# Subcellular Biochemistry

Volume 106

The book series SUBCELLULAR BIOCHEMISTRY is a renowned and well recognized forum for disseminating advances of emerging topics in Cell Biology and related subjects. All volumes are edited by established scientists and the individual chapters are written by experts on the relevant topic. The individual chapters of each volume are fully citable and indexed in Medline/Pubmed to ensure maximum visibility of the work.

Swetha Vijayakrishnan • Yaming Jiu •
J. Robin Harris
Editors

# Virus Infected Cells

Springer

*Editors*
Swetha Vijayakrishnan
Centre for Virus Research (CVR)
University of Glasgow
Glasgow, UK

Yaming Jiu
Shanghai Institute of Immunity and Infection
(Formerly Institut Pasteur of Shanghai)
Chinese Academy of Sciences
Shanghai, China

University of Chinese Academy of Sciences
Beijing, China

Unit of Cell Biology and Imaging Study of Pathogen Host Interaction, The Center for Microbes, Development and Health, Key Laboratory of Molecular Virology and Immunology, Institut Pasteur of Shanghai
Chinese Academy of Sciences
Shanghai, China

J. Robin Harris
Institute for Molecular Physiology
University of Mainz
Mainz, Germany

ISSN 0306-0225 ISSN 2542-8810 (electronic)
Subcellular Biochemistry
ISBN 978-3-031-40085-8 ISBN 978-3-031-40086-5 (eBook)
https://doi.org/10.1007/978-3-031-40086-5



This Springer imprint is published by the registered company Springer Nature Switzerland AG
The registered company address is: Gewerbestrasse 11, 6330 Cham, Switzerland

Paper in this product is recyclable.

# Preface

This muti-author book was conceived at a time of massive worldwide academic, medical and public interest in virus infections. Indeed, the book supplements previous volumes in the *Subcellular Biochemistsry* series devoted to viruses (Vol. 15, 1989, Virally Infected Cells; Vol. 68. 2013, Structure and Physics of Viruses; Vol. 88, 2018, Virus Protein and Nucleoprotein Complexes). Also, individual chapters relating to viruses appear elsewhere within volumes of the series. The chapters in this book encompass a diverse collection of interesting topics covering both structural and functional aspects of viral infection within cells.

The book is largely split into two sections. The first group of chapters showcases general aspects and state-of-the-art technical advancements applied currently in studying virus infected cells. The second group of chapters focus primarily on specific viruses covering a broad range of human and animal RNA and DNA viruses.

The 16 chapters here have been contributed by eminent international scientists and together provide a good overview of the subject. The initial book project commissioned more than 20 chapters, but several were lost along the way to publication, no doubt due to the increasing pressures of work placed upon scientists working in research institutes and academia through recent years. Rather than selecting a few chapters for comment, we simply refer the reader to the Contents list, immediately following.

This unique collection of chapters highlight the recent advances made in cellular virology. We hope that this book will be of interest and value to the broad international community of virologists and cellular biologists and to others who wish to delve in some depth into this fascinating and rapidly advancing field of study.

Glasgow, UK — Swetha Vijayakrishnan
Shanghai, China — Yaming Jiu
Mainz, Germany — J. Robin Harris

# Contents

# Part I
# General/Technical Aspects

# Chapter 1
# In Situ Imaging of Virus-Infected Cells by Cryo-Electron Tomography: An Overview

**Swetha Vijayakrishnan**

**Abstract** Cryo-electron tomography (cryo-ET) has emerged as a powerful tool in structural biology to study viruses and is undergoing a resolution revolution. Enveloped viruses comprise several RNA and DNA pleomorphic viruses that are pathogens of clinical importance to humans and animals. Considerable efforts in cryogenic correlative light and electron microscopy (cryo-CLEM), cryogenic focused ion beam milling (cryo-FIB), and integrative structural techniques are helping to identify virus structures within cells leading to a rise of in situ discoveries shedding light on how viruses interact with their hosts during different stages of infection. This chapter reviews recent advances in the application of cryo-ET in imaging enveloped viruses and the structural and mechanistic insights revealed studying the viral infection cycle within their eukaryotic cellular hosts, with particular attention to viral entry, replication, assembly, and egress during infection.

**Keywords** Virus infection · Cryo-electron tomography · In situ cellular imaging · Enveloped virus · Correlative light and electron microscopy · 3D structural determination

## Introduction

Being the most abundant life form on the planet, viruses are present everywhere, from plants to organisms deep within the ocean. Studies in marine ecology indicate a gigantic number ($10^{30}$) of viruses inhabiting the biosphere, exhibiting astounding

---

The original version of the chapter has been revised. A correction to this chapter can be found at https://doi.org/10.1007/978-3-031-40086-5_17

---

S. Vijayakrishnan (✉)
MRC-University of Glasgow Centre for Virus Research, Sir Michael Stoker Building, Garscube Campus, Glasgow, Scotland, UK
e-mail: Swetha.Vijayakrishnan@glasgow.ac.uk



S. Vijayakrishnan et al. (eds.), *Virus Infected Cells*, Subcellular Biochemistry 106,
https://doi.org/10.1007/978-3-031-40086-5_1

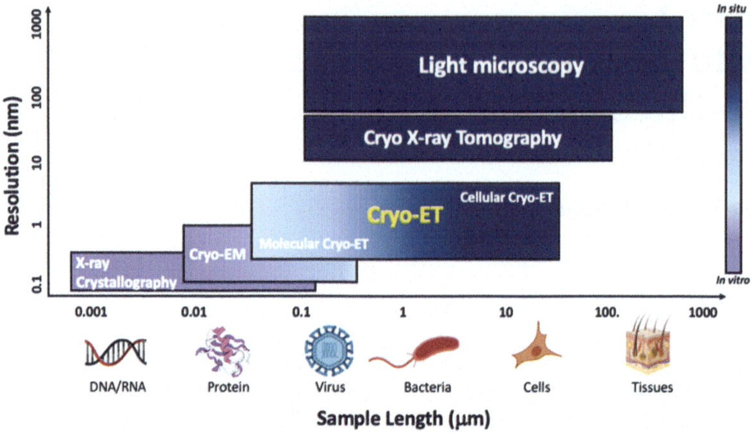


**Fig. 1.1 Correlation of structural biology and imaging methods with sample size and resolution.** Sample length, resolution, and the extent of nativeness (in vitro or in situ) are the main determinants when choosing structural biology and imaging methods. Sample size and resolution ranges of the most common methods are shown for comparison, highlighting the suitability of cryo-ET for imaging a wide range of molecular and cellular samples including protein assemblies, viruses, and cells. In combination with correlative light-electron microscopy (CLEM) and cryo-focused ion beam milling (cryo-FIB), cryo-ET resolves native structures of protein at relatively high resolution as well as provides abundant information on protein–protein, protein–membrane, or protein–nucleic acids interactions in an in situ setting. Figure adapted from (Li S, 2022) under CC BY 4.0. The various sample types used in this figure were created with Biorender.com

levels of genetic diversity (Jian et al. 2021; Suttle 2007). Viruses need suitable host cells to replicate and release progeny for subsequent cycles of productive infection. Interactions of viruses with their hosts are complex and diverse, encompassing a balanced interplay between viral and host factors that regulate viral replication and host responses to infection. Dissecting the molecular mechanisms involved is critical to our understanding of virus infection and the changes they bring about in the cellular environment of their hosts (Rodrigues et al. 2017). Knowledge gained from these crucial studies is important for implementing strategies to control virus infection and transmission, predicting their impact on the environment, zoonotic, and human populations as well as the design of effective antiviral and vaccine interventions against current and emerging diseases.

Many different techniques are employed to image biological processes at different length scales ranging from tissues (mm) and cells (microns) to nanometre scale subcellular structures (Fig. 1.1). At one end of the spectrum, light microscopy images whole cells and tissues at low resolution (120–250 nm), while at the other end, three main techniques—X-ray crystallography, nuclear magnetic resonance (NMR), and cryo-electron microscopy (cryo-EM)—enable high-resolution imaging (0.1–0.5 nm) and structural determination of proteins and macromolecular assemblies such as viruses (Campbell 2002; Li 2022). While the former two methods have helped push structural biology to the forefront by resolving structures of isolated proteins and complexes at close-to-atomic resolutions, they are time-consuming and occasionally prone to misinterpretation due to lack of the native environment in which such proteins and complexes normally exist. On the other hand, cryo-EM is a versatile technique that resolves three-dimensional (3D) structures of macromolecular assemblies of various sizes in multiple functional states under near-physiological conditions at 0.2–0.5 nm resolutions. Further, samples are in an intact hydrated state and are not fixed, stained, or manipulated into a crystalline state. The most widespread application of cryo-EM is the study of purified samples by the "single particle analysis" method. However, with the cryo-EM resolution revolution era, cryo-electron tomography (cryo-ET) has emerged as a powerful method in structural biology, specifically for viruses, bridging the resolution range (sub-nanometre) achievable between light microscopy and cryo-EM (Fig 1.1). It allows imaging of purified non-symmetric pleomorphic viruses such as HIV-1 and Influenza A as well as large complex viruses like poliovirus and herpes simplex virus (HSV-1) (Calder et al. 2010; Cyrklaff et al. 2005; Grunewald et al. 2003; Liu et al. 2010; Mangala Prasad et al. 2022b). But perhaps the most advantageous application of cryo-ET over other structural methods is the ability to visualize cells and the macromolecular complexes housed within it at 3–5 nm resolutions.

Enveloped viruses are surrounded by a lipid membrane and comprise a large number of RNA and DNA viruses, the majority of which are pathogenic to animals or humans. They are highly pleomorphic proving difficult for structure determination by traditional structural biology methods including cryo-EM. Asymmetric elements associated with these viruses such as locally ordered helical symmetry of the matrix protein layer in influenza A virus (IAV) and Ebola virus, absence of global symmetry in capsid cores (HIV-1), and flexibility of glycoproteins (SARS-CoV-2) are uniquely suited for investigation by cryo-ET. This chapter will examine sample preparation methods, data collection and image processing pipelines, and integrative approaches used to aid cryo-ET imaging of enveloped viruses. Further, the role of cryo-ET in expanding the structural biology of viruses and their interactions within the cellular host during entry, receptor interactions, replication, and assembly will be highlighted.

## Sample Preparation

For imaging samples by cryo-ET, the primary step is to plunge freeze the specimen by a method known as vitrification (Fig. 1.2). Based on sample thickness there are two ways to plunge freeze. Samples less than 10 μm such as purified proteins, viruses, or thin cells can be vitrified directly by plunge freezing into a cryogen with high heat capacity and thermal conductivity such as liquid ethane or propane (Dubochet et al. 1988). For purified preparations of proteins and viruses, typically 2–5 μl is applied to holey carbon copper grids, and then blotted with filter paper to leave behind a thin layer of liquid (typically 50–100 nm). For thin cells directly cultured on grids, gold grids are used to avoid the toxicity of copper. After blotting, grids are then vitrified in ethane and stored for future use.

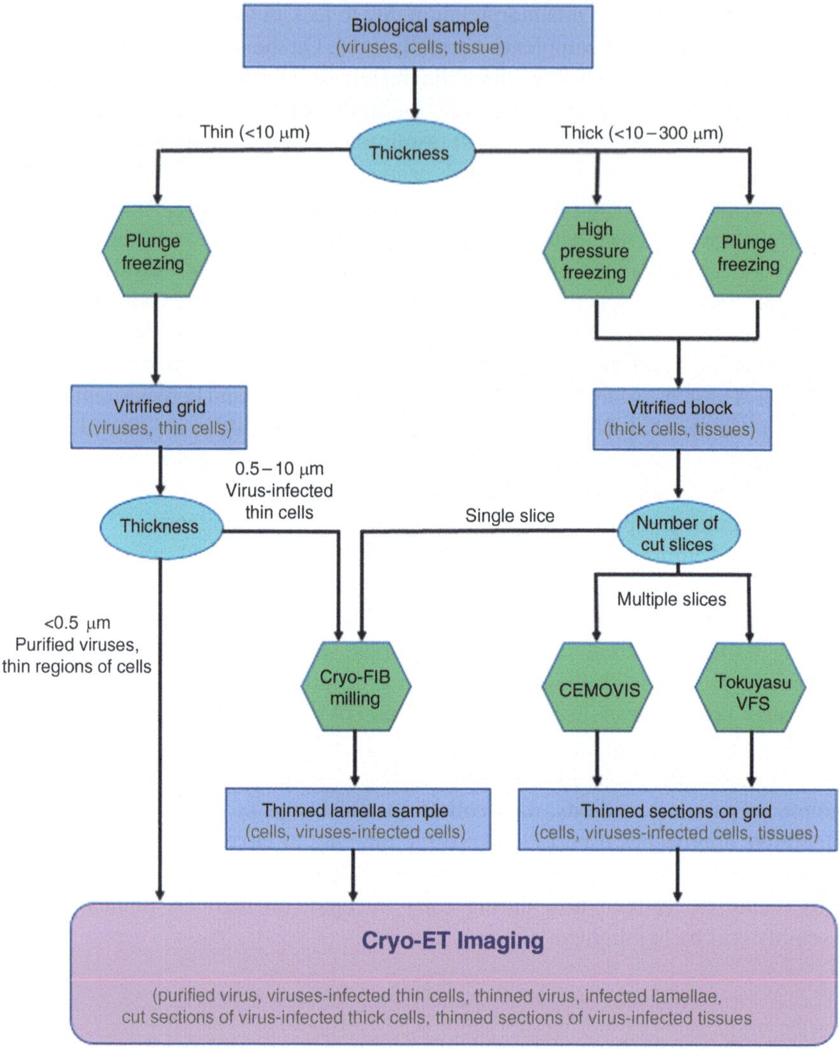


**Fig. 1.2 Sample processing for cryo-ET imaging**. Flowchart illustrating sample preparation for cryo-ET imaging where the choices for preparation (cyan ellipses), appropriate techniques used (green hexagons), and the desired structures to visualize (purple rectangles) are depicted. Samples thinner than 10 μm such as viral proteins and thin cells are vitrified by plunge freezing, while thicker samples like cell pellets or tissue blocks/biopsies are mainly cryo-fixed by high pressure freezing (HPF). The only exception is thick, chemically fixed, gelatin-embedded, sucrose-infiltrated Tokuyasu cell pellet or tissue samples, where they are fixed by plunge freezing in liquid nitrogen. Samples thicker than ~500 nm (e.g., cells and tissue) are thinned by cryo-ultramicrotomy (CEMOVIS, Tokuyasu vitrified frozen sections [VFS]) or cryo-FIB milling prior to imaging in order to make them electron transparent. Finally, all vitrified cryo-specimens—thinner than 500 nm samples or cryo-ultramicrotomy/cryo-FIB thinned samples are imaged by cryo-ET, acquiring tilt series for 3D tomogram generation and/or sub-tomogram averaging

Samples greater than 10 μm in thickness but less than 300 μm like cell pellets and tissues are frozen at high pressure, commonly known as "high-pressure freezing" (HPF) (McDonald and Auer 2006). When high pressure (200 MPa) is applied using a liquid nitrogen jet at −196 °C to biological samples, expansion of water is counteracted, promoting the formation of vitreous ice, and samples are frozen within ~15 ms in a reproducible manner. After vitrification, thick HPF samples are thinned by cryo-ultramicrotomy or cryo-focused ion beam (cryo-FIB) milling prior to imaging (Fig. 1.2).

Cryo-ultramicrotomy is performed at cryogenic temperatures (−180 °C) using methods like cryo-electron microscopy of vitrified sections (CEMOVIS) or Tokuyasu sectioning, where the sample is thinned using a diamond knife to slice the frozen sample into thin sections (50–200 nm) (Al-Amoudi et al. 2004, 2005; Tokuyasu 1973). Both CEMOVIS and Tokuyasu combine cryo-preservation and high-resolution imaging of thin sections of thick samples such as vitrified cell pellets. While both techniques are technically demanding, CEMOVIS causes knife-related artefacts, sample distortion, and compression, and Tokuyasu involves chemical fixation of samples prior to sectioning (Al-Amoudi et al. 2005; Tokuyasu 1973). Single or multiple cut sections can be transferred onto an EM grid and imaged by cryo-ET.

Alternatively, unwanted material can be removed from vitrified specimens and thinned down by ablation to a specified thickness employing a cryo-FIB by using heavy metal ions such as Gallium. Cryo-FIB provides artefact-free thinning of specimens and has been successfully used to section samples prepared by HPF and direct plunge freezing (such as cells) (Hayles et al. 2010; Marko et al. 2007; Rigort et al. 2012a). HPF samples can be cut by the ion beam and imaged by the electron beam as serial sections; within the cryo-FIB scanning electron microscope, plunge-frozen cellular samples on EM grids are milled by cryo-FIB to produce electron-transparent lamellae for subsequent cryo-ET imaging (Rigort et al. 2012a; Klein et al. 2020; Mendonca et al. 2021a; Berger et al. 2023). Although this technique enables peering into the interior of the cell, it is low in throughput often resulting in lamellae that are broken, have irregularities, or are contaminated by frost during transfer to the cryo-transmission electron microscope (Rigort and Plitzko 2015). The development of integrated cryo-FIB machines housing fluorescence light microscope systems along with the design of high-vacuum transfer systems and cryo-shutters will help mitigate these problems and streamline the workflow for automated cryo-FIB milling.

An attractive and easier alternative is using vitrified frozen sections (VFS), where Tokuyasu sections of cellular samples cut at cryogenic conditions are thawed and then re-vitrified by plunge freezing for imaging by cryo-ET. Using Tokuyasu VFS coupled with CLEM and STA allows for successful in situ imaging and high-resolution 3D structure determination of viruses from within cells (Tokuyasu 1973; Vijayakrishnan et al. 2020; Sabanay et al. 1991). Further, this method offers several advantages such as the availability of less technically challenging, relatively inexpensive equipment, higher throughput, and the addition of CLEM to the pipeline, broadening its scope to the study of low-frequency events with a greater likelihood of successful imaging. Using automated data collection procedures on 300 KV microscopes fitted with high-speed direct electron detectors and streamlined STA procedures will further improve the applicability of this method and enable the determination of in situ near-atomic resolution structures.

## Data Collection and Image Processing

Biological materials undergo damage by radiation within the electron microscope, and therefore, it is crucial to minimize electron exposure of the sample. In cryo-ET, plunge-frozen samples are rotated around an axis over a tilt range, typically –60° to +60° and two-dimensional (2D) projection images are recorded at each tilt to generate a "tilt series". As images are collected on the same region of the sample at multiple angles (or tilts), electron exposure is minimized by reducing the overall electron dose, typically not greater than 78–100 $e^{-}/\text{Å}^{2}$ (to achieve nanometre resolution) and distributing (fractionating) it over the number of images recorded in each tilt series. Consequently, the fractionated electron dose results in low-contrast and heavily noisy images, making accurate alignment of images in the tilt series challenging. To overcome this problem, gold fiducial markers (diameter 5–20 nm) are added to the samples prior to vitrification. As these fiducials are high contrast, they can be tracked successfully through the images, helping increase the alignment accuracy of the tilt series.

Computational analysis of the aligned tilt series produces a 3D volume of the sample known as a "tomogram", which provides detailed ultrastructure of the size, shape, and organization of macromolecules and cells at 3–5 nm resolutions (Wan and Briggs 2016). Repeating structures within tomograms can be extracted from tomograms (sub-tomograms/sub-volumes) and subject to extensive computational processing with iterative steps of alignment and averaging by a process called sub-tomogram averaging (STA), to obtain high-resolution structures at 0.5–2 nm. Identification and extraction of sub-tomograms is executed by template matching to a known structure or generating a template by manually picking sub-tomograms. Alignment of sub-tomograms is carried out with one or multiple references based on cross-correlation methods that inform on the similarity of the structures. Due to the restricted tilting range (–60° to +60°) and mechanical limitations of the stage, data cannot be acquired at all angles, causing missing information in reconstructed tomograms and sub-tomograms, called the "missing wedge". However, cross-correlation-based sub-tomogram alignments and classification procedures are sensitive to the missing wedge information. Using a wedge-shaped mask for the reference as well as application of defocus-gradient corrections or contrast transfer function-corrected sub-tomograms helps mitigate these problems, allowing near-atomic 3D structure determination (Tegunov et al. 2021; Xiong et al. 2009; Zanetti et al. 2009). In recent years, major microscope hardware developments such as phase plates, energy filters, and direct electron detectors (DEDs) have significantly increased the quality of cryo-ET images enabling visualization and study of macromolecules in situ (Vijayakrishnan et al. 2020; Tegunov et al. 2021; Bammes et al. 2012; Danev et al. 2014; Haney et al. 2022; Schur et al. 2016; Vijayakrishnan et al. 2013; Zila et al. 2021). Additionally, ongoing improvements in software development and processing pipelines have allowed us to reach near-atomic resolution (~0.4 nm) for cryo-ET data by STA (Tegunov et al. 2021; Schur et al. 2016).

To obtain insights into the biological processes, interpretation of the 3D volume data is important. To achieve this, specifically in tomography samples of cells and tissues, biologically relevant features within the 3D-reconstructed tomograms are surface rendered and pseudo-coloured by a process known as segmentation to aid in the interpretation and visualization of the 3D volume. In practice, structures of interest are manually drawn by hand into the volume, being highly time intensive and subjective. On the other hand, automated segmentation can speed up the process and provide a reproducible and objective way for the interpretation and quantification of data. Software development implementing faster and better algorithms using template matching, neural networks-based AI, and 3D deep learning are making this a reality, revolutionizing automated segmentation to visualize proteins and larger structures such as actin filaments, microtubules, and vesicles inside cellular tomograms (Klein et al. 2020; Chen et al. 2017; Rigort et al. 2012b; Rusu et al. 2012).

## Integrative Techniques

In addition to cryo-ET, several newly developing accessory techniques such as cryo serial block face scanning electron microscopy (cryo-SBF SEM), cryo-soft X-ray tomography (CSXT), and cryo-correlative light and electron microscopy (cryo-CLEM) enable 3D imaging of cells and tissues in close-to-native environments. Combining these methods alongside cryo-ET can be a powerful and informative approach to obtain insightful data at multi-scale resolutions (6–200 nm).

### *Cryo-serial Block Face Scanning Electron Microscopy*

Cryo-SBF SEM combines serial sectioning of thick samples coupled with successive imaging by SEM. Typically one flat surface of a HPF cellular or tissue specimen is repeatedly imaged by SEM alongside removal of consecutive slices sectioned by a focused ion beam (FIB). Successive imaging and sectioning are performed automatically several hundreds of times to reconstruct a 3D volume of the sample. A major advantage of this technique is the capability to image thick and large sized samples that are not amenable for direct imaging by cryo-ET. Cryo-SBF SEM has been successfully applied to study the process of cells transforming into mineralized tissues in the tail fin of zebrafish, and more recently in the study of SARS-CoV-2 assembly (Mendonca et al. 2021a, b; Vidavsky et al. 2016).

## *Cryo-soft X-Ray Tomography*

In CSXT, X-rays are used for imaging vitrified biological samples. As X-rays have a larger penetration depth, thicker specimens can be used such as whole cells without the need for sectioning, a fundamental and crucial difference from cryo-ET. Soft X-rays have energies (284–543 eV) that fall within the so-called "water window", allowing for selective absorption between carbon-rich elements and water. This absorption difference provides significant contrast in vitrified biological specimens that are largely rich in carbon and oxygen. Typically, samples are tilted (–60° to +60° with EM grids or a full 360° with capillary tubes) and projection images captured at each tilt are subsequently reconstructed to obtain a 3D tomographic volume. In CSXT, the X-ray beam focusing on the sample is detected by zone plates that act as detectors. Unlike lenses or mirrors, zone plates define the resolution limit and use diffraction instead of refraction or reflection. Current resolutions of CSXT range in the order of 25–40 nm. Several studies have been performed using CSXT to explore the changes in ultrastructure of whole cells upon virus infection (Fig. 1.3) (Harkiolaki et al. 2018; Perez-Berna et al. 2016). Hepatitis C virus (HCV) infection in liver cells showed significant changes in the endoplasmic reticulum (ER) and cellular morphology of the cytoplasm, with formation of a membranous web of tubules and vesicles of variable size and orientation, similar to a sponge-like appearance (Fig. 1.3a) (Perez-Berna et al. 2016). Furthermore, treatment of HCV-infected cells with clinically approved antivirals reversed completely the virus-induced ultrastructure to an uninfected state indistinguishable from control samples, demonstrating the potential for using CSXT as a preclinical tool for drug efficacy testing, selection, and development. Recent quantitative analysis of SARS-CoV-2 infected cells showed large cellular remodelling induced by the virus, specifically changes in the mitochondrial network, increase in lipid droplets, and convoluted membrane structures and vesicles, overall indicating cell apoptosis (Mendonca et al. 2021a, b; Loconte et al. 2021). CSXT imaging has also been extended to DNA viruses such as Herpes simplex virus-1 (HSV-1) infection showing unique nuclear bodies and virus capsid aggregates (JH Chen et al. 2022). In combination with fluorescence light microscopy (LM), CSXT has been used to image in a correlative manner to link morphology, location, and function in virus-infected cells. Correlative CSXT with LM revealed that HSV-1 induces organelle remodelling, elongation of mitochondria, and influencing dynamics during fission/fusion (Fig. 1.3b) (Nahas et al. 2022). The highly complex virus maturation and morphogenesis pathway of the large poxvirus vaccinia virus was elucidated in detail by imaging the various forms of the virus—immature and mature within the cytoplasm of whole cells by CSXT (Fig. 1.3c) (Chichon et al. 2012).

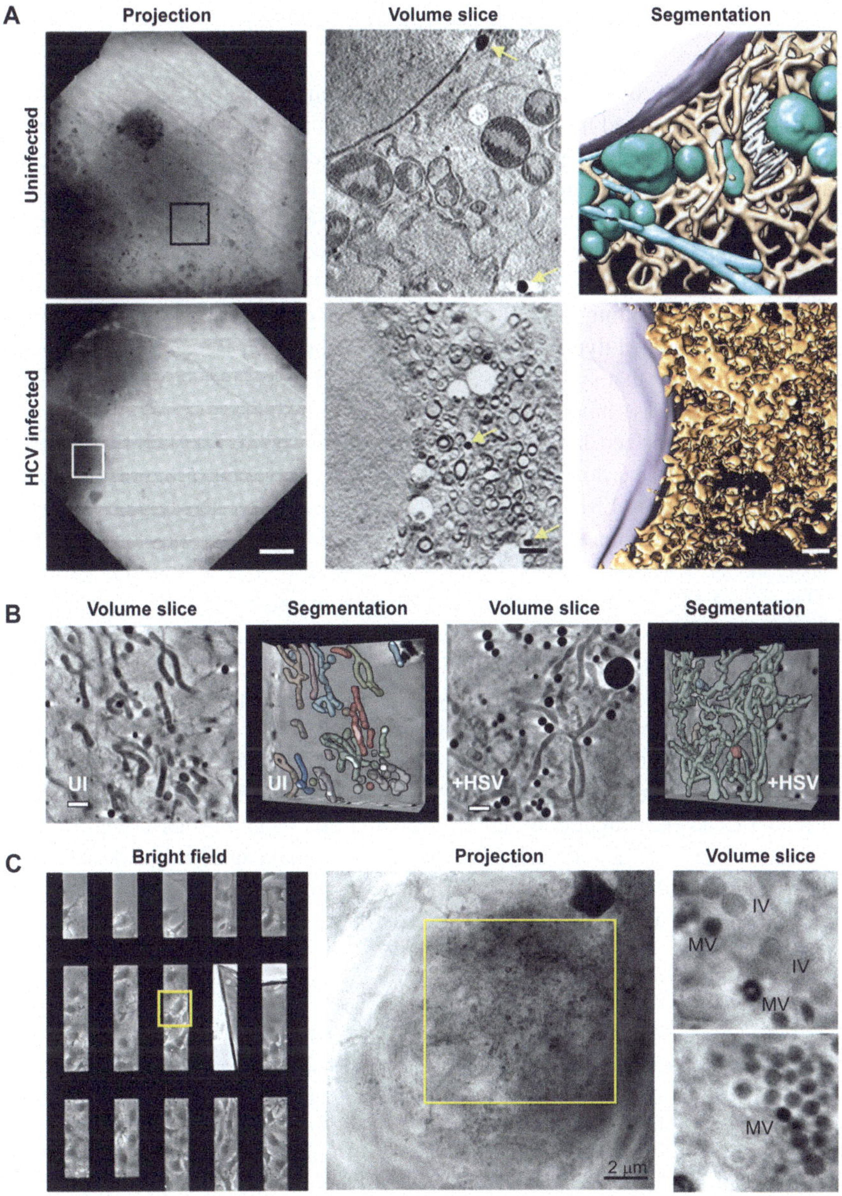


**Fig. 1.3 Whole-cell imaging by cryo-soft X-ray tomography**. CSXT enables 3D imaging of vitrified whole cells, without the need for sectioning, providing overall cellular architecture at mesoscale resolution (25–40 nm). Three viral systems HCV (Perez-Berna et al. 2016), HSV-1 (Nahas et al. 2022), and VACV (Chichon et al. 2012) imaged by CSXT are depicted demonstrating detection of virus particles and remodelling of cellular organelles that is triggered upon infection. (**a**) Tiled projection images, 3D tomographic volumes, and annotation by segmentation of whole-cell volumes of control and HCV-infected cells are shown. ROI selected for CSXT imaging from the projection images (black and white squares) and their corresponding volume slices and segmentation images show significant alteration of ER forming a membranous web network upon HCV infection. The various organelles, nucleus (pink), mitochondria (green), normal ER (light

## *Cryo-correlative Light and Electron Microscopy*

Although cryo-ET provides high-resolution tomograms of the cellular landscape, the high magnifications used limit the area that can be imaged. As the interior of the cell is extremely dense, identifying regions or macromolecules of interest within the cell is an arduous and challenging task, similar to searching for a "needle in a haystack". Cryo-CLEM overcomes this problem by combining fluorescent light microscopy (LM) imaging to target the same specific sites with cryo-ET, increasing the efficiency of subcellular localization with single molecule precision. This is achieved by using fluorescent labelling techniques like genetic tagging (using GFP) or chemical labelling of specialized dyes targeting structures of interest. Recently, a genetically encoded protein tag that synthesizes gold ions into beads and DNA based signpost origami tags (SPOTs) have been developed, offering attractive options for both intracellular and surface labelling of proteins (Jiang et al. 2020; Silvester et al. 2021). Imaging can be done via a two-step process, selecting a region of interest (ROI) and imaging by LM first followed by cryo-EM/cryo-ET or via an integrated set-up having an integrated LM within a transmission electron microscope (TEM) (Faas et al. 2013; Hampton et al. 2017; Wang et al. 2017). Although the resolution gap between LM (250 nm) and cryo-ET (5 nm) is massive, currently a localization precision of ~40 nm can be achieved using cryo-LM. Targeting smaller molecules within the crowded cell can be performed by cryo-CLEM employing super-resolution LM (Wolff et al. 2016). Cryo-CLEM studies have been performed to identify virus and virus–host interactions in cells (Fig. 1.4). Cryo-CLEM has elegantly shown the transport of HSV-1 in neurons (Fig. 1.4a) and offered structural insights on HSV-1 capsid interaction with its associated tegument proteins within the cell nuclei (Vijayakrishnan et al. 2020; Ibiricu et al. 2011). Furthermore, in HIV-1 infected cells, cryo-CLEM revealed the inhibitory mechanism of virus release during egress under the influence of tetherin, a host restriction factor (Fig. 1.4b) (Hampton et al. 2017; Jun et al. 2011; Strauss et al. 2016).

**Fig. 1.3** (continued) brown), altered ER (yellow), cytoskeleton actin (light blue), and Golgi (grey), are depicted. Yellow arrows indicate putative lysosomes/lipid droplets. Scale bars are 10 μm (projection images) and 0.5 μm (volume slices). Figure was adapted from Perez-Berna et al. (2016), copyright © 2016, American chemical society, (**b**) Comparison of uninfected (UI) and HSV-1 infected cells (+HSV) by CSXT (volume slices and segmentation) highlights extensive morphological changes in mitochondria, with the 3D geometry shifting from short and long mitochondria in uninfected control cells to an elongated, branched dense network upon HSV-1 infection. Scale bars = 1 μm. Figure adapted from Nahas et al. (2022) under CC BY 4.0. (**c**) ROI selected (yellow square) from a bright field image of VACV-infected cells was imaged by CSXT (projection image) to obtain tomograms. Single volume slices of tomograms provided insights into the assembly pathway identifying both immature (IV) and mature (MV) viral particles of VACV, present close to each other in densely packed clusters within the cell cytosol. Scale bar represents 100 μm and 0.5 μm in bright field and volume slice images, respectively. The figure was adapted from Chichon et al. (2012), copyright © 2012, Elsevier

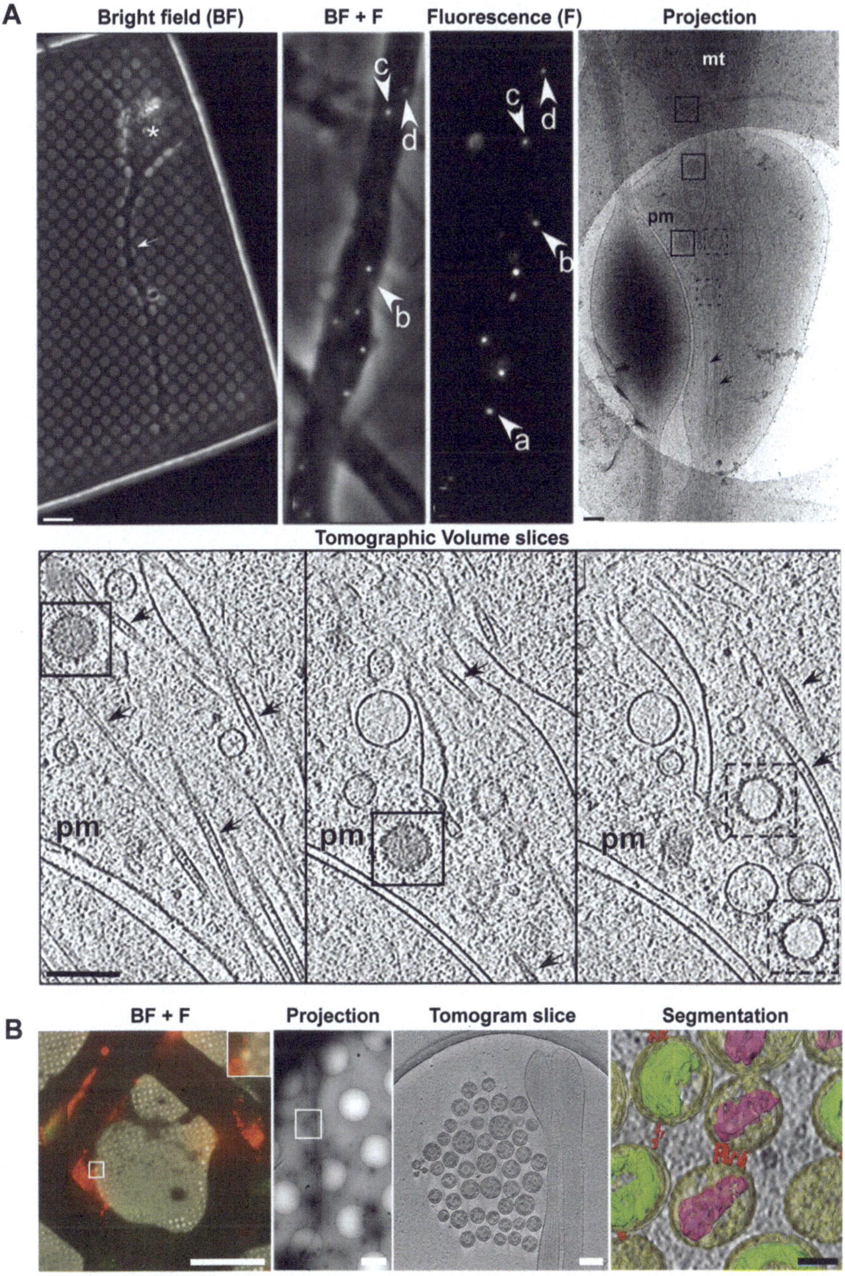


**Fig. 1.4 Virus–host interactions revealed by cryo-CLEM and cryo-ET.** Cryo-CLEM imaging provides information on spatial dynamics, specifically in locating the "needle in a haystack", i.e., protein or virus within the dense environment of the cell. However, when coupled with cryo-ET, it offers an extremely powerful approach to obtain structural insights into virus–protein interactions in situ. Interaction of two viruses HSV-1 (Ibiricu et al. 2011) and HIV-1 (Hampton et al. 2017) with their hosts is shown. (**a**) Cryo-CLEM and cryo-ET imaging of HSV-1 infected neurons detected the transport of virus capsids during egress. Neurons observed by bright field imaging (BF) revealed

## Cryo-ET Studies of Enveloped Viruses and Their Interactions Within the Host Cell

Cryo-ET in combination with STA provides detailed 3D ultrastructure and insights into in situ structure and assembly of macromolecular assemblies. However, sub-optimal throughput in data acquisition coupled with relatively low resolution of STA has limited its potential. But recent advances in these areas are steering the resolution revolution of cryo-ET.

Obtaining high-resolution structures from cryo-ET is determined by the quality of high-throughput data collected in minimal time. To obtain a reasonable level of signal-to-noise ratio (SNR) in the tilt series, the sample is exposed to a high cumulative dose (2–3 times more than in SPA). However, the earliest images collected during the tilt series have the lowest exposure of electrons and therefore high SNR. This problem was elegantly overcome by employing a dose-symmetric scheme that minimized the accumulated electron dose by starting the data acquisition at low tilts (0°) following a walk through to higher tilts with the dose symmetrically spread across the tilt range (Hagen et al. 2017). Hybrid STA-SPA approaches have further maximized SNR with successful determination of retroviral capsid and tobacco mosaic virus (TA Bharat et al. 2012; Sanchez et al. 2020).

Although the dose-symmetric tilt scheme retains high-resolution information, it comes at the cost of low throughput. Several newly developed tilt schemes incorporating beam-tilt induced image shift methods such as beam image-shift electron cryo-tomography (BISECT), parallel cryo-electron tomography (PACE-tomo), and multishot-tomography have significantly increased the throughput of tilt series acquisition (Bouvette et al. 2021; Eisenstein et al. 2023; Khavnekar et al. 2023). These methods exploit the dose symmetric acquisition scheme by collecting them in

**Fig. 1.4** (continued) neurite (white arrow) and cell body (white asterisk) structures. The fluorescence (F) and overlay (BF and F) images show bright punctate spots, denoting individual virus particles (white arrows) travelling along the axon. Detail is observed in the projection image of the intact axon including HSV-1 genome-filled C-capsids (black squares) and empty A-capsids (black dashed square). Tomographic volume slices of these selected regions (squares in projection image) clearly denote the virus capsids transporting across the axon close to the plasma membrane (pm) and microtubules (black arrows). Scale bar represents 6 µm, 5 µm, and 200 nm in the BF, F and BF+F volume slice images, respectively. Figure was adapted from Ibiricu et al. (2011) under CC BY 4.0. (**b**) Structural insights on tetherin binding to HIV-1, preventing virus release, were obtained by successful cryo-CLEM and cryo-ET imaging. The montage map overlay of BF+F was used to select ROI (white square) representing colocalization of mCherry-Gag (red) and EGFP-tetherin (green). The ROI selected by CLEM was imaged by cryo-EM to produce a low montage projection map that was further used to select areas (white squares) for cryo-ET data acquisition. Tomographic slice and segmentation images depict a cluster of HIV-1 virions (capsid cores—pink and virion envelope—yellow) connected by tetherin (elongated structures—red) near a cell extension. Scale is 50 µm, 2 µm, and 200 nm for the BF+F overlay, projection, and tomogram slice images, respectively. The figure was adapted from Hampton et al. (2017), copyright © 2016, Springer Nature

multiple areas in parallel, reducing the acquisition time substantially from 45–70 min to 3–4 min for each tilt series, yielding high-resolution structures. The beam image-shift scheme has also been used to acquire montage tomograms by splitting a large field of view into overlapping tiles that can be subsequently stitched computationally, particularly useful for imaging cellular samples by cryo-ET such as respiratory syncytial virus-infected cells (Yang et al. 2022; Peck et al. 2022). While BISECT performs effectively with purified specimens, PACE-tomo and multishot tomography have been tested on both in vitro and FIB-milled in situ lamellae with tremendous success.

In cryo-EM analysis, the mathematical equation termed the contrast transfer function (CTF) defines how aberrations within the TEM influence the image of the sample. Therefore, CTF correction is crucial for obtaining high-resolution structural information using cryo-EM and cryo-ET. During cryo-ET tilt series acquisition, CTF estimation is required for accurate determination of defocus, to enable extraction of high-resolution information from the data. Major improvements in microscope hardware, data acquisition schemes, and software algorithms including improving CTF estimation and correction have gradually evolved paving the way for the cryo-ET resolution revolution from 20 to 8 Å and currently to near-atomic 3–4 Å resolution (Xiong et al. 2009; Zanetti et al. 2009; Schur et al. 2016; Schur et al. 2013; Bharat et al. 2015; Ni et al. 2021).

## *Cryo-ET of Purified Enveloped Viruses*

Several hundreds of pleomorphic, enveloped viruses are responsible for a wide range of human diseases. Over the past 20 years, cryo-ET has provided astonishing structural insights into the assembly of several of these viruses including HIV-1, influenza A virus (IAV), SARS-CoV-2, RSV, HSV-1, Ebola virus, Lassa virus (LASV), Marburg virus (MARV), and Rift valley fever virus (RVFV) (Conley et al. 2022; Halldorsson et al. 2018; Li et al. 2016; McElwee et al. 2018; Peukes et al. 2020; Schur et al. 2016; Vankadari et al. 2002; Wan et al. 2017, 2020; Yao et al. 2020). In this chapter, we will focus on a subset of viruses mentioned above with special attention to HIV-1 and SARS-CoV-2 (Fig. 1.5).

### Human Immunodeficiency Virus (HIV-1)

HIV-1 is a single-stranded RNA retrovirus and causative factor of acquired immunodeficiency syndrome (AIDS). Upon infection of its host, HIV-1 integrates its genome into the host DNA, and weakens the host immune system, resulting in a gradual decline of immune cells (such as CD4+ T cells, macrophages) over time, causing AIDS. Despite antiretroviral therapy being an effective treatment to lower mortality rates due to AIDS, nearly 9.7 million people do not have access to it, especially in developing countries (WHO 2021).

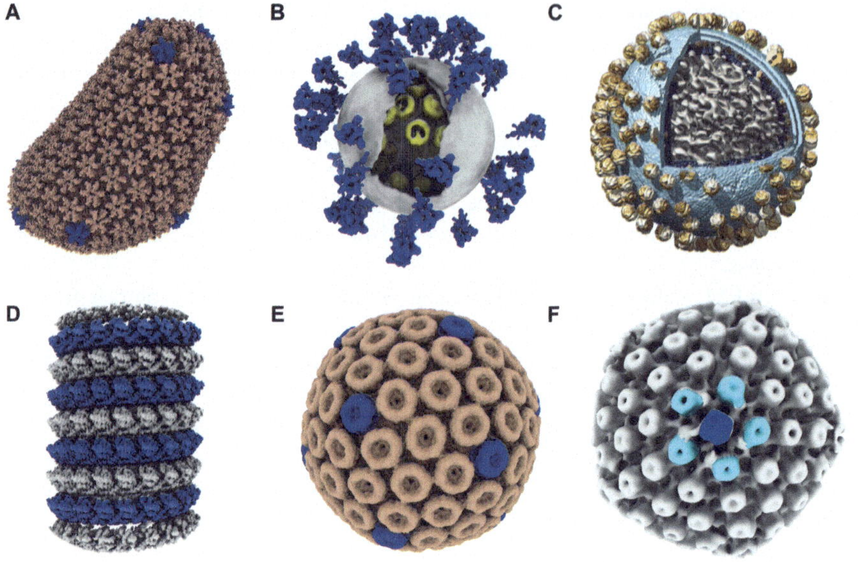


**Fig. 1.5 Structures of purified enveloped viruses resolved by cryo-ET and STA.** A selection of high-resolution structures of representative enveloped purified viruses determined by cryo-ET and STA is shown. (**a**) Structure of mature HIV-1 capsid (PDB: 3J3Y) colour coded with CA hexamers (brown) and pentamers (blue). (**b**) Ultrastructure of the SARS-CoV-2 virion (EMD-30430); lipid envelope (grey), S protein (blue), and RNPs (yellow). (**c**) The STA reconstructed prefusion LASV glycoprotein (GP) trimers (yellow) embedded on the viral envelope (blue) and filled with internal material (grey) segmented from the tomogram. Figure reproduced from Li et al. (2016) under CC BY 4.0. (**d**) Structure of the helical IAV M1 matrix protein (grey and blue) (EMD-22384). (**e**) Structure of RVFV prefusion GP pentamers (blue) and hexamers (brown) reprojected on the viral envelope (EMDB-4197). (**f**) Structure of mature HSV-1 capsid (EMD-5452) denoting the five-fold portal structure (blue) and surrounding hexamers (cyan). Figure was adapted from (Vankadari et al. 2002) under CC BY 4.0 using publicly available atomic coordinates and maps from the protein data bank (PDB) and Electron Microscopy Data Bank (EMDB) and using UCSF ChimeraX (Pettersen et al. 2021)

The mature HIV-1 virion is composed of a capsid core made of capsid protein (CA) enclosing the dimeric RNA genome. Upon entry and membrane fusion of the CA core, viral RNA is reverse transcribed to double-stranded DNA (dsDNA), which in turn traffics to the nucleus to integrate into the host chromosome. HIV-1 morphogenesis is a structured process composed of three steps—assembly, budding, and maturation (Sundquist and Krausslich 2012; Sundquist and Krug 2012). In association with the membrane, the 55 kDa Gag polyprotein that comprises the matrix (MA), CA, nucleocapsid (NC), SP1, and p6 domains mediates virus membrane curvature via self-multimerization to form a Gag shell layer during HIV-1 assembly. Post-assembly the viral envelope encoding the immature Gag shell undergoes scission from the cell membrane via the ESCRT-III complex, resulting in the budding of spherical immature HIV-1 particles. Subsequently, the envelope of released HIV-1 virions containing immature Gag is cleaved at the MA–CA, CA–NC, SP1-NC, CA-SP1, and NC–p6 sites by the virally encoded protease. This produces major structural rearrangements including disassembly and re-organization of the conical CA core as well as condensation of the RNA genome, forming mature virions that are now infectious and ready to enter and replicate in a nearby cell.

Studies on HIV-1 capsids over more than a decade have greatly impacted in situ structural virology by pushing cryo-ET and STA development, leading the cryo-ET resolution revolution towards achieving sub-nanometre and near-atomic resolutions (Mangala Prasad et al. 2022b; Schur et al. 2016; Li et al. 2020; Mattei et al. 2016; Schur et al. 2015; Zanetti et al. 2006; Zhao et al. 2013). Cryo-ET imaging of purified HIV-1 virions has revealed fascinating insights into the structure of CA and the molecular mechanisms driving assembly and maturation. In immature virions, CA forms discontinuous lattices with partial hexamers at the lattice edge, making the overall surface very elastic (Schur et al. 2015; Briggs et al. 2009). This arrangement of CA in the lattice is vital for the regulation of cleavage and the production of mature infectious virions. Determination of the CA structure in immature HIV-1 particles provided a comprehensive view of the proteolytic mechanism revealing that cleavage at the other sites caused CA lattice disturbance allowing for the final step of CA-SP1 proteolysis prior to maturation (Schur et al. 2016; Mendonca et al. 2021a, b). The lattice structure of CA also helped identify the destabilization of the CA-SP1 helix bundle as being the key factor for CA maturation (Mattei et al. 2018). An integrative approach combining cryo-ET, cryo-EM, and all-atom molecular dynamics was used to determine the first high-resolution structure of a mature HIV-1 capsid core at 3 nm (Zhao et al. 2013). The CA core in mature particles is cone-shaped and heterogeneous, comprising 186 hexamers and 12 pentamers (Fig. 1.5a). The heterogeneity is attributed to the relative movements of the N- and C-terminal domains that result in tilt and twist of hexamer pairs that in turn dictates variable curvature throughout the entire CA cone (Mattei et al. 2016). Further, in mature HIV-1 particles, the matrix protein MA is packed as a hexagonal lattice that facilitates the distribution and fusion of the surface glycoprotein, Env. These studies employed a dose-symmetric tilt acquisition scheme and 3D CTF correction, showcasing the full power of cryo-ET coupled with STA to achieve near-atomic resolutions (Qu et al. 2021). Put together, these observations suggest that in HIV-1, the Gag lattice not only modulates the assembly of CA and MA but also regulates the functional properties of the viral envelope and Env.

## Severe Acute Respiratory Syndrome Coronavirus 2 (SARS-CoV-2)

Belonging to the family of *Coronaviridae*, SARS-CoV-2, the causative agent of the COVID-19 pandemic, is a positive sense single-stranded RNA virus that primarily infects the respiratory tract causing respiratory illness, and in severe cases pneumonia and acute respiratory distress syndrome (ARDS). SARS-CoV-2 uses its surface glycoprotein, spike protein "S", to bind the receptor, angiotensin converting enzyme

2 (ACE2), to enter cells. The unprecedented public health and economic impact caused by the COVID-19 pandemic has been met with an equally extraordinary scientific response. The majority of this response has been focused on extensive structural and cellular analysis by cryo-ET and cryo-EM to understand the multi-step entry process of SARS-CoV-2, including the structure of the S protein conformations, required for SARS-CoV-2 to associate with receptor ACE2, engagement of the receptor-binding domain of S with ACE2, proteolytic activation of S, endocytosis, and membrane fusion (Yao et al. 2020; Barnes et al. 2020; Hoffmann et al. 2020; Jackson et al. 2022; Ke et al. 2020; Liu et al. 2020; Turonova et al. 2020; Wrapp et al. 2020).

Cryo-ET of purified, inactivated clinical strains of SARS-CoV2 virions exhibit pleomorphic morphology, with 20–40 randomly distributed S proteins on their surface (Yao et al. 2020; Ke et al. 2020; Liu et al. 2020; Turonova et al. 2020). In SARS-CoV-2 virions, the RNA genome is encapsulated by the nucleoprotein, N, forming ribonucleoprotein complexes (RNPs). In situ cryo-ET coupled with STA provided insights into the detailed structures of prefusion and postfusion S protein as well as RNPs (Fig. 1.5b). Specifically of note was the inverse G-shaped architecture (~15 nm in diameter) of RNP obtained from 2D classification and averaging within virions. Further, the overall 3D ultrastructures of RNPs were observed to be similar to a bead on a string following a specific configuration, a hexagonal "eggs in a nest" or a tetrahedral "pyramid" arrangement (Yao et al. 2020). The flexibility present in the stalk of the trimeric S spike drives the conformational switching of the receptor-binding domain (RBD) allowing it to adopt an open (1 RBD up or 2 RBDs up) or closed conformation in the prefusion form (Ke et al. 2020; Turonova et al. 2020). Elegant STA workflows in these studies confirmed that the open and closed conformations of S protein on the virus surface prime the RBDs for association with the ACE2 receptor as well as modulating antibody binding. Interestingly, the S on purified virions was mostly present in the prefusion conformation and observed to be extensively glycosylated, protecting S from antibody access and binding. STA also revealed the presence of three hinges in the stalk domain of S, allowing the head domain significant orientational freedom, permitting it to flex as high as 90°. Crucially, these studies indicated that it was important to consider how virus samples were prepared for vaccine development, as inactivation and purification procedures altered the distribution and ratio of the S open and closed conformations. This, in turn, can modify their immune sensitivity, and their ability to induce a neutralizing antibody response, therefore being an effective vaccine immunogen. Crucially, cryo-ET coupled with STA demonstrated unequivocally the power of this technique to examine the structure of new emerging viruses and its capability in providing tools to screen and assess the effectiveness of vaccine candidates for therapeutic development.

## Other Enveloped Viruses

This section will highlight how cryo-ET has advanced our understanding of other pleomorphic and clinically important viruses such as LASV, IAV, RVFV, and HSV-1 (Fig. 1.5).

Lassa virus (LASV) is a zoonotic pathogen belonging to the family of arenaviruses and causes haemorrhagic fever. Endemic in West Africa, with nearly half a million people being affected annually, it is spread and transmitted via aerosols, therefore posing a major health risk and burden. LASV encodes for the heavily glycosylated membrane glycoprotein (GP) that comprises the receptor-binding GP1 subunit and the fusion GP2 subunit. Cryo-ET studies of LASV at low pH revealed the trimeric structure of GP on the virion surface and the conformational changes it undergoes for receptor binding and viral entry (Fig. 1.5c) (Li et al. 2016). As LASV is a containment level-4 pathogen, samples were chemically fixed prior to imaging. Although the structure of GP from a non-infectious virus-like particles differed in morphology, it shared a high degree of structural similarity with the virion GP. This enabled the use of low pH to probe the structural states of unbound GP and GP in complex with lysosomal receptor LAMP1, similar to that encountered during LASV endosomal entry and membrane fusion (Li et al. 2016).

Spherical viruses like LASV have a discontinuous and loosely structured matrix layer beneath the lipid envelope of the virion, proving its investigation a huge challenge. Cryo-ET and STA showed interaction of the cytoplasmic tail of LASV GP with the underlying matrix protein (Z) of LASV. While a hexameric assembly of Z dimers was obtained by X-ray crystal structures, it was not observed in the tomograms, possibly due to the lower stiffness of oligomerized Z beneath the viral membrane. Not only did this study resolve the first high-resolution in situ structure of viral glycoprotein, but it also provided snapshots of structural rearrangements that occur during host cell entry and membrane fusion.

Influenza A virus (IAV) is a pleomorphic RNA virus, which causes seasonal epidemics and occasional pandemics infecting millions of people worldwide. IAV exhibits spherical and filamentous morphology, with the latter form believed to be the clinical phenotype. The virus envelope is studded with glycoproteins haemagglutinin and neuraminidase which are critical for virus entry and egress. Beneath the viral envelope is the matrix protein (M) layer that plays a key role in virus fusion, assembly, and morphogenesis. Recent studies on intact IAV filaments and virus-like particles (VLP) suggest dimerization of M through binding of the C-terminal domain (CTD) with the N-terminal domain (NTD) of a neighbouring M via electrostatic interactions to form a polymeric helical assembly (Fig. 1.5d) (Peukes et al. 2020; Selzer et al. 2020). A positively charged cluster of five histidine residues present at the interface of three M dimers serves as a trigger for pH-mediated disassembly. These structural data imply that the M dimer represents the basic assembly unit for filamentous enveloped viruses, providing a greater understanding of the IAV assembly and disassembly process.

The GPs of some enveloped viruses form oligomers covering the entire surface of the virus. Such architectures are the result of viruses that frequently lack the M proteins and rely on the GPs for assembly and budding. For example, in bunyaviruses, GPs can form diverse oligomers ranging from monomers on nairovirus to tetramers of dimers on phleboviruses, like Rift Valley fever virus (RVFV). Transmitted by mosquitoes, RVFV is a clinically important pathogen that infects humans and livestock causing viral haemorrhagic fever. The structure of an early fusion intermediate of RVFV with liposomes at pH 5 determined by cryo-ET/STA pipelines revealed the insertion of GP pentamers into the target liposome membrane forming an intermediate sandwich structure comprising GP pentamer-viral membrane-target membrane (Halldorsson et al. 2018). Interestingly, this intermediate was not observed on pentamers unbound to a target membrane or on any hexamers, suggesting greater stability of the pentamers. Further, the arrangement of GP oligomers in a continuous higher-order lattice covering the entire virion surface (Fig. 1.5e) reduced the overall solvent-exposed area on the GPs, aiding RVFV to evade antibody recognition. This work was the first structure of a higher oligomeric GP, demonstrating the virus fusion mechanism in action using cryo-ET and STA.

Belonging to the alphaherpesvirus subfamily, herpes simplex virus, HSV-1 is a double-stranded DNA (dsDNA) virus that is contagious and causes cold sores and rare but severe encephalitis. The dsDNA genome is packed into the icosahedral ($T = 16$) capsid in a liquid crystalline arrangement (Booy et al. 1991). Release and packaging of the genome during viral disassembly and assembly are mediated through a special structure in the capsid called the portal (Fig. 1.5f) (McElwee et al. 2018; Chang et al. 2007; Schmid et al. 2012). Imaging of HSV-1 virions by cryo-ET has uncovered several asymmetric elements including polarization of the nucleocapsid at one end within virions and the disproportionate arrangement of the tegument (the proteinaceous region between the capsid and the viral envelope) cap-like structure, which contained 40 nm long-filaments ($\sim$7 nm wide) apposed to the membrane (Grunewald et al. 2003). In recent times, cryo-ET/STA has been used to determine the structure of the glycoprotein B prefusion complex to 9 Å resolution, demonstrating the power and potential of this technique to study uncharacterized protein structures of large viruses at high resolution, which serve as immunogens for vaccine development (Vollmer et al. 2020).

Overall, these studies demonstrate that cryo-ET has a tremendous capacity to improve our understanding of the molecular architecture of purified viruses at high resolution, as well as their mechanisms of entry, disassembly, and assembly, highlighting its ability to reveal novel drug targets and the development of potential therapeutics and vaccines.

## In Situ Cellular Cryo-ET Enables Peering into the Cell

Cellular cryo-ET is emerging as a major structural tool to study virus-host interactions in situ (within cells and tissues). In combination with other integrative techniques like cryo-CLEM, cryo-FIB, and CSXT, cryo-ET is making great strides, overcoming the significant gap in length scales and resolution between viruses and cells. This section of the chapter will describe how cryo-ET is aiding studies of enveloped virus behaviour within infected cells, focusing on recent discoveries that have expanded our knowledge of the viral life cycle, cellular interactions, and viral activities within eukaryotic cells at nanometre to sub-nanometre resolution.

### *Probing Enveloped Viral Infections and Molecular Mechanisms of Virus-Host Interactions*

For viruses to enter cells and replicate successfully, it is imperative that they carry out several steps in an ordered manner. Being the most abundant species in the biosphere (~1 quintillion), they adopt diverse strategies for the key steps of their life cycle such as cell attachment and entry, viral disassembly and genome release, replication and translation, assembly, and virus budding, with the common goal for survival via multiplication and evolution. Until now, high-resolution in situ visualization of these viral activities have been hindered by the massive gap in length scales between cells and viruses and the thickness of the cellular samples. For a long time, hypothesis generation, functional experiments and observations, and classical plastic resin embedding EM have driven our understanding of the viral life cycle, devoid of any direct structural evidence for proposed virus–host interactions. Resin-embedded plastic cell sections provide striking contrast images of viruses infecting cells, but at the cost of sample preservation, quality, and resolution. In recent years, the combination of cryo-CLEM, EM-specific labelling, cryo-FIB, Tokuyasu VFS, cellular cryo-ET, and STA techniques has provided unprecedented insights into infected cells at nanometre resolution (Fig. 1.6). This section will highlight the identification of various activities of enveloped viruses inside cells based on the key steps of the viral life cycle.

#### Viral Disassembly and Genome Release

Enveloped viruses attach and enter cells by binding to receptors. This often involves using the endocytotic machinery for membrane fusion and subsequent uncoating to release their genome, all of which happens in the cytoplasm of the cell. Protein-mediated membrane fusion is a critical step in enveloped virus infection and a fundamental process that underpins many cellular functions. Extensive cryo-ET studies on virus binding to liposomes have provided a simplistic model to extract

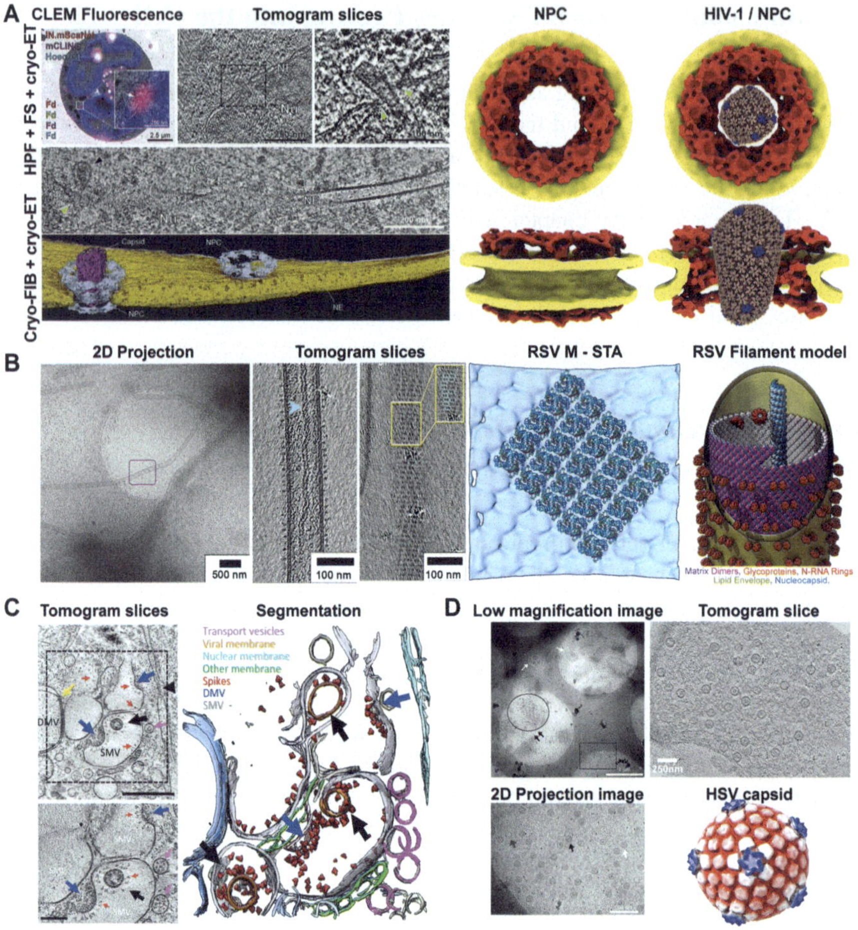


**Fig. 1.6 In situ imaging of viral processes in cells by cryo-ET and STA.** In combination with cryo-CLEM, CEMOVIS, Tokuyasu VFS, and cryo-FIB milling, cryo-ET has provided important structural insights for various steps of the viral life cycle. (**a**) Overlay of a fluorescence HPF CEMOVIS section of a cell with the correlated EM image showing staining for HIV-1 (mScarlet, red), NPC (mCLING, pink), nucleus (Hoescht, blue), and decorated with multi-fluorescent fiducials (Fd). Enlarged region (white square) displays the mScarlet HIV capsid signal at the nuclear envelope (white arrow). A slice through a tomogram of the CEMOVIS section at the CLEM correlated position shows an HIV capsid lodged at the nuclear envelope (NE). Both cytosol (Cy) and the nucleus (Nu) regions are on either side of NE. The enlarged image (black square) of the tomogram slice shows an intact HIV-1 capsid (black arrowhead) docked deep inside the central channel of the NPC (green arrowheads) with its narrow end towards the nucleoplasm. Slice through a tomogram of a cryo-FIB-milled lamella infected with HIV-1 corroborates the data, illustrating a capsid (black arrowhead) localized inside of the NPC (green arrowhead). Isosurface rendering via segmentation is shown. Capsid (magenta); NE (yellow); NPC (purple). Top and side views of the NPC structures (EMDB-11967; red; NE—yellow) without and with superposition of the HIV-1 capsid (PDB: 3J3Y, cut-open side view) revealing dilation of the NPC wide enough to transport intact HIV-1 capsids to the nucleus. The CLEM overlay and cryo-ET images were adapted from Zila et al. (2021), copyright © 2021, Elsevier. Structural models of NPC with and without HIV-1 capsids were made using publicly available PDB (3J3Y) and EMDB (11967) coordinates. (**b**) Budding RSV filaments studied by cryo-ET (2D projection image) revealed preservation of intact

important knowledge on fusion activation, intermediate conformations of glycoproteins interacting with the membrane, and the sequence of events resulting in membrane remodelling, fusion, and genome release (Halldorsson et al. 2018; Calder and Rosenthal 2016; Ward et al. 2020; Mangala Prasad et al. 2022a). IAV fusion studies have revealed important insights into the fusion process. Membrane fusion of viral and endosomal membranes at low pH is driven by conformational changes in the IAV glycoprotein haemagglutinin, HA. Cholesterol-mediated membrane fusion proceeds via an intermediate called hemifusion that is induced by HA, resulting in the formation of a hemifusion stalk junction leading to the formation of a fusion pore (Chlanda et al. 2016). A sequence of membrane-centric remodelling events at pH 5.5–6.0 leads to the formation of this fusion pore. This includes the formation of a "dimple structure" in the target membrane created by a localized HA cluster, scission of the target membrane with the viral membrane being unperturbed, creation of an open mouth funnel structure that inserts into the viral membrane, which in turn leads to apposition, and a hemifusion intermediate with a constrained pore that is maintained by the intact IAV matrix protein layer (Lee, 2010). When pH is lowered to 5.0 as in the case of late endosomes, the matrix disintegrates or weakens, causing dilation of the pore and release of the genome. These studies suggest that by

**Fig. 1.6** (continued) filamentous morphology and a complex envelope. Tomographic slice of the enlarged region (black square) showed the viral envelope of the filament lined with the matrix protein M below acting as a scaffold. RSV filaments also displayed dense packing of outer glycoprotein spikes exhibiting a picket-fence-like appearance. Virions packed helical nucleocapsids having characteristic herringbone morphology (blue arrow). Helical ordering of the glycoproteins as well as their clustering in pairs (blue ovals in enlarged image—yellow rectangle) is clearly evident looking at the top slice of the tomogram. STA of the viral envelope reveals the helical arrangement of the matrix layer (blue), and it is packing as a curved lattice of M dimers upon docking with the X-ray structure of M (PDB: 4V23). A cut-open view of an RSV filament model incorporating structural information pertaining to all its viral proteins. Figure adapted from Conley et al. (2022) under CC BY 4.0. (**c**) In situ cryo-ET of cryo-FIB lamella depicts the SARS-CoV-2 assembly process. Tomographic slice of the lamella (scale = 300 nm) and the enlarged inset (black square, scale = 100 nm) show DMV portals (yellow arrow), assembling virus (blue arrow), assembled virus (black arrow), viral spikes on SMV membranes (red arrows), transporting vesicles around the assembly site (pink arrow), and a nuclear pore (black arrowhead). 3D volume rendering by segmentation illustrates three virus particles (black arrows) and two assembly sites (blue arrows). Figure reproduced from Mendonca et al. (2021a) under CC BY 4.0. (**d**) In situ cryo-ET of Tokuyasu VFS revealed critical information on the HSV-1 assembly pathway inside the nucleus. Low magnification cryo-EM image of the VFS showed well-preserved structure and an abundance of capsids within the nucleus (black rectangle) and cytoplasm (black circle) of the cell separated by the nuclear membrane (black arrows). Other membranous and vacuolar structures (white arrows) are also observed within the dense cytoplasm. High-magnification 2D cryo-EM projection image and a central tomographic slice display uniform dispersal of the three different types of capsids, A-type (empty capsids, grey arrow), B-type (scaffold-containing capsids, white arrow), and C-type (DNA-containing capsids, black arrow). HSV-1 capsid structure determined by STA showed additional pronounced star-like density (blue) above the pentons on the five-fold capsid vertex corresponding to the tegument protein complex CATC, confirming its addition to the capsid prior to nuclear egress during HSV assembly. Figure adapted from Vijayakrishnan et al. (2020) under CC BY 4.0

concentrating on membrane deformation of only the target membrane during the early steps of fusion, IAV does not risk the premature release of its RNPs into the degradative lumens of mature endosomes and lysosomes.

Fusion intermediates of RVFV, Chikungunya virus (CHIKV), and SARS-CoV-2 have also been observed by cryo-ET and STA. Using liposomes and varying conditions of pH as well as reaction time points, the ultrastructure of CHIKV, its class-II glycoprotein conformations, membrane re-organization, and the fusion pathway were mapped at high resolution, demonstrating that the process is markedly different from fusion mediated by class-I fusion systems such as IAV. Further, the surface glycoprotein-nucleocapsid interaction is highly dependent on the pH and plays a regulatory role in CHIKV fusion (Mangala Prasad et al. 2022a). Recent high-resolution in situ structural studies of SARS-CoV-2 virions provided evidence for the molecular basis of enhanced infectivity and attenuated dependency on cellular proteins during the fusion of the Delta variant, suggesting the correlation of fusogenicity with pathogenicity (Song et al. 2022).

Postfusion, some enveloped viruses like retroviruses and orthomyxoviruses need to import their genome into the nucleus of the host cell for replication, a process that is not well understood. It was believed that HIV-1 capsids disassemble at the nuclear membrane prior to crossing the nuclear pore complex (NPC) owing to the small size of NPC. However, in a ground-breaking study using in situ cryo-CLEM, CEMOVIS, cryo-FIB, cryo-ET, and STA workflows, it was elegantly shown that in human CD4+ T cells, intact HIV-1 capsid cores dock at the nuclear membrane and initiate NPC dilation instead to accommodate complete viral capsids during translocation (Fig. 1.6a) (Zila et al. 2021). Once the capsids enter the nucleus, they undergo partial disruption leading to uncoating and genome release. This strengthens the model proposing HIV-1 uncoating occurring at the nuclear pore rather than it happening upon entry, i.e., cytoplasmic uncoating. Crucially, this indicates significant differences in structural information obtained between in vitro and in situ studies, arguing for the importance of studying in situ systems when interpreting data to understand the function and molecular mechanisms of viral infection in native-like environments such as cells and tissues.

## Viral Factories Are Sites of Viral Replication

After release, the genome has to be replicated and translated for packaging into progeny virions. The majority of RNA viruses (positive and negative strands) replicate in the cytosol, modifying intracellular membranes to form virus-induced vesicles. These vesicles serve as viral replication factories by concentrating all the necessary ingredients prior to assembly and budding whilst evading intracellular detection and host immune defences.

### Enveloped Positive-Strand RNA Viruses

In positive-strand RNA viruses (+RNA), the viral antigenome is used for replication and transcription, but not directly for protein translation or packaging progeny viruses. Cryo-ET of SARS-CoV-2 identified endoplasmic reticulum (ER)-derived double-membrane vesicles (DMV) encompassing branched double-stranded RNA filaments (Klein et al. 2020; Mendonca et al. 2021a; Knoops et al. 2008; Miller and Krijnse-Locker 2008). It was further realized that SARS-CoV-2 uses membranous viral replication compartments like DMVs to protect the genome, adopting these structures as safehouses during replication to evade the innate antiviral response. A similar strategy has been observed in cells infected with alphaviruses, which form single membrane, balloon-shaped replication complex structures, called spherules, composed of several non-structural proteins (NSP) and house RNA intermediates (Laurent et al. 2022; Tan et al. 2022). Similar to the NPC, +RNA viruses form highly oligomerized pore complexes that import and export material through the viral replication complex. In alphaviruses, these pores are present on the neck of spherules and found to be a 12-fold symmetric crown-shaped complex by STA, comprising NSP4, the RNA-dependent RNA polymerase facing the spherule side (Laurent et al. 2022; Tan et al. 2022). Similar pores were observed in cells infected with coronaviruses (mouse hepatitis virus and SARS-CoV-2) by GFP and CLEM labelling. This answered the long-standing question of how RNAs in the DMVs access the cytosol for RNP encapsidation and budding. Using cryo-FIB milling coupled with cryo-ET, pores spanning the DMV membranes were observed to be hexameric 6-fold symmetric structures, comprising NSP3 as the central component. These pores formed an exit pathway for the newly made viral genomic RNA from the DMV interior to cytosolic sites for encapsidation (Wolff et al. 2020).

While NSPs have several important roles during viral infection, a crucial one being suppression and evasion of cellular antiviral defences, their functions depend on coordinated interactions with a number of cellular proteins, membranes, and multiple NSPs. An in-depth understanding of their function requires knowledge of their residing cellular context. High-resolution in situ structures of NSPs by cryo-ET offer the key to deciphering defence strategies adopted by viruses, promising to be exciting targets for drug screening and therapeutic development.

### Enveloped Negative-Strand RNA Viruses

In negative-strand RNA viruses (-RNA), the viral antigenome functions for replication as well as protein coding, requiring exposure of both antigenome and genome to the cytosol. RNA viruses like vesicular stomatitis virus, measles virus, and RSV form unique virus-induced, membrane-less organelles called inclusion bodies (IBs), serving as sites of replication, i.e. virus factories (Heinrich et al. 2018; Rincheval et al. 2017; Su et al. 2021; Zhou et al. 2019). These IBs are suggested to be stress-induced liquid phase-separated condensates and their architecture remains largely obscured. A recent study integrating light microscopy, proteomics, and in situ cryo-

ET on cryo-FIB-milled lamellae of mumps virus-infected cells provided some insight into identifying IBs as liquid-like condensates that are enriched in nucleocapsids, and resolved the nucleocapsid structure to 6.5 Å resolution (Zhang et al. 2022).

## Viral Assembly and Egress

In enveloped viruses, the replicated genome along with translated viral proteins will need to gather to assemble into progeny virions and bud from the cell membrane. Recently, these events have been captured by in situ cryo-ET imaging on various clinically relevant human viruses revealing overall virus morphology and confirmations of the viral structural proteins involved.

Although respiratory viruses are largely pleomorphic and enveloped, budding and egress of the clinically relevant filamentous morphology of IAV and RSV from human cells have been well studied by cryo-ET. Using light microscopy and cryo-ET, intact budding IAV filaments attached or in proximity to the cell membrane were visualized by performing viral infections directly on EM grids prior to imaging. This impactful study helped develop sample preparation methods for imaging the ultrastructure of fragile filamentous viruses budding at the membrane, which are otherwise ruptured during harsh purification preparations. Budding IAV filaments were extremely long (greater than 20 μm) and often attached to the cell membrane. In addition to smaller bacilliform particles, three distinct classes of viral filaments (empty filaments, RNP-containing filaments, and Archetti bodies) were identified and proposed to have different functions, with empty filaments devoid of RNPs suggested to play a role in immune evasion (Vijayakrishnan et al. 2013; Dadonaite et al. 2016).

Using a similar approach, Cryo-ET and STA of RSV filaments egressing from cells revealed the helical curved lattice architecture of the matrix protein M that forms an endoskeleton below the viral envelope (Fig. 1.6b) (Conley et al. 2022). Further, the helical packing of M dimers coordinates the helical ordering of the surface glycoproteins that cluster in pairs as doublets. In addition to packing multiple helical RNPs, virions also package an abundance of ring assemblies formed of nucleocapsid N and RNA. During respiratory virus infections, the respiratory tract constitutes an ecological niche where taxonomically different respiratory viruses can coinfect and co-exist. In a landmark study, employing multimodal imaging approaches, budding infectious hybrid virus particles (HVPs) from coinfection of taxonomically different, clinically important viruses IAV and RSV were identified. HVPs are composed of structural, genomic, and functional components of both parental viruses, and exhibit altered antigenicity and expanded tropism, a crucial factor impacting pathogenesis and disease outcome (Haney et al. 2022).

Coronaviruses bud and egress at different sites. This was confirmed by cryo-ET of cryo-FIB-milled MHV and SARS-CoV-2 infected cells, where virus assembly was observed at ER-Golgi intermediate compartment (ERGIC)-like vesicles. Further, several assembly intermediates were captured including trimeric S protein inside vesicles, RNPs outside the vesicles, and smaller adjacent vesicles comprising inward pointing S trimers (Fig. 1.6c) (Klein et al. 2020; Mendonca et al. 2021a, b; Wolff et al. 2020). The RNPs are proposed to assemble following encapsidation of newly synthesized RNA, exported via the replication and transcription complex at the gate of the DMV-spanning pore, by accumulated nucleoprotein N (Wolff et al. 2020). This is further validated by the presence of ordered RNP assemblies in isolated SARS-CoV-2 virions (Yao et al. 2020). In addition, egress at the plasma membrane suggests two egressing pathways, through extended tunnels via exocytosis-like release or through membrane rupture (Mendonca et al. 2021a, b). Further, these studies confirmed that in vitro characterization of structures and distribution of S trimers in purified virions were comparable to in situ intracellular virions.

In some enveloped viruses, the crucial steps of assembly and budding take place concomitantly. In CHIKV-infected cells, cryo-ET and STA captured the various snapshots of assembly and budding at nanometre resolution, providing mechanistic insights into the whole process. Structures on budding intermediates revealed a progressive glycoprotein spike-driven nucleocapsid morphogenesis, whereupon initial convergence of membrane-embedded spikes and immature non-icosahedral nucleocapsid-like particles (NLPs) occurs at the cell surface (Chmielewski et al. 2022). Subsequently, the assembly of the icosahedral spike lattice enwraps the NLPs and reorganizes it from an asymmetric structure to an icosahedral nucleocapsid through a series of sequential steps. Further, the structure of a neutralizing antibody C9, crosslinking with CHIKV spikes on the infected cell surface and inhibiting virus budding, provided a mechanistic model for the development of therapeutics.

The measles virus (MeV) remains an important human pathogen with no antivirals to treat infection. Whole-cell cryo-ET of MeV provided direct evidence for a detailed assembly pathway and release of MeV virions. The matrix protein M drives virus assembly by coalescing with the host membrane at the budding site, concentrating the glycoproteins and RNPs via direct protein–protein interactions (Ke et al. 2018). Budding and release are driven by membrane remodelling mediated by M that forms a 2D paracrystalline lattice, which in turn induces an F array formation. This study elegantly demonstrated that in the paramyxovirus family, oligomerization of the scaffold viral protein M into ordered arrays regulates the spatial arrangement of surface glycoproteins and the RNP complex for viral assembly and budding.

Some enveloped viruses like DNA viruses undergo multiple assembly steps at various sites in the cell. Herpesviruses replicate and assemble the capsid inside the nucleus. After nuclear egress, the mature nucleocapsid gathers its envelope, accessory tegument proteins, and glycoproteins during budding at the plasma membrane to egress as mature enveloped virions. CLEM and cryo-ET imaging along with STA analysis of Tokuyasu VFS of HSV-1-infected cells revealed that different forms of capsids (A-type, B-type, and C-type) assemble at the nucleus with varying scaffold proteins and packaged DNA (Vijayakrishnan et al. 2020). The capsid-associated tegument complex (CATC) comprises three tegument proteins and is observed on matured enveloped HSV-1 virions. For a long time, the precise site of CATC addition, i.e. in the nucleus or the cytoplasm, had been unclear. Importantly, this study resolved the structure of CATC on intranuclear HSV-1 capsids (Fig. 1.6d), confirming that CATC binds capsids in the nuclei prior to egress, thereby exemplifying the potential of CLEM/cryo-ET/STA workflows in imaging and resolving structure within crowded intracellular environments like the nucleus.

Using multimodal imaging including CEMOVIS, HSV-1 maturation and assembly at the nuclear membrane was investigated, revealing that a protein coat of the nuclear egress complex (NEC) is added to the HSV-1 capsids prior to egress from the inner nuclear membrane (C Hagen et al. 2015). Further, in infected cells, the NEC coat displayed a curved double-layered hexagonal lattice composed of viral protein pUL34 that mediates the curvature and size of the coat to match the size of HSV-1 capsids.

## Concluding Remarks and Perspectives

Understanding the molecular architecture of different viruses and the structural and mechanistic basis of their infection, replication and transmission are crucial to understanding the fundamental biology of how viruses work. While X-ray crystallography and NMR yield atomic structures of protein complexes, they are limited to purified specimens. Super-resolution light microscopy and CSXT are helpful to study large proteins and cells in situ but at lower resolutions (25–250 nm). Cryo-ET with STA is uniquely suited to image both in vitro and in situ homogeneous and pleomorphic samples, bridging the resolution gap and achieving 0.5–5 nm resolution. Advances in microscopes, detectors, and data processing software have enabled cryo-ET to resolve structures of macromolecular assemblies at near-atomic resolution, increasing its application in the field of structural biology. All these technical developments built on earlier cryo-ET studies were quickly applied to intact SARS-CoV-2 viruses during the pandemic, providing vital structural information for vaccine candidates and the development of antivirals. In combination with integrative techniques and other imaging modalities such as cryo-CLEM, CSXT, cryo-FIB, and VFS, cryo-ET has provided important insights into structure and mechanisms driving entry, replication, assembly, maturation, and egress during infection of several pleomorphic enveloped viruses.

Although cryo-ET and ancillary techniques have allowed in situ visualization of structures and life cycles of many important viruses, steps associated with data collection and processing are challenging and are relatively low throughput. The cryo-ET resolution revolution has made possible several innovations pertaining to tilt series (TS) acquisition, data processing, and hardware, but general throughput and resolution are still incomparable to SPA. Recent progress in TS acquisition developments has incorporated tilt schemes and instrument innovations, aiming to achieve both precision and throughput in practical cryo-ET applications. Further, the lack of a streamlined approach to cryo-ET data processing, involving the execution of several steps with a range of software applications, makes the computational processing time intensive and tedious. Integrating STA applications into a single package will provide a flexible choice of tools and speed up the whole process. In addition, as we aim to approach higher resolutions, STA methods need further development and evaluation to better extract high-frequency information from low-resolution SNR images. Further, artificial intelligence-based semi-automated template matching, image segmentation, and surface-based rendering techniques are necessary to circumvent time-consuming and subjective manual drawing of structures to speed up the entire process for 3D visualization, annotation and data interpretation.

Locating a protein or virus within the dense environment of the cell is comparable to finding a needle in a haystack. Advances in cryo-CLEM coupled with cryo-ET have helped visualize several steps of the viral cycle by localizing proteins of interest, albeit at low resolution (~50–100 nm). The precise location of proteins/viruses requires the development of super-resolution CLEM to improve resolutions to 10–40 nm along the lateral Z-axis. The success of such technical development alongside high throughput will be crucial for routine imaging of intracellular viral events at sub-nm resolution. Alternatively, the development of EM tags will prove useful when correlating data across imaging methods. High-contrast EM tags that are genetically coded and pose minimal toxicity to cells with negligible influence on protein structure such as dual nano tags made of fluorescent gold or fluorescent silver, quantum dots, and DNA origami offer viable options for EM labelling.

Thinning down cellular and tissue samples by cryo-FIB milling is a key step for in situ imaging by cryo-ET. A major drawback currently is the low throughput in milling due to challenges in controlling cell growth on grids, targeted milling, minimizing lamella damage during or after milling, lamella contamination during milling and specimen transfer, and processing thick samples such as tissues or cell blocks. Recently, several approaches have been implemented to mitigate these problems. Although at an early stage, the development and optimization of integrated correlative light and cryo-FIB microscopes will allow simultaneous localization and milling of ROI, improving the efficiency of targeted milling. Furthermore, the development of several techniques to improve lamella throughput and quality will enable automated pipelines for routine efficient cryo-milling of samples from organisms to macromolecules in the future. This includes increasing sample throughput and optimizing cell distribution procedures by micropatterning of EM grids, designing devices that minimize lamella contamination such as high-vacuum transfer systems and cryo-shields, improving milling protocols like the Waffle method to increase milling speed and lamella stability, and implementing the cryo-FIB lift-out technique to increase milling efficiency and aid milling of bulky specimens. Synchronized development of all of the above will critically help streamline the workflow for automated milling, facilitating the ambitious vision for the future of correlated cryo-FIB.

As more and more processes pertaining to the viral life cycle are being discovered by in situ cryo-ET within cells, mechanistic insights on some key steps such as viral entry, membrane fusion and genome release, the architecture of IBs, and replication and transcription complexes are currently lacking, and therefore deserving of attention to fill the knowledge gaps for currently well-studied enveloped viruses. But more importantly, in situ cryo-ET will be a critical technique to understand virus and host interplay during infection for the huge spectrum of other pathogenic enveloped viruses and novel emerging viruses that are yet to be investigated and may potentially be sources of future epidemics/pandemics.

With cryo-ET rapidly evolving, it is without a doubt that it will become an indispensable and sought-after structural biology tool to study complex processes of viral infections in their hosts within the native environment of cells and tissues at extraordinary resolution. With improvements in sample and data throughput, sample quality, and data processing pipelines, sub-nm resolution structures are expected to be routinely determined for purified proteins and macromolecular assemblies within the cell. This paves the way for exciting future innovations such as time-resolved cryo-ET to study dynamic molecular processes and imaging large volume clinical samples that in turn will expand the capability of cryo-ET for drug discovery and vaccine development.

Looking ahead into the future, given how cryo-ET has advanced in such a short time, there are bound to be many more surprising and exciting technological advancements as well as breathtaking scientific discoveries in the years to come. In situ cryo-ET imaging is heading towards a grand and attainable reality: a 3D reconstruction of a virus-infected cell in its native state, providing unprecedented insights on protein interactions and molecular mechanisms.

**Acknowledgements** The author apologizes to all colleagues whose significant contributions could not be covered in this chapter due to space limitations.

## References

Al-Amoudi A, Norlen LP, Dubochet J (2004) Cryo-electron microscopy of vitreous sections of native biological cells and tissues. J Struct Biol 148(1):131–135. https://doi.org/10.1016/j.jsb.2004.03.010

Al-Amoudi A, Studer D, Dubochet J (2005) Cutting artefacts and cutting process in vitreous sections for cryo-electron microscopy. J Struct Biol 150(1):109–121. https://doi.org/10.1016/j.jsb.2005.01.003

Bammes BE, Rochat RH, Jakana J, Chen DH, Chiu W (2012) Direct electron detection yields cryo-EM reconstructions at resolutions beyond 3/4 Nyquist frequency. J Struct Biol 177(3):589–601. https://doi.org/10.1016/j.jsb.2012.01.008

Barnes CO, West AP Jr, Huey-Tubman KE, Hoffmann MAG, Sharaf NG, Hoffman PR et al (2020) Structures of human antibodies bound to SARS-CoV-2 spike reveal common epitopes and recurrent features of antibodies. Cell 182(4):828–842 e816. https://doi.org/10.1016/j.cell.2020.06.025

Berger C, Dumoux M, Glen T, Yee NB, Mitchels JM, Patakova Z et al (2023) Plasma FIB milling for the determination of structures in situ. Nat Commun 14(1):629. https://doi.org/10.1038/s41467-023-36372-9

Bharat TA, Davey NE, Ulbrich P, Riches JD, de Marco A, Rumlova M et al (2012) Structure of the immature retroviral capsid at 8 A resolution by cryo-electron microscopy. Nature 487(7407):385–389. https://doi.org/10.1038/nature11169

Bharat TAM, Russo CJ, Lowe J, Passmore LA, Scheres SHW (2015) Advances in single-particle electron cryomicroscopy structure determination applied to sub-tomogram averaging. Structure 23(9):1743–1753. https://doi.org/10.1016/j.str.2015.06.026

Booy FP, Newcomb WW, Trus BL, Brown JC, Baker TS, Steven AC (1991) Liquid-crystalline, phage-like packing of encapsidated DNA in herpes simplex virus. Cell 64(5):1007–1015. https://doi.org/10.1016/0092-8674(91)90324-r

Bouvette J, Liu HF, Du X, Zhou Y, Sikkema AP, da Fonseca Rezende EMJ et al (2021) Beam image-shift accelerated data acquisition for near-atomic resolution single-particle cryo-electron tomography. Nat Commun 12(1):1957. https://doi.org/10.1038/s41467-021-22251-8

Briggs JA, Riches JD, Glass B, Bartonova V, Zanetti G, Krausslich HG (2009) Structure and assembly of immature HIV. Proc Natl Acad Sci U S A 106(27):11090–11095. https://doi.org/10.1073/pnas.0903535106

Calder LJ, Rosenthal PB (2016) Cryomicroscopy provides structural snapshots of influenza virus membrane fusion. Nat Struct Mol Biol 23(9):853–858. https://doi.org/10.1038/nsmb.3271

Calder LJ, Wasilewski S, Berriman JA, Rosenthal PB (2010) Structural organization of a filamentous influenza A virus. Proc Natl Acad Sci U S A 107(23):10685–10690. https://doi.org/10.1073/pnas.1002123107

Campbell ID (2002) Timeline: the march of structural biology. Nat Rev Mol Cell Biol 3(5):377–381. https://doi.org/10.1038/nrm800

Chang JT, Schmid MF, Rixon FJ, Chiu W (2007) Electron cryotomography reveals the portal in the herpesvirus capsid. J Virol 81(4):2065–2068. https://doi.org/10.1128/JVI.02053-06

Chen M, Dai W, Sun SY, Jonasch D, He CY, Schmid MF et al (2017) Convolutional neural networks for automated annotation of cellular cryo-electron tomograms. Nat Methods 14(10):983–985. https://doi.org/10.1038/nmeth.4405

Chen JH, Vanslembrouck B, Ekman A, Aho V, Larabell CA, Le Gros MA et al (2022) Soft X-ray tomography reveals HSV-1-induced remodeling of human B cells. Viruses 14(12). https://doi.org/10.3390/v14122651

Chichon FJ, Rodriguez MJ, Pereiro E, Chiappi M, Perdiguero B, Guttmann P et al (2012) Cryo X-ray nano-tomography of vaccinia virus infected cells. J Struct Biol 177(2):202–211. https://doi.org/10.1016/j.jsb.2011.12.001

Chlanda P, Mekhedov E, Waters H, Schwartz CL, Fischer ER, Ryham RJ et al (2016) The hemifusion structure induced by influenza virus haemagglutinin is determined by physical properties of the target membranes. Nat Microbiol 1(6):16050. https://doi.org/10.1038/nmicrobiol.2016.50

Chmielewski D, Schmid MF, Simmons G, Jin J, Chiu W (2022) Chikungunya virus assembly and budding visualized in situ using cryogenic electron tomography. Nat Microbiol 7(8):1270–1279. https://doi.org/10.1038/s41564-022-01164-2

Conley MJ, Short JM, Burns AM, Streetley J, Hutchings J, Bakker SE et al (2022) Helical ordering of envelope-associated proteins and glycoproteins in respiratory syncytial virus. EMBO J 41(3):e109728. https://doi.org/10.15252/embj.2021109728

Cyrklaff M, Risco C, Fernandez JJ, Jimenez MV, Esteban M, Baumeister W et al (2005) Cryo-electron tomography of vaccinia virus. Proc Natl Acad Sci U S A 102(8):2772–2777. https://doi.org/10.1073/pnas.0409825102

Dadonaite B, Vijayakrishnan S, Fodor E, Bhella D, Hutchinson EC (2016) Filamentous influenza viruses. J Gen Virol 97(8):1755–1764. https://doi.org/10.1099/jgv.0.000535

Danev R, Buijsse B, Khoshouei M, Plitzko JM, Baumeister W (2014) Volta potential phase plate for in-focus phase contrast transmission electron microscopy. Proc Natl Acad Sci U S A 111(44):15635–15640. https://doi.org/10.1073/pnas.1418377111

Dubochet J, Adrian M, Chang JJ, Homo JC, Lepault J, McDowall AW et al (1988) Cryo-electron microscopy of vitrified specimens. Q Rev Biophys 21(2):129–228. https://doi.org/10.1017/s0033583500004297

Eisenstein F, Yanagisawa H, Kashihara H, Kikkawa M, Tsukita S, Danev R (2023) Parallel cryo electron tomography on in situ lamellae. Nat Methods 20(1):131–138. https://doi.org/10.1038/s41592-022-01690-1

Faas FG, Barcena M, Agronskaia AV, Gerritsen HC, Moscicka KB, Diebolder CA et al (2013) Localization of fluorescently labeled structures in frozen-hydrated samples using integrated light electron microscopy. J Struct Biol 181(3):283–290. https://doi.org/10.1016/j.jsb.2012.12.004

Grunewald K, Desai P, Winkler DC, Heymann JB, Belnap DM, Baumeister W et al (2003) Three-dimensional structure of herpes simplex virus from cryo-electron tomography. Science 302(5649):1396–1398. https://doi.org/10.1126/science.1090284

Hagen C, Dent KC, Zeev-Ben-Mordehai T, Grange M, Bosse JB, Whittle C et al (2015) Structural basis of vesicle formation at the inner nuclear membrane. Cell 163(7):1692–1701. https://doi.org/10.1016/j.cell.2015.11.029

Hagen WJH, Wan W, Briggs JAG (2017) Implementation of a cryo-electron tomography tilt-scheme optimized for high resolution subtomogram averaging. J Struct Biol 197(2):191–198. https://doi.org/10.1016/j.jsb.2016.06.007

Halldorsson S, Li S, Li M, Harlos K, Bowden TA, Huiskonen JT (2018) Shielding and activation of a viral membrane fusion protein. Nat Commun 9(1):349. https://doi.org/10.1038/s41467-017-02789-2

Hampton CM, Strauss JD, Ke Z, Dillard RS, Hammonds JE, Alonas E et al (2017) Correlated fluorescence microscopy and cryo-electron tomography of virus-infected or transfected mammalian cells. Nat Protoc 12(1):150–167. https://doi.org/10.1038/nprot.2016.168

Haney J, Vijayakrishnan S, Streetley J, Dee K, Goldfarb DM, Clarke M et al (2022) Coinfection by influenza A virus and respiratory syncytial virus produces hybrid virus particles. Nat Microbiol 7(11):1879–1890. https://doi.org/10.1038/s41564-022-01242-5

Harkiolaki M, Darrow MC, Spink MC, Kosior E, Dent K, Duke E (2018) Cryo-soft X-ray tomography: using soft X-rays to explore the ultrastructure of whole cells. Emerg Top Life Sci 2(1):81–92. https://doi.org/10.1042/ETLS20170086

Hayles MF, de Winter DA, Schneijdenberg CT, Meeldijk JD, Luecken U, Persoon H et al (2010) The making of frozen-hydrated, vitreous lamellas from cells for cryo-electron microscopy. J Struct Biol 172(2):180–190. https://doi.org/10.1016/j.jsb.2010.07.004

Heinrich BS, Maliga Z, Stein DA, Hyman AA, Whelan SPJ (2018) Phase transitions drive the formation of vesicular stomatitis virus replication compartments. mBio 9(5). https://doi.org/10.1128/mBio.02290-17

Hoffmann M, Kleine-Weber H, Schroeder S, Kruger N, Herrler T, Erichsen S et al (2020) SARS-CoV-2 cell entry depends on ACE2 and TMPRSS2 and is blocked by a clinically proven protease inhibitor. Cell 181(2):271–280 e278. https://doi.org/10.1016/j.cell.2020.02.052

Ibiricu I, Huiskonen JT, Dohner K, Bradke F, Sodeik B, Grunewald K (2011) Cryo electron tomography of herpes simplex virus during axonal transport and secondary envelopment in primary neurons. PLoS Pathog 7(12):e1002406. https://doi.org/10.1371/journal.ppat.1002406

Jackson CB, Farzan M, Chen B, Choe H (2022) Mechanisms of SARS-CoV-2 entry into cells. Nat Rev Mol Cell Biol 23(1):3–20. https://doi.org/10.1038/s41580-021-00418-x

Jian H, Yi Y, Wang J, Hao Y, Zhang M, Wang S et al (2021) Diversity and distribution of viruses inhabiting the deepest ocean on Earth. ISME J 15(10):3094–3110. https://doi.org/10.1038/s41396-021-00994-y

Jiang Z, Jin X, Li Y, Liu S, Liu XM, Wang YY et al (2020) Genetically encoded tags for direct synthesis of EM-visible gold nanoparticles in cells. Nat Methods 17(9):937–946. https://doi.org/10.1038/s41592-020-0911-z

Jun S, Ke D, Debiec K, Zhao G, Meng X, Ambrose Z et al (2011) Direct visualization of HIV-1 with correlative live-cell microscopy and cryo-electron tomography. Structure 19(11):1573–1581. https://doi.org/10.1016/j.str.2011.09.006

Ke Z, Strauss JD, Hampton CM, Brindley MA, Dillard RS, Leon F et al (2018) Promotion of virus assembly and organization by the measles virus matrix protein. Nat Commun 9(1):1736. https://doi.org/10.1038/s41467-018-04058-2

Ke Z, Oton J, Qu K, Cortese M, Zila V, McKeane L et al (2020) Structures and distributions of SARS-CoV-2 spike proteins on intact virions. Nature 588(7838):498–502. https://doi.org/10.1038/s41586-020-2665-2

Khavnekar S, Wan W, Majumder P, Wietrzynski W, Erdmann PS, Plitzko JM (2023) Multishot tomography for high-resolution in situ subtomogram averaging. J Struct Biol 215(1):107911. https://doi.org/10.1016/j.jsb.2022.107911

Klein S, Cortese M, Winter SL, Wachsmuth-Melm M, Neufeldt CJ, Cerikan B et al (2020) SARS-CoV-2 structure and replication characterized by in situ cryo-electron tomography. Nat Commun 11(1):5885. https://doi.org/10.1038/s41467-020-19619-7

Knoops K, Kikkert M, Worm SH, Zevenhoven-Dobbe JC, van der Meer Y, Koster AJ et al (2008) SARS-coronavirus replication is supported by a reticulovesicular network of modified endoplasmic reticulum. PLoS Biol 6(9):e226. https://doi.org/10.1371/journal.pbio.0060226

Laurent T, Kumar P, Liese S, Zare F, Jonasson M, Carlson A et al (2022) Architecture of the chikungunya virus replication organelle. Elife 11. https://doi.org/10.7554/eLife.83042

Lee KK (2010) Architecture of a nascent viral fusion pore. EMBO J 29(7):1299–1311. https://doi.org/10.1038/emboj.2010.13

Li S (2022) Cryo-electron tomography of enveloped viruses. *Trends Biochem Sci 47*(2):173–186. https://doi.org/10.1016/j.tibs.2021.08.005

Li S, Sun Z, Pryce R, Parsy ML, Fehling SK, Schlie K et al (2016) Acidic pH-induced conformations and LAMP1 binding of the lassa virus glycoprotein spike. PLoS Pathog 12(2):e1005418. https://doi.org/10.1371/journal.ppat.1005418

Li Z, Li W, Lu M, Bess J Jr, Chao CW, Gorman J et al (2020) Subnanometer structures of HIV-1 envelope trimers on aldrithiol-2-inactivated virus particles. Nat Struct Mol Biol 27(8):726–734. https://doi.org/10.1038/s41594-020-0452-2

Liu J, Wright ER, Winkler H (2010) 3D visualization of HIV virions by cryoelectron tomography. Methods Enzymol 483:267–290. https://doi.org/10.1016/S0076-6879(10)83014-9

Liu C, Mendonca L, Yang Y, Gao Y, Shen C, Liu J et al (2020) The architecture of inactivated SARS-CoV-2 with postfusion spikes revealed by cryo-EM and cryo-ET. Structure 28(11): 1218–1224 e1214. https://doi.org/10.1016/j.str.2020.10.001

Loconte V, Chen JH, Cortese M, Ekman A, Le Gros MA, Larabell C et al (2021) Using soft X-ray tomography for rapid whole-cell quantitative imaging of SARS-CoV-2-infected cells. Cell Rep Methods 1(7):100117. https://doi.org/10.1016/j.crmeth.2021.100117

Mangala Prasad V, Blijleven JS, Smit JM, Lee KK (2022a) Visualization of conformational changes and membrane remodeling leading to genome delivery by viral class-II fusion machinery. Nat Commun 13(1):4772. https://doi.org/10.1038/s41467-022-32431-9

Mangala Prasad V, Leaman DP, Lovendahl KN, Croft JT, Benhaim MA, Hodge EA et al (2022b) Cryo-ET of Env on intact HIV virions reveals structural variation and positioning on the Gag lattice. Cell 185(4):641–653 e617. https://doi.org/10.1016/j.cell.2022.01.013

Marko M, Hsieh C, Schalek R, Frank J, Mannella C (2007) Focused-ion-beam thinning of frozen-hydrated biological specimens for cryo-electron microscopy. Nat Methods 4(3):215–217. https://doi.org/10.1038/nmeth1014

Mattei S, Glass B, Hagen WJ, Krausslich HG, Briggs JA (2016) The structure and flexibility of conical HIV-1 capsids determined within intact virions. Science 354(6318):1434–1437. https://doi.org/10.1126/science.aah4972

Mattei S, Tan A, Glass B, Muller B, Krausslich HG, Briggs JAG (2018) High-resolution structures of HIV-1 Gag cleavage mutants determine structural switch for virus maturation. Proc Natl Acad Sci U S A 115(40):E9401–E9410. https://doi.org/10.1073/pnas.1811237115

McDonald KL, Auer M (2006) High-pressure freezing, cellular tomography, and structural cell biology. Biotechniques 41(2):137, 139, 141 passim. https://doi.org/10.2144/000112226

McElwee M, Vijayakrishnan S, Rixon F, Bhella D (2018) Structure of the herpes simplex virus portal-vertex. PLoS Biol 16(6):e2006191. https://doi.org/10.1371/journal.pbio.2006191

Mendonca L, Howe A, Gilchrist JB, Sheng Y, Sun D, Knight ML et al (2021a) Correlative multi-scale cryo-imaging unveils SARS-CoV-2 assembly and egress. Nat Commun 12(1):4629. https://doi.org/10.1038/s41467-021-24887-y

Mendonca L, Sun D, Ning J, Liu J, Kotecha A, Olek M et al (2021b) CryoET structures of immature HIV Gag reveal six-helix bundle. Commun Biol 4(1):481. https://doi.org/10.1038/s42003-021-01999-1

Miller S, Krijnse-Locker J (2008) Modification of intracellular membrane structures for virus replication. Nat Rev Microbiol 6(5):363–374. https://doi.org/10.1038/nrmicro1890

Nahas KL, Connor V, Scherer KM, Kaminski CF, Harkiolaki M, Crump CM et al (2022) Near-native state imaging by cryo-soft-X-ray tomography reveals remodelling of multiple cellular organelles during HSV-1 infection. PLoS Pathog 18(7):e1010629. https://doi.org/10.1371/journal.ppat.1010629

Ni T, Zhu Y, Yang Z, Xu C, Chaban Y, Nesterova T et al (2021) Structure of native HIV-1 cores and their interactions with IP6 and CypA. Sci Adv 7(47):eabj5715. https://doi.org/10.1126/sciadv.abj5715

Peck A, Carter SD, Mai H, Chen S, Burt A, Jensen GJ (2022) Montage electron tomography of vitrified specimens. J Struct Biol 214(2):107860. https://doi.org/10.1016/j.jsb.2022.107860

Perez-Berna AJ, Rodriguez MJ, Chichon FJ, Friesland MF, Sorrentino A, Carrascosa JL et al (2016) Structural changes in cells imaged by soft X-ray cryo-tomography during hepatitis C virus infection. ACS Nano 10(7):6597–6611. https://doi.org/10.1021/acsnano.6b01374

Pettersen EF, Goddard TD, Huang CC, Meng EC, Couch GS, Croll TI et al (2021) UCSF ChimeraX: Structure visualization for researchers, educators, and developers. Protein Sci 30(1):70–82. https://doi.org/10.1002/pro.3943

Peukes J, Xiong X, Erlendsson S, Qu K, Wan W, Calder LJ et al (2020) The native structure of the assembled matrix protein 1 of influenza A virus. Nature 587(7834):495–498. https://doi.org/10.1038/s41586-020-2696-8

Qu K, Ke Z, Zila V, Anders-Osswein M, Glass B, Mucksch F et al (2021) Maturation of the matrix and viral membrane of HIV-1. Science 373(6555):700–704. https://doi.org/10.1126/science.abe6821

Rigort A, Plitzko JM (2015) Cryo-focused-ion-beam applications in structural biology. Arch Biochem Biophys 581:122–130. https://doi.org/10.1016/j.abb.2015.02.009

Rigort A, Bauerlein FJ, Villa E, Eibauer M, Laugks T, Baumeister W et al (2012a) Focused ion beam micromachining of eukaryotic cells for cryoelectron tomography. Proc Natl Acad Sci U S A 109(12):4449–4454. https://doi.org/10.1073/pnas.1201333109

Rigort A, Gunther D, Hegerl R, Baum D, Weber B, Prohaska S et al (2012b) Automated segmentation of electron tomograms for a quantitative description of actin filament networks. J Struct Biol 177(1):135–144. https://doi.org/10.1016/j.jsb.2011.08.012

Rincheval V, Lelek M, Gault E, Bouillier C, Sitterlin D, Blouquit-Laye S et al (2017) Functional organization of cytoplasmic inclusion bodies in cells infected by respiratory syncytial virus. Nat Commun 8(1):563. https://doi.org/10.1038/s41467-017-00655-9

Rodrigues RAL, Andrade A, Boratto PVM, Trindade GS, Kroon EG, Abrahao JS (2017) An anthropocentric view of the virosphere-host relationship. Front Microbiol 8:1673. https://doi.org/10.3389/fmicb.2017.01673

Rusu M, Starosolski Z, Wahle M, Rigort A, Wriggers W (2012) Automated tracing of filaments in 3D electron tomography reconstructions using Sculptor and Situs. J Struct Biol 178(2):121–128. https://doi.org/10.1016/j.jsb.2012.03.001

Sabanay I, Arad T, Weiner S, Geiger B (1991) Study of vitrified, unstained frozen tissue sections by cryoimmunoelectron microscopy. J Cell Sci 100(Pt 1):227–236. https://doi.org/10.1242/jcs.100.1.227

Sanchez RM, Zhang Y, Chen W, Dietrich L, Kudryashev M (2020) Subnanometer-resolution structure determination in situ by hybrid subtomogram averaging—single particle cryo-EM. Nat Commun 11(1):3709. https://doi.org/10.1038/s41467-020-17466-0

Schmid MF, Hecksel CW, Rochat RH, Bhella D, Chiu W, Rixon FJ (2012) A tail-like assembly at the portal vertex in intact herpes simplex type-1 virions. PLoS Pathog 8(10):e1002961. https://doi.org/10.1371/journal.ppat.1002961

Schur FK, Hagen WJ, de Marco A, Briggs JA (2013) Determination of protein structure at 8.5A resolution using cryo-electron tomography and sub-tomogram averaging. J Struct Biol 184(3):394–400. https://doi.org/10.1016/j.jsb.2013.10.015

Schur FK, Hagen WJ, Rumlova M, Ruml T, Muller B, Krausslich HG et al (2015) Structure of the immature HIV-1 capsid in intact virus particles at 8.8 A resolution. Nature 517(7535):505–508. https://doi.org/10.1038/nature13838
Schur FK, Obr M, Hagen WJ, Wan W, Jakobi AJ, Kirkpatrick JM et al (2016) An atomic model of HIV-1 capsid-SP1 reveals structures regulating assembly and maturation. Science 353(6298): 506–508. https://doi.org/10.1126/science.aaf9620
Selzer L, Su Z, Pintilie GD, Chiu W, Kirkegaard K (2020) Full-length three-dimensional structure of the influenza A virus M1 protein and its organization into a matrix layer. PLoS Biol 18(9): e3000827. https://doi.org/10.1371/journal.pbio.3000827
Silvester E, Vollmer B, Prazak V, Vasishtan D, Machala EA, Whittle C et al (2021) DNA origami signposts for identifying proteins on cell membranes by electron cryotomography. Cell 184(4): 1110–1121 e1116. https://doi.org/10.1016/j.cell.2021.01.033
Song Y, Yao H, Wu N, Xu J, Zhang Z, Peng C et al (2022) *In situ* architecture and membrane fusion of SARS-CoV-2 Delta variant. bioRxiv. https://doi.org/10.1101/2022.05.13.491759
Strauss JD, Hammonds JE, Yi H, Ding L, Spearman P, Wright ER (2016) Three-dimensional structural characterization of HIV-1 tethered to human cells. J Virol 90(3):1507–1521. https://doi.org/10.1128/JVI.01880-15
Su JM, Wilson MZ, Samuel CE, Ma D (2021) Formation and function of liquid-like viral factories in negative-sense single-stranded RNA virus infections. Viruses 13(1). https://doi.org/10.3390/v13010126
Sundquist WI, Krausslich HG (2012) HIV-1 assembly, budding, and maturation. Cold Spring Harb Perspect Med 2(7):a006924. https://doi.org/10.1101/cshperspect.a006924
Sundquist WI, Krug RM (2012) Assemble, replicate, remodel and evade. Curr Opin Virol 2(2): 111–114. https://doi.org/10.1016/j.coviro.2012.02.008
Suttle CA (2007) Marine viruses—major players in the global ecosystem. Nat Rev Microbiol 5(10): 801–812. https://doi.org/10.1038/nrmicro1750
Tan YB, Chmielewski D, Law MCY, Zhang K, He Y, Chen M et al (2022) Molecular architecture of the Chikungunya virus replication complex. Sci Adv 8(48):eadd2536. https://doi.org/10.1126/sciadv.add2536
Tegunov D, Xue L, Dienemann C, Cramer P, Mahamid J (2021) Multi-particle cryo-EM refinement with M visualizes ribosome-antibiotic complex at 3.5 A in cells. Nat Methods 18(2):186–193. https://doi.org/10.1038/s41592-020-01054-7
Tokuyasu KT (1973) A technique for ultracryotomy of cell suspensions and tissues. J Cell Biol 57(2):551–565. https://doi.org/10.1083/jcb.57.2.551
Turonova B, Sikora M, Schurmann C, Hagen WJH, Welsch S, Blanc FEC et al (2020) In situ structural analysis of SARS-CoV-2 spike reveals flexibility mediated by three hinges. Science 370(6513):203–208. https://doi.org/10.1126/science.abd5223
Vankadari N, Shepherd DC, Carter SD, Ghoshal D (2002) Three-dimensional insights into human enveloped viruses in vitro and in situ. *Biochem Soc Trans 50*(1):95–105. https://doi.org/10.1042/BST20210433
Vidavsky N, Akiva A, Kaplan-Ashiri I, Rechav K, Addadi L, Weiner S et al (2016) Cryo-FIB-SEM serial milling and block face imaging: Large volume structural analysis of biological tissues preserved close to their native state. J Struct Biol 196(3):487–495. https://doi.org/10.1016/j.jsb.2016.09.016
Vijayakrishnan S, Loney C, Jackson D, Suphamungmee W, Rixon FJ, Bhella D (2013) Cryotomography of budding influenza A virus reveals filaments with diverse morphologies that mostly do not bear a genome at their distal end. PLoS Pathog 9(6):e1003413. https://doi.org/10.1371/journal.ppat.1003413
Vijayakrishnan S, McElwee M, Loney C, Rixon F, Bhella D (2020) In situ structure of virus capsids within cell nuclei by correlative light and cryo-electron tomography. Sci Rep 10(1):17596. https://doi.org/10.1038/s41598-020-74104-x
Vollmer B, Prazak V, Vasishtan D, Jefferys EE, Hernandez-Duran A, Vallbracht M et al (2020) The prefusion structure of herpes simplex virus glycoprotein B. Sci Adv 6(39). https://doi.org/10.1126/sciadv.abc1726

Wan W, Briggs JA (2016) Cryo-electron tomography and subtomogram averaging. Methods Enzymol 579:329–367. https://doi.org/10.1016/bs.mie.2016.04.014
Wan W, Kolesnikova L, Clarke M, Koehler A, Noda T, Becker S et al (2017) Structure and assembly of the Ebola virus nucleocapsid. Nature 551(7680):394–397. https://doi.org/10.1038/nature24490
Wan W, Clarke M, Norris MJ, Kolesnikova L, Koehler A, Bornholdt ZA et al (2020) Ebola and Marburg virus matrix layers are locally ordered assemblies of VP40 dimers. Elife 9. https://doi.org/10.7554/eLife.59225
Wang S, Li S, Ji G, Huang X, Sun F (2017) Using integrated correlative cryo-light and electron microscopy to directly observe syntaphilin-immobilized neuronal mitochondria in situ. Biophys Rep 3(1):8–16. https://doi.org/10.1007/s41048-017-0035-x
Ward AE, Kiessling V, Pornillos O, White JM, Ganser-Pornillos BK, Tamm LK (2020) HIV-cell membrane fusion intermediates are restricted by Serincs as revealed by cryo-electron and TIRF microscopy. J Biol Chem 295(45):15183–15195. https://doi.org/10.1074/jbc.RA120.014466
WHO (2021) Global progress report on HIV, viral hepatitis and sexually transmitted infections, 2021: accountability for the global health sector strategies 2016–2021: actions for impact. World Health Organization
Wolff G, Hagen C, Grunewald K, Kaufmann R (2016) Towards correlative super-resolution fluorescence and electron cryo-microscopy. Biol Cell 108(9):245–258. https://doi.org/10.1111/boc.201600008
Wolff G, Limpens R, Zevenhoven-Dobbe JC, Laugks U, Zheng S, de Jong AWM et al (2020) A molecular pore spans the double membrane of the coronavirus replication organelle. Science 369(6509):1395–1398. https://doi.org/10.1126/science.abd3629
Wrapp D, Wang N, Corbett KS, Goldsmith JA, Hsieh CL, Abiona O et al (2020) Cryo-EM structure of the 2019-nCoV spike in the prefusion conformation. Science 367(6483):1260–1263. https://doi.org/10.1126/science.abb2507
Xiong Q, Morphew MK, Schwartz CL, Hoenger AH, Mastronarde DN (2009) CTF determination and correction for low dose tomographic tilt series. J Struct Biol 168(3):378–387. https://doi.org/10.1016/j.jsb.2009.08.016
Yang JE, Larson MR, Sibert BS, Kim JY, Parrell D, Sanchez JC et al (2022) Correlative cryogenic montage electron tomography for comprehensive in-situ whole-cell structural studies, bioRxiv. https://doi.org/10.1101/2021.12.31.474669
Yao H, Song Y, Chen Y, Wu N, Xu J, Sun C et al (2020) Molecular architecture of the SARS-CoV-2 virus. Cell 183(3):730–738 e713. https://doi.org/10.1016/j.cell.2020.09.018
Zanetti G, Briggs JA, Grunewald K, Sattentau QJ, Fuller SD (2006) Cryo-electron tomographic structure of an immunodeficiency virus envelope complex in situ. PLoS Pathog 2(8):e83. https://doi.org/10.1371/journal.ppat.0020083
Zanetti G, Riches JD, Fuller SD, Briggs JA (2009) Contrast transfer function correction applied to cryo-electron tomography and sub-tomogram averaging. J Struct Biol 168(2):305–312. https://doi.org/10.1016/j.jsb.2009.08.002
Zhang X, Sridharan S, Zagoriy L, Oegema CE, Ching C, Pflaesterer T et al (2022) Molecular mechanisms of stress-induced reactivation in mumps virus condensates. bioRxiv. https://doi.org/10.1101/2021.07.10.451879
Zhao G, Perilla JR, Yufenyuy EL, Meng X, Chen B, Ning J et al (2013) Mature HIV-1 capsid structure by cryo-electron microscopy and all-atom molecular dynamics. Nature 497(7451):643–646. https://doi.org/10.1038/nature12162
Zhou Y, Su JM, Samuel CE, Ma D (2019) Measles virus forms inclusion bodies with properties of liquid organelles. J Virol 93(21). https://doi.org/10.1128/JVI.00948-19
Zila V, Margiotta E, Turonova B, Muller TG, Zimmerli CE, Mattei S et al (2021) Cone-shaped HIV-1 capsids are transported through intact nuclear pores. Cell 184(4):1032–1046 e1018. https://doi.org/10.1016/j.cell.2021.01.025

# Chapter 2
# Approaches to Evaluating Necroptosis in Virus-Infected Cells

**Crystal A. Lawson, Derek J. Titus, and Heather S. Koehler**

**Abstract** The immune system functions to protect the host from pathogens. To counter host defense mechanisms, pathogens have developed unique strategies to evade detection or restrict host immune responses. Programmed cell death is a major contributor to the multiple host responses that help to eliminate infected cells for obligate intracellular pathogens like viruses. Initiation of programmed cell death pathways during the early stages of viral infections is critical for organismal survival as it restricts the virus from replicating and serves to drive antiviral inflammation immune recruitment through the release of damage-associated molecular patterns (DAMPs) from the dying cell. Necroptosis has been implicated as a critical programmed cell death pathway in a diverse set of diseases and pathological conditions including acute viral infections. This cell death pathway occurs when certain host sensors are triggered leading to the downstream induction of mixed-lineage kinase domain-like protein (MLKL). MLKL induction leads to cytoplasmic membrane disruption and subsequent cellular destruction with the release of DAMPs. As the role of this cell death pathway in human disease becomes apparent, methods identifying necroptosis patterns and outcomes will need to be further developed. Here, we discuss advances in our understanding of how viruses counteract necroptosis, methods to quantify the pathway, its effects on viral pathogenesis, and its impact on cellular signaling.

**Keywords** Programmed cell death · Necroptosis · Poxviruses · Herpesviruses · Influenza viruses · E3L · Z-nucleic acid binding protein · vIRA

C. A. Lawson · H. S. Koehler (✉)
School of Molecular Biosciences, Washington State University, Pullman, WA, USA

Center for Reproductive Biology, Washington State University, Pullman, WA, USA
e-mail: caputnam@wsu.edu; heather.koehler@wsu.edu

D. J. Titus
Providence Sacred Heart, Spokane Teaching Health Center, Spokane, WA, USA
e-mail: derek.titus1@providence.org

 
S. Vijayakrishnan et al. (eds.), *Virus Infected Cells*, Subcellular Biochemistry 106,
https://doi.org/10.1007/978-3-031-40086-5_2

## Abbreviations

| | |
|---|---|
| AIM2 | Absent in melanoma 2 |
| ATP | Adenosine triphosphate |
| DAMPs | Damage-associated molecular patterns |
| DNA | Deoxy-ribose nucleic acid |
| dsRNA | Double-stranded RNA |
| hpi | Hours post-infection |
| IFN | Type I interferon |
| ISG | IFN-stimulated gene |
| LPS | Lipopolysaccharide |
| MEF | Mouse embryo fibroblasts |
| MLKL | Mixed-lineage kinase-like |
| PAMPs | Pathogen-associated molecular patterns |
| PKR | Double-stranded RNA-dependent protein kinase |
| PRRs | Pattern recognition receptors |
| RHIM | RIP homotypic interaction motif |
| RIPK | Receptor-interacting protein kinase |
| RNase | Ribonuclease |
| serpins | Serine protease inhibitors |
| ssRNA | Single-stranded ribonucleic acid |
| TNF | Tumor necrosis factor |
| TRIF/TICAM-1 | TIR domain-containing adaptor inducing interferon-β/TIR domain-containing adaptor molecule 1 |
| VACV | Vaccinia virus |
| wt | Wild type |
| Za | Z-nucleic acid binding domain |
| ZBP1 | Z-nucleic acid binding protein 1 |
| Z-NA | Z-form nucleic acid |

## Introduction

Viruses are obligate intracellular parasites that depend on host cell machinery for replication. As viruses evolve and increase pathogenicity, infected hosts counter these changes by developing immune strategies that protect the host. Mammalian host immune systems, in particular, have evolved defense mechanisms that increase resistance to infection as well as strategies to recognize viruses and target infected cells for elimination. In a perpetual arms race, the viral pathogens respond to these host defense mechanisms by developing gene products that antagonize or allow the virus to evade detection via the immune system. While the strategies used to subvert innate immunity are recognized as crucial to ensuring the propagation and transmission of infections, our full understanding of their role in disease pathogenesis remains limited. Part of what researchers have been able to discover is that many viruses manage to overcome the full impact of a host antiviral state, involving the interferon (IFN) family and other inflammatory cytokines, by counteracting pathogen pattern recognition receptors (PRRs) and downstream signaling cascades (Pestka et al. 2004; Amarante-Mendes et al. 2018). Without employing these pathogenic strategies, elements produced from viral infections typically are recognized by PRRs which trigger an inflammatory response causing the upregulation of inflammatory cytokines along with programmed cell death pathways. Programmed cell death of infected cells has long been recognized as a highly effective mechanism to restrict pathogens by eliminating host machinery needed for viral replication and subsequent spread. The critical process has been exemplified by the abundance of virus-encoded inhibitors that have been found to suppress programmed cell death pathways(Mocarski et al. 2015; Jorgensen et al. 2017; Koehler and Jacobs 2021; Guo et al. 2022a). Here we discuss a specific type of programmed cell death known as necroptosis.

Necroptosis is part of a subgroup of programmed cell death known as necrosis. Necrosis was previously believed to be a dysregulated or accidental type of cell death that results from a toxic insult or physical stressors. Now it has become recognized to include a multitude of distinct regulated pathways with similar morphologies including necroptosis, pyroptosis, ferroptosis, pyronecrosis, parthanatos, oxytosis, and NETosis. Of these morphologies, necroptosis is characterized by a gain in cell volume, swelling of organelles, plasma membrane rupture, and subsequent loss of intracellular contents (Berghe et al. 2014; Green and Llambi 2015; Cotsmire et al. 2021). Unlike apoptosis and pyroptosis, it is independent of caspases (Vercammen et al. 1998; Holler et al. 2000) and generated from death receptor activation (TNF superfamily) due to ligand sensing by PPRs along with innate sensors. Necroptosis has also been implicated to contribute to numerous non-infectious pathological medical conditions including Alzheimer's disease, Parkinson's disease, age-related macular degeneration, skin autoimmunity/inflammation, tumor invasion to chemotherapeutics during colorectal cancer treatment, and many more (Bonnet et al. 2011; Moriwaki et al. 2015; Hanus et al. 2015; Bozec et al. 2016; Zhang et al. 2017). Non-infectious diseases and their relationship to necroptosis have been reviewed at length previously in "Necroptosis in the Pathophysiology of Disease" (Khoury et al. 2020). Since the pathway is also so critical for the early control of both DNA and RNA viruses, this work will focus on necroptosis as it relates to viral pathogens. We will explore the viruses that induce necroptosis, the strategies viruses employ to disrupt it, and the ways this death pathway has been studied.

## Necroptosis Pathway and Players

Programmed cell death is a genetically regulated process leading to the death of cells that was synonymous with apoptosis for some time. However, over the past several years, an array of alternative programmed cell death pathways have been identified and established as important regulators of the innate immune response to viral infections. Through these discoveries, the field of virus-induced, programmed cellular death rapidly evolved along with apoptosis, necroptosis, and pyroptosis, representing key pathways involved in the innate response to intracellular pathogens. Out of these pathways, necroptosis is characterized as an inflammatory form of programmed cell death believed to function as an alternative to extrinsic apoptosis, observed when caspases are inhibited (Vercammen et al. 1998; Li and Beg 2000; Holler et al. 2000; Mocarski et al. 2011). Activation of necroptosis leads to a signal cascade dependent on the receptor-interacting serine/threonine protein kinase (RIPK)3 and the downstream pseudokinase executer, MLKL (Sun et al. 2002; Murphy et al. 2013; Rodriguez et al. 2016) (Fig. 2.1). Because necroptosis is a caspase-independent programmed cell death, it is thought to play a critical role in the host defense mechanism against viruses that encode anti-apoptotic genes (Mocarski et al. 2011, 2015; Koehler and Jacobs 2021).

Much of our current understanding of the necroptosis pathway arises from studies investigating TNF-α signaling in the presence of a caspase inhibitor (de Almagro et al. 2017). Through these studies, it is now known that binding of a TNF-α homotrimer with TNF-R1 promotes the formation of a membrane-associated signaling complex termed complex I through scaffolding of TRADD, RIPK1, TRAF2, and cIAP1/cIAP2 (Li et al. 2012; Chen et al. 2019). Formation of complex I leads to two distinct outcomes depending on cosigning and other cellular factors. In cellular conditions that promote the polyubiquitination of RIPK1 via cIAP1/cIAP2 and LUBAC, complex I drives pro-survival signaling via NFκB and MAP kinase (O'Donnell et al. 2007; Declercq et al. 2009; Dikic et al. 2009). In contrast, when RIPK1 undergoes deubiquitination, complex I dissociates from the membrane and becomes cytoplasmic in order to recruit FADD to form complex II which favors death signaling (Li et al. 2012). by inducing the dimerization and activation of caspase 8 causing apoptosis. This activation of caspase 8 not only initiates extrinsic apoptosis but also precludes necroptosis since it cleaves essential necroptotic mediators including RIPK1, RIPK3, and CYLD (Oberst et al. 2011). However, in situations where caspase activity is restricted, necroptosis proteins are not inactivated by caspase 8, but instead, RIPK1 recruits RIPK3 to complex II, thereby forming the necrosome. This can occur with viral caspase inhibitors like serpin or chemical inhibitors such as zVAD-*fmk*. This interaction of RIPK1 with RIPK3 drives large amyloid-like complexes via RIP homotypic interaction motifs (RHIMs) (Li et al. 2012). The combination of blocking caspase 8 activity and the formation of the necrosome by RIPK1 recruitment of RIPK3 is the molecular switch that turns cell death from apoptosis to necroptosis. Necrosome formation then results in the recruitment of MLKL via its interaction with RIPK3 which causes MLKL to

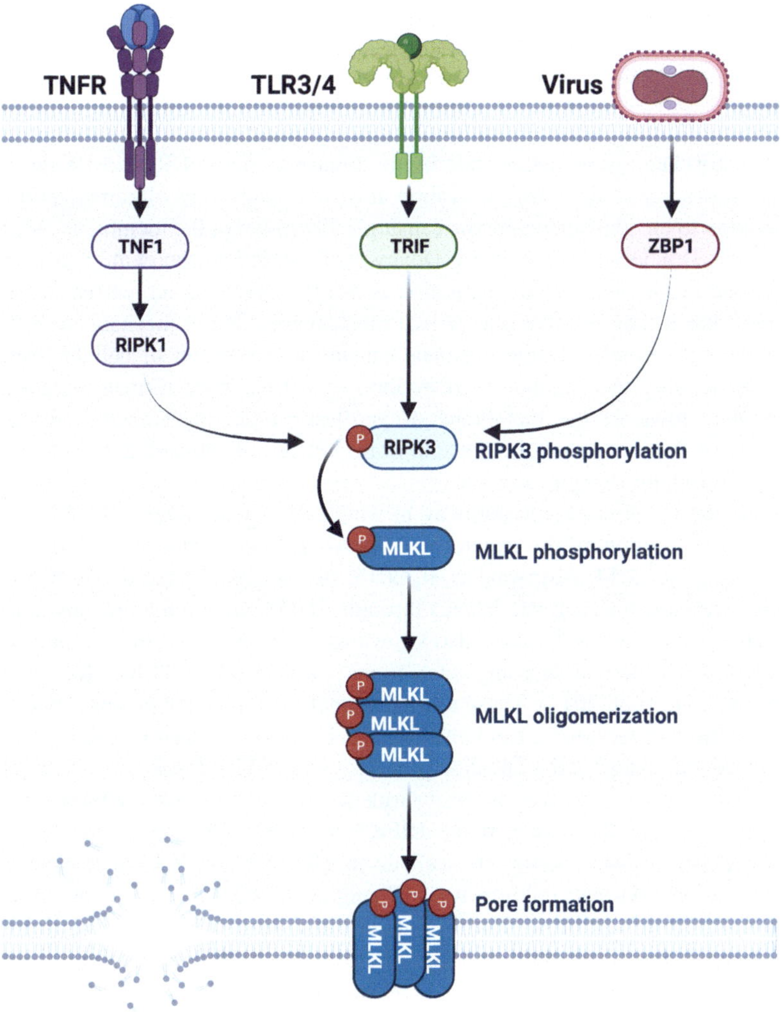


**Fig. 2.1** Necroptosis pathway and the three adaptors. Necroptosis is a form of programmed cell death that is initiated in response to various stimuli, such as viral infection. The signaling pathway of necroptosis involves three major adaptor proteins, namely receptor-interacting protein kinase 1 (RIPK1), TIR domain-containing adapter-inducing interferon-β (TRIF), and Z-DNA binding protein 1 (ZBP1). *RIPK1*: Various stimuli such as tumor necrosis factor (TNF) binding to TNF receptor 1 (TNFR1) activate RIPK1. RIPK1 has a death domain that enables it to interact with other death domain-containing proteins. When caspase-8 is inhibited, RIPK1 undergoes autophosphorylation and recruits other proteins to form a complex called the necrosome. *TRIF:* The second adaptor protein involved in the necroptosis pathway is TRIF. TRIF is a signaling protein that plays a role in the immune response to viral infection through the activation of TLR3. When TRIF is activated by a viral infection, it recruits RIPK1 to form a complex that leads to necroptosis. *ZBP1:* The third adaptor protein involved in the necroptosis pathway is ZBP1. ZBP1 is a cytoplasmic Z RNA sensor. When ZBP1 recognizes Z-RNA, it undergoes a conformational change and recruits RIPK3 to form a complex that leads to necroptosis. Activation of any of these three adaptors results in the recruitment and activation of RIPK3. Activated RIPK3 then phosphorylates and

**Fig. 2.1** (continued) activates MLKL, which translocates to the plasma membrane, where it oligomerizes and forms pores, leading to necroptotic cell death

be phosphorylated to its active state by the kinase activity of RIPK3 (Wang et al. 2014a; Rodriguez et al. 2016). Activation of MLKL leads to its oligomerization and localization to the plasma membrane causing a disruption of the membrane (Murphy et al. 2013; Dondelinger et al. 2014; Dovey et al. 2018; Samson et al. 2020). While the precise mechanism of this disruption is not fully understood, current evidence suggests that it may involve one of two mechanisms. These mechanisms involve either the activation of cation channels causing an alteration of cellular osmotic pressures leading to a membrane disruption or MLKL pore formation causing a direct breakdown of the membrane. Regardless of the mechanism, subsequent rupturing of the membrane then results in the release of intracellular DAMPs that drive local inflammatory processes.

While the cytokine-driven pathway in which RIPK1 activates RIPK3 has been fundamental to defining the general necroptosis pathway (Sun et al. 2002, 2012; Polykratis et al. 2014; Moriwaki et al. 2015; de Almagro et al. 2017), two other proteins are known to activate RIPK3 through RHIM domain interactions to cause necroptotic cell death. TICAM1, also known as TRIF, is one of these proteins that activate RIPK3 after undergoing interactions with TLR3 or TLR4 (Kaiser et al. 2008, 2013; He et al. 2011). These toll-like receptors initiate the process after being activated through receptor–ligand interactions. TLR3 is activated by dsRNA known to be present in many viral replication processes, while TLR4 recognizes LPS which is a common gram-negative bacterial virulence factor. The other RHIM domain-containing protein that can activate RIPK3 is ZBP1, formerly called DAI and DLM1. ZBP1 triggers necroptotic cell death after sensing Z-form nucleic acids (Z-NA) released during viral infections (Zhang et al. 2020; Koehler et al. 2021).

Overall, RIPK3 can be activated to initiate necroptotic cell death through RHIM domain interactions with RIPK1, TRIF, or ZBP1. Regardless of the trigger, activation of RIPK3 and subsequent MLKL phosphorylation lead to a rupturing of the plasma membrane and release of pro-inflammatory DAMPs which drives a local inflammatory response.

## Necroptosis Inhibitors

Many necroptotic inhibitors have been developed over the past several years that target different components of the cell death pathway including RIPK1, RIPK3, and MLKL (Fig. 2.2). Notably, RIPK1 inhibitors include necrostatin (Nec)-1 and GSK2982772. Nec-1 is a tryptophan-based inhibitor that works by stabilizing RIPK1 into its inactive form (Degterev et al. 2013). RIPK1 inhibitors: the most studied and well-characterized RIPK1 inhibitors are kinase inhibitors that bind to the

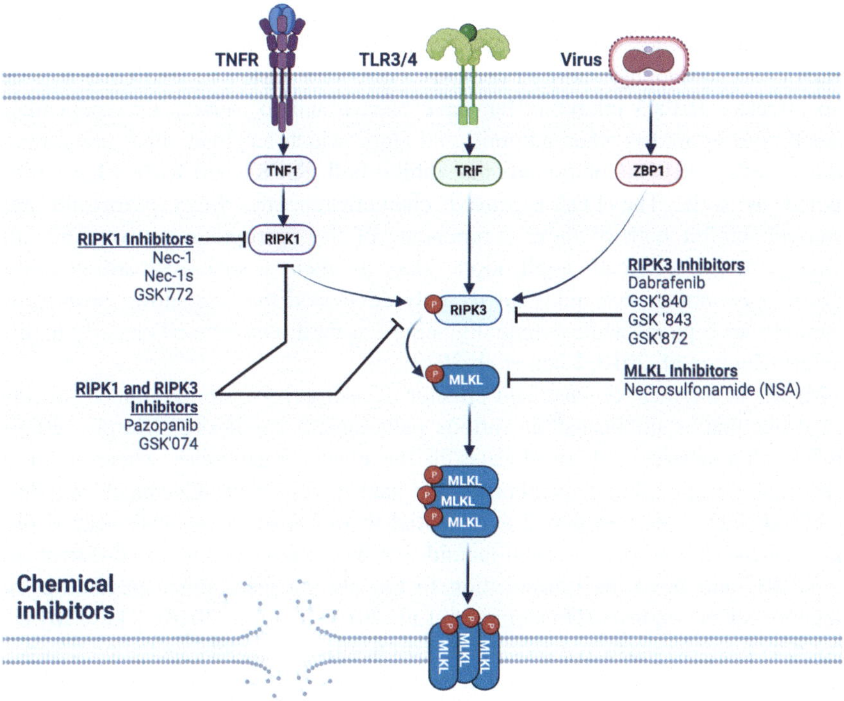


**Fig. 2.2** The chemical inhibitors of necroptosis and their targets. The necroptosis signaling pathway involves a series of protein–protein interactions and post-translational modifications that ultimately lead to the activation of the executioner protein, MLKL, and subsequent cell death. The signaling pathway of necroptosis involves three major adaptor proteins, namely RIPK1, TRIF, and ZBP1. Chemical inhibitors can target specific upstream adaptors of RIPK3, RIPK3, or the downstream executer MLKL. This figure summarizes the chemical inhibitors and their specific targets

ATP-binding pocket of the kinase domain and block its activity. GSK'963, for example, has been shown to inhibit RIPK1 kinase activity and prevent the formation of the necrosome complex, which is composed of RIPK1, RIPK3, and MLKL (Berger et al. 2015). Other RIPK1 inhibitors, such as Nec-1s and Nec-1h, have been shown to stabilize RIPK1 in a non-phosphorylated state, thereby preventing its recruitment to the necrosome (Degterev et al. 2013; Cao and Mu 2021). A derivative of Nec-1, Nec-1s, has also been developed and shows increased specificity for RIPK1 binding along with better stabilization of RIPK1 into its inactive conformation compared to Nec-1. While both of these compounds have been utilized to study a multitude of disease processes, poor pharmacokinetic properties have made their potential limited.

RIPK3 inhibitors include dabrafenib and other small-molecule inhibitors (Martens et al. 2020). Dabrafenib has been traditionally used in the treatment of melanoma where it acts as a B-RAF inhibitor (Rheault et al. 2013). Recently though, it has also been found to compete for an ATP-binding site on RIPK3, thereby inhibiting RIPK3's kinase capability (Li et al. 2014). This compound has been implemented in studies using human hepatocytes where it has shown promise in reducing acetaminophen-induced necrosis. GSK'840, GSK'843, and GSK'872 are also effective RIPK3 inhibitors but have been found to cause a dose-dependent induction of apoptosis when administered alone which has made their use limited (Mandal et al. 2014). Compounds that inhibit both RIPK1 and RIPK3 have been studied as well. The FDA-approved chemotherapeutic drugs pazopanib and ponatinib inhibit both of these components of the necroptotic pathway but are unlikely to have broad implications due to their associated cardiotoxicity. GSK'074 is another dual inhibitor recently developed that has shown promise in inhibiting necroptosis while potentially being less cardiotoxic based on early mouse models (Zhou et al. 2019; Chen et al. 2022).

MLKL is a critical downstream effector of necroptosis, a programmed form of cell death that is implicated in various pathological conditions (Murphy 2020). MLKL oligomerizes and translocates to the plasma membrane, where it forms pores that disrupt cellular membranes and lead to cell death. Chemical inhibitors of MLKL have been developed as potential therapies for necroptosis-related diseases. One such inhibitor necrosulfonamide has been shown to bind to a different site on MLKL and block its translocation to the plasma membrane, preventing the execution of necroptosis (Dondelinger et al. 2014; Su et al. 2014). These MLKL inhibitors have demonstrated efficacy in various disease models, including myocardial infarction, ischemic stroke, and acute pancreatitis, suggesting their potential as novel therapies for necroptosis-related disorders (Dondelinger et al. 2014; Zhang et al. 2017, 2022). However, further research is needed to optimize their efficacy and safety in humans. Another MLKL inhibitor that has shown promise as a potential therapeutic agent is NSA (necrosulfonamide-α), which is a derivative of necrosulfonamide. NSA binds specifically to the MLKL protein and inhibits its oligomerization and translocation to the plasma membrane, thereby preventing necroptosis. In addition to its potent anti-necroptotic activity, NSA has also been shown to have anti-inflammatory effects in various disease models, including colitis and sepsis (Jiao et al. 2020). These findings suggest that NSA could be a promising therapeutic agent for necroptosis-related disorders with an added anti-inflammatory benefit. Further preclinical studies and clinical trials will be needed to assess the efficacy and safety of NSA in humans.

In summary, many necroptotic inhibitors have been developed that target a variety of critical components of the cell death pathway. The complexity of signaling involved in necroptosis has contributed to the pros and cons associated with each inhibitor choice currently available. Challenges include unwanted signaling overlap with the apoptotic death pathway, a lack of target specificity, and unwanted tissue toxicity. While these challenges do represent hurdles in working to develop necroptosis-based therapeutics, current progress suggests that continued research into this intricate cell death pathway could yield significant insights into treating a multitude of disease processes in the future.

## In Vitro Necroptosis in Cultured Cells

Necroptotic death was very difficult to identify and study for several years since only a few specific cell lines actually had the necessary machinery to elicit necroptosis as a result of epigenetic regulation (Koo et al. 2015). Of these necroptotic-capable cells, the L929 and SVEC4-10 lines have been utilized the most due to ease of use and long-established protocols. HT29 cells are used frequently as well but need to have ZBP1 artificially expressed due to the limited presence of the protein in the line. Other necroptotic-capable lines include 3T3-SA, BMDM, HMEC-1, IMR-90, J2, MDM, MEF, U937, and THP-1 cells. Of these, the THP-1 line has been found to have variable necroptotic expression, based on differentiation status. Other lines that do not have the necessary necroptotic machinery have been adapted to have necroptotic capabilities through either transduction or ectopic expression of missing components. These lines include NIH3T3, HEK293T, HF, and HT29 cells. A full list of cell lines utilized to study necroptotic death can be found in Table 2.1. This is not an exclusive list as not all cells have been evaluated. However, prior to initiating studies, there should be a careful evaluation of a culture systems ability to undergo necroptosis and the suitability for the specific virus.

## In Vivo Models of Necroptosis

Concerns over the complicated delivery of chemical necroptotic inhibitors with potential off-target effects have led to the development of a multitude of genetically modified in vivo models to study the cell death pathway. Many of these genetically modified models have worked to alter necroptotic steps involved in the RIPK1-RIPK3-MLKL and ZBP1-RIPK3-MLKL pathways. A list of necroptosis-deficient mice and the genetic targets utilized to study the role of necroptotic death in viral pathogenesis can be found in Table 2.2.

The $RIPK1^{-/-}$ mice generated have not been able to be implemented in studying necroptosis since they demonstrate perinatal lethality due to apoptosis and RIPK1-independent necroptosis (Berger et al. 2014; Polykratis et al. 2014). As such, RIPK1 knock-in models have been developed in which the native RIPK1 gene is replaced with a genetically modified version that alters the kinase capabilities of RIPK1 making it catalytically inactive. $Ripk1^{K45A}$ and $Ripk1^{D138N/D138N}$ are two of these catalytically inactive RIPK1 knock-in models that produce viable mice while inhibiting RIPK1-dependent necroptosis (Berger et al. 2014; Polykratis et al. 2014). This suggests that the kinase activity of RIPK1 is needed for a necroptotic response, but not necessary for embryonic development. A RIPK1 scaffold mechanism has been proposed which suggests that RIPK1 prevents ZBP1-induced necroptotic death. It is interesting to also note that RIPK1 knockout mice can be rescued when crossed with RIPK3/Caspase-8 double knockout mice or $ZBP1^{-/-}$

**Table 2.1** Genetic summary of mice used in the studies of virus-induced necroptosis

| Protein | Function | Mice | Description | Sources |
|---|---|---|---|---|
| RIPK1 | RIPK1 recruits and activates RIPK3<br>Potential dual role in apoptosis and necroptosis | $RIPK1^{-/-}$ | • Not typically used due to perinatal lethality<br>• Lethality can be rescued when crossed with Caspase-8 knockout | https://doi.org/10.1073/pnas.1401857111<br>https://doi.org/10.1016/j.cell.2020.02.050 (IAV) |
| | | $RIPK1^{K45A/K45A}$ | • Catalytically inactive. Knock-in inactivates kinase activity but retains activity for development | https://doi.org/10.1073/pnas.1401857111<br>https://doi.org/10.1007/s00430-015-0410-5 (HSV1)<br>https://doi.org/10.1126/sciimmunol.aag2045 (IAV) |
| | | $RIPK1^{D138N/D138N}$ | • Catalytically inactive. Knock-in inactivates kinase activity but retains activity for development | https://doi.org/10.4049/jimmunol.1400590 |
| RIPK3 | Interacts with RIPK1 via RHIM domain<br>Phosphorylates MLKL | $RIPK3^{-/-}$ | Allows examination of RIPK3 mediated cell death | https://doi.org/10.1016/j.cell.2009.05.037<br>https://doi.org/10.4049/jimmunol.1400590 (VACV)<br>https://doi.org/10.1073/pnas.1700999114 (VACV)<br>https://doi.org/10.1073/pnas.1412767111 (HSV1)<br>https://doi.org/10.1007/s00430-015-0410-5 (HSV1)<br>https://doi.org/10.1038/s41419-018-0868-3 (HSV1) |

(continued)

**Table 2.1** (continued)

| Protein | Function | Mice | Description | Sources |
|---|---|---|---|---|
| | | | | https://doi.org/10.1016/j.chom.2012.01.016 (MCMV) https://doi.org/10.1126/sciimmunol.aag2045 (IAV) https://doi.org/10.1016/j.cell.2020.02.050 (IAV) |
| | | $RIPK3^{K51A/K51A}$ | | https://doi.org/10.1007/s00430-015-0410-5 (HSV1) |
| | | $RIPK3^{\Delta R/\Delta R}$ | | https://doi.org/10.1101/gad.223321.113 https://doi.org/10.1016/j.immuni.2020.11.020 |
| ZBP1 | Recognizes Z-form nucleic acid, activates RIPK3 via RHIM interaction | $ZBP1^{-/-}$ | Used to study necroptotic death triggered by Z-form nucleic acid | https://doi.org/10.1073/pnas.1700999114 (VACV) https://doi.org/10.1038/s41419-018-0868-3 (HSV1) https://doi.org/10.1016/j.chom.2012.01.016 (MCMV) https://doi.org/10.1126/sciimmunol.aag2045 (IAV) https://doi.org/10.1016/j.cell.2020.02.050 (IAV) |

(continued)

**Table 2.1** (continued)

| Protein | Function | Mice | Description | Sources |
|---|---|---|---|---|
| MLKL | Activated by RIPK3<br>Oligomerizes and translocates to membrane for pore formation<br>Executioner of necroptosis | $MLKL^{-/-}$ | Used to study impact of necroptotic death | https://doi.org/10.1016/j.immuni.2013.06.018 (origin)<br>https://doi.org/10.1038/s41419-018-0868-3 (HSV1)<br>https://doi.org/10.1084/jem.20191259 (IAV)<br>https://doi.org/10.1126/sciimmunol.aag2045 (IAV)<br>https://doi.org/10.1016/j.cell.2020.02.050 (IAV) |
| Caspase8 | Activates apoptosis, inhibits necroptosis | $Casp8^{DA}$ | Knock in point mutant of caspase 8 | https://doi.org/10.1084/jem.20191259 (IAV)<br>https://doi.org/10.1016/j.cell.2020C.02.050 (IAV)<br>https://doi.org/10.1126/sciimmunol.aag2045 (IAV) |

This table is a summary of small animal models used in the studies of virus-induced necroptosis to investigate the role necroptosis play in viral pathogenesis. Column one designated the gene/protein target. The column provides a brief functional role of the protein. Column 3 describes the specific genetic mutation and designated identification of the mutation. Column 4 describes the specific function and general rationale for mutation use. The column provided the link to the reference and viruses used in each study

mice which suggests RIPK1's dual role in both apoptotic and necroptotic pathways. Domain-specific ZBP1 mutants have been developed but have not been applied to viral research, but would likely give significant insights to advance our understanding of virus-induced necroptosis in vivo.

While these animal models are useful in studying RIPK1-induced necroptosis, they are not able to prevent all forms of necroptosis such as those involved in RIPK3 activation via TRIP or ZBP1. Consequently, catalytically inactive RIPK3 knock-in mice have also been developed (Newton et al. 2014). While these lines can prevent necroptotic death, they are subject to death by RIPK3-mediated apoptosis, evident in influenza A viral and HSV models (Nogusa et al. 2016b; Kuriakose et al. 2016; Guo et al. 2022b). This suggests that RIPK3 also has a dual role in both apoptotic and necroptotic pathways. The dual necroptotic and apoptotic roles both RIPK1 and RIPK3 play help to represent the complexity of these regulators involved with multiple immunologic responses. Due to this complexity, MLKL-knockout mice have also been produced and are currently the best option for studying the impact of necroptotic cell death in various diseases (Wu et al. 2013; Moerke et al. 2019; Tovey Crutchfield et al. 2021; Cao et al. 2022). This is evident in studies showing intact pro-inflammatory cytokine, NF-kB, and MAPK signaling post-LPS or TNF treatment in MLKL$^{-/-}$ mouse models immune to necroptotic death (Wu et al. 2013).

Other in vivo models have looked at TLR-induced necroptosis through *TLR3*$^{-/-}$ mice or Z-NA-induced necroptosis through *ZBP1*$^{-/-}$ mice. In *TLR3*$^{-/-}$ mice, TLR3 ligand sensing of dsRNA cannot occur to activate TRIF (Chen et al. 2021; Kaiser et al. 2013). Research utilizing this model has shown that *TLR3*$^{-/-}$ mice infected with various viruses show greater viral loads and disease progression compared to controls (Vercammen et al. 2008; Zhang et al. 2013; Perales-Linares and Navas-Martin 2013). This is interesting since no viral infection has been able to produce TLR3-dependent necroptosis in vitro with the pathway only being studied with synthetic RNAs designed to mimic viral nucleic acids (Kaiser et al. 2013). Future research will need to delve into whether or not certain viruses actually utilize this pathway to stimulate necroptotic death in vitro. *ZBP1*$^{-/-}$ mice on the other hand have been able to produce evidence of necroptotic cell death when infected with various viruses that correlates with in vitro studies (Thapa et al. 2016; Koehler et al. 2017; Upton et al. 2019). In these mice, ZBP1 cannot sense Z-NA released during viral infections to stimulate necroptotic death. Since many viruses have developed inhibitors to counteract necroptotic cell death, work utilizing this model will be interesting to follow as it could lead to the development of novel therapeutics designed to treat a variety of viral infections.

## Cellular Patterns and Methods to Evaluate Necroptosis

Since cell death pathways are complex in nature and share a multitude of components with one another, distinctly identifying necroptotic processes from other cell death pathways has been challenging. Therefore, it comes as no surprise that necroptosis identification tends to require a multifaceted approach that assesses many different interactions within the pathway. Additionally, in the setting of viral

**Table 2.2** Cells used in the studies of virus-induced necroptosis

| Cells | Use | Origin | Sources |
|---|---|---|---|
| 3T3 | Used for transduction studies | Mouse NIH/Swiss embryo fibroblast. NIH3T3 ATCC number CRL-1658<br>3T3-SA ATCC CCL92 | https://doi.org/10.1007/s00430-015-0410-5 (HSV1)<br>https://doi.org/10.3390/v13112134 (HCMV)<br>https://doi.org/10.1074/jbc.M115.646042 (HCMV)<br>https://doi.org/10.1016/j.chom.2010.03.006 (MCMV)<br>https://doi.org/10.1074/jbc.C800051200 (MCMV)<br>https://doi.org/10.15252/embr.201743947 (MCMV)<br>https://doi.org/10.1016/j.chom.2012.01.016 (MCMV) |
| A549 | In vitro studies for viruses that impact lung tissue | Human airway epithelium cell line<br>ATCC CCL-185 | https://doi.org/10.1016/j.cell.2020.02.050 (IAV) |
| BMDM | Cell death assessment | Bone marrow-derived macrophages | https://doi.org/10.1101/2022.04.20.488871 (IAV) TLR2$^{-/-}$ and TLR4$^{-/-}$<br>https://doi.org/10.1126/sciimmunol.aag2045 (IAV) (ZBP1$^{-/-}$) |
| HEK293T | Used for ectopic or lentivirus transduction to reconstitute necroptosis machinery | ATCC<br>Invitrogen Cat#: R70007 | https://doi.org/10.1016/j.immuni.2020.11.020 (CPXV)<br>https://doi.org/10.1073/pnas.1700999114 (VACV)<br>https://doi.org/10.1074/jbc.M115.646042 (HCMV)<br>https://doi.org/10.1016/j.chom.2010.03.006 (MCMV)<br>https://doi.org/10.1074/jbc.C800051200 (MCMV)<br>https://doi.org/10.1016/j.chom.2016.09.014 (IAV) |

(continued)

**Table 2.2** (continued)

| Cells | Use | Origin | Sources |
|---|---|---|---|
| HFFFs-TERT | | TERT-immortalized primary human fetal foreskin fibroblasts | https://doi.org/10.1073/pnas.2001887117 (HCMV) |
| HF and ihf-ei1.3 | Lentivirus transduction | Newborn foreskin fibroblasts | https://doi.org/10.1074/jbc.M115.646042 (HCMV) |
| HMEC-1 | Cell death signaling assessment | Human microvascular endothelial cells | https://doi.org/10.1074/jbc.M115.646042 (HCMV) |
| HT29 | Necroptosis inducible cells | Epithelial colorectal adenocarcinoma cell line ATCC Cat# HTB-38 | https://doi.org/10.1016/j.celrep.2019.08.055 (Poxviruses) https://doi.org/10.1016/j.chom.2015.01.003 (HSV1) EV, ICP6 https://doi.org/10.1074/jbc.RA118.004651 (HSV1) https://doi.org/10.3390/v13112134 (HCMV) https://doi.org/10.1074/jbc.M115.646042 (HCMV) https://doi.org/10.15252/embr.201743947 (MCMV) https://doi.org/10.1016/j.cell.2020.02.050 (IAV) |
| IMR-90 | Cell death assessment | Human fetal fibroblasts | https://doi.org/10.1074/jbc.M115.646042 (HCMV) |
| J2 | Cell death signaling assessment | Murine J2 virus transformed macrophages | https://doi.org/10.1016/j.immuni.2020.11.020 (CPXV) |
| Jurkat | Used to study cell signaling events | Human T lymphocyte cell line | https://doi.org/10.1016/j.cell.2009.05.037 (VACV) https://doi.org/10.1074/jbc.M305633200 (VACV) |
| L929 | Necroptosis inducible cells | Mouse-derived connective tissue cell line ATCC Cat# CCL-1 | https://doi.org/10.1016/j.cell.2009.05.037 (VACV) https://doi.org/10.1073/pnas.1700999114 (VACV) https://doi.org/10.1016/j.immuni.2020.11.020 (CPXV) |

(continued)

**Table 2.2** (continued)

| Cells | Use | Origin | Sources |
|---|---|---|---|
| | | | https://doi.org/10.1016/j.chom.2010.03.006 (MCMV) |
| LET1 AEC | In vitro studies for viruses that impact lung tissue | Mouse airway epithelial cell line | https://doi.org/10.1016/j.chom.2016.09.014 (IAV)<br>https://doi.org/10.1016/j.cell.2020.02.050 (IAV) |
| MDF | Transduction and cell death signaling assessment | Mouse dermal fibroblasts | https://doi.org/10.1016/j.celrep.2019.08.055 (poxviruses) |
| MDM | Cell death analysis | Human monocyte-derived macrophages | https://doi.org/10.1101/2022.04.20.488871 (IAV) |
| MEF | Cell death analysis | Mouse embryonic fibroblasts | https://doi.org/10.1016/j.immuni.2020.11.020 (CPXV)<br>https://doi.org/10.1016/j.chom.2021.05.009 (VACV) (Pkr deficient)<br>https://doi.org/10.4049/jimmunol.1400590 (VACV)<br>https://doi.org/10.1038/s41419-018-0868-3 (HSV1)<br>https://doi.org/10.1073/pnas.2001887117 (HCMV)<br>https://doi.org/10.1016/j.chom.2010.03.006 (MCMV)<br>https://doi.org/10.1016/j.chom.2012.01.016 (MCMV)<br>ZBP1$^{-/-}$<br>https://doi.org/10.1084/jem.20191259 (IAV)<br>MLKL$^{-/-}$<br>https://doi.org/10.1016/j.chom.2016.09.014 (IAV)<br>https://doi.org/10.1016/j.cell.2020.02.050 (IAV) |
| SVEC4-10 | Necroptosis inducible cells | Mouse endothelial cell line<br>ATCC Cat# CRL-2181 | https://doi.org/10.1016/j.chom.2021.05.009 (VACV) (ZBP1 deficient and WT<br>https://doi.org/10.1038/s41419-018-0868-3 (HSV1) (ZBP1 KO) |

(continued)

**Table 2.2** (continued)

| Cells | Use | Origin | Sources |
|---|---|---|---|
| | | | https://doi.org/10.3390/v13112134 (HCMV) https://doi.org/10.1016/j.chom.2010.03.006 (MCMV) (RIP1$^{-/-}$ transformed) https://doi.org/10.1074/jbc.C800051200 (MCMV) https://doi.org/10.15252/embr.201743947 (MCMV) https://doi.org/10.1016/j.chom.2012.01.016 (MCMV) |
| THP-1 | Undifferentiated cells sensitive to necroptosis | Human leukemia monocyte cell line ATCC Cat# TIB-202 | https://doi.org/10.1016/j.immuni.2020.11.020 (CPXV) |
| U937 | Cell death assays | Human histiocytic lymphoma cell line ATCC Cat# CRL-1593.2 | https://doi.org/10.1016/j.celrep.2019.08.055 (poxviruses) |

Table of common cell types used as tools to study virus-induced necroptosis. 1st column is the common name of the cell; the column contains a brief categorical description of their use; column 3 is the type of cell and ATCC catalogue number (is listed) and the column is the reference and viruses used in the studies

infections, it is important to note that viruses have been found to inhibit necroptosis at different levels which makes it critical to assess which component a particular virus of interest may be targeted to inhibit the pathway. This is especially important in the development of therapeutics which require specific targets to be efficacious.

## *Metabolic Viability Assays*

Metabolic viability assays are an important method to evaluate cell death that utilizes markers of mitochondrial activity to detect changes manifested in viable or nonviable cells. Here we describe three assays that can be used to measure metabolic changes in cells exposed to viral infections:

***ATP Assay*** Adenosine triphosphate (ATP) is an organic, energy-carrying molecule that is essential to fueling many cellular processes. Produced in the mitochondria by metabolically active cells, ATP can be assayed to determine the changes in cell viability with nonviable cells showing diminished ATP levels from a lack of production. While not specific to only necroptotic death, diminished ATP levels help us to broadly determine if some type of cellular demise is taking place for a given disease process. Most commonly this is done by using a bioluminescent assay that uses the two-part firefly luciferase reaction. Lysing cells to release intracellular ATP in the presence of a luciferin enzyme and substrates can effectively measure the concentrations of cellular ATP. The available ATP in the lysate is used to adenylate luciferin forming luciferyl-adenylate. In the second step of the reaction, oxidative decarboxylation of luciferyl-adenylate forms an excited oxyluciferin. When this oxyluciferin returns to the ground state, light is emitted which can be measured by a luminometer. The measurement of emitted light from this reaction can be correlated to the amount of ATP in the sample. In utilizing this technique, it is important to note that in cases of programmed cell death, ATP breakdown is delayed from death execution steps and may produce a false negative limiting the use as an immediate indicator of a necroptotic death. This makes using appropriate time points essential when using the technique to study necroptosis. Additionally, pairing the assay with small-molecule or genetic modification approaches can verify if the cell death events observed are in fact a result of necroptosis.

***MTT Assay*** The MTT assay is a very common colorimetric assay to measure cell viability or cytotoxicity in viral infections. Viable, metabolically active cells produce mitochondrial dehydrogenases that can reduce MTT, a yellow tetrazolium salt, to form purple formazan crystals. Cells that are no longer metabolically active or producing mitochondrial dehydrogenases are dead or dying and these cannot complete the reduction reaction to generate the purple formazan. The purple color correlates to the number of metabolically active cells and the intensity of the color can be spectrophotometrically measured: the darker the color, the greater the number of viable cells.

***Resazurin Assay*** Similar to the MTT assay, the resazurin assay is also an indicator dye assay that measures cell viability based on the reduction of blue resazurin by mitochondrial reductases to red fluorescent resorufin. Cells that are dead or dying will not have the necessary reductases to convert the blue resazurin to red resorufin. The measurement of the resorufin produced is proportional to the number of viable cells in the sample, the greater red fluorescence signal corresponds to an increase in the number of metabolically active cells, and, conversely, more blue fluorescence indicates more cell death, including cells undergoing necroptosis.

These cell viability assays are not specific indicators of necroptosis, but they are indicators of cell death that can be paired with other assays to determine if necroptosis is occurring. Additionally, these assays are not accurate indicators of the timing of cell death, and they do require careful control of cell number and an understanding that states or conditions that cause metabolic senescence should be avoided.

## Membrane Integrity Assays

Cell membrane integrity assays are crucial to assessing virus-induced necroptotic cell death, the hallmark of which is cell swelling and rupture due to plasma membrane damage.

***LDH Assay*** The lactate dehydrogenase (LDH) assay uses LDH to measure a loss in cell membrane integrity. In viable or uninfected cells, the LDH enzyme is present only inside the cells, but infection or other cytotoxic insults can degrade the plasma membrane and allow the release of this enzyme into the media. In a damaged cell, LDH catalyzes the reduction of NAD+ to NADH and the oxidation of lactate to pyruvate. In the second step, the NADH is used to catalyze the reduction of an indicator dye, like luciferin or resazurin. The resulting amount of fluorescence or color measured is an indicator of membrane integrity loss. This assay is not a specific measure of necroptosis but should be used with other assays such as inhibitors or genetic knockouts to verify that an increase in LDH levels is the result of necroptosis.

***Membrane Impermeant DNA Dyes*** A simple method for quantifying cell death is to use DNA-binding fluorescent dyes that are impermeant to healthy cell membranes. These fluorescent nucleic acid-binding dyes can penetrate the compromised membranes of dead or dying cells and the fluorescence of the nucleic acid-bound dye can be measured using live cell imaging. This method gives a real-time measurement of membrane permeability to help establish the timing of signaling events in cells that are dying. This is an effective tool to use for cells undergoing necroptosis where the pathway results in pore formation and the dyes can access the DNA and can be utilized as a real-time measurement of loss of membrane integrity in live cell imaging to give kinetic insights into death (Upton et al. 2019). Many dye systems are available including SYTOX from Invitrogen and Incucyte Cytotox Dyes from Sartorius.

## Necroptosis Morphology

Various imaging techniques have been implemented to assess the morphological characteristics of cells undergoing necroptosis. These techniques have been used to show evidence of typical characteristics of cells undergoing necroptotic death, including a gain in cell volume, swelling of organelles, and a rupturing of the plasma membrane (Newton et al. 2014; Cotsmire et al. 2021) (Fig. 2.3). Since this process of cellular swelling with subsequent membrane rupture generates fairly fragile specimens, techniques that rely on fixation like transmission microscopy and confocal imaging have been unsuccessful. However, techniques that work by freezing the specimen in place before fixation such as cryo-electron microscopy have shown moderate success (Daley-Bauer et al. 2017). Additionally, utilizing necroptotic

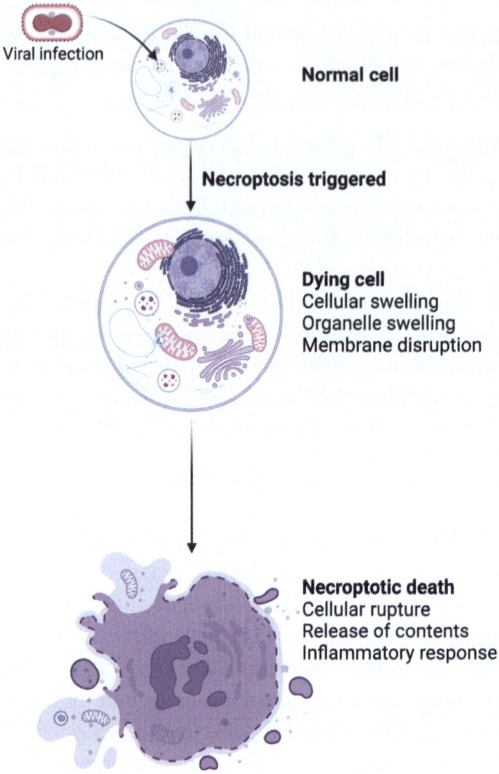


**Fig. 2.3** Diagram of common morphology of a virus-infected cell undergoing necroptosis. During necroptosis, cells undergo a series of morphological changes that ultimately lead to cell death. At the early stage of necroptosis, cells undergo swelling and rounding, followed by the formation of blebs on the cell surface. The blebs are caused by the disintegration of the plasma membrane and the release of cytoplasmic contents. The cells also exhibit a loss of plasma membrane integrity, which leads to the leakage of intracellular contents into the extracellular space. As the process progresses, the cells become progressively more swollen and eventually burst open, releasing their contents into the extracellular environment. This final stage of necroptosis is characterized by the formation of large membrane pores, which allow the influx of water and ions into the cell and lead to osmotic lysis. Overall, the morphology of cells undergoing necroptosis is characterized by swelling, membrane blebbing, loss of membrane integrity, and osmotic lysis. These changes are distinct from other forms of programmed cell death, such as apoptosis, which is characterized by shrinkage, condensation of chromatin, and fragmentation of the nucleus

inhibitors cells deficient in necroptotic machinery that halt cell death has allowed more traditional imaging techniques to observe activation of the necroptosis pathway for unaffected proteins at discrete time points. Ultimately though, given the quick onset and fragility of the necroptosis process, live cell imaging has proven to be the most successful in studying the pathway and provides critical kinetic information (Koehler et al. 2017, 2021; Upton et al. 2019). This technique allows a researcher to observe the necroptotic process in real time to differentiate morphological manifestations from those observed in other cellular death pathways.

## Immuno-techniques

Immuno-techniques utilized to study necroptosis rely on antibody–antigen interactions to isolate and characterize various components pertinent to the necroptotic pathway. These techniques include immunohistochemistry, immunoprecipitation, and western blots.

***Immunohistochemistry*** This technique utilizes chromogen-tagged antibodies that attach to antigens on critical necroptotic components. After allowing time for these antibodies to adhere to their respective antigens, a substrate is added that reacts with the tagged antibody complex to produce a color change. This color change correlates to the amount of a particular necroptotic component present. While this technique makes theoretical sense for studying the pathway, antibody cross-reactivity has made it difficult to assess if the changes observed are accurate. As such, the technique has not been utilized too much in present studies but has potential for future research endeavors as more specific antibodies are being developed.

***Immunoprecipitation*** These techniques work by precipitating out necroptotic components of interest using immobilized antibodies. Direct immunoprecipitation techniques involve antibodies immobilized on density beads that can directly bind to antigens of interest. Indirect immunoprecipitation techniques are performed by exposing a solution containing necroptotic components of interest to antibodies directed toward specific antigens on these components. After allowing time for these antibodies to bind to their antigens, beads designed to capture antibodies are added to the solution and bind to the antibody–antigen complexes. After necroptotic components of interest have been immobilized by either technique, the remaining solution is washed away. This allows these components to be isolated for further analysis with other techniques such as immunoblotting or proteomic analysis. This approach is useful in determining which necroptotic components are being affected by various disease processes and protein interactions that propagate necroptosis signaling. Careful consideration should be taken to minimize off-targeting effects of non-specific antibody interactions. The use and development of engendered cell lines that express epitope-tagged necroptotic proteins has drastically mitigated the off-targeting effects observed and facilitated key advances in our understanding of necroptosis signaling (Sridharan et al. 2017; Upton et al. 2019).

***Western Blots*** Western blots are common experiments used to study various cell signaling events including necroptosis protein activation states. These experiments work by separating different proteins of interest by size with denaturing gel electrophoresis. This allows for the identification and semi-quantification of the levels of proteins in a given sample. A multitude of antibodies have been developed that specifically target activated or phosphorylated necroptotic proteins including MLKL and RIPK3 which have been used to document necroptosis signaling in virus-infected cells. This helps to garner evidence that a particular component of interest is present and activated by controlling for both size and phosphorylation state.

## The Viruses That Activate or Inhibit Necroptosis

Much like apoptosis, necroptosis plays out as a cell-autonomous innate immune mechanism to kill virus-infected cells and cut short productive infection. This potent host defense mechanism limits viral replication and spread by eliminating infected cells (Koehler et al. 2017; Upton et al. 2019). The significance of necroptosis as a host defense mechanism has been reinforced by the identification of virus-encoded inhibitors that block specific steps in the pathway (Fig. 2.4). These viral inhibitors sustain infection allowing the virus to replicate and spread. A series of reports demonstrated the significance of necroptosis in restricting viral pathogenesis and how ZBP1 is a critical mediator of virus-induced necroptosis observed during innate immune sensing of viruses in both DNA and RNA viruses.

### *DNA Viruses*

#### Herpesviruses

Herpesviruses are a large family of double-stranded DNA viruses. The *Herpesviridae* family consists of more than 100 viruses, nine of which primarily infect humans where infection and replication take place in the epithelial cells and some of which can cause lifelong latent infections. Necroptosis studies utilizing herpes viruses have included investigations with murine cytomegalovirus (MCMV), human cytomegalovirus (HCMV), herpes simplex virus 1 (HSV1), and herpes simplex virus 2 (HSV2).

#### Murine Cytomegalovirus

MCMV is a species-specific double-stranded DNA virus that is a natural pathogen in wild mice populations. While MCMV infection triggers necroptotic cell death, MCMV has also evolved viral inhibitors to counteract the host defense mechanisms to ensure viral replication. Early evaluation of a mutant MCMV transposon library identified M45 as an important gene for endothelial and macrophage cell survival

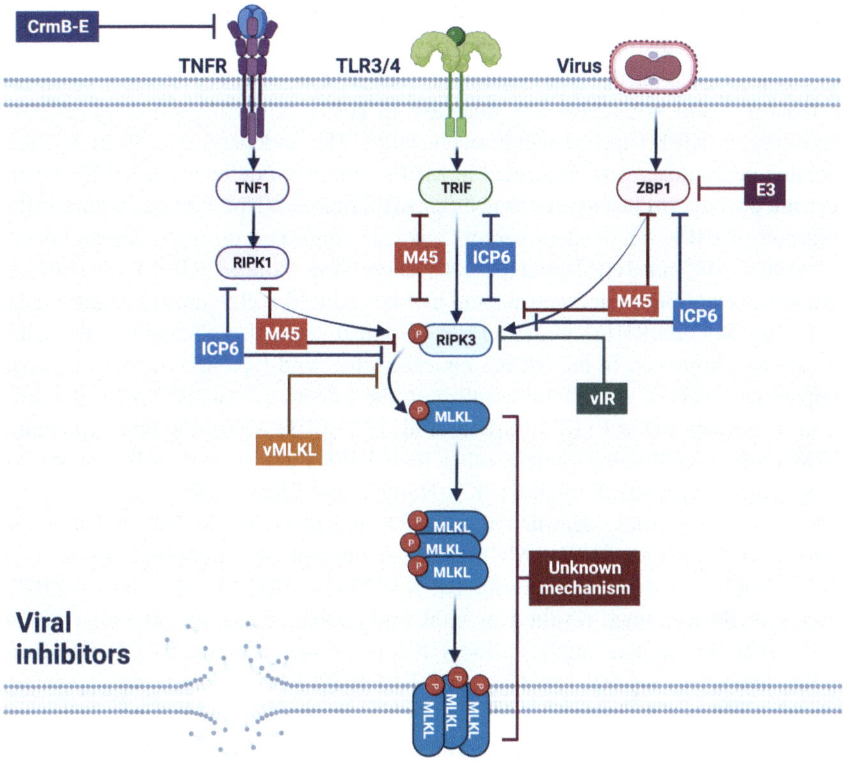


**Fig. 2.4** Viral inhibitors of necroptosis and their cellular targets. Crm (light purple) proteins are found in orthopoxviruses but are best described in the cowpox virus. These proteins are competitive inhibitors of TNF receptor signaling and prevent RIPK1-mediated necroptosis. ICP6(blue) is the HSV1 and HSV2 RHIM-containing inhibitor that targets host RHIM interactions of adaptors and RIPK3. M45(Red) are found in MCMV and are an RHIM-containing protein similar to ICP6 in HSV. M45 blocks necroptosis by preventing RHIM interactions of the adaptor proteins and RIPK3, thereby blocking all three pathways of necroptosis. vIR (green) is found is poxviruses and functions to enhance the turnover of RIPK3 preventing it from activating MLKL. vMLKL is a truncated MLKL homologue and functions to restrict necroptosis by acting as a decoy preventing host MLKL from getting phosphorylated. E3(dark purple) is a Z-RNA competitive inhibitor found in orthopoxviruses and prevents the activation of ZBP1. HCMV has an unidentified inhibitor of MLKL (burgundy)

in vitro and viral replication (Brune et al. 2001; Lembo et al. 2004). Delving further to elucidate M45-encoded protein function, the sequence was analyzed and found to contain an N-terminal RHIM domain facilitating binding to RHIM-containing receptor-interacting kinases, RIPK1 and RIPK3 (Kaiser et al. 2008). To examine the impact of this protein on necroptosis, cells were infected with recombinant MCMV engineered to have a nonfunctional M45 RHIM domain by mutations of four amino acids in the RHIM sequence (Kaiser et al. 2008). Cells infected with the M45-deficient MCMV (M45*Mut*RHIM) had decreased cell viability compared to wild-type M45. The M45-encoded viral inhibitor of RIP activation (vIRA) was found to impede the RIPK1 and RIPK3 complex from forming via RHIM domains. To evaluate this disruption, the roles of these receptor interaction kinases were individually examined. Treatment of cells with RIPK1-specific inhibitor Necrostatin-1 did not reveal any decrease in cell survival (Upton et al. 2010a), suggesting a RIPK1-independent mechanism. The independence from RIPK1-mediated necroptosis was confirmed in MEFs from mice deficient in RIPK1 where these mice were also insensitive to necroptosis induction (Upton et al. 2010a). vIRA inhibition of RIPK3-dependent necroptosis was confirmed through various knock-out studies. M45*mut*RHIM-infected cells expressing reduced RIPK3 via shRNAs were not susceptible to necroptosis and had no reduced cell viability. Additionally, the mutant M45*mut*RHIM virus did not demonstrate viral replication in the wild-type animal; however, in the RIPK3 knockout, the wild-type and mutant virus had comparable levels of replication confirming the importance of RIPK3 for the inhibition of necroptosis via vIRA (Upton et al. 2010a). This was the first time that a RHIM domain-containing protein other than RIPK1 was found to be capable of stimulating necroptosis through RIPK3 (Nailwal and Chan 2019).

More recent studies identified that ZBP1 interacts with RIPK3 and that this complex is the target for the MCMV vIRA necroptosis inhibitor (Upton et al. 2012). Similar to experiments with RIPK3, M45mutRHIM infection of ZBP1-deficient cells and mice resulted in viral replication of the mutant virus (Upton et al. 2012). These data implicate the vIRA in preventing the ZBP1 RHIM from binding and activating RIP3, halting the ZBP1/RIPK3/MLKL necroptotic pathway and protecting the infected cell, allowing it to continue replication and pathogenesis (Upton et al. 2012). Further studies identified viral transport factor immediate early protein 3 (IE3) as a pivotal player in ZBP1-dependent necroptosis in MCMV infection. Recombinant viruses for wild-type MCMV and M45*mut*RHIM were generated that contained an IE3 protein fused to the C-terminus (Sridharan et al. 2017). Under normal conditions, the addition of the destabilizing domain triggers rapid proteosomal degradation. Chemical ligand Shield-1 stabilizes the destabilization domain and allows the transcriptional functions to persist. Results from these studies demonstrated that M45*mut*RHIM-infected cells would undergo ZBP1-mediated necroptosis in the presence of Shield-1 but would be resistant to necroptotic death in its absence. This showed that the early transcription of viral RNA via IE3 is required for necroptosis and is the ligand for ZBP1-mediated necroptosis. One last interesting MCMV study found that when MCMV's viral inhibitor of caspase 8-induced apoptosis (vICA) is altered to be nonfunctional, cells show evidence of RIPK3, MLKL, and caspase 8 activation. This showed that apoptotic and necroptotic processes could occur simultaneously.

## Human Cytomegalovirus

HCMV studies reveal significant species-specific differences in the human virus compared to the murine strain. HCMV functions to block necroptosis after the phosphorylation of RIPK3 and MLKL. This resistance to necroptotic death is dependent on viral transcriptional processes regulated by the viral protein IE1; however, necroptosis-resistant cells still show evidence of MLKL phosphorylation (Omoto et al. 2015). It has been shown that the HCMV UL36 protein is a key player in inhibiting necroptosis by interacting with MLKL and inducing its demise (Fletcher-Etherington et al. 2020). UL36, a dual cell death pathway inhibitor also able to inhibit apoptosis by binding caspase 8, is capable of inhibiting necroptosis in a species-specific manner and does not inhibit necroptosis in murine cells.

### Herpes Simplex Viruses

Studies investigating herpes simplex viruses (HSV1 and HSV2) have also found that these viruses are able to inhibit necroptosis in virus-infected human cells (Guo et al. 2015, 2018). This mechanism of cell death inhibition is similar to that of MCMV, where vIRA blocks the necrosome formation by RHIM interaction of RIPK3 and prevents further execution of the necroptotic pathway through RIPK3 and MLKL (Upton et al. 2010a, b, 2012). In HSV1 infection of human cells, ICP6 acts as an inhibitor that contains an N-terminal RHIM domain allowing it to competitively bind and sequester RIPK3 from the other interacting proteins, blocking necrosome formation and inhibiting the activation of RIPK3/MLKL-dependent necroptosis. Strangely enough, ICP6 has been found to induce necroptotic death in HSV1-infected murine cells through interactions of the RHIM domains on ICP6 and RIPK3 (Wang et al. 2014b). Later studies revealed HSV1 infection triggers necroptosis through the ZBP1/RIPK3-dependent pathway. Evaluation of ZBP1-deficient MEFs and mouse cell lines found the ZBP1-deficient cells resistant to necroptosis, confirming the importance of ZBP1 in HSV1 infection in mouse cells (Guo et al. 2018). Evaluation of MLKL phosphorylation during WT and ICP6 RHIM mutant virus infection revealed that the ICP6 mutant virus promotes the formation of a necrosome-like complex involving ZBP1/RIPK3/MLKL to promote MLKL phosphorylation and necroptosis in mouse cells (Guo et al. 2018). To analyze if ZBP1 also contributes to the initiation of necroptosis in human cells, ZBP1 expressing HT29 and ZBP1 deficient cells were infected with HSV1 ICP6 RHIM mutant virus. As in the mouse cells, the ZBP1-expressing human cells showed an induction of cell death, but the ZBP1-deficient cells did not. Using various inhibitors of necroptosis pathway players, RIPK3 and MLKL and caspase inhibitor zVAD, it was determined that the HSV ICP6 RHIM mutant virus-induced cell death through ZBP1/RIPK3/MLKL-mediated necroptosis. WT HSV1 infection did not induce cell death in the human cells, showing that the ICP6 RHIM was required for HSV1 necroptosis inhibition in its natural host species. These studies highlight species-specific differences in how viral inhibitors act on cells, suggesting a viral restriction in non-natural host infection. In addition to necroptotic inhibition in human cells, ICP6 has demonstrated an ability to bind to and inhibit caspase 8 (Langelier et al. 2002; Esaki et al. 2010; Dufour et al. 2011). This means that the viral protein ICP6 can inhibit both necroptosis and apoptosis in human cells.

## Poxviruses

Poxviruses are a double-stranded DNA family of viruses. Most notable among these viruses may be the *Orthopoxvirus* genus due to the infamy of the variola virus (smallpox virus). The *Orthopoxvirus* genus includes vaccinia virus (VACV), cowpox virus (CPXV), Mpox (formerly monkeypox), and ectromelia (ECTV). Like other dsDNA viruses, this family encodes numerous accessory virulence factors that inhibit cell death.

Vaccinia virus (VACV) is the prototypic model for orthopoxvirus research and the poxvirus strain utilized in smallpox vaccinations. Early VACV studies identified a novel necrotic cell death pathway that eliminated virus infected cells. Researchers found that VACV encodes a caspase-8 inhibitor known as CrmA-like ortholog B13R (aka Spi2), which blocks caspase-dependent cell death pathways and sensitizes infected cells for this necrotic death pathway through TNF signaling (Li and Beg 2000). TNF signaling in VACV infection was further clarified using VACV-infected cells that are normally sensitive to necroptosis. RIPK1-deficient cells showed a reduction in necrosis, confirming RIPK1 as an essential player in the pathway. Further, TNFR-2 was also shown to be essential for this pathway in an in vivo infection by creating TNFR-2-deficient mice and observing reduced inflammation in the liver (Chan et al. 2003). RNA interference screens identified RIPK3 as another pivotal player in the necroptosis pathway. Wild-type and RIP3K-deficient cells were infected with VACV showing the deficient cells resistant to TNF-α-induced necroptosis. As with the RIPK1-deficient mice, RIPK3-deficient mice infected with VACV showed much less necrosis and liver inflammation than infected wild-type mice (Cho et al. 2009), showing that RIPK3 is part of the host defense mechanism protecting against VACV. Together, during VACV infection, these two protein kinases form a RIPK1–RIPK3 necrosome complex through RHIM interactions, RIPK1 thereby activating RIPK3 to phosphorylate and subsequently activate MLKL to execute necroptosis (Sun et al. 2002, 2012; Rodriguez et al. 2016). Later studies would look at another RHIM-containing stimulator of necroptosis known as ZBP1. In addition to two RHIM domains (RHIM-A and RHIM-B), this protein contains two Z-NA (Zα) binding domains located at the N-terminus. These Z-NA binding domains can be triggered by Z-NA released during VACV infections. Once triggered, type I IFN-signaling increases along with further ZBP1 expression. Utilizing RHIM domain interactions, ZBP1 can then instigate necroptotic cell death via RIPK3/MLKL activation. This necroptotic pathway is not observed in wild-type VACV infections, however, since the virus also encodes a protein with a Zα-binding domain known as E3 which serves to inhibit necroptosis (Koehler et al. 2017).

Studies examining how VACV utilizes E3 to prevent necroptotic death have produced some interesting findings. In vivo studies have shown that mice infected with a mutant VACV that encodes a truncated E3L lacking a Zα-binding domain did not exhibit severe pathology comparable to mice infected with the wt virus. However, pathogenicity was restored in both $ZBP1^{-/-}$ and $RIPK3^{-/-}$ mice. Additionally, follow-up in vitro studies demonstrated that knocking out the Zα-binding domains of

ZBP1 was sufficient for increasing viral pathogenesis. The culmination of these and other studies provided evidence that VACV's E3 is a competitive inhibitor of ZBP1/RIPK3/MLKL-induced necroptosis (Koehler et al. 2021). This provides another example of how viruses have co-evolved with us to evade our innate immune system.

Another virus that utilizes necroptotic inhibitors is cowpoxvirus (CPXV). Cowpox is known to block caspase activation by CrmA (homologue of VACV B13R) making the cells susceptible to necroptosis by triggering RIPK1/RIPK3 (Kettle et al. 1997; Li and Beg 2000; Chan et al. 2003). Most poxviruses, like VACV, also encode various amounts of TNF inhibitors to block signaling to restrict TNF-induced necroptosis. CPXV has been found to encode many inhibitors: CrmB, CrmC, CrmD, and CrmE (Cunnion 1999). Additionally, CPXV has been found to produce a CD30 homologue with exclusive TNFSF8 binding whose role in pathogenesis is still unknown (Panus et al. 2002). Aside from these TNF inhibitors, a targeted small interfering RNA (siRNA) screen was able to identify a novel inhibitor of RIPK3 generated by CPXV termed viral inducer of RIPK3 degradation (vIRD) (Liu et al. 2021). vIRD binds to RIPK3, causing RIPK3 ubiquitination and subsequent proteasome degradation as detected in western blots. Mice infected with mutant CPXV in which vIRD had been deleted showed significant decreases in many pathogenic processes including mortality. These pathogenic processes were able to be reversed in mice deficient in RIPK3 and MLKL. This evidence shows vIRD was able to inhibit all known forms of necroptotic cell death by directly targeting RIPK3 for degradation.

Other poxviruses have been co-opted to study the vIRD inhibitor including VACV, ectromelia virus (ECTV), and leporipoxvirus myxoma virus (MYXV). VACV encodes a defective vIRD that has been truncated. When a functional version of vIRD is integrated into VACV, mice infected with the mutant virus show increased viral titers. MYXV, which also lacks a functional vIRD, has been shown to be more pathogenic in RIPK3-deficient hosts, which makes sense since these hosts would have resistance to necroptotic death. ECTV, which also has an intact vIRD ortholog, induced RIPK3 degradation and resistance to TNF-induced necroptosis (Liu et al. 2021). ECTV, however, does not fit the typical necroptosis story where the crucial players, RIPK3 and MLKL, do not seem to play the same role in controlling viral infections (Montoya et al. 2023). Studying these inhibitors and comparing them to other viruses can help us to understand more about how viruses have continued to evolve survival mechanisms within their hosts.

Another method used in some poxviruses to inhibit necroptotic death is to mimic viral proteins. BeAn58058 and Cotia poxviruses encode proteins with high enough homology to competitively bind and sequester RIPK3 and out-compete host MLKL and disrupt necroptosis (Petrie et al. 2019). These mimic proteins have been identified in several clades of poxviruses, including avipoxviruses and leporipoxviruses.

## *RNA Viruses*

### Influenza A

Influenza A (IAV) is a negative, enveloped single-stranded RNA virus from the *Orthomyxoviridae* family that causes seasonal and sometimes severe respiratory tract infections in humans worldwide. Current and commonly circulating subtypes of influenza are A(H1N1) and A(H3N2). IAV infection typically affects human lung epithelial cells and triggers cell death. IAV was originally thought to mainly activate the apoptotic cell death pathway. After the discovery of RIPK3 as a crucial factor for cell death in IAV infections, it was recognized that the cell types most commonly used in viral and medical research were not appropriate to study cell death pathways for IAV infection because they did not express RIPK3 (Nogusa et al. 2016a).

Research utilizing the influenza A virus P8 strain model has helped to illuminate the dual role ZBP1 and RIPK3 play in both apoptosis and necroptosis. Early on, it was found that a ZBP1/RIPK3-dependent signaling axis induced cell death in human lung epithelial cells and MEFs following IAV infection. MEFs from RIPK3-deficient mice were infected with PR8 and found to be resistant to cell death and cytopathic effects, whereas wild-type mice showed susceptibility to the infection resulting in cell death (Nogusa et al. 2016a). However, MEFs generated to be deficient in MLKL, the executioner of necroptotic death, were determined to be just as susceptible as wild types to cell death (Nogusa et al. 2016b). This suggested that ZBP1 was capable of stimulating a non-necroptotic form of cell death post-IAV infection. Further genetic analyses revealed the absence of MLKL-activated ZBP1 and could trigger apoptosis in IAV-infected MEFs. Consequently, it was found that both the apoptosis regulator Fadd and the necroptosis executor MLKL needed to be knocked out in order to rescue IAV-infected MEF survival. Looking further into this ZBP1-stimulated apoptotic pathway demonstrated some interesting findings. After ZBP1 is activated post-IAV infection, RIPK3 is needed for the early onset of apoptosis, but not required at later time points. Additionally, the kinase activity of RIPK3 has been shown to be needed for necroptotic but not apoptotic cell death. As such, catalytically inactive RIPK3-expressing mice are subject to early caspase-8-dependent apoptosis, while those deficient in RIPK3 altogether ($RIPK3^{-/-}$) are subject to later onset apoptotic cell death.

Additional studies verified that IAV was able to generate Z-RNA and activate ZBP1 to initiate the activation of RIPK3 and cell death pathways (Zhang et al. 2020). By transfecting labeled antiserum to Z-NA, it was found that the antiserum co-localized with Z-RNAs in the nucleus of infected cells. This Z-RNA was then shown to co-localize with ZBP1 and activate RIPK3 death signaling culminating in cell death.

While program cell death can be advantageous as a host defense mechanism, necroptosis, by way of its inflammatory nature, can also have severe pathogenic disadvantages when triggered by IAV infection. This was evaluated by examining MLKL-, ZBP1-, and RIPK3-deficient animals after IAV infection. Again, MLKL did not show any increased lethality or viral load, while *ZPB1*$^{-/-}$ and *RIPK3*$^{-/-}$ mice had significantly higher pathogenesis than their wild-type counterparts (Zhang et al. 2020).

## Reovirus

Reoviruses are non-enveloped RNA viruses with ten linear double-stranded genomic segments (Danthi et al. 2013). Four main serotypes of mammalian ReoV have been identified. These serotypes are represented by prototype strains Type 1 Lang (T1L), Type 2 Jones (T2J), Type 3 Dearing (T3D), and Type 4 Ndelle (T4N) (Attoui et al. 2001).

It was previously believed that the primary PCD pathway triggered by ReoV infection was apoptosis. However, recent studies of T3D-infected L929 cells demonstrated that necroptosis is an important innate mechanism triggered in response to infection. L929 cells infected with the T3D strain underwent a rapid induction of death even in the presence of a caspase inhibitor (Berger and Danthi 2013). The cell death was characterized to be necroptosis as a result of loss of membrane integrity that was inhibited by necrostatins. A follow-up study then demonstrated that IFN-β production, cellular incorporation of viral genomic RNA, and de novo synthesis of viral dsRNA were required to induce necroptosis (Berger et al. 2017). Together, these results suggest that incoming genomic RNA is detected by RLRs in the cytoplasm of the infected cell, which then signals via the adaptor protein MAVS to produce IFN-β followed by the synthesis of viral RNA establishing a pathway to upregulate and detect the likely ligand of dsRNA. Additionally, it was found that knockdown of the dsRNA binding protein σ3 resulted in enhanced necroptosis which was attributed to RLR-mediated increase in IFN-β production (Roebke et al. 2020). Further support of multifactorial death requirements is indicated by the enhanced rate of necroptosis when the ReoV μ1 protein is knocked down, which is associated with an increase of accumulated progeny dsRNA and viral protein synthesis. The specific mechanism of necroptosis induction by ReoV remains unclear. One proposed mechanism is the downregulation of the cIAP family of E3 ubiquitin ligase ReoV (Kominsky et al. 2002). cIAPs are known to polyubiquinate RIPK1 and RIPK3 resulting in kinase inhibition and therefore the decreased cIAP could result in more RIPK1 and RIPK3 and thus more necroptosis (Annibaldi et al. 2018). This model, however, does not address the need for de novo dsRNA synthesis required for necroptosis activation (Jiffry et al. 2021). It is also unclear if the induction of necroptosis, as in many other virus models, contributes to viral clearance or pathogenesis. RIPK3 deficiency does result in an increase in ReoV progeny virions suggesting potential for a protective mechanism indicative of a potentially protective role for necroptosis in ReoV clearance (Yue and Shatkin 1997; Berger et al. 2017). Both cultured BMDMs and L929 fibroblasts undergo necroptosis within a few days post-infection (DeBiasi et al. 2004; Xing et al. 2016; Gummersheimer and Danthi 2020). It remains unclear if ReoV-induced necroptosis occurs within biologically relevant cell types in the gut, CNS, or heart and, furthermore, whether

Reo-mediated necroptosis is protective or deleterious due to inflammation during other RNA viral infections like IAV (Balachandran and Rall 2020). More studies are warranted to further clarify the role of necroptosis in ReoV infections and its role in enhancing or restricting pathogenesis.

## Respiratory Syncytial Virus

Human respiratory syncytial virus (RSV) is a nonsegmented negative-stranded RNA virus from the *Paramyxoviridae* family (Collins et al. 2013). This enveloped virus is a significant cause of acute bronchiolitis in infants, the outcomes and severity of which are associated with necroptosis. It is hypothesized that infants suffering acute RSV infection may later be predisposed to respiratory complications, such as asthma and wheezing, later in life. While much of the data is clinical, reports of RSV-triggered necroptosis are relevant to the focus of this chapter. Early RSV necroptosis studies used selective inhibitors of necroptosis pathway players to determine that RSV infection activated the necroptotic pathway RIPK1/RIPK3/MLKL (Muraro et al. 2018; Simpson et al. 2020). Further studies demonstrated that RSV infection may involve TNF-mediated macrophage necroptosis which can cause increased lung damage during RSV infections. This is supported by findings showing that *Tnfr1*$^{-/-}$ mice demonstrate decreased numbers of necrotic macrophages in lung tissue with less RIPK3 and MLKL expression, while elevated levels of TNF in RSV-infected infants correlate with increased disease severity (Santos et al. 2021). RSV-infected mice demonstrate an increased expression of activated RIPK1 and MLKL with no increase in activated caspase-3 expression which correlates with neutrophilic inflammation and airway epithelial cell sloughing (Simpson et al. 2020). When MLKL or RIPK1 were taken out of the equation using pharmacological inhibitors, inflammation and airway remodeling were prevented and there was a decreased viral load. Furthermore, limiting necroptosis in RSV infection in neonatal mice was associated with decreased viral or allergen-triggered asthma later on in life. These findings argue that necroptotic processes stimulated by RSV infection are actually harmful by increasing viral pathogenesis and making those who become infected more susceptible to developing asthma. This is important to note in the development of future therapeutics as researchers will need to weigh the benefits versus harms of utilizing an inflammatory form of cell death in necroptosis.

## SARS-CoV-2

SARS-CoV-2 is a positive single-stranded RNA virus. SARS-CoV-2 infection has been shown to induce a strong inflammatory response that results in severe lung disease in critically ill patients. SARS-CoV-2 infection in lung epithelial cells revealed activation of NF-κB, upregulation of inflammatory cytokines, upregulation of ZBP1, and significant cell death. It has been reported that SARS-CoV-2 triggers the production of viral Z-RNA and initiates the ZBP1-RIPK3-MLKL necroptosis pathway. In ZBP1 CRISPR knockout epithelial cells, a reduced MLKL phosphorylation confirmed the role of necroptosis. Additionally, a significant reduction in inflammatory cytokines released in the ZBP1-deficient cells. This showed that SARS-CoV-2 infection activates necroptosis and induces pro-inflammatory signaling (Li et al. 2023). It was shown that ZBP1 deficiency in mice also reduced immune cell infiltration and lung damage, demonstrating that in the mouse model, ZBP1 is important for SARS-CoV-2 pathogenesis (Li et al. 2023). Further down the pathway, RIPK3 knockdown cells did not appear to impact viral load, but a significant reduction of inflammatory cytokines and chemokines was demonstrated. This depletion of MLKL did not reduce the inflammatory response. Inhibition of the RIPK3 kinase activity also did not appear to impact the inflammatory response but only the necroptosis activation; thus, the scaffolding of RIPK3 was determined to be crucial for triggering the inflammatory response resulting in severe lung damage.

## Flaviviruses

Flaviviruses are a family of positive, single-stranded, enveloped RNA viruses. This genus of viruses includes mosquito-transmitted viruses such as yellow fever, Japanese encephalitis, Dengue fever, West Nile, and Zika virus (Pierson and Diamond 2020). Ticks are also responsible for some flavivirus transmission, causing encephalitis and hemorrhagic fever viruses. These include Palm Creek virus, Parramatta River virus, Kyasanur Forest disease, Alkhurma disease, and Omsk hemorrhagic fever.

Zika virus (ZIKV) is a flavivirus transmitted by mosquitoes that can cause various neurological outcomes in humans. ZIKV infection in pregnancy can result in the virus crossing the placenta resulting in severe fetal abnormalities like microcephaly (Coyne and Lazear 2016; Costa and Ko 2018). Infection in adulthood can also result in neurological injury but is rare (Mehta et al. 2018). Early studies conducted found that neurons infected with ZIKV activated a cell-restricting pathway. Using mice deficient in the crucial necroptotic players, ZBP1, RIPK3, or the RIPK1 kinase activity, researchers found that the incidence of paresis to be rapidly increased during CNS infection demonstrating the protective role of necroptosis in neuron ZIKV infection. After activation of the ZBP1 necroptosis pathway, instead of completing necroptosis, RIPK signaling in ZIKV-infected neurons causes them to enter a metabolic state that suppresses viral genome replication (Daniels et al. 2019). This change in metabolic state is evidenced by the identification of IRG1 production of itaconate that inhibits succinate dehydrogenase activity in the mitochondria, restricting viral replication (Daniels et al. 2019). Intracranially infected mice deficient in IRG1 had a greater viral burden, but injection of itaconate rescued the IRG1-deficient mice from this viral burden. Later studies, however, found evidence of necroptosis occurring in ZIKV-infected human astrocytes and resulted in a protective effect on glial cells. This is of interest since astrocytes are known to support neurons in many ways including providing key components to fuel neuronal metabolism. It is important to also note that the necroptotic cell death observed was RIPK1 independent and associated with an upregulation of ZBP1 (Wen et al. 2021). While these are some fascinating findings, ultimately, studies using ZIKV-infected MLKL-knockout mice have shown no differences in mortality compared to wild types infected with the virus which suggests that necroptosis does not play a major role in pathogenesis (Daniels et al. 2019).

Necroptotic processes in West Nile virus (WNV) models have not been thoroughly studied. What has been demonstrated is that microarray analysis of brain tissues harvested from WNV-infected mice shows an upregulation of necroptotic genes (Peng and Wang 2019). While these genes are upregulated, other in vivo studies have shown that RIPK3-deficient mice had increased mortality after WNV infection; however, in mice deficient in MLKL, the executioner of necroptosis, no impact on mortality was observed. These data suggest RIPK3 is involved in coordinating other immune responses including chemokine expression and lymphocyte recruitment (Daniels et al. 2017). This further demonstrates the complexity of key necroptotic components and the multiple roles RIPK3 plays in the host defense.

### Rotavirus

Rotavirus (RV) is a double-stranded RNA virus of the Reoviridae family and a leading cause of gastroenteritis in young children. Rotavirus infection mainly impacts enterocyte cells in the small intestine where infection concludes in cell death. RV induces necroptosis and apoptosis, the latter being well characterized. In early RV-induced necroptosis studies, RV was demonstrated to trigger MLKL phosphorylation, oligomerization of MLKL, and translocation to the plasma membrane for pore formation (Mukhopadhyay et al. 2022) and shown to act through RIPK1-dependent pathway. Interestingly, a cooperative element between apoptosis and necroptosis appears to be a part of RV infection. When necroptosis is inhibited, apoptosis is activated and cell viability is also increased, resulting in an antiviral state where infected cells are phagocytized and RV replication reduces. When apoptosis is restricted in vivo, necroptosis becomes the death pathway of choice and a proviral response occurs. As a consequence of necroptosis, infected cells burst and the viral contents are released to promote infection (Soliman et al. 2022).

## Conclusion

Viruses have invested substantial resources in the inhibition of programmed cell death by the host, highlighting the importance of cell death in the arms race between hosts and pathogens. While caspase-dependent cell death is traditionally thought of as the default host cell death pathway, numerous virus-encoded gene products have been identified that inhibit caspase-dependent cell death. However, inhibition of caspase-dependent cell death sensitizes cells to MLKL-dependent necroptotic cell death. Thus, viruses have evolved multiple mechanisms to inhibit necroptotic cell death that can be engaged to gain further understanding of the consequences of unleashing or restricting this death pathway. Viral inhibitors target RIPK3 degradation or redistribution, ZBP1/DAI antagonist, restriction of RHIM activation, TNFR activation, and MLKL antagonist. Antagonism of necroptosis is important for poxviral resistance to interferon, both in cells in culture and in animal models, and for pathogenesis in animal models. Thus, inhibition of necroptosis appears to be an important means for viruses to maintain an edge in the arms race between viruses and their hosts.

## References


Amarante-Mendes GP, Adjemian S, Branco LM et al (2018) Pattern recognition receptors and the host cell death molecular machinery. Front Immunol 9:2379

Annibaldi A, Wicky John S, Vanden Berghe T et al (2018) Ubiquitin-mediated regulation of RIPK1 kinase activity independent of IKK and MK2. Mol Cell 69:566–580.e5. https://doi.org/10.1016/j.molcel.2018.01.027

Attoui H, Biagini P, Stirling J et al (2001) Sequence characterization of Ndelle virus genome segments 1, 5, 7, 8, and 10: evidence for reassignment to the genus Orthoreovirus, family Reoviridae. Biochem Biophys Res Commun 287:583–588. https://doi.org/10.1006/bbrc.2001.5612

Balachandran S, Rall GF (2020) Benefits and perils of necroptosis in influenza virus infection. J Virol 94:e01101-19. https://doi.org/10.1128/JVI.01101-19

Berger AK, Danthi P (2013) Reovirus activates a caspase-independent cell death pathway. mBio 4: e00178-00113. https://doi.org/10.1128/mBio.00178-13

Berger SB, Kasparcova V, Hoffman S et al (2014) RIP1 kinase activity is dispensable for normal development but is a key regulator of inflammation in SHARPIN-deficient mice. J Immunol Baltim Md 1950 192:5476–5480. https://doi.org/10.4049/jimmunol.1400499

Berger SB, Harris P, Nagilla R et al (2015) Characterization of GSK'963: a structurally distinct, potent and selective inhibitor of RIP1 kinase. Cell Death Discov 1:15009. https://doi.org/10.1038/cddiscovery.2015.9

Berger AK, Hiller BE, Thete D et al (2017) Viral RNA at two stages of reovirus infection is required for the induction of necroptosis. J Virol 91:e02404-16. https://doi.org/10.1128/JVI.02404-16

Berghe TV, Linkermann A, Jouan-Lanhouet S et al (2014) Regulated necrosis: the expanding network of non-apoptotic cell death pathways. Nat Rev Mol Cell Biol 15:135–147. https://doi.org/10.1038/nrm3737

Bonnet MC, Preukschat D, Welz P-S et al (2011) The adaptor protein FADD protects epidermal keratinocytes from necroptosis *in vivo* and prevents skin inflammation. Immunity 35:572–582. https://doi.org/10.1016/j.immuni.2011.08.014

Bozec D, Iuga AC, Roda G et al (2016) Critical function of the necroptosis adaptor RIPK3 in protecting from intestinal tumorigenesis. Oncotarget 7:46384–46400. https://doi.org/10.18632/oncotarget.10135

Brune W, Ménard C, Heesemann J, Koszinowski UH (2001) A ribonucleotide reductase homolog of cytomegalovirus and endothelial cell tropism. Science 291:303–305. https://doi.org/10.1126/science.291.5502.303

Cao L, Mu W (2021) Necrostatin-1 and necroptosis inhibition: Pathophysiology and therapeutic implications. Pharmacol Res 163:105297. https://doi.org/10.1016/j.phrs.2020.105297

Cao T, Ni R, Ding W et al (2022) MLKL-mediated necroptosis is a target for cardiac protection in mouse models of type-1 diabetes. Cardiovasc Diabetol 21:165. https://doi.org/10.1186/s12933-022-01602-9

Chan FK-M, Shisler J, Bixby JG et al (2003) A role for tumor necrosis factor receptor-2 and receptor-interacting protein in programmed necrosis and antiviral responses. J Biol Chem 278: 51613–51621. https://doi.org/10.1074/jbc.M305633200

Chen J, Kos R, Garssen J, Redegeld F (2019) Molecular insights into the mechanism of necroptosis: the necrosome as a potential therapeutic target. Cells 8:1486. https://doi.org/10.3390/cells8121486

Chen Y, Lin J, Zhao Y et al (2021) Toll-like receptor 3 (TLR3) regulation mechanisms and roles in antiviral innate immune responses. J Zhejiang Univ Sci B 22:609–632. https://doi.org/10.1631/jzus.B2000808

Chen L, Zhang X, Ou Y et al (2022) Advances in RIPK1 kinase inhibitors. Front Pharmacol 13: 976435

Cho Y, Challa S, Moquin D et al (2009) Phosphorylation-driven assembly of RIP1-RIP3 complex regulates programmed necrosis and virus-induced inflammation. Cell 137:1112–1123. https://doi.org/10.1016/j.cell.2009.05.037

Collins PL, Fearns R, Graham BS (2013) Respiratory syncytial virus: virology, reverse genetics, and pathogenesis of disease. Curr Top Microbiol Immunol 372:3–38. https://doi.org/10.1007/978-3-642-38919-1_1

Costa F, Ko AI (2018) Zika virus and microcephaly: where do we go from here? Lancet Infect Dis 18:236–237. https://doi.org/10.1016/S1473-3099(17)30697-7

Cotsmire SM, Szczerba M, Jacobs BL (2021) Detecting necroptosis in virus-infected cells. Methods Mol Biol Clifton NJ 2225:199–216. https://doi.org/10.1007/978-1-0716-1012-1_11

Coyne CB, Lazear HM (2016) Zika virus—reigniting the TORCH. Nat Rev Microbiol 14:707–715. https://doi.org/10.1038/nrmicro.2016.125

Cunnion KM (1999) Tumor necrosis factor receptors encoded by poxviruses. Mol Genet Metab 67: 278–282. https://doi.org/10.1006/mgme.1999.2878

Daley-Bauer LP, Roback L, Crosby LN et al (2017) Mouse cytomegalovirus M36 and M45 death suppressors cooperate to prevent inflammation resulting from antiviral programmed cell death pathways. Proc Natl Acad Sci U S A 114:E2786–E2795. https://doi.org/10.1073/pnas.1616829114

Daniels BP, Snyder AG, Olsen TM et al (2017) RIPK3 restricts viral pathogenesis via cell death-independent neuroinflammation. Cell 169:301–313.e11. https://doi.org/10.1016/j.cell.2017.03.011

Daniels BP, Kofman SB, Smith JR et al (2019) The nucleotide sensor ZBP1 and kinase RIPK3 induce the enzyme IRG1 to promote an antiviral metabolic state in neurons. Immunity 50:64–76.e4. https://doi.org/10.1016/j.immuni.2018.11.017

Danthi P, Holm GH, Stehle T, Dermody TS (2013) Reovirus receptors, cell entry, and proapoptotic signaling. In: Pöhlmann S, Simmons G (eds) Viral entry into host cells. Springer, New York, pp 42–71

de Almagro MC, Goncharov T, Izrael-Tomasevic A et al (2017) Coordinated ubiquitination and phosphorylation of RIP1 regulates necroptotic cell death. Cell Death Differ 24:26–37. https://doi.org/10.1038/cdd.2016.78

DeBiasi RL, Robinson BA, Sherry B et al (2004) Caspase inhibition protects against reovirus-induced myocardial injury *in vitro* and *in vivo*. J Virol 78:11040–11050. https://doi.org/10.1128/JVI.78.20.11040-11050.2004

Declercq W, Vanden Berghe T, Vandenabeele P (2009) RIP kinases at the crossroads of cell death and survival. Cell 138:229–232. https://doi.org/10.1016/j.cell.2009.07.006

Degterev A, Maki JL, Yuan J (2013) Activity and specificity of necrostatin-1, small-molecule inhibitor of RIP1 kinase. Cell Death Differ 20:366. https://doi.org/10.1038/cdd.2012.133

Dikic I, Wakatsuki S, Walters KJ (2009) Ubiquitin-binding domains—from structures to functions. Nat Rev Mol Cell Biol 10:659–671. https://doi.org/10.1038/nrm2767

Dondelinger Y, Declercq W, Montessuit S et al (2014) MLKL compromises plasma membrane integrity by binding to phosphatidylinositol phosphates. Cell Rep 7:971–981. https://doi.org/10.1016/j.celrep.2014.04.026

Dovey CM, Diep J, Clarke BP et al (2018) MLKL requires the inositol phosphate code to execute necroptosis. Mol Cell 70:936–948.e7. https://doi.org/10.1016/j.molcel.2018.05.010

Dufour F, Bertrand L, Pearson A et al (2011) The ribonucleotide reductase R1 subunits of herpes simplex virus 1 and 2 protect cells against poly(I · C)-induced apoptosis. J Virol 85:8689–8701. https://doi.org/10.1128/JVI.00362-11

Esaki S, Goshima F, Katsumi S et al (2010) Apoptosis induction after herpes simplex virus infection differs according to cell type *in vivo*. Arch Virol 155:1235–1245. https://doi.org/10.1007/s00705-010-0712-2

Fletcher-Etherington A, Nobre L, Nightingale K et al (2020) Human cytomegalovirus protein pUL36: A dual cell death pathway inhibitor. Proc Natl Acad Sci U S A 117:18771–18779. https://doi.org/10.1073/pnas.2001887117

Green DR, Llambi F (2015) Cell death signaling. Cold Spring Harb Perspect Biol 7:a006080. https://doi.org/10.1101/cshperspect.a006080

Gummersheimer SL, Danthi P (2020) Reovirus core proteins λ1 and σ2 promote stability of disassembly intermediates and influence early replication events. J Virol 94:e00491-20. https://doi.org/10.1128/JVI.00491-20

Guo H, Omoto S, Harris PA et al (2015) Herpes simplex virus suppresses necroptosis in human cells. Cell Host Microbe 17:243–251. https://doi.org/10.1016/j.chom.2015.01.003

Guo H, Gilley RP, Fisher A et al (2018) Species-independent contribution of ZBP1/DAI/DLM-1-triggered necroptosis in host defense against HSV1. Cell Death Dis 9:816. https://doi.org/10.1038/s41419-018-0868-3

Guo H, Koehler HS, Dix RD, Mocarski ES (2022a) Programmed cell death-dependent host defense in ocular herpes simplex virus infection. Front Microbiol 13:869064. https://doi.org/10.3389/fmicb.2022.869064

Guo H, Koehler HS, Mocarski ES, Dix RD (2022b) RIPK3 and caspase 8 collaborate to limit herpes simplex encephalitis. PLoS Pathog 18:e1010857. https://doi.org/10.1371/journal.ppat.1010857

Hanus J, Anderson C, Wang S (2015) RPE necroptosis in response to oxidative stress and in AMD. Ageing Res Rev 24:286–298. https://doi.org/10.1016/j.arr.2015.09.002

He S, Liang Y, Shao F, Wang X (2011) Toll-like receptors activate programmed necrosis in macrophages through a receptor-interacting kinase-3-mediated pathway. Proc Natl Acad Sci U S A 108:20054–20059. https://doi.org/10.1073/pnas.1116302108

Holler N, Zaru R, Micheau O et al (2000) Fas triggers an alternative, caspase-8-independent cell death pathway using the kinase RIP as effector molecule. Nat Immunol 1:489–495. https://doi.org/10.1038/82732

Jiao J, Wang Y, Ren P et al (2020) Necrosulfonamide ameliorates neurological impairment in spinal cord injury by improving antioxidative capacity. Front Pharmacol 10:1538

Jiffry J, Thavornwatanayong T, Rao D et al (2021) Oncolytic reovirus (pelareorep) induces autophagy in KRAS-mutated colorectal cancer. Clin Cancer Res Off J Am Assoc Cancer Res 27:865–876. https://doi.org/10.1158/1078-0432.CCR-20-2385

Jorgensen I, Rayamajhi M, Miao EA (2017) Programmed cell death as a defence against infection. Nat Rev Immunol 17:151–164. https://doi.org/10.1038/nri.2016.147

Kaiser WJ, Upton JW, Mocarski ES (2008) Receptor-interacting protein homotypic interaction motif-dependent control of NF-kappa B activation via the DNA-dependent activator of IFN regulatory factors. J Immunol Baltim Md 1950 181:6427–6434. https://doi.org/10.4049/jimmunol.181.9.6427

Kaiser WJ, Sridharan H, Huang C et al (2013) Toll-like receptor 3-mediated necrosis via TRIF, RIP3, and MLKL. J Biol Chem 288:31268–31279. https://doi.org/10.1074/jbc.M113.462341

Kettle S, Alcamí A, Khanna A et al (1997) Vaccinia virus serpin B13R (SPI-2) inhibits interleukin-1beta-converting enzyme and protects virus-infected cells from TNF- and Fas-mediated apoptosis, but does not prevent IL-1beta-induced fever. J Gen Virol 78(Pt 3):677–685. https://doi.org/10.1099/0022-1317-78-3-677

Khoury MK, Gupta K, Franco SR, Liu B (2020) Necroptosis in the pathophysiology of disease. Am J Pathol 190:272–285. https://doi.org/10.1016/j.ajpath.2019.10.012

Koehler HS, Jacobs BL (2021) Subversion of programed cell death by poxviruses. Curr Top Microbiol Immunol. https://doi.org/10.1007/82_2020_229

Koehler H, Cotsmire S, Langland J et al (2017) Inhibition of DAI-dependent necroptosis by the Z-DNA binding domain of the vaccinia virus innate immune evasion protein, E3. Proc Natl Acad Sci U S A 114:11506–11511. https://doi.org/10.1073/pnas.1700999114

Koehler H, Cotsmire S, Zhang T et al (2021) Vaccinia virus E3 prevents sensing of Z-RNA to block ZBP1-dependent necroptosis. Cell Host Microbe 29:1266–1276.e5. https://doi.org/10.1016/j.chom.2021.05.009

Kominsky DJ, Bickel RJ, Tyler KL (2002) Reovirus-induced apoptosis requires mitochondrial release of Smac/DIABLO and involves reduction of cellular inhibitor of apoptosis protein levels. J Virol 76:11414–11424. https://doi.org/10.1128/jvi.76.22.11414-11424.2002

Koo G-B, Morgan MJ, Lee D-G et al (2015) Methylation-dependent loss of RIP3 expression in cancer represses programmed necrosis in response to chemotherapeutics. Cell Res 25:707–725. https://doi.org/10.1038/cr.2015.56

Kuriakose T, Man SM, Malireddi RKS et al (2016) ZBP1/DAI is an innate sensor of influenza virus triggering the NLRP3 inflammasome and programmed cell death pathways. Sci Immunol 1: aag2045. https://doi.org/10.1126/sciimmunol.aag2045

Langelier Y, Bergeron S, Chabaud S et al (2002) The R1 subunit of herpes simplex virus ribonucleotide reductase protects cells against apoptosis at, or upstream of, caspase-8 activation. J Gen Virol 83:2779–2789. https://doi.org/10.1099/0022-1317-83-11-2779

Lembo D, Donalisio M, Hofer A et al (2004) The ribonucleotide reductase R1 homolog of murine cytomegalovirus is not a functional enzyme subunit but is required for pathogenesis. J Virol 78: 4278–4288. https://doi.org/10.1128/jvi.78.8.4278-4288.2004

Li M, Beg AA (2000) Induction of necrotic-like cell death by tumor necrosis factor alpha and caspase inhibitors: novel mechanism for killing virus-infected cells. J Virol 74:7470–7477. https://doi.org/10.1128/jvi.74.16.7470-7477.2000

Li J, McQuade T, Siemer AB et al (2012) The RIP1/RIP3 necrosome forms a functional amyloid signaling complex required for programmed necrosis. Cell 150:339–350. https://doi.org/10.1016/j.cell.2012.06.019

Li J-X, Feng J-M, Wang Y et al (2014) The B-Raf(V600E) inhibitor dabrafenib selectively inhibits RIP3 and alleviates acetaminophen-induced liver injury. Cell Death Dis 5:e1278. https://doi.org/10.1038/cddis.2014.241

Li S, Zhang Y, Guan Z et al (2023) SARS-CoV-2 Z-RNA activates the ZBP1-RIPK3 pathway to promote virus-induced inflammatory responses. Cell Res 33:201–214. https://doi.org/10.1038/s41422-022-00775-y

Liu Z, Nailwal H, Rector J et al (2021) A class of viral inducer of degradation of the necroptosis adaptor RIPK3 regulates virus-induced inflammation. Immunity 54:247–258.e7. https://doi.org/10.1016/j.immuni.2020.11.020

Mandal P, Berger SB, Pillay S et al (2014) RIP3 induces apoptosis independent of pronecrotic kinase activity. Mol Cell 56:481–495. https://doi.org/10.1016/j.molcel.2014.10.021

Martens S, Hofmans S, Declercq W et al (2020) Inhibitors targeting RIPK1/RIPK3: old and new drugs. Trends Pharmacol Sci 41:209–224. https://doi.org/10.1016/j.tips.2020.01.002

Mehta R, Soares CN, Medialdea-Carrera R et al (2018) The spectrum of neurological disease associated with Zika and chikungunya viruses in adults in Rio de Janeiro, Brazil: A case series. PLoS Negl Trop Dis 12:e0006212. https://doi.org/10.1371/journal.pntd.0006212

Mocarski ES, Upton JW, Kaiser WJ (2011) Viral infection and the evolution of caspase 8-regulated apoptotic and necrotic death pathways. Nat Rev Immunol 12:79–88. https://doi.org/10.1038/nri3131

Mocarski ES, Guo H, Kaiser WJ (2015) Necroptosis: The Trojan horse in cell autonomous antiviral host defense. Virology 479–480:160–166. https://doi.org/10.1016/j.virol.2015.03.016

Moerke C, Bleibaum F, Kunzendorf U, Krautwald S (2019) Combined knockout of RIPK3 and MLKL reveals unexpected outcome in tissue injury and inflammation. Front Cell Dev Biol 7:19. https://doi.org/10.3389/fcell.2019.00019

Montoya B, Knudson CJ, Melo-Silva CR et al (2023) Resistance to poxvirus lethality does not require the necroptosis proteins RIPK3 or MLKL. J Virol 97:e01945-22. https://doi.org/10.1128/jvi.01945-22

Moriwaki K, Bertin J, Gough PJ et al (2015) Differential roles of RIPK1 and RIPK3 in TNF-induced necroptosis and chemotherapeutic agent-induced cell death. Cell Death Dis 6: e1636. https://doi.org/10.1038/cddis.2015.16

Mukhopadhyay U, Patra U, Chandra P et al (2022) Rotavirus activates MLKL-mediated host cellular necroptosis concomitantly with apoptosis to facilitate dissemination of viral progeny. Mol Microbiol 117:818–836. https://doi.org/10.1111/mmi.14874

Muraro SP, De Souza GF, Gallo SW et al (2018) Respiratory syncytial virus induces the classical ROS-dependent NETosis through PAD-4 and necroptosis pathways activation. Sci Rep 8: 14166. https://doi.org/10.1038/s41598-018-32576-y

Murphy JM (2020) The killer pseudokinase mixed lineage kinase domain-like protein (MLKL). Cold Spring Harb Perspect Biol 12:a036376. https://doi.org/10.1101/cshperspect.a036376

Murphy JM, Czabotar PE, Hildebrand JM et al (2013) The pseudokinase MLKL mediates necroptosis via a molecular switch mechanism. Immunity 39:443–453. https://doi.org/10.1016/j.immuni.2013.06.018

Nailwal H, Chan FK-M (2019) Necroptosis in anti-viral inflammation. Cell Death Differ 26:4–13. https://doi.org/10.1038/s41418-018-0172-x

Newton K, Dugger DL, Wickliffe KE et al (2014) Activity of protein kinase RIPK3 determines whether cells die by necroptosis or apoptosis. Science 343:1357–1360. https://doi.org/10.1126/science.1249361

Nogusa S, Slifker MJ, Ingram JP et al (2016a) RIPK3 is largely dispensable for RIG-I-like receptor- and type I interferon-driven transcriptional responses to influenza A virus in murine fibroblasts. PLoS One 11:e0158774. https://doi.org/10.1371/journal.pone.0158774

Nogusa S, Thapa RJ, Dillon CP et al (2016b) RIPK3 activates parallel pathways of MLKL-driven necroptosis and FADD-mediated apoptosis to protect against influenza A virus. Cell Host Microbe 20:13–24. https://doi.org/10.1016/j.chom.2016.05.011

Oberst A, Dillon CP, Weinlich R et al (2011) Catalytic activity of the caspase-8-FLIP(L) complex inhibits RIPK3-dependent necrosis. Nature 471:363–367. https://doi.org/10.1038/nature09852

O'Donnell MA, Legarda-Addison D, Skountzos P et al (2007) Ubiquitination of RIP1 regulates an NF-kappaB-independent cell-death switch in TNF signaling. Curr Biol CB 17:418–424. https://doi.org/10.1016/j.cub.2007.01.027

Omoto S, Guo H, Talekar GR et al (2015) Suppression of RIP3-dependent necroptosis by human cytomegalovirus. J Biol Chem 290:11635–11648. https://doi.org/10.1074/jbc.M115.646042

Panus JF, Smith CA, Ray CA et al (2002) Cowpox virus encodes a fifth member of the tumor necrosis factor receptor family: A soluble, secreted CD30 homologue. Proc Natl Acad Sci U S A 99:8348–8353. https://doi.org/10.1073/pnas.122238599

Peng B-H, Wang T (2019) West Nile virus induced cell death in the central nervous system. Pathog Basel Switz 8:215. https://doi.org/10.3390/pathogens8040215

Perales-Linares R, Navas-Martin S (2013) Toll-like receptor 3 in viral pathogenesis: friend or foe? Immunology 140:153–167. https://doi.org/10.1111/imm.12143

Pestka S, Krause CD, Walter MR (2004) Interferons, interferon-like cytokines, and their receptors. Immunol Rev 202:8–32. https://doi.org/10.1111/j.0105-2896.2004.00204.x

Petrie EJ, Sandow JJ, Lehmann WIL et al (2019) Viral mlkl homologs subvert necroptotic cell death by sequestering cellular RIPK3. Cell Rep 28:3309–3319.e5. https://doi.org/10.1016/j.celrep.2019.08.055

Pierson TC, Diamond MS (2020) The continued threat of emerging flaviviruses. Nat Microbiol 5: 796–812. https://doi.org/10.1038/s41564-020-0714-0

Polykratis A, Hermance N, Zelic M et al (2014) Cutting edge: RIPK1 Kinase inactive mice are viable and protected from TNF-induced necroptosis *in vivo*. J Immunol Baltim Md 1950 193: 1539–1543. https://doi.org/10.4049/jimmunol.1400590

Rheault TR, Stellwagen JC, Adjabeng GM et al (2013) Discovery of dabrafenib: a selective inhibitor of Raf kinases with antitumor activity against B-Raf-driven tumors. ACS Med Chem Lett 4:358–362. https://doi.org/10.1021/ml4000063

Rodriguez DA, Weinlich R, Brown S et al (2016) Characterization of RIPK3-mediated phosphorylation of the activation loop of MLKL during necroptosis. Cell Death Differ 23:76–88. https://doi.org/10.1038/cdd.2015.70

Roebke KE, Guo Y, Parker JSL, Danthi P (2020) Reovirus σ3 protein limits interferon expression and cell death induction. J Virol 94:e01485-20. https://doi.org/10.1128/JVI.01485-20

Samson AL, Zhang Y, Geoghegan ND et al (2020) MLKL trafficking and accumulation at the plasma membrane control the kinetics and threshold for necroptosis. Nat Commun 11:3151. https://doi.org/10.1038/s41467-020-16887-1

Santos LD, Antunes KH, Muraro SP et al (2021) TNF-mediated alveolar macrophage necroptosis drives disease pathogenesis during respiratory syncytial virus infection. Eur Respir J 57. https://doi.org/10.1183/13993003.03764-2020

Simpson J, Loh Z, Ullah MA et al (2020) Respiratory syncytial virus infection promotes necroptosis and HMGB1 release by airway epithelial cells. Am J Respir Crit Care Med 201:1358–1371. https://doi.org/10.1164/rccm.201906-1149OC

Soliman M, Seo J-Y, Baek Y-B et al (2022) Opposite effects of apoptotic and necroptotic cellular pathways on rotavirus replication. J Virol 96:e0122221. https://doi.org/10.1128/JVI.01222-21

Sridharan H, Ragan KB, Guo H et al (2017) Murine cytomegalovirus IE3-dependent transcription is required for DAI/ZBP1-mediated necroptosis. EMBO Rep 18:1429–1441. https://doi.org/10.15252/embr.201743947

Su L, Quade B, Wang H et al (2014) A plug release mechanism for membrane permeation by MLKL. Structure 22:1489–1500. https://doi.org/10.1016/j.str.2014.07.014

Sun X, Yin J, Starovasnik MA et al (2002) Identification of a novel homotypic interaction motif required for the phosphorylation of receptor-interacting protein (RIP) by RIP3. J Biol Chem 277:9505–9511. https://doi.org/10.1074/jbc.M109488200

Sun L, Wang H, Wang Z et al (2012) Mixed lineage kinase domain-like protein mediates necrosis signaling downstream of RIP3 kinase. Cell 148:213–227. https://doi.org/10.1016/j.cell.2011.11.031

Thapa RJ, Ingram JP, Ragan KB et al (2016) DAI senses influenza A virus genomic RNA and activates RIPK3-dependent cell death. Cell Host Microbe 20:674–681. https://doi.org/10.1016/j.chom.2016.09.014

Tovey Crutchfield EC, Garnish SE, Hildebrand JM (2021) The role of the key effector of necroptotic cell death, MLKL, in mouse models of disease. Biomolecules 11:803. https://doi.org/10.3390/biom11060803

Upton JW, Kaiser WJ, Mocarski ES (2010a) Pathogen subversion of RIP3-dependent necrosis. Cell Host Microbe 7:302–313. https://doi.org/10.1016/j.chom.2010.03.006

Upton JW, Kaiser WJ, Mocarski ES (2010b) Virus inhibition of RIP3-dependent necrosis. Cell Host Microbe 7:302–313. https://doi.org/10.1016/j.chom.2010.03.006

Upton JW, Kaiser WJ, Mocarski ES (2012) DAI complexes with RIP3 to mediate virus-induced programmed necrosis that is targeted by murine cytomegalovirus vIRA. Cell Host Microbe 11:290–297. https://doi.org/10.1016/j.chom.2012.01.016

Upton JW, Kaiser WJ, Mocarski ES (2019) DAI/ZBP1/DLM-1 complexes with RIP3 to mediate virus-induced programmed necrosis that is targeted by murine cytomegalovirus vIRA. Cell Host Microbe 26:564. https://doi.org/10.1016/j.chom.2019.09.004

Vercammen D, Beyaert R, Denecker G et al (1998) Inhibition of caspases increases the sensitivity of L929 cells to necrosis mediated by tumor necrosis factor. J Exp Med 187:1477–1485. https://doi.org/10.1084/jem.187.9.1477

Vercammen E, Staal J, Beyaert R (2008) Sensing of viral infection and activation of innate immunity by toll-like receptor 3. Clin Microbiol Rev 21:13–25. https://doi.org/10.1128/CMR.00022-07

Wang H, Sun L, Su L et al (2014a) Mixed lineage kinase domain-like protein MLKL causes necrotic membrane disruption upon phosphorylation by RIP3. Mol Cell 54:133–146. https://doi.org/10.1016/j.molcel.2014.03.003
Wang X, Li Y, Liu S et al (2014b) Direct activation of RIP3/MLKL-dependent necrosis by herpes simplex virus 1 (HSV-1) protein ICP6 triggers host antiviral defense. Proc Natl Acad Sci U S A 111:15438–15443. https://doi.org/10.1073/pnas.1412767111
Wen C, Yu Y, Gao C et al (2021) RIPK3-dependent necroptosis is induced and restricts viral replication in human astrocytes infected with Zika virus. Front Cell Infect Microbiol 11:637710. https://doi.org/10.3389/fcimb.2021.637710
Wu J, Huang Z, Ren J et al (2013) Mlkl knockout mice demonstrate the indispensable role of Mlkl in necroptosis. Cell Res 23:994–1006. https://doi.org/10.1038/cr.2013.91
Xing J, Weng L, Yuan B et al (2016) Identification of a role for TRIM29 in the control of innate immunity in the respiratory tract. Nat Immunol 17:1373–1380. https://doi.org/10.1038/ni.3580
Yue Z, Shatkin AJ (1997) Double-stranded RNA-dependent protein kinase (PKR) is regulated by reovirus structural proteins. Virology 234:364–371. https://doi.org/10.1006/viro.1997.8664
Zhang S-Y, Herman M, Ciancanelli MJ et al (2013) TLR3 immunity to infection in mice and humans. Curr Opin Immunol 25:19–33. https://doi.org/10.1016/j.coi.2012.11.001
Zhang S, Tang M-B, Luo H-Y et al (2017) Necroptosis in neurodegenerative diseases: a potential therapeutic target. Cell Death Dis 8:e2905. https://doi.org/10.1038/cddis.2017.286
Zhang T, Yin C, Boyd DF et al (2020) Influenza virus Z-RNAs induce ZBP1-mediated necroptosis. Cell 180:1115–1129.e13. https://doi.org/10.1016/j.cell.2020.02.050
Zhang X, Zhang Y, Wang F et al (2022) Necrosulfonamide alleviates acute brain injury of intracerebral hemorrhage via inhibiting inflammation and necroptosis. Front Mol Neurosci 15: 916249. https://doi.org/10.3389/fnmol.2022.916249
Zhou T, Wang Q, Phan N et al (2019) Identification of a novel class of RIP1/RIP3 dual inhibitors that impede cell death and inflammation in mouse abdominal aortic aneurysm models. Cell Death Dis 10:226. https://doi.org/10.1038/s41419-019-1468-6

# Chapter 3
# Apoptosis and Phagocytosis as Antiviral Mechanisms

**Firzan Nainu, Youdiil Ophinni, Akiko Shiratsuchi, and Yoshinobu Nakanishi**

**Abstract** Viruses are infectious entities that make use of the replication machinery of their hosts to produce more progenies, causing disease and sometimes death. To counter viral infection, metazoan hosts are equipped with various defense mechanisms, from the rapid-evoking innate immune responses to the most advanced adaptive immune responses. Previous research demonstrated that cells in fruit flies and mice infected with Drosophila C virus and influenza, respectively, undergo apoptosis, which triggers the engulfment of apoptotic virus-infected cells by phagocytes. This process involves the recognition of eat-me signals on the surface of virus-infected cells by receptors of specialized phagocytes, such as macrophages and neutrophils in mice and hemocytes in fruit flies, to facilitate the phagocytic elimination of virus-infected cells. Inhibition of phagocytosis led to severe pathologies and death in both species, indicating that apoptosis-dependent phagocytosis of virus-infected cells is a conserved antiviral mechanism in multicellular organisms. Indeed, our understanding of the mechanisms underlying apoptosis-dependent phagocytosis of virus-infected cells has shed a new perspective on how hosts defend themselves

F. Nainu (✉)
Department of Pharmacy, Faculty of Pharmacy, Hasanuddin University, Makassar, Indonesia
e-mail: firzannainu@unhas.ac.id

Y. Ophinni
Division of Clinical Virology, Center for Infectious Diseases, Kobe University Graduate School of Medicine, Kobe, Japan

Laboratory of Host Defense, Immunology Frontier Research Center, Osaka University, Osaka, Japan
e-mail: yophinni@panda.kobe-u.ac.jp

A. Shiratsuchi
Center for Medical Education, Sapporo Medical University, Sapporo, Japan

Division of Biological Function and Regulation, Graduate School of Medicine, Sapporo Medical University, Sapporo, Japan
e-mail: ashira@sapmed.ac.jp

Y. Nakanishi
Faculty of Pharmaceutical Sciences, Kanazawa University, Kanazawa, Japan
e-mail: nakanaka@p.kanazawa-u.ac.jp



S. Vijayakrishnan et al. (eds.), *Virus Infected Cells*, Subcellular Biochemistry 106,
https://doi.org/10.1007/978-3-031-40086-5_3

against viral infection. This chapter explores the mechanisms of this process and its potential for developing new treatments for viral diseases.

**Keywords** Viruses · Apoptosis-dependent phagocytosis · Innate immunity · Host defense mechanisms · Antiviral

## Introduction

Viruses are tiny infectious agents composed of a small amount of genetic material (either DNA or RNA), surrounded by a protein coat called the capsid. Viruses were originally considered pathogenic agents that can pass a common filter used in the isolation and detection of bacteria (Louten 2016). Soon enough, we later appreciated the existence of these tiny pathogenic entities, mostly with caution, and learned to live with them. Structurally, a virus is an infectious particle composed of genomic materials, enclosed in physically connected proteins to form capsid. The capsid protects the viral genome and helps the virus to enter the host cells. The combination of genomic materials and capsid proteins is termed nucleocapsid. Some viruses are surrounded by lipid membranes, designated as the viral envelope, derived from the host cell membranes (Madigan et al. 2017; Louten 2016).

In order to replicate, a virus must enter a host cell and use the host's cellular machinery to produce progenies. This process can cause defects in protein synthesis in the host cell, as the cell's normal functions may be disrupted by the presence of the virus (Walsh et al. 2013; Walsh and Mohr 2011). The virus may also produce proteins that interfere with the host's normal processes, further disrupting the cell's functions (Walsh et al. 2013). As a result, the host cell may be damaged or killed in response to the viral infection, which can have significant consequences for the overall health of the host organism (Walsh and Mohr 2011).

To mitigate viral infection, the host employs multilayer innate and adaptive immune responses (Aoshi et al. 2011). The orchestration of cellular and humoral mechanisms in those two arms of immunity has been reported to be able to alleviate viral propagation either directly by responding to the virus particles or indirectly via the elimination of virus-infected cells (Aoshi et al. 2011; Nainu et al. 2017). In this chapter, we aim to outline recent progress and insights on the role of apoptosis-dependent phagocytosis of virus-infected cells as one of the evolutionarily conserved mechanisms to eliminate virus-infected cells from the host. To achieve this purpose, we will delineate knowledge on what we have so far regarding the biological process and significance of apoptosis-dependent phagocytosis as an antiviral innate immune response in both invertebrates and vertebrates. Hopefully, by addressing this particular topic, we can improve our understanding of the potential targets for antiviral drug discovery, aiming toward the best treatment and safety of the patient.

## Virus Life Cycle, Classification, and Infection

Viruses are obligate intracellular parasites; henceforth they are fully dependent on the replicative machinery of the host cells to produce new infectious particles (virions). To start its life cycle, a virus needs to bind to an appropriate receptor at the surface of the host cell and further promote a safe passage to deliver its genetic materials into the cytoplasm of the host cell (Louten 2016). Later, the viral genetic materials will be properly processed to produce new virions that are ready to be released to infect new healthy host cells (Madigan et al. 2017). A simplified diagram to illustrate the life cycle of a virus is presented in Fig. 3.1.

Viruses produce multiple copies of virions, or progenies, by replicating their genetic material and constructing new viral particles using the host cell's machinery (Sumbria et al. 2020). The process of viral replication can occur in several ways, depending on the type of virus. Some viruses, such as retroviruses, replicate using a reverse transcription process, in which the virus's RNA genome is converted into DNA, which can then be integrated into the host cell's genome (Madigan et al. 2017). Other viruses, such as herpesviruses, replicate using a process called rolling circle replication, in which the viral genome is replicated as a circle and then cut into pieces, which are packaged into new viral particles (Madigan et al. 2017).

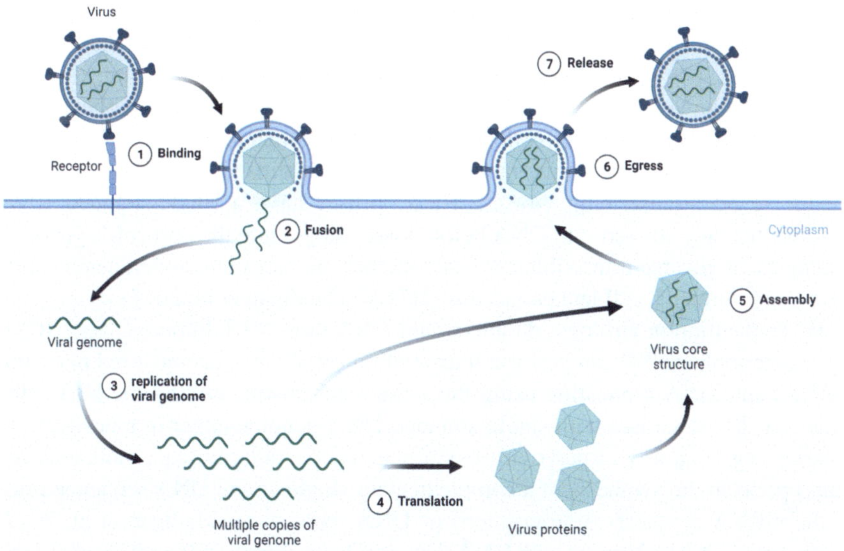


**Fig. 3.1** Life cycle of a virus. Virus starts its life cycle by infecting a healthy cell to carry out replication of its genome as well as the production of virus proteins. The newly synthesized viral proteins are then assembled with the viral genome to form new virus particles prior to being released into the extracellular regions. Adapted from "Generic Viral Life Cycle" by BioRender.com (2023). Retrieved from https://app.biorender.com/biorender-templates

Regardless of the specific mechanism of replication, the outcome of viral infection is the production of new viral particles that are able to infect other cells and continue the cycle of infection. This process can be very efficient, with some viruses able to produce hundreds of progenies per infected cell in a matter of hours (Mateu 2013).

Viruses can be classified based on their shape, size, genomic structure, and the type of disease they cause (Mateu 2013). Some common shapes include spherical, rod-shaped, and filamentous (Madigan et al. 2017; Mateu 2013). Most viruses are so small that they can only be seen with the help of an electron microscope. However, some viruses, termed as giant viruses, can more easily be seen as their size is larger than bacteria. Nevertheless, in general, viruses are equipped with simple and efficient genomic components, which can be in the form of either DNA or RNA, structured as a single- or double-strand. Hence, based on their genomic structures, viruses can be classified into either DNA viruses or RNA viruses (Madigan et al. 2017). The genome of a DNA virus can be as small as 1.75 kilobase (kb) in the form of single-stranded and as big as 2.5 megabase (Mb) in the form of double-stranded (Madigan et al. 2017). On the other hand, the genome of RNA viruses is typically smaller in size than that of DNA viruses.

Viruses can be classified into three main groups based on the type of genome they possess: DNA viruses, RNA viruses, and reverse-transcribing viruses (Madigan et al. 2017). This classification scheme, widely known as the Baltimore Classification System, can be seen in Fig. 3.2. Common DNA viruses have a double-stranded DNA genome and can replicate their genome using the host cell's machinery and can produce proteins using the host cell's ribosomes. Examples of DNA viruses include herpesviruses, poxviruses, and adenoviruses (AdVs). On the other hand, common RNA viruses have a single-stranded RNA genome. These viruses can replicate their genome using their own RNA-dependent RNA polymerase. Examples of RNA viruses include influenza virus, rabies virus, and hepatitis C virus (HCV). Lastly are the reverse-transcribing viruses. These viruses have a single-stranded RNA genome that they convert into DNA before integrating it into the host cell's genome. Examples of reverse-transcribing viruses include human immunodeficiency virus (HIV) and human T-cell leukemia virus (HTLV) (Madigan et al. 2017).

By convention in virology, double-strand DNA viruses (Baltimore Class (BC) I, e.g., bacteriophage T4) can perform transcription (viral DNA is converted into viral mRNA) and DNA replication using the same mechanisms used by the host cells. However, BC II viruses with single-stranded DNA genome either in a positive (+) polarity (e.g., majority of parvoviruses) or negative (–) polarity (e.g., mink enteritis virus) perform the synthesis of a complementary single-strand DNA sequence prior to the mRNA synthesis (transcription) or DNA replication (Madigan et al. 2017; Koonin et al. 2021). Viruses with RNA genomes have slightly different mechanisms to produce their new infectious virions. The formation of new RNA genomes for their newly produced virions is carried out using a dedicated RNA-dependent RNA polymerase. For double-strand RNA viruses (BC III, e.g., rotavirus) and single negative-strand RNA viruses (BC V, e.g., Influenza virus), the negative (–) strand serves as the template to produce a new viral genome. However, unlike BC III and V, the genome of RNA viruses of BC IV (positive-strand RNA genome, e.g.,

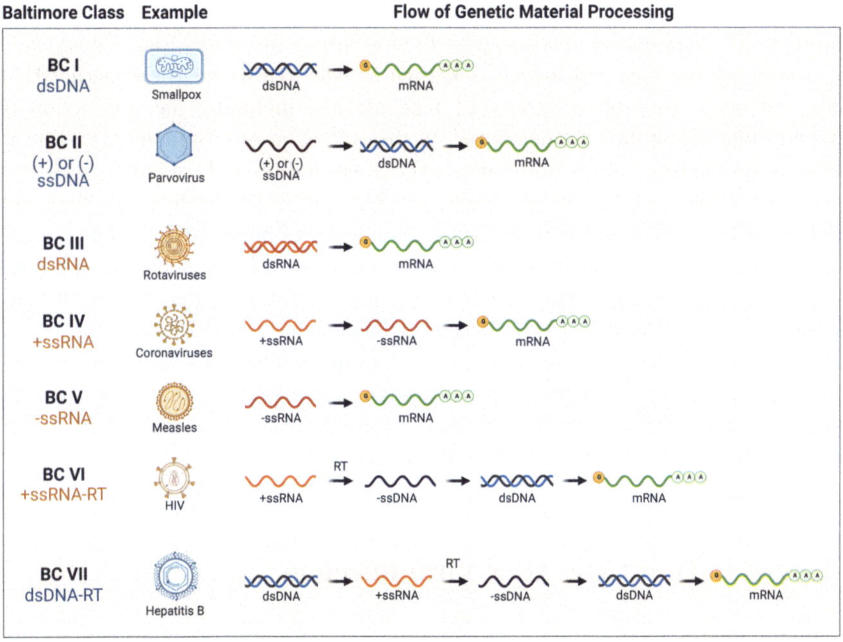


**Fig. 3.2** The Baltimore classification system. This system classifies viruses into seven different groups based on the nature of genetic materials and the way their genetic materials are processed to generate messenger RNAs. +, positive sense; –, negative sense; BC, Baltimore Class; ss, single strand; ds, double strand; mRNA, messenger RNA; RT, reverse transcriptase. Adapted from "The Baltimore Classification of Viruses" by BioRender.com (2023). Retrieved from https://app.biorender.com/biorender-templates

Poliovirus) can undergo a direct translation into proteins. For its genome replication, BC IV RNA viruses need to synthesize a complementary negative-strand RNA to accommodate its genomic replication (Madigan et al. 2017; Koonin et al. 2021).

The last two classes are positive-sense RNA viruses equipped with reverse transcriptase (RT) that promote the replication via a DNA intermediate (BC VI) and viruses that package a genomic dsDNA (despite the fact that one of the strands is usually incomplete) that make use of RT to replicate via an RNA intermediate (BC VII). In general, viruses classified into these two latter classes, BC VI and BC VII, have a distinct role of RT in their replication cycle, hence separated from other viruses in BC I (dsDNA) and BC IV (positive-sense ssRNA). For more details on the virus classification and their replication, please refer to Fig. 3.2.

Upon successful replication, viruses can escape from infected cells through a variety of mechanisms. Some viruses, for example, varicella-zoster virus (VZV) and dengue virus (DENV), use the host cell's exocytosis machinery to be released from the cell (Buckingham et al. 2016; Islam et al. 2021). Exocytosis is a process by which cells release substances through the fusion of vesicles with the cell membrane (Südhof and Rizo 2011). In exocytosis, the virus is enclosed in a vesicle, which fuses with the cell membrane and releases the virus into the extracellular space (Südhof and Rizo 2011). However, it is also possible for viruses to escape from infected cells by destroying the host cell, which is known as lysis (Bird and Kirkegaard 2015). This can occur through a variety of mechanisms, including the production of enzymes that digest the cell membrane, or the activation of programmed cell death pathways in the host cell. In addition to exocytosis and lysis, there are several other ways that viruses use to escape from infected host cells, such as budding and envelopment (Chazal and Gerlier 2003). Budding is employed by some viruses, such as HIV. Using this mechanism, the virus forms small buds on the surface of the host cell and eventually breaks off from the host cell, taking the virus with them (Votteler and Sundquist 2013). In another mechanism, termed envelopment, some viruses, such as influenza virus, wrap themselves in the host cell's membrane as they are released, taking the membrane with them, and protecting the virus from the immune system (Sun and Whittaker 2003). Overall, the specific mechanism of viral escape depends on the type of virus and the host cell.

## Changes in Host Cells After Viral Infection

In the event of viral infection, host cells initiate transcriptional changes leading to the induction of intrinsic immune responses for the resolution of infection (Aoshi et al. 2011). Alternatively, the virus will escape from the intrinsic immunity and subsequently infect the host cell to initiate a variety of changes that may favor the viral invader. In the latter event, the host cells become infected and subsequently manipulated to produce viral progenies (Andreu-Moreno et al. 2020). While both events lead to different outcomes, they may drive molecular and/or phenotypical changes in the infected cells. Obviously, either events will impact the global transcriptional state of infected host cells. This section is dedicated to exploring what are the changes identified in the host cells and why such changes occurred in response to viral infection.

### *Alteration of Gene Expression*

Introduction of foreign entities such as virus particles/genomes into the intrinsic environment of host cells can lead to a variety of changes in the cell, including robust alteration of gene expression that may either facilitate viral replication or initiate the intrinsic immune responses (Sumbria et al. 2020; Walsh et al. 2013). In addition, viral infection may drive specific modifications of host cell structure to accommodate the formation of viral envelopes, enabling the assemble of newly generated viral particles (Omasta and Tomaskova 2022; Villanueva et al. 2005).

Some of the most common changes that are related to the modulation of gene expression in the infected host cells are mainly stimulated by replication of the viral genome and initiation of viral protein synthesis. The replication of the viral genome using either the host cell's genetic machinery or the virus's own proteins leads to the production of many copies of the virus (Rampersad and Tennant 2018). In accordance with such an invasive event, viral proteins are subsequently synthesized by the host cell's ribosomes and other host cellular machinery. These newly synthesized viral genomes and proteins are then assembled into new virions, ready to infect new healthy host cells (Rampersad and Tennant 2018).

In general, a virus is a metabolically inert entity. However, during infection, viruses may alter the metabolism of the host cells to support their replication (Sumbria et al. 2020). There are several virus-mediated cell metabolism modifications that have been reported. For example, the upregulation of glycolysis by AdV has been shown essential to support its optimal replication in the primary lung epithelial cells (Thai et al. 2014). Another example is the activation of glutamine-mediated, instead of glucose-mediated, production of ATP in the host cells by human cytomegalovirus (Chambers et al. 2010). For a detailed explanation of the impact of virus infection on the metabolic state of host cells, please refer to the published literature (Sumbria et al. 2020; Thaker et al. 2019).

## *Changes in Cell Surface Structures*

During a viral infection, the structure of the cell surface can undergo various changes. Some viruses, such as the influenza virus, can cause changes in the glycosylation patterns of the cell surface (York et al. 2019; Alymova et al. 2022). While others, such as HIV, can alter the expression of cell surface proteins (Speth and Dierich 1999; Sugden et al. 2016). Additional changes can also occur in the cell surface structure during viral infection. Some examples include the appearance of viral surface proteins (Villanueva et al. 2005), the downregulation of host cell surface proteins (Depierreux et al. 2022), and the upregulation of immune checkpoint receptors (Saheb Sharif-Askari et al. 2021). These changes can be infection-specific, as different viruses can have different effects on the cell surface.

In addition to the above-mentioned changes, virus infection can trigger the death of virus-infected cells, mostly in a process termed apoptosis (Verburg et al. 2022; Nainu et al. 2017). However, it is important to note that changes present on the cell surface during viral infection do not necessarily occur in an apoptosis-dependent manner (Thomson 2001). Apoptosis is a separate process that can occur concurrently with changes in the cell surface structure during viral infection. Nevertheless, induction of apoptosis in response to viral infection has been widely reported (Majchrzak and Poręba 2022; Rex et al. 2022; Turpin et al. 2022; Holzerland et al. 2020), and these apoptotic virus-infected cells are subjected to evolutionarily conserved phagocytic elimination by immune cells such as neutrophils and macrophages (Hashimoto et al. 2007; Nainu et al. 2015), which will be described later in detail.

The sequence leading to the engulfment of apoptotic cells typically starts with the release of "find-me" signals from the apoptotic cells to recruit phagocytic cells, recognition via expression of "eat-me" signals, internalization and cytoskeletal rearrangement by phagocytic cells, and finally, digestion, phagocytic response, and cytokine release. Secretion of the "find-me" signals creates the chemotactic gradient surrounding the apoptotic cells that may attract chemotaxis and diapedesis of local and distant phagocytes (Ravichandran 2003). The classic example is chemokines, such as fractalkine (CXC3CL1), which binds to CX3CR1 expressed on numerous lymphoid and myeloid effector immune cells (Imai et al. 1997). CX3CR1 was initially described as an HIV-1 fusion co-receptor and implicated in faster progression to AIDS (Faure et al. 2000; Combadiere et al. 1998), and indeed, fractalkine expression is increased in HIV-1 neuro-infection (Pereira et al. 2001). MIP1α and MCP-1 are other chemoattractants for nearby macrophages, which are upregulated in influenza virus and HCV infection in macrophages before cell death (Hofmann et al. 1997; Liu et al. 2017). Later discovered "find-me" signals include lipids, such as lysophosphatidylcholine (Lauber et al. 2003) and nucleotides (Elliott et al. 2009).

## *Apoptosis and Viral Resistance to Apoptosis*

Physiological homeostasis is one of the most important properties to maintain. To achieve this, a delicate balance of the formation of new cells via the cell cycle and the destruction of damaged/dead cells via programmed cell death needs to be strictly maintained (Arandjelovic and Ravichandran 2015). Apoptosis is one type of programmed cell death process that occurs in cells during development and in the maintenance of tissue homeostasis (Arandjelovic and Ravichandran 2015). In the event of viral infection, virus-infected host cells may undergo apoptotic cell death (Nainu et al. 2017). The antiviral properties of virus-induced apoptosis have been widely reported (Barber 2001; Clarke and Tyler 2009; Orzalli and Kagan 2017; Rex et al. 2022; Upton and Chan 2014). Induction of apoptosis is essential to maintain tissue homeostasis (Taylor et al. 2008; Fuchs and Steller 2011), as it allows the body to rid itself of unnecessary or damaged cells (Arandjelovic and Ravichandran 2015), including the ones that are infected with viruses (Nainu et al. 2017). However, some viruses are known to resist apoptosis in the infected cells in order to maintain their replication state and spread to new healthy cells (Thomson 2001; Clarke and Tyler 2009).

Apoptosis is an evolutionarily conserved process and has been intensively studied in invertebrate model organisms (*Caenorhabditis elegans* and *Drosophila melanogaster*) and vertebrate animals (the mouse *Mus musculus*). While the detail of mechanisms is slightly differed across species, the main components and fundamental pathways that regulate the induction of apoptosis remain similar (Fuchs and Steller 2011; Hay et al. 2004). For example, the role of initiator and effector cysteine proteases, simply termed caspases, in the activation of apoptosis has been reported in

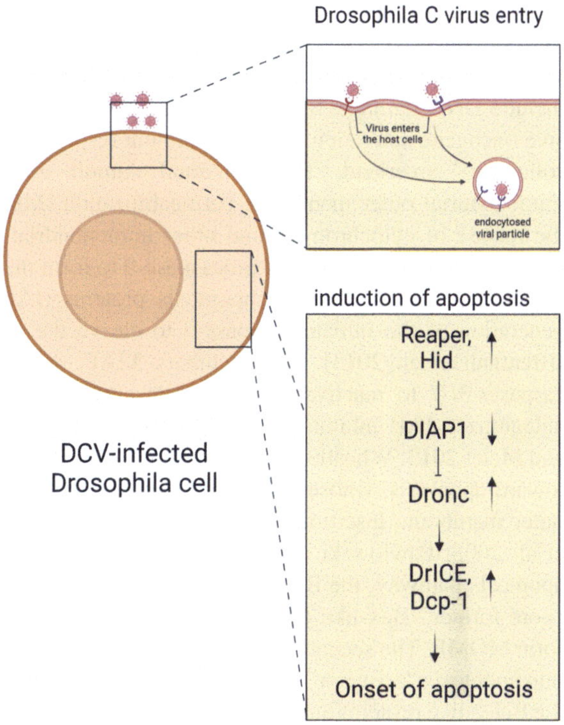


**Fig. 3.3** Induction of caspase-mediated apoptosis in *Drosophila* upon viral infection. Apoptosis is induced upon DCV infection possibly via depletion of DIAP1 leading to the activation of initiation (Dronc) and effector (DrICE, Dcp-1) caspases. Dcp-1, death caspase-1; DCV, *Drosophila* C virus; DIAP1, *Drosophila* inhibitor of apoptosis protein 1; DrICE, death-related ICE-like caspase; Dronc, death regulator Nedd2-like caspase. Created with BioRender.com

model organisms (Taylor et al. 2008; Shi 2004; Parrish et al. 2013). In addition, similar cellular changes such as DNA fragmentation, chromatin condensation, and cell shrinkage were observed in the apoptotic cells of worms, flies, and mice (Xu et al. 2009; Saraste and Pulkki 2000).

We previously demonstrated that *Drosophila* C virus (DCV), a picorna-like RNA virus that naturally infects *Drosophila*, was able to induce the activation of caspase-mediated apoptosis in *Drosophila* (Nainu et al. 2015). Hallmarks of apoptotic cells such as chromatin condensation, DNA fragmentation, and caspase activation were evident in the DCV-infected *Drosophila* S2 cells, suggesting that these DCV-infected cells underwent apoptotic cell death (Nainu et al. 2015). We observed that DCV-infected cells experienced depletion of *Drosophila* inhibitor of apoptosis protein 1 (DIAP1) which led to the onset of apoptosis (Nainu et al. 2015). Since DIAP1 acts to inhibit caspases, and DIAP1 has been known to be regulated by the pro-apoptotic protein reaper, hid, and grim, we speculated that the upregulated expression of the *reaper*, *hid*, and *grim* at the upstream of DIAP1 plays an important role in the depletion of DIAP1, resulting in the subsequent action of initiator and effector caspases to induce apoptosis (Fig. 3.3). While detailed mechanisms remain unrevealed, prior data from our group and others indicate the existence of DIAP1-regulated, caspase-mediated apoptosis in the DCV-infected cells (Nainu et al. 2015; Liu et al. 2013).

In mammals, there are two main paths toward apoptosis: the mitochondria-initiated intrinsic pathway and the death receptor-mediated extrinsic signaling pathway. The intrinsic pathway can be induced by various intracellular stimuli, which include DNA damage, oxidative stress, endoplasmic reticulum (ER) stress, excessive oncogene activation, unfolded protein response, cytokine deprivation, or cytosolic $Ca^{2+}$ overload. These internal stimuli trigger sequential processes of mitochondrial outer membrane permeabilization (MOMP) (Tait and Green 2010), the release of cytochrome *c* and other mitochondrial factors (e.g., Smac) into the cytosol to bind to Apaf-1 and procaspase-9 to form the apoptosome (Li et al. 1997), degradation of inhibitors of apoptosis proteins (IAPs) (Vince et al. 2018), and generation of the initiator caspase-9 to cleave the executioner caspases-3 and -7 (Brentnall et al. 2013). IAP members XIAP, c-IAP1, and C-IAP2 can bind to caspases-3/-7 to inactivate them, while survivin inhibits apoptosome formation, antagonizes IAP inhibitor Smac/DIABLO, and inhibits caspases directly (Silke and Meier 2013; Wheatley and Altieri 2019). MOMP signifies the point-of-no-return toward apoptosis, whose formation is dependent on the homooligomerization and outer membrane insertion of the pro-apoptotic Bcl-2 family of proteins (Dewson et al. 2008; Pawlowski and Kraft 2000). As the master regulator of the intrinsic apoptosis pathway, the Bcl-2 family can be classified into three classes. First is the "pore-former" Bax-like pro-apoptotic factors (Bax, Bak) that can oligomerize to form MOMP. The second class is the BH3-only factor, which can further be divided into apoptosis "activator" (Bid, Bim, Puma) that activates the pore formers or inhibit the Bcl-2-like proteins and apoptosis "sensitizer" (Hrk, Noxa, Bad, Bmf, Bik, Blm-s) that releases the activators or inhibit the anti-apoptotic counterparts. The last is the Bcl-2-like anti-apoptotic factors (Bcl-2, Bcl-$X_L$, Mcl-1, Bcl-w, Bcl-B, Bfl-1) that can sequester and inhibit both the pore-former and the activator BH3-only proteins. Membrane interaction via the C-terminal transmembrane domains (TMD) is essential for the function of the Bcl-2 family of proteins (Kale et al. 2018).

Some viruses encode proteins that inhibit apoptosis, allowing the infected cells to survive longer and provide a favorable environment for viral replication (Clarke and Tyler 2009; Kaminskyy and Zhivotovsky 2010). For example, herpes simplex virus (HSV) can express viral protein ICP34.5 to prevent the activation of caspases, enzymes that play a key role in initiating apoptosis (Pasieka et al. 2006). Another herpesvirus, Epstein-Barr virus (EBV), expresses the viral protein BHRF1 that can bind to and inactivate the pro-apoptotic proteins expressed in the host cells (Kvansakul et al. 2010; Desbien et al. 2009). Inhibition of the apoptotic mechanism was also evident in other viruses. The viral protein E6 of human papillomavirus (HPV) can bind to and inactivate the tumor suppressor protein p53, which plays a key role in promoting apoptosis in response to DNA damage (Shimada et al. 2020), and the viral protein HBx of hepatitis B virus (HBV) induces anti-apoptotic pathways via activation of the Akt signaling pathway (Rawat and Bouchard 2015). In addition, in the realm of retroviruses, HTLV type 1 (HTLV-1) has been shown to be able to generate viral protein Tax that binds to and activates the anti-apoptotic protein NF-*k*B (Kannian and Green 2010). These are some examples of viral proteins that have been shown to inhibit or resist apoptosis, but it is important to note that not all viruses have the same mechanisms to resist apoptosis. Examples of viral proteins with anti-apoptotic activity can be seen in Fig. 3.4.

Viruses have evolutionarily developed strategies to modulate the intrinsic pathway, which include disruption of the intracellular sensors, signaling cascade, caspases, and viral mimicry of cellular apoptotic factors. A well-studied viral factor is vBcl-2, the viral homolog of cellular Bcl-2, which can hijack intrinsic pathway processes to prevent the premature death of host cells (Cuconati and White 2002). Similar to cBcl-2, vBcl-2 possesses TMD and can homo-oligomerize and interact with cellular Bcl-2 TMDs within the membranes of ER, mitochondria, and nuclear envelope (García-Murria et al. 2020). Membrane-associated vBcl2 proteins include AdV E1B 19K (White et al. 1984), EBV BHRF1 (Henderson et al. 1993), murine gammaherpesvirus-68 (MHV-68) M11 (Wang et al. 1999), African swine fever virus (ASFV) A179L (Hernaez et al. 2013), and Frog virus 3 (FV3) 97R (Ring et al. 2013). Functionally, vBcl-2 bound to apoptosis activator and pore-forming cBcl2 can disturb their functions: AdV E1B 19K engages with, at that time, the newly discovered Bak, Bax, and Bik (Farrow et al. 1995; Han et al. 1996; Boyd et al. 1995); Herpesvirus saimiri (HVS) ORF16 binds to Bax and Bak (Nava et al. 1997); cytomegalovirus (CMV) vMIA (UL37) inhibits Bax and Bak oligomerization and upregulates Mcl-1 before shifting to Bcl-2 (Goldmacher et al. 1999; Collins-McMillen et al. 2015); EBV BHRF1 resembles Bcl-2 and sequesters to Bim and Bak (Flanagan and Letai 2008); Kaposi's sarcoma-associated herpesvirus (KSHV) Ks-Bcl-2 is more related to Mcl-1 and binds to Bim, Bid, and Noxa (Flanagan and Letai 2008); MHV-68 M11 binds to Bak, Bim, Noxa, Bid, Bmf, and Puma (Ku et al. 2008); vaccinia virus (VACV) F1L binds to Bim and Bax and replaces Mcl-1 to inhibit Bak activation (Taylor et al. 2006); variola virus (VARV) F1L binds to Bid, Bak, Bax, and inhibits Bax independent of Bim (Marshall et al. 2015); myxoma virus (MYXV) M11L engages with Bak, Bax, Bim, and Bid (Wang et al. 2004; Kvansakul et al. 2007); ASFV A179L interacts with Bax, Bak, Bid, and Noxa (Galindo et al. 2008); orf virus (ORFV) ORFV125 interacts with Bik, Puma, Hrk, Noxa, and Bim (Westphal et al. 2009); canarypox virus (CNPV) CNP058 engages with Bax, BimL, and Bik to prevent Bax activation (Banadyga et al. 2009); sheeppox virus (SPPV) SPPV14 binds to Bam, Bak, Bim, Bid, Bmf, Hrk, and Puma (Okamoto et al. 2012); deerpox virus (DPV) DPV022 binds to Bax, Bak, and Bim (Burton et al. 2015); and fowlpox virus (FPV) FPV039 interacts with Bax, Bak, and all BH3-only proteins (Anasir et al. 2017).

The extrinsic apoptotic pathway is induced by the binding of the TNF family of death receptors (DRs) with their respective ligands: TNFR1 with TNF (Chen and Goeddel 2002), Fas (CD95/Apo-1) with Fas ligand (FasL) (CD95L) (Lacana and D'Adamio 1999), DR4 (TRAIL-R1) and DR5 (TRAIL-R2) with TRAIL (Apo2L) (Wang and El-Deiry 2003), DR3 with TL1A (Migone et al. 2002), and DR6 possibly with N-APP (Pan et al. 1998). DRs are transmembrane proteins consisting of extracellular domains containing 2–4 cysteine-rich repeats for ligand binding and an intracellular death domain to recruit adaptors required for downstream signaling. DR-induced cascades can be divided into two: TNFR1 and DR3 recruit the adaptor molecule TRADD, which subsequently forms the death-inducing signaling complex

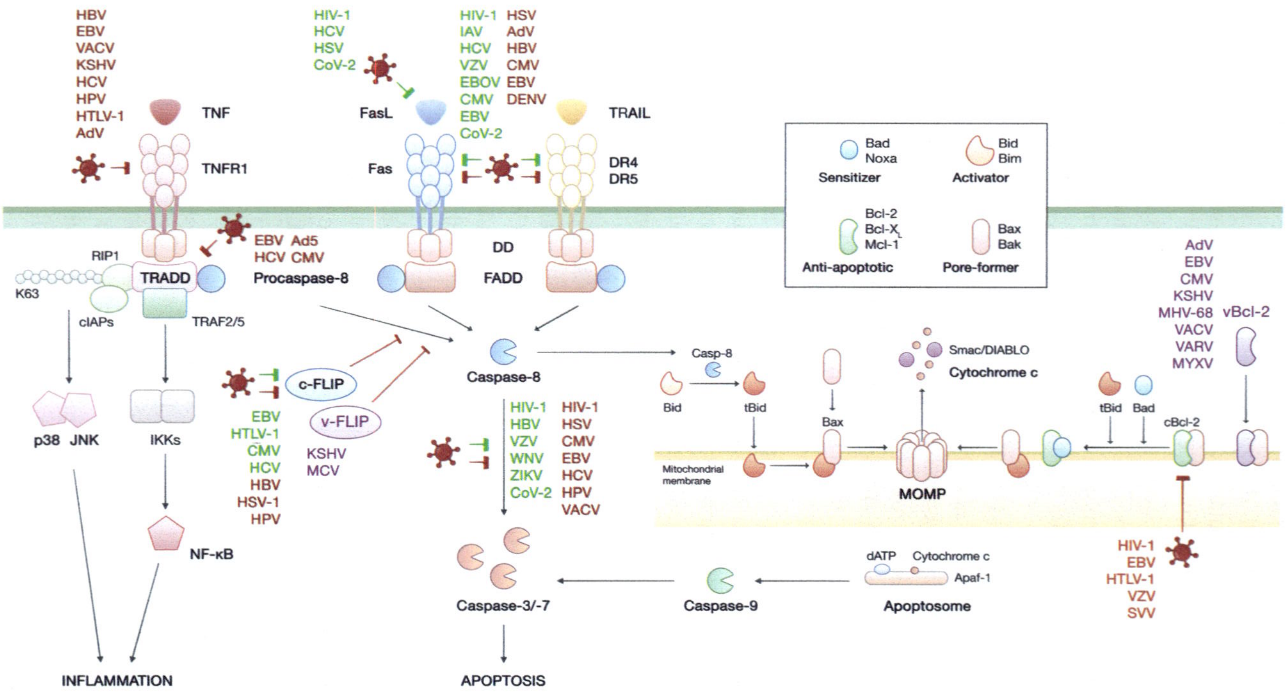


**Fig. 3.4** Biological activities of viral proteins in apoptotic pathways. Viruses may affect the extrinsic apoptotic pathway via manipulation of the death receptors (TNFR1, Fas, and DR4/5), FasL, adaptor molecule TRADD, caspase-8 regulator c-FLIP, or caspase-3/-7 maturation. The intrinsic apoptotic pathways are mainly affected by the vBcl-2 family of proteins, which mimics the cellular cBcl-2. Viruses may also directly inhibit cBcl-2. AdV, adenovirus; Ad5, adenovirus type 5; CMV, cytomegalovirus; CoV-2, severe acute respiratory syndrome coronavirus-2; DENV, dengue virus; EBOV, Ebola virus; EBV, Epstein-Barr virus; HBV, hepatitis B virus; HCV, hepatitis C virus; HIV-1, human immunodeficiency virus-1; HTLV-1, human T-lymphotropic virus 1; HPV, human papillomavirus; HSV, herpes simplex virus; IAV, influenza A virus; KSHV, Kaposi's sarcoma-associated herpesvirus; MCV, molluscum contagiosum virus; MHV-68, murine gammaherpesvirus 68; MYXV, myxoma virus; SVV, simian varicella virus; VACV, vaccinia virus; VARV, variola virus; VZV, varicella-zoster virus; WNV, West Nile virus; ZIKV, Zika virus

(DISC) by involving FADD together with RIP and TRAF2 (Pobezinskaya et al. 2008); the latter recruits E3 ubiquitin ligase cIAPs to ubiquitinate RIP1 (Mahoney et al. 2008). The outcome of TRADD-recruiting DR signaling is the activation of NF-κB, c-Jun amino-terminal kinase (JNK), and p38 mitogen-activated protein kinases (MAPKs), and thus is inflammatory, while cell death is a secondary outcome due to inadequate or aberrant signaling (Natoli et al. 1998). On the other hand, Fas and TRAIL-Rs recruit FADD as the adaptor protein together with procaspase-8, procaspase-10, and the caspase-8/-10 regulator c-FLIP to form the DISC. Downstream of DISC is either cascade activation of the initiator caspases-8/-10 to cleave the executioner caspases-3/-7 to prompt apoptosis (Aouad et al. 2004) or caspase-8 cleavage of Bid to produce tBid and subsequently trigger MOMP, thus creating linkage to the intrinsic apoptosis pathway (Li et al. 1998). Caspase-8 and -9 activation signify the extrinsic and intrinsic pathways, respectively, and both pathways converge to the executioner caspases-3/-7. However, the caspase cascade in the intrinsic pathway is not required for apoptosis but instead to dampen cytokine production of the dying cell (McArthur and Kile 2018).

Besides the perforin/granzyme pathway, DR stimulation can also be used by innate and adaptive immune cells to facilitate cytolysis and combat viral infections. The Fas/FasL-binding is utilized by cytotoxic T cells (CTLs) to induce the death of virus-infected and bystander cells in both MHC-II- and MHC-I-restricted lymphocytic choriomeningitis virus (LCMV) infection (Zajac et al. 1996; Medana et al. 2000), HCV-infected cells (Gremion et al. 2004), and EBV-transformed B cells (Wilson et al. 1998). TRAIL also facilitates virus-specific CD8+ T-cell killing of influenza A virus (IAV)-infected pulmonary epithelial cells (Brincks et al. 2008). DRs may also exert immunoregulatory effects; for example, the Fas/FasL pathway may regulate mucosal neutrophil infiltration in HSV-2 infection (Krzyzowska et al. 2011) and M1 macrophage differentiation and neuroinflammation in HSV-1 (Krzyzowska et al. 2021).

On the other hand, viruses can upregulate DRs in host cells to induce apoptosis, which is beneficial during the lytic stage to disseminate viral progeny with minimal host immune response. One of the most well-known examples is Fas; HIV-1 infection, via Tat, Env, and Vpu proteins, upregulates both Fas on the surface of CD4+ T cells and macrophages, which leads to apoptosis and depletion of T cells (Aries et al. 1995; Badley et al. 1996). Fas upregulation has also been noted to mediate apoptosis of Madin-Darby canine kidney (MDCK) cells and HeLa cells in influenza virus infection (Takizawa et al. 1993), hepatocytes in HCV and HIV–HCV coinfection (Pianko et al. 2001; Macias et al. 2005), lymphocytes in VZV and Zaire ebolavirus (ZEBOV) (Ito et al. 1995; Baize et al. 1999), hematopoietic progenitor cells in murine CMV (Mori et al. 1997), CD4+ T and B cells in EBV (Tanner and Alfieri 1999; Le Clorennec et al. 2006), airway epithelial cells in respiratory syncytial virus (RSV) (O'Donnell et al. 1999), human umbilical vein endothelial cells (HUVECs) in DENV (Liao et al. 2010), PK-15 cells in transmissible gastroenteritis virus (TGEV) (Ding et al. 2012), neurons in CHPV (Ghosh et al. 2013), astrocytes and microglia in HSV-1 (Krzyzowska et al. 2021), and peripheral T cells in SARS-CoV-2 (Bellesi et al. 2020). FasL instead is upregulated in effector cells, such as mononuclear cells in HCV (Mita et al. 1994), macrophages in HIV-1 (Badley et al. 1996), and monocytes in HSV-1 (Iannello et al. 2011). Increased FasL, but not Fas, DR4/5 has also been observed in apoptotic LNCaP cells after SARS-CoV-2 infection (Johnson et al. 2022). Moreover, Fas also mediates bystander cell death, such as in HIV-1 (Algeciras-Schimnich et al. 2002), and surface or soluble FasL from infected cells can also promote Fas upregulation in CTL for elimination and thus weaken cellular immunity, as observed in EBV (Contini et al. 2000), HIV (Mueller et al. 2001), HSV-1 (Iannello et al. 2011), rabies virus (RABV) (Baloul et al. 2004), and AdV infection of the liver (Liu et al. 2001). Augmented co-expression of Fas and FasL in influenza-infected HeLa cells induces apoptosis from cell-to-cell contact (Fujimoto et al. 1998). In HSV-2 infection, an apoptotic tendency from Fas/FasL upregulation is balanced by anti-apoptotic factors, such as Bcl-2, Akt, and NF-κB (Krzyzowska et al. 2011).

Similarly, increased susceptibility to TRAIL-induced apoptosis due to upregulation of DR4/5 or TRAIL is noted in fibroblasts in CMV and reovirus infection (Sedger et al. 1999; Clarke et al. 2000), resting memory CD4+ T cells and macrophages in HIV-1 (Lum et al. 2001), T cells in HTLV (Rivera-Walsh et al. 2001), primary airway cells in RSV and IAV (Kotelkin et al. 2003; Ishikawa et al. 2005), HepG2 cells in HBV in part via Hbx protein (Janssen et al. 2003), hepatocytes in HCV (Deng et al. 2012), neuroblastoma cells in lyssaviruses (Kassis et al. 2004), primary keratinocytes in Sendai virus (SeV) (Kirshner et al. 2005), and HeLa cells in Newcastle disease virus (NDV) (Liao et al. 2017). SARS-CoV-2 ORF7b can induce IFN-β, TNF-α, and IRF3 phosphorylation, thus promoting TNF-mediated apoptosis in 293T and Vero cells (Yang et al. 2021). TRAIL induction also depends on type-I interferon and IRF3 (Kirshner et al. 2005; Herbeuval et al. 2006). DR4/5 upregulation also contributes to HIV-1 pathogenesis by inducing apoptosis of bystander T cells, neurons, and B cells (Herbeuval et al. 2005; Miura et al. 2003; van Grevenynghe et al. 2011). The HBx protein of HBV also increased TRAIL-induced apoptosis through Bax or E3 ubiquitin ligase (Liang et al. 2007; Zhang et al. 2015).

Further downstream, the c-FLIP anti-apoptotic function can be disrupted by HBV HBx protein in HepG2 cells (Kim and Seong 2003), HSV-1 in dendritic cells (Kather et al. 2010), and HPV E2 (Wang et al. 2011). Caspase-8 maturation has been observed in T cells in HIV-1 via Env gp160 (Algeciras-Schimnich et al. 2002) and in Middle East respiratory syndrome coronavirus (MERS-CoV) infection (Chu et al. 2016), melanoma cells in VZV (Brazeau et al. 2010), mouse brain tissue in West Nile virus (WNV) (Clarke et al. 2014), and neural progenitor cells in Zika virus (ZIKV) (Souza et al. 2016). SARS-CoV-2 ORF3a also increased caspase-8-activation, BID cleavage, and cytochrome c release, indicating an external apoptotic entry pathway that leads to mitochondrial damage (Ren et al. 2020).

On the contrary, DR downregulation to subvert apoptosis in viral infection is associated with latency establishment and chronic disease progression. Fas is downregulated in AdV infection by the adenoviral E3 protein 10.4K/14.5K complex (Shisler et al. 1997), human CMV (HCMV) in myeloid progenitors (Seirafian et al. 2014), and HBV in hepatocytes by the core protein (Liu et al. 2015). NF-κB activation protects against Fas-mediated apoptosis, which can be induced by HSV-1 gD protein (Medici et al. 2003) and HCV E2 (Chen et al. 2011). Caspase-8 activation blockade can be enacted by HIV-1 Nef (Yoon et al. 2001), the R1 subunits of HSV-1 and HSV-2 (Dufour et al. 2011; Langelier et al. 2002), vICA (UL36) of HCMV and murine CMV (MCMV) (Skaletskaya et al. 2001; McCormick et al. 2003), and EBV LMP1 (Snow et al. 2006). The HBc protein of HBV represses the DR5 promoter and prevents TRAIL-induced apoptosis contributing to chronic hepatitis development (Du et al. 2009). DR4/5 are also downregulated in AdV infection by E3 10.4K/14.5K/6.7K (Benedict et al. 2001) and DENV-2-infected HUVECs (Liao et al. 2010). Additionally, the TNFR1 apoptotic pathway can also be disrupted by EBV BZLF1 protein via TNFR1 downregulation (Morrison et al. 2004), HCV NS5A via TRADD interaction (Majumder et al. 2002), adenovirus type 5 (Ad5) E3-14.7K via DISC formation blockade (Schneider-Brachert et al. 2006), and MCMV M45 via RIP1 interference (Mack et al. 2008). c-FLIP is induced in EBV infection of B cells and in nasopharyngeal carcinoma cells via LMP1 protein (Snow et al. 2006; Li et al. 2011), HTLV-I in T cells via Tax oncoprotein (Okamoto et al. 2006), HCMV in retinal epithelial cells via IE2 (Chiou et al. 2006), and HCV via core protein (Ray et al. 1998). The viral homolog of c-FLIP is the v-FLIP proteins of gammaherpesviruses, such as KSHV K13 protein, and poxviruses, such as molluscum contagiosum virus (MCV) MC159 and M160, which act by competitively inhibiting interaction with FADD and blocking procaspase-8 maturation, thus subverting apoptosis via all three of Fas-, DR4/5-, and TNFR1-mediated pathways (Thome et al. 1997; Bélanger et al. 2001). HCMV vMIA also inhibits Fas induction of the intrinsic apoptotic pathway downstream of the Bid cleavage (Goldmacher et al. 1999).

## *Cellular and Humoral Immune Responses Against Virus Infection and the Possible Induction of Cytokine Storm*

The immune system is a complex network of cells, tissues, and organs that work together to defend the body against foreign invaders, such as viruses and bacteria. There are two main types of immune responses in the innate and adaptive arms of immunity: cellular immune responses and humoral immune responses (Marshall et al. 2018). Cellular immune responses involve the activation of immune cells, for example, macrophages and dendritic cells (innate immune cells) as well as T cells (adaptive immune cells), which can directly attack infected cells and destroy them. These immune cells can also release cytokines, either with pro-inflammatory or anti-inflammatory signatures, which help to coordinate the immune responses (Marshall et al. 2018). When a virus infects a cell, it can often evade the immune system by hiding inside the cell and replicating (Beachboard and Horner 2016; Simmons et al. 2013). However, the immune system has various mechanisms for detecting virus-infected cells and eliminating them (Beachboard and Horner 2016; Carty et al. 2021; Nainu et al. 2017).

One way that the immune system can detect infected cells is through the presence of viral proteins on the surface of the cell. When a virus infects a cell, it often expresses viral proteins on the cell surface, which can be recognized by the immune system as foreign materials. Once infected cells are detected, immune cells such as phagocytes and T cells can be activated to attack and destroy them (Herbert and Panagiotou 2022). There are two main types of T cells: CD8+ T cells and CD4+ T cells. CD8+ T cells, also known as CTLs, can directly attack and kill infected cells by releasing toxic substances. CD4+ T cells, also known as helper T cells, can help to coordinate the immune response by releasing cytokines and activating other immune cells (Herbert and Panagiotou 2022).

Another way that the immune system can detect infected cells is through the presence of viral RNA or DNA inside the cell (Iwasaki 2012). Cells employ the pathogen-associated molecular pattern (PAMP) recognition receptor (PRR) system that can detect viral RNA or DNA and trigger an immune response. When a cell is infected with a virus, it can release pro-inflammatory cytokines and interferons, which are signaling molecules that can alert nearby cells to the presence of the virus and help to activate the immune response (Neufeldt et al. 2022; Kopitar-Jerala 2017). Overall, the immune system has multiple mechanisms for detecting and eliminating virus-infected cells, including the activation of T cells and the interferon system.

Humoral immune responses involve the activation of immune cells called B cells, which produce antibodies that can recognize and neutralize foreign substances in the body (Herbert and Panagiotou 2022; Aoshi et al. 2011). When a virus infects a cell, it can trigger the activation of B cells, which can produce antibodies that are specific to the virus. Once activated, B cells divide and differentiate into two main types of cells: plasma cells and memory B cells. Plasma cells produce large amounts of antibodies, which are proteins that can recognize and bind to specific viral proteins. These antibodies can neutralize the virus by blocking its ability to infect cells or by marking it for destruction by other immune cells (Herbert and Panagiotou 2022; Aoshi et al. 2011). Memory B cells, on the other hand, are long-lived cells that remain in the body after the initial infection has been cleared. These cells can "remember" the specific viral proteins that they encountered, and can rapidly produce antibodies if the same virus is encountered again in the future (Dörner and Radbruch 2007). Such a convenient principle has been applied in the vaccination with a weakened or inactivated form of a virus, which can provide long-lasting immunity to the corresponding virus (Pollard and Bijker 2021). Overall, the humoral immune response involves the production of antibodies that can recognize and neutralize virus-infected cells, and it can provide long-lasting immunity to future infections with the same virus.

A cytokine storm is a phenomenon that occurs when the immune system releases a large number of cytokines in response to an infection or other stimulus (Tisoncik et al. 2012). This can lead to inflammation and damage to healthy cells and tissues, and can sometimes be life-threatening (Tisoncik et al. 2012; Fajgenbaum and June

2020). Cytokines are signaling molecules that are produced by immune cells and other cells in the body. They play a crucial role in coordinating the immune response and helping to eliminate foreign substances, such as viruses and bacteria (Marshall et al. 2018). However, if the immune system releases too many cytokines at once, or if the cytokines are not properly regulated, it can lead to a cytokine storm (Tisoncik et al. 2012; Fajgenbaum and June 2020). Several factors can contribute to the development of a cytokine storm. One factor is the severity of the infection or other stimulus. More severe infections or other stimuli can trigger a larger immune response and a greater release of cytokines (Tisoncik et al. 2012). Another factor is the overall health of the individual. Individuals with compromised immune systems or underlying health conditions, including aging, may be more likely to experience a cytokine storm (Nidadavolu and Walston 2021). Finally, genetic factors can also play a role in the development of a cytokine storm, causing some individuals maybe more prone to developing a cytokine storm due to their genetic makeup (Vakil et al. 2022). Overall, a cytokine storm is a complex process that is triggered by the immune system in response to an infection or other stimulus. It can lead to inflammation and damage to healthy cells and tissues, and it can sometimes be life-threatening.

## Roles for Phagocytosis in the Prevention and Progression of Viral Diseases

The emergence of virus-related proteins on the surface of infected cells can be a sign of an ongoing immune response to the infection. These proteomic biomarkers may be infection-specific, meaning that they are only expressed in response to a particular type of infection, or they may be apoptosis-specific, meaning that they are only expressed on cells that are undergoing apoptosis. Infection-specific biomarkers are molecules that are expressed on the surface of infected cells in response to a particular type of infection. These biomarkers can be used to identify the progression of the infection and may also be involved in the immune response against infection (Shapiro et al. 2022). Apoptotic biomarkers, on the other hand, are molecules that are expressed on the surface of cells undergoing apoptotic cell death, which can be used to identify apoptotic cells and may be involved in the regulation of the apoptotic process (Ward et al. 2008). Overall, the emergence of these markers on the surface of infected cells can provide important information about the immune response to the infection and the status of the infected cells.

One of the most important viral-related biomarkers is viral antigens themselves. Viral antigens are molecules that are produced by viruses in infected cells and can be recognized and targeted by the immune system. These molecules may be expressed on the surface of virus-infected cells and can be used to identify and track the site of infection (Selvarajan et al. 2021; Tsao et al. 2020). These molecules are usually identified as PAMPs, molecules that are characteristic of pathogens, including viruses, and can be recognized by the immune system (Thompson et al. 2011). However, in some cases, certain molecules that are produced by cells in response to damage or stress, including viral infection, can be recognized by the immune system (Land 2021). These molecules are collectively termed damage-associated molecular patterns (DAMPs). DAMPs may be expressed on the surface of virus-infected cells and can be used to stimulate an immune response to the infection (Land 2021). Overall, these markers can be used to identify and track cells undergoing apoptosis in response to viral infection, and are useful in understanding the role of apoptosis in viral infection.

The emergence of virus-related markers on the surface of virus-infected cells is a prime subject for immunological responses. In addition to the swift response, T cells and phagocytic immune cells, such as neutrophils and macrophages, can respond to the alarming presence of virus-infected cells (Nainu et al. 2017). These latter cells have been reported to be able to eliminate virus-infected cells via apoptosis-dependent phagocytosis of virus-infected cells leading to the mitigation of viral infection (Nainu et al. 2015; Hashimoto et al. 2007).

## *Phagocytosis of Virus-Infected Cells as Innate Immune Response*

Metazoan species, which include animals such as insects, mammals, and birds, have evolved a variety of mechanisms to recognize and destroy virus-infected cells (Herbert and Panagiotou 2022). For example, mammals have several types of immune pathways that can be activated in response to viral infections, including the innate immune pathway and the adaptive immune pathway (Herbert and Panagiotou 2022). The innate immune pathway is activated by the presence of PAMPs that are present on the surface or in the interior of the viruses (Lee 2013; Mogensen 2009). Subsequent interaction of PAMPs with PRRs activates the production of pro-inflammatory cytokines, antimicrobial peptides, and the recruitment of immune cells to the site of the infection (Herbert and Panagiotou 2022; Mogensen 2009). The adaptive immune pathway, on the other hand, involves the activation of specific types of T cells and B cells, which can recognize and target specific viruses and help to eliminate them from the body (Herbert and Panagiotou 2022). In addition to the activation of immune pathways, metazoan species can also use other mechanisms to recognize and destroy virus-infected cells. For example, some species have developed mechanisms for detecting the presence of abnormal proteins or changes in gene expression that are characteristic of virus-infected cells, and they can use these mechanisms to identify and eliminate infected cells. Additionally, vertebrates have evolved specialized immune cells called CTLs that can directly recognize and kill virus-infected cells (Herbert and Panagiotou 2022). Overall, the ability to recognize and destroy virus-infected cells is a critical component of the immune systems of metazoan species, and it plays a key role in protecting them from viral infections (Herbert and Panagiotou 2022).

Phagocytosis of virus-infected cells is an important mechanism by which the immune system can eliminate viral infections and prevent the spread of the virus (Nainu et al. 2017; Fujimoto et al. 2000; Hashimoto et al. 2007). Phagocytes, such as neutrophils and macrophages, are important immune cells that can recognize and engulf virus-infected cells (Hashimoto et al. 2007). Once inside the phagocyte, the engulfed virus-infected cell is exposed to a range of antimicrobial substances, such as reactive oxygen and nitrogen species, which can kill the virus and prevent it from replicating (Paiva and Bozza 2014; Molteni et al. 2014). Subsequently, the engulfed virus-infected cell is broken down and its components, including the virus, are degraded and recycled (Nainu et al. 2017). Many types of immune cells, including neutrophils, monocytes, and macrophages, are capable of phagocytosis and use this process to help eliminate pathogens and infected cells from the body (Hashimoto et al. 2007; Nainu et al. 2017).

In *Drosophila* or fruit flies, phagocytosis is a process by which immune cells called hemocytes or blood cells engulf and digest foreign substances, including virus particles (Zhu and Zhang 2013) and virus-infected cells (Nainu et al. 2015). It is an important part of the immune response in *Drosophila*, as it helps to remove infected cells and prevent the spread of the virus (Nainu et al. 2015; Zhu and Zhang 2013). When a virus-infected cell is detected, hemocytes can bind to and engulf it (Nainu et al. 2015). During phagocytosis, the engulfed materials are subjected to a digestion process within a specialized structure called the lysosome, which contains enzymes that break down foreign substances (Melcarne et al. 2019). In addition to phagocytosis, *Drosophila* is also equipped with other mechanisms to fight against viral infection, including the activation of autophagy, the production of antiviral proteins, and the activation of RNA interference pathways (Mussabekova et al. 2017; Xu and Cherry 2014). Together, these mechanisms help to protect the fly from viral infections and maintain the health of the fly's immune system. Phagocytosis of virus-infected cells in *Drosophila* is mediated by two types of phagocytosis receptors, namely, Draper and integrin (Nainu et al. 2015). Draper is a transmembrane protein that is expressed on the surface of phagocytes, such as hemocytes (Manaka et al. 2004). It plays a key role in the recognition and engulfment of apoptotic cells (Manaka et al. 2004; Zheng et al. 2017). Integrin, also a transmembrane protein, is generally involved in cell adhesion and signaling (Moreira et al. 2013). Our study demonstrated that *Drosophila* integrins can play a role in the phagocytosis of virus-infected cells (Nainu et al. 2015). Both Draper and integrin receptors are activated upon binding to specific ligands, most likely phosphatidylserine (PS), on the surface of infected cells, which leads to the activation of signaling pathways that trigger phagocytosis (Nainu et al. 2015). Specific ligands and signaling pathways involved in Draper- and integrin-mediated phagocytosis may vary depending on the types of viruses and host immune responses, but no studies have been done to explore this. Nevertheless, our initial study confirmed the antiviral role of phagocytosis of virus-infected cells in *Drosophila* (Fig. 3.5), and such an event plays a critical role in the control of viral infections and the maintenance of immune homeostasis.

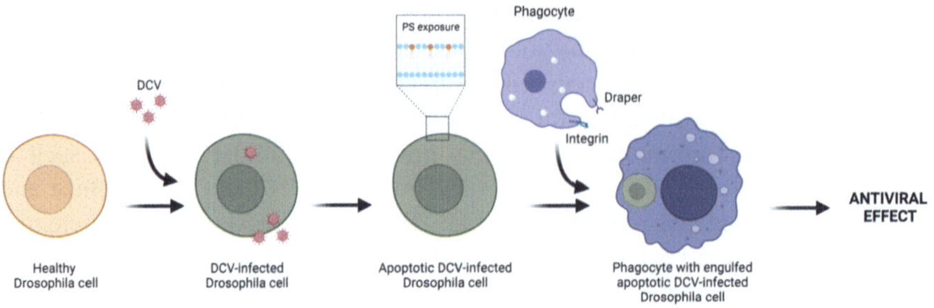


**Fig. 3.5** Induction of apoptosis upon DCV infection and apoptosis-dependent phagocytosis of DCV-infected cells in *Drosophila*. Cells with apoptotic signatures, for example, increased levels of PS exposure, are targets for phagocytic elimination by *Drosophila* phagocytes. Adapted from Nainu et al. (2017). DCV, *Drosophila* C virus; PS, phosphatidylserine. Created with BioRender.com

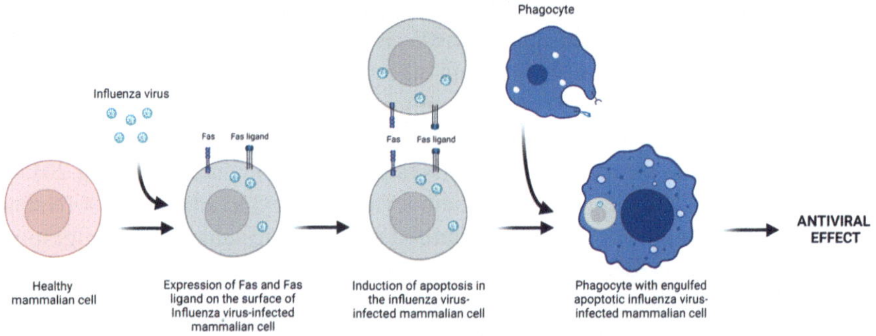


**Fig. 3.6** Mechanisms of apoptosis-dependent phagocytosis of virus-infected cells in the mammalian host. Cells with apoptotic signatures, for example, increased levels of Fas and Fas ligands, are targets for phagocytic elimination by mammalian phagocytes. Adapted from Nainu et al. (2017). Created with BioRender.com

In the mammalian host, virus-infected cells are also targeted for removal from the host body by the apoptosis-dependent phagocytosis mechanism (Fig. 3.6) (Fujimoto et al. 2000; Hashimoto et al. 2007). Previous in vitro data have suggested that HeLa cells and MDCK cells underwent apoptosis in the presence of influenza A virus infection (Takizawa et al. 1993), most likely as the outcome of the increased levels of Fas and FasL (Wada et al. 1995; Fujimoto et al. 1998; Takizawa et al. 1993). Subsequent studies further revealed that influenza A virus-infected cells were targeted for phagocytic elimination by macrophages in both the in vitro (Fujimoto et al. 2000) and the in vivo (Hashimoto et al. 2007) settings. Preliminary analysis using HeLa cells suggested that apoptotic cells were subjected to engulfment in a manner dependent on the presence of PS on the surface of virus-infected cells (Shiratsuchi et al. 2000; Watanabe et al. 2002). While the biological implication of such an event was convincingly stated: the inhibition of viral propagation (Fujimoto et al. 2000) that leads to the mitigation of virus-related pathologies (Hashimoto et al. 2007), it is important to note that the find-me signals involved in the recruitment of phagocytes to the infected sites are still unidentified. In addition to that, detailed mechanisms of apoptosis-dependent phagocytosis of virus-infected cells remain unexplored, which would be scientifically suitable for future studies.

## *Consequences of Apoptosis-Dependent Phagocytosis of Virus-Infected Cells*

Phagocytosis is the process by which immune or non-immune cells engulf and digest foreign particles or microorganisms (Gordon 2016; Uribe-Querol and Rosales 2020). Phagocytosis can help remove viruses from the body by engulfing and digesting them. This can help prevent the spread of the virus and reduce the severity of the infection (Tay et al. 2019; Nainu et al. 2017). Thus, phagocytosis is an important defense mechanism used by the immune system to remove pathogens, including viruses, from the body. When a phagocyte encounters a virus, it will extend pseudopodia (projections) around the virus and engulf it, forming a phagosome. The phagosome then fuses with a lysosome, which contains hydrolytic enzymes that can digest and destroy the virus (Gordon 2016; Tay et al. 2019). This process helps to prevent the virus from replicating and spreading and can help to reduce the severity of the infection (Tay et al. 2019).

In the context of viral infections, the outcome of phagocytosis may extend to the presentation of viral antigens, removal of viruses, and mitigation of viral infection (Nainu et al. 2017). When viruses are phagocytosed, they are broken down into smaller components that may be recognized by immune cells. These components, collectively referred to as viral antigens, can be presented on the surface of immune cells, such as dendritic cells, to stimulate the development of adaptive immune responses (Nainu et al. 2017). This process, known as antigen presentation, is an important way that the immune system can recognize and respond to viral infections. When a dendritic cell presents viral antigens to a T cell, the T cell can recognize the antigens and becomes activated. Activated T cells can then secrete cytokines and other signaling molecules to stimulate other immune cells to respond to the infection, and can also directly kill infected cells. This process helps to clear the virus from the body and can also help to establish immunity to the virus, protecting against future infections (Nainu et al. 2017; Tay et al. 2019).

However, in some cases, phagocytosis may not be effective at eliminating viruses, and the infection may progress. While phagocytosis can be an effective way for the immune system to eliminate viruses and other pathogens, it is not always successful in every case. Some viruses are able to evade the immune system and replicate inside cells, despite the efforts of phagocytic cells to destroy them. This can lead to the progression of the viral infection and the development of associated diseases. For example, HIV is able to evade the immune system and replicate inside immune cells, including CD4+ T cells and macrophages. Despite the efforts of these cells to remove the virus through phagocytosis, HIV is able to replicate and spread, leading to the progression of AIDS (Balasubramaniam et al. 2019).

## Perspectives

Phagocytosis can play both a helpful and harmful role in virus-induced diseases, depending on the specific virus and the context in which it is operating. On the one hand, phagocytosis can be a useful defense mechanism for the immune system to remove viruses and other pathogens from the body. By engulfing and digesting viruses, phagocytic cells can help prevent the spread of the virus and reduce the severity of the infection. However, in some cases, phagocytosis can also contribute to the progression of viral diseases. Some viruses evade digestion and achieve successful replication in phagocytes after engulfment, and continue the infection process.

There are currently no medical treatments that are specifically based on phagocytosis for the treatment of viral diseases, including COVID-19. However, some treatments for viral infections, including antiviral medications and vaccines, may indirectly enhance phagocytosis as part of the immune response to the virus. Antiviral medications work by inhibiting the replication of viruses, which can help to reduce the severity of the infection and the load of the virus in the body. This can make it easier for the immune system, including phagocytic cells, to clear the virus from the body. Vaccines, on the other hand, stimulate the immune system to develop immunity to a particular virus, without causing illness. When a person is vaccinated against a virus, their immune system is exposed to viral antigens, which can stimulate the development of adaptive immune responses, including the production of antibodies and the activation of T cells. These immune responses can help to clear the virus from the body and protect against future infections.

Overall, while phagocytosis is an important defense mechanism for the immune system to remove viruses and other pathogens from the body, it is not currently a primary focus of medical treatments for viral diseases. Serious attention should now be given to this research area, as phagocytosis has been widely shown to play important roles in many aspects of viral diseases. Successful navigation on the prospective yet uncharted mechanisms of targeted phagocytic elimination of viral particles and virus-infected cells may lead to the discovery of generic targets presented on the surface of virus-infected cells and/or universal phagocytic receptors that can target viral particles and virus-infected cells. Therefore, instead of targeting different viruses with different medicines, generic targeting of virus-specific antigens and/or virus-infected cells with enhanced phagocytosis activity in the patients' bodies may provide a universal and hopefully better disease management in the future.

**Acknowledgments** Research carried out in FN's lab is supported by PDUPT Research Grant 2022 (No.020/E5/PG.02.00PT/2022) from the Directorate General of Higher Education, Ministry of

Education, Culture, Research, and Technology, Indonesia. This study was also partly supported by Grant-in-Aid for Scientific Research (KAKENHI) from the Japan Society for the Promotion of Science (grant number 19 K17925 to YO). We are sincerely grateful to the members of UNHAS Fly Research Group (UFRG) for their support during the preparation of the manuscript.

## References

Algeciras-Schimnich A, Vlahakis SR, Villasis-Keever A, Gomez T, Heppelmann CJ, Bou G, Paya CV (2002) CCR5 mediates Fas- and caspase-8 dependent apoptosis of both uninfected and HIV infected primary human CD4 T cells. AIDS 16(11):1467–1478. https://doi.org/10.1097/00002030-200207260-00003

Alymova IV, Cipollo JF, Parsons LM, Music N, Kamal RP, Tzeng WP, Goldsmith CS, Contessa JN, Hartshorn KL, Wilson JR, Zeng H, Gansebom S, York IA (2022) Aberrant cellular glycosylation may increase the ability of influenza viruses to escape host immune responses through modification of the viral glycome. mBio 13(2):e0298321. https://doi.org/10.1128/mbio.02983-21

Anasir MI, Caria S, Skinner MA, Kvansakul M (2017) Structural basis of apoptosis inhibition by the fowlpox virus protein FPV039. J Biol Chem 292(22):9010–9021. https://doi.org/10.1074/jbc.M116.768879

Andreu-Moreno I, Bou J-V, Sanjuán R (2020) Cooperative nature of viral replication. Sci Adv 6(49):eabd4942. https://doi.org/10.1126/sciadv.abd4942

Aoshi T, Koyama S, Kobiyama K, Akira S, Ishii KJ (2011) Innate and adaptive immune responses to viral infection and vaccination. Curr Opin Virol 1(4):226–232. https://doi.org/10.1016/j.coviro.2011.07.002

Aouad SM, Cohen LY, Sharif-Askari E, Haddad EK, Alam A, Sekaly R-P (2004) Caspase-3 is a component of Fas death-inducing signalling complex in lipid rafts and its activity is required for complete caspase-8 activation during Fas-mediated cell death. J Immunol 172(4):2316–2323. https://doi.org/10.4049/jimmunol.172.4.2316

Arandjelovic S, Ravichandran KS (2015) Phagocytosis of apoptotic cells in homeostasis. Nat Immunol 16(9):907–917. https://doi.org/10.1038/ni.3253

Aries SP, Schaaf B, Müller C, Dennin RH, Dalhoff K (1995) Fas (CD95) expression on CD4+ T cells from HIV-infected patients increases with disease progression. J Mol Med 73(12):591–593. https://doi.org/10.1007/BF00196352

Badley AD, McElhinny JA, Leibson PJ, Lynch DH, Alderson MR, Paya CV (1996) Upregulation of Fas ligand expression by human immunodeficiency virus in human macrophages mediates apoptosis of uninfected T lymphocytes. J Virol 70(1):199–206. https://doi.org/10.1128/JVI.70.1.199-206.1996

Baize S, Leroy EM, Georges-Courbot MC, Capron M, Lansoud-Soukate J, Debré P, Fisher-Hoch SP, McCormick JB, Georges AJ (1999) Defective humoral responses and extensive intravascular apoptosis are associated with fatal outcome in Ebola virus-infected patients. Nat Med 5(4):423–426. https://doi.org/10.1038/7422

Balasubramaniam M, Pandhare J, Dash C (2019) Immune control of HIV. J Life Sci (Westlake village) 1(1):4–37

Baloul L, Camelo S, Lafon M (2004) Up-regulation of Fas ligand (FasL) in the central nervous system: a mechanism of immune evasion by rabies virus. J Neurovirol 10(6):372–382. https://doi.org/10.1080/13550280490521122

Banadyga L, Veugelers K, Campbell S, Barry M (2009) The fowlpox virus BCL-2 homologue, FPV039, interacts with activated Bax and a discrete subset of BH3-only proteins to inhibit apoptosis. J Virol 83(14):7085–7098. https://doi.org/10.1128/JVI.00437-09

Barber GN (2001) Host defense, viruses and apoptosis. Cell Death Differ 8(2):113–126. https://doi.org/10.1038/sj.cdd.4400823

Beachboard DC, Horner SM (2016) Innate immune evasion strategies of DNA and RNA viruses. Curr Opin Microbiol 32:113–119. https://doi.org/10.1016/j.mib.2016.05.015

Bélanger C, Gravel A, Tomoiu A, Janelle ME, Gosselin J, Tremblay MJ, Flamand L (2001) Human herpesvirus 8 viral FLICE-inhibitory protein inhibits Fas-mediated apoptosis through binding and prevention of procaspase-8 maturation. J Hum Virol 4(2):62–73

Bellesi S, Metafuni E, Hohaus S et al (2020) Increased CD95 (Fas) and PD-1 expression in peripheral blood T lymphocytes in COVID-19 patients. Br J Haematol 191(2):207–211

Benedict CA, Norris PS, Prigozy TI, Bodmer JL, Mahr JA, Garnett CT, Martinon F, Tschopp J, Gooding LR, Ware CF (2001) Three adenovirus E3 proteins cooperate to evade apoptosis by tumor necrosis factor-related apoptosis-inducing ligand receptor-1 and -2. J Biol Chem 276(5): 3270–3278. https://doi.org/10.1074/jbc.M008218200

Bird SW, Kirkegaard K (2015) Escape of non-enveloped virus from intact cells. Virology 479–480: 444–449. https://doi.org/10.1016/j.virol.2015.03.044

Boyd JM, Gallo GJ, Elangovan B, Houghton AB, Malstrom S, Avery BJ, Ebb RG, Subramanian T, Chittenden T, Lutz RJ (1995) Bik, a novel death-inducing protein shares a distinct sequence motif with Bcl-2 family proteins and interacts with viral and cellular survival-promoting proteins. Oncogene 11(9):1921–1928

Brazeau E, Mahalingam R, Gilden D, Wellish M, Kaufer BB, Osterrieder N, Pugazhenthi S (2010) Varicella-zoster virus-induced apoptosis in MeWo cells is accompanied by down-regulation of Bcl-2 expression. J Neurovirol 16(2):133–140. https://doi.org/10.3109/13550281003682547

Brentnall M, Rodriguez-Menocal L, De Guevara RL, Cepero E, Boise LH (2013) Caspase-9, caspase-3 and caspase-7 have distinct roles during intrinsic apoptosis. BMC Cell Biol 14:32. https://doi.org/10.1186/1471-2121-14-32

Brincks EL, Katewa A, Kucaba TA, Griffith TS, Legge KL (2008) CD8 T cells utilize TRAIL to control influenza virus infection. J Immunol 181(7):4918–4925. https://doi.org/10.4049/jimmunol.181.7.4918

Buckingham EM, Jarosinski KW, Jackson W, Carpenter JE, Grose C (2016) Exocytosis of Varicella-Zoster virus virions involves a convergence of endosomal and autophagy pathways. J Virol 90(19):8673–8685. https://doi.org/10.1128/jvi.00915-16

Burton DR, Caria S, Marshall B, Barry M, Kvansakul M (2015) Structural basis of Deerpox virus-mediated inhibition of apoptosis. Acta Crystallogr D Biol Crystallogr 71(Pt 8):1593–1603. https://doi.org/10.1107/S1399004715009402

Carty M, Guy C, Bowie AG (2021) Detection of viral infections by innate immunity. Biochem Pharmacol 183:114316. https://doi.org/10.1016/j.bcp.2020.114316

Chambers JW, Maguire TG, Alwine JC (2010) Glutamine metabolism is essential for human cytomegalovirus infection. J Virol 84(4):1867–1873. https://doi.org/10.1128/jvi.02123-09

Chazal N, Gerlier D (2003) Virus entry, assembly, budding, and membrane rafts. Microbiol Mol Biol Rev 67(2):226–237. https://doi.org/10.1128/MMBR.67.2.226-237.2003

Chen G, Goeddel DV (2002) TNF-R1 signalling: a beautiful pathway. Science 296(5573): 1634–1635. https://doi.org/10.1126/science.1071924

Chen Z, Zhu Y, Ren Y, Tong Y, Hua X, Zhu F, Huang L, Liu Y, Luo Y, Lu W, Zhao P, Qi Z (2011) Hepatitis C virus protects human B lymphocytes from Fas-mediated apoptosis via E2-CD81 engagement. PLoS One 6(4):e18933. https://doi.org/10.1371/journal.pone.0018933

Chiou S-H, Yang Y-P, Lin J-C, Hsu C-H, Jhang H-C, Yang Y-T, Lee C-H, Ho LLT, Hsu W-M, Ku H-H, Chen S-J, Chen SSL, Chang MDT, Wu C-W, Juan L-J (2006) The immediate early 2 protein of human cytomegalovirus (HCMV) mediates the apoptotic control in HCMV retinitis through up-regulation of the cellular FLICE-inhibitory protein expression. J Immunol 177(9): 6199–6206. https://doi.org/10.4049/jimmunol.177.9.6199

Chu H, Zhou J, Wong BH-Y, Li C, Chan JF-W, Cheng Z-S, Yang D, Wang D, Lee AC-Y, Li C, Yeung M-L, Cai J-P, Chan IH-Y, Ho W-K, To KK-W, Zheng B-J, Yao Y, Qin C, Yuen K-Y (2016) Middle East respiratory syndrome coronavirus efficiently infects human primary T lymphocytes and activates the extrinsic and intrinsic apoptosis pathways. J Infect Dis 213(6): 904–914. https://doi.org/10.1093/infdis/jiv380

Clarke P, Tyler KL (2009) Apoptosis in animal models of virus-induced disease. Nat Rev Microbiol 7(2):144–155. https://doi.org/10.1038/nrmicro2071

Clarke P, Meintzer SM, Gibson S, Widmann C, Garrington TP, Johnson GL, Tyler KL (2000) Reovirus-induced apoptosis is mediated by TRAIL. J Virol 74(17):8135–8139. https://doi.org/10.1128/jvi.74.17.8135-8139.2000

Clarke P, Leser JS, Quick ED, Dionne KR, Beckham JD, Tyler KL (2014) Death receptor-mediated apoptotic signalling is activated in the brain following infection with West Nile virus in the absence of a peripheral immune response. J Virol 88(2):1080–1089. https://doi.org/10.1128/JVI.02944-13

Collins-McMillen D, Kim JH, Nogalski MT, Stevenson EV, Chan GC, Caskey JR, Cieply SJ, Yurochko AD (2015) Human cytomegalovirus promotes survival of infected monocytes via a distinct temporal regulation of cellular Bcl-2 family proteins. J Virol 90(5):2356–2371. https://doi.org/10.1128/JVI.01994-15

Combadiere C, Salzwedel K, Smith ED, Tiffany HL, Berger EA, Murphy PM (1998) Identification of CX3CR1. A chemotactic receptor for the human CX3C chemokine fractalkine and a fusion coreceptor for HIV-1. J Biol Chem 273(37):23799–23804. https://doi.org/10.1074/jbc.273.37.23799

Contini P, Ghio M, Merlo A, Brenci S, Filaci G, Indiveri F, Puppo F (2000) Soluble HLA class I/CD8 ligation triggers apoptosis in EBV-specific CD8+ cytotoxic T lymphocytes by Fas/Fas-ligand interaction. Hum Immunol 61(12):1347–1351. https://doi.org/10.1016/s0198-8859(00)00212-3

Cuconati A, White E (2002) Viral homologs of BCL-2: role of apoptosis in the regulation of virus infection. Genes Dev 16(19):2465–2478. https://doi.org/10.1101/gad.1012702

Deng Z, Yan H, Hu J, Zhang S, Peng P, Liu Q, Guo D (2012) Hepatitis C virus sensitizes host cells to TRAIL-induced apoptosis by up-regulating DR4 and DR5 via a MEK1-dependent pathway. PLoS One 7(5):e37700. https://doi.org/10.1371/journal.pone.0037700

Depierreux DM, Altenburg AF, Soday L, Fletcher-Etherington A, Antrobus R, Ferguson BJ, Weekes MP, Smith GL (2022) Selective modulation of cell surface proteins during vaccinia infection: A resource for identifying viral immune evasion strategies. PLoS Pathog 18(6): e1010612. https://doi.org/10.1371/journal.ppat.1010612

Desbien AL, Kappler JW, Marrack P (2009) The Epstein–Barr virus Bcl-2 homolog, BHRF1, blocks apoptosis by binding to a limited amount of Bim. Proc Natl Acad Sci USA 106(14): 5663–5668. https://doi.org/10.1073/pnas.0901036106

Dewson G, Kratina T, Sim HW, Puthalakath H, Adams JM, Colman PM, Kluck RM (2008) To trigger apoptosis, Bak exposes its BH3 domain and homodimerizes via BH3:groove interactions. Mol Cell 30(3):369–380. https://doi.org/10.1016/j.molcel.2008.04.005

Ding L, Xu X, Huang Y, Li Z, Zhang K, Chen G, Yu G, Wang Z, Li W, Tong D (2012) Transmissible gastroenteritis virus infection induces apoptosis through FasL- and mitochondria-mediated pathways. Vet Microbiol 158(1-2):12–22. https://doi.org/10.1016/j.vetmic.2012.01.017

Dörner T, Radbruch A (2007) Antibodies and B cell memory in viral immunity. Immunity 27(3): 384–392. https://doi.org/10.1016/j.immuni.2007.09.002

Du J, Liang X, Liu Y, Qu Z, Gao L, Han L, Liu S, Cui M, Shi Y, Zhang Z, Yu L, Cao L, Ma C, Zhang L, Chen Y, Sun W (2009) Hepatitis B virus core protein inhibits TRAIL-induced apoptosis of hepatocytes by blocking DR5 expression. Cell Death Differ 16(2):219–229. https://doi.org/10.1038/cdd.2008.144

Dufour F, Sasseville AM-J, Chabaud S, Massie B, Siegel RM, Langelier Y (2011) The ribonucleotide reductase R1 subunits of herpes simplex virus types 1 and 2 protect cells against TNFα- and FasL-induced apoptosis by interacting with caspase-8. Apoptosis 16(3):256–271. https://doi.org/10.1007/s10495-010-0560-2

Elliott MR, Chekeni FB, Trampont PC, Lazarowski ER, Kadl A, Walk SF, Park D, Woodson RI, Ostankovich M, Sharma P, Lysiak JJ, Harden TK, Leitinger N, Ravichandran KS (2009) Nucleotides released by apoptotic cells act as a find-me signal to promote phagocytic clearance. Nature 461(7261):282–286. https://doi.org/10.1038/nature08296

Fajgenbaum DC, June CH (2020) Cytokine storm. N Engl J Med 383(23):2255–2273. https://doi.org/10.1056/NEJMra2026131

Farrow SN, White JH, Martinou I, Raven T, Pun KT, Grinham CJ, Martinou JC, Brown R (1995) Cloning of a bcl-2 homologue by interaction with adenovirus E1B 19K. Nature 374(6524): 731–733. https://doi.org/10.1038/374731a0

Faure S, Meyer L, Costagliola D, Vaneensberghe C, Genin E, Autran B, Delfraissy JF, McDermott DH, Murphy PM, Debré P, Théodorou I, Combadière C (2000) Rapid progression to AIDS in HIV+ individuals with a structural variant of the chemokine receptor CX3CR1. Science 287(5461):2274–2277. https://doi.org/10.1126/science.287.5461.2274

Flanagan AM, Letai A (2008) BH3 domains define selective inhibitory interactions with BHRF-1 and KSHV BCL-2. Cell Death Differ 15(3):580–588. https://doi.org/10.1038/sj.cdd.4402292

Fuchs Y, Steller H (2011) Programmed cell death in animal development and disease. Cell 147(4): 742–758. https://doi.org/10.1016/j.cell.2011.10.033

Fujimoto I, Takizawa T, Ohba Y, Nakanishi Y (1998) Co-expression of Fas and Fas-ligand on the surface of influenza virus-infected cells. Cell Death Differ 5(5):426–431. https://doi.org/10.1038/sj.cdd.4400362

Fujimoto I, Pan J, Takizawa T, Nakanishi Y (2000) Virus clearance through apoptosis-dependent phagocytosis of influenza a virus-infected cells by macrophages. J Virol 74(7):3399–3403. https://doi.org/10.1128/JVI.74.7.3399-3403.2000

Galindo I, Hernaez B, Díaz-Gil G, Escribano JM, Alonso C (2008) A179L, a viral Bcl-2 homologue, targets the core Bcl-2 apoptotic machinery and its upstream BH3 activators with selective binding restrictions for Bid and Noxa. Virology 375(2):561–572. https://doi.org/10.1016/j.virol.2008.01.050

García-Murria MJ, Duart G, Grau B, Diaz-Beneitez E, Rodríguez D, Mingarro I, Martínez-Gil L (2020) Viral Bcl2s' transmembrane domain interact with host Bcl2 proteins to control cellular apoptosis. Nat Commun 11(1):6056. https://doi.org/10.1038/s41467-020-19881-9

Ghosh S, Dutta K, Basu A (2013) Chandipura virus induces neuronal death through Fas-mediated extrinsic apoptotic pathway. J Virol 87(22):12398–12406. https://doi.org/10.1128/JVI.01864-13

Goldmacher VS, Bartle LM, Skaletskaya A, Dionne CA, Kedersha NL, Vater CA, Han JW, Lutz RJ, Watanabe S, Cahir McFarland ED, Kieff ED, Mocarski ES, Chittenden T (1999) A cytomegalovirus-encoded mitochondria-localized inhibitor of apoptosis structurally unrelated to Bcl-2. Proc Natl Acad Sci U S A 96(22):12536–12541. https://doi.org/10.1073/pnas.96.22.12536

Gordon S (2016) Phagocytosis: an immunobiologic process. Immunity 44(3):463–475. https://doi.org/10.1016/j.immuni.2016.02.026

Gremion C, Grabscheid B, Wölk B, Moradpour D, Reichen J, Pichler W, Cerny A (2004) Cytotoxic T lymphocytes derived from patients with chronic hepatitis C virus infection kill bystander cells via Fas-FasL interaction. J Virol 78(4):2152–2157. https://doi.org/10.1128/jvi.78.4.2152-2157.2004

Han J, Sabbatini P, Perez D, Rao L, Modha D, White E (1996) The E1B 19K protein blocks apoptosis by interacting with and inhibiting the p53-inducible and death-promoting Bax protein. Genes Dev 10(4):461–477. https://doi.org/10.1101/gad.10.4.461

Hashimoto Y, Moki T, Takizawa T, Shiratsuchi A, Nakanishi Y (2007) Evidence for phagocytosis of influenza virus-infected, apoptotic cells by neutrophils and macrophages in Mice1. J Immunol 178(4):2448–2457. https://doi.org/10.4049/jimmunol.178.4.2448

Hay BA, Huh JR, Guo M (2004) The genetics of cell death: approaches, insights and opportunities in Drosophila. Nat Rev Genet 5(12):911–922. https://doi.org/10.1038/nrg1491

Henderson S, Huen D, Rowe M, Dawson C, Johnson G, Rickinson A (1993) Epstein-Barr virus-coded BHRF1 protein, a viral homologue of Bcl-2, protects human B cells from programmed cell death. Proc Natl Acad Sci U S A 90(18):8479–8483. https://doi.org/10.1073/pnas.90.18.8479

Herbert JA, Panagiotou S (2022) Immune response to viruses. In: Rezaei N (ed) Encyclopedia of infection and immunity, vol 1. Elsevier, Oxford, pp 429–444

Herbeuval J-P, Grivel J-C, Boasso A, Hardy AW, Chougnet C, Dolan MJ, Yagita H, Lifson JD, Shearer GM (2005) CD4+ T-cell death induced by infectious and noninfectious HIV-1: role of type 1 interferon-dependent, TRAIL/DR5-mediated apoptosis. Blood 106(10):3524–3531. https://doi.org/10.1182/blood-2005-03-1243

Herbeuval J-P, Nilsson J, Boasso A, Hardy AW, Kruhlak MJ, Anderson SA, Dolan MJ, Dy M, Andersson J, Shearer GM (2006) Differential expression of IFN-α and TRAIL/DR5 in lymphoid tissue of progressor versus nonprogressor HIV-1-infected patients. Proc Natl Acad Sci USA 103(18):7000–7005. https://doi.org/10.1073/pnas.0600363103

Hernaez B, Cabezas M, Muñoz-Moreno R, Galindo I, Cuesta-Geijo MA, Alonso C (2013) A179L, a new viral Bcl2 homolog targeting Beclin 1 autophagy related protein. Curr Mol Med 13(2): 305–316

Hofmann P, Sprenger H, Kaufmann A, Bender A, Hasse C, Nain M, Gemsa D (1997) Susceptibility of mononuclear phagocytes to influenza A virus infection and possible role in the antiviral response. J Leukoc Biol 61(4):408–414. https://doi.org/10.1002/jlb.61.4.408

Holzerland J, Fénéant L, Banadyga L, Hölper JE, Knittler MR, Groseth A (2020) BH3-only sensors Bad, Noxa and Puma are key regulators of tacaribe virus-induced apoptosis. PLoS Pathog 16(10):e1008948. https://doi.org/10.1371/journal.ppat.1008948

Iannello A, Debbeche O, El Arabi R, Samarani S, Hamel D, Rozenberg F, Heveker N, Ahmad A (2011) Herpes simplex virus type 1-induced FasL expression in human monocytic cells and its implications for cell death, viral replication, and immune evasion. Viral Immunol 24(1):11–26. https://doi.org/10.1089/vim.2010.0083

Imai T, Hieshima K, Haskell C, Baba M, Nagira M, Nishimura M, Kakizaki M, Takagi S, Nomiyama H, Schall TJ, Yoshie O (1997) Identification and molecular characterization of fractalkine receptor CX3CR1, which mediates both leukocyte migration and adhesion. Cell 91(4):521–530. https://doi.org/10.1016/s0092-8674(00)80438-9

Ishikawa E, Nakazawa M, Yoshinari M, Minami M (2005) Role of tumor necrosis factor-related apoptosis-inducing ligand in immune response to influenza virus infection in mice. J Virol 79(12):7658–7663. https://doi.org/10.1128/JVI.79.12.7658-7663.2005

Islam MT, Quispe C, Herrera-Bravo J, Sarkar C, Sharma R, Garg N, Fredes LI, Martorell M, Alshehri MM, Sharifi-Rad J, Daştan SD, Calina D, Alsafi R, Alghamdi S, Batiha GE-S, Cruz-Martins N (2021) Production, transmission, pathogenesis, and control of dengue virus: a literature-based undivided perspective. BioMed Res Int 2021:4224816. https://doi.org/10.1155/2021/4224816

Ito M, Watanabe M, Ihara T, Kamiya H, Sakurai M (1995) Fas antigen and bcl-2 expression on lymphocytes cultured with cytomegalovirus and varicella-zoster virus antigen. Cell Immunol 160(2):173–177. https://doi.org/10.1016/0008-8749(95)80024-d

Iwasaki A (2012) A virological view of innate immune recognition. Annu Rev Microbiol 66:177–196. https://doi.org/10.1146/annurev-micro-092611-150203

Janssen HLA, Higuchi H, Abdulkarim A, Gores GJ (2003) Hepatitis B virus enhances tumor necrosis factor-related apoptosis-inducing ligand (TRAIL) cytotoxicity by increasing TRAIL-R1/death receptor 4 expression. J Hepatol 39(3):414–420. https://doi.org/10.1016/s0168-8278(03)00265-4

Johnson BD, Zhu Z, Lequio M, Powers CGD, Bai Q, Xiao H, Fajardo E, Wakefield MR, Fang Y (2022) SARS-CoV-2 spike protein inhibits growth of prostate cancer: a potential role of the COVID-19 vaccine killing two birds with one stone. Med Oncol 39(3):32. https://doi.org/10.1007/s12032-021-01628-1

Kale J, Osterlund EJ, Andrews DW (2018) BCL-2 family proteins: changing partners in the dance towards death. Cell Death Differ 25(1):65–80. https://doi.org/10.1038/cdd.2017.186

Kaminskyy V, Zhivotovsky B (2010) To kill or be killed: how viruses interact with the cell death machinery. J Intern Med 267(5):473–482. https://doi.org/10.1111/j.1365-2796.2010.02222.x

Kannian P, Green PL (2010) Human T lymphotropic virus type 1 (HTLV-1): molecular biology and oncogenesis. Viruses 2(9):2037–2077. https://doi.org/10.3390/v2092037

Kassis R, Larrous F, Estaquier J, Bourhy H (2004) Lyssavirus matrix protein induces apoptosis by a TRAIL-dependent mechanism involving caspase-8 activation. J Virol 78(12):6543–6555. https://doi.org/10.1128/JVI.78.12.6543-6555.2004

Kather A, Raftery MJ, Devi-Rao G, Lippmann J, Giese T, Sandri-Goldin RM, Schönrich G (2010) Herpes simplex virus type 1 (HSV-1)-induced apoptosis in human dendritic cells as a result of downregulation of cellular FLICE-inhibitory protein and reduced expression of HSV-1 antiapoptotic latency-associated transcript sequences. J Virol 84(2):1034–1046. https://doi.org/10.1128/JVI.01409-09

Kim K-H, Seong BL (2003) Pro-apoptotic function of HBV X protein is mediated by interaction with c-FLIP and enhancement of death-inducing signal. EMBO J 22(9):2104–2116. https://doi.org/10.1093/emboj/cdg210

Kirshner JR, Karpova AY, Kops M, Howley PM (2005) Identification of TRAIL as an interferon regulatory factor 3 transcriptional target. J Virol 79(14):9320–9324. https://doi.org/10.1128/JVI.79.14.9320-9324.2005

Koonin EV, Krupovic M, Agol VI (2021) The Baltimore classification of viruses 50 years later: how does it stand in the light of virus evolution? Microbiol Mol Biol Rev 85(3):e0005321. https://doi.org/10.1128/mmbr.00053-21

Kopitar-Jerala N (2017) The role of interferons in inflammation and inflammasome activation. Front Immunol 8:873. https://doi.org/10.3389/fimmu.2017.00873

Kotelkin A, Prikhod'ko EA, Cohen JI, Collins PL, Bukreyev A (2003) Respiratory syncytial virus infection sensitizes cells to apoptosis mediated by tumor necrosis factor-related apoptosis-inducing ligand. J Virol 77(17):9156–9172. https://doi.org/10.1128/jvi.77.17.9156-9172.2003

Krzyzowska M, Shestakov A, Eriksson K, Chiodi F (2011) Role of Fas/FasL in regulation of inflammation in vaginal tissue during HSV-2 infection. Cell Death Dis 2(3):e132. https://doi.org/10.1038/cddis.2011.14

Krzyzowska M, Kowalczyk A, Skulska K, Thörn K, Eriksson K (2021) Fas/FasL contributes to HSV-1 brain infection and neuroinflammation. Front Immunol 12:714821. https://doi.org/10.3389/fimmu.2021.714821

Ku B, Woo J-S, Liang C, Lee K-H, Hong H-S, Xiaofei E, Kim K-S, Jung JU, Oh B-H (2008) Structural and biochemical bases for the inhibition of autophagy and apoptosis by viral BCL-2 of murine gamma-herpesvirus 68. PLoS Pathog 4(2):e25. https://doi.org/10.1371/journal.ppat.0040025

Kvansakul M, van Delft MF, Lee EF, Gulbis JM, Fairlie WD, Huang DCS, Colman PM (2007) A structural viral mimic of prosurvival Bcl-2: a pivotal role for sequestering proapoptotic Bax and Bak. Mol Cell 25(6):933–942. https://doi.org/10.1016/j.molcel.2007.02.004

Kvansakul M, Wei AH, Fletcher JI, Willis SN, Chen L, Roberts AW, Huang DC, Colman PM (2010) Structural basis for apoptosis inhibition by Epstein-Barr virus BHRF1. PLoS Pathog 6(12):e1001236. https://doi.org/10.1371/journal.ppat.1001236

Lacana E, D'Adamio L (1999) Regulation of Fas ligand expression and cell death by apoptosis-linked gene 4. Nat Med 5(5):542–547. https://doi.org/10.1038/8420

Land WG (2021) Role of DAMPs in respiratory virus-induced acute respiratory distress syndrome—with a preliminary reference to SARS-CoV-2 pneumonia. Genes Immun 22(3): 141–160. https://doi.org/10.1038/s41435-021-00140-w

Langelier Y, Bergeron S, Chabaud S, Lippens J, Guilbault C, Sasseville AM-J, Denis S, Mosser DD, Massie B (2002) The R1 subunit of herpes simplex virus ribonucleotide reductase protects cells against apoptosis at, or upstream of, caspase-8 activation. J Gen Virol 83(Pt 11): 2779–2789. https://doi.org/10.1099/0022-1317-83-11-2779

Lauber K, Bohn E, Kröber SM, Xiao Y-J, Blumenthal SG, Lindemann RK, Marini P, Wiedig C, Zobywalski A, Baksh S, Xu Y, Autenrieth IB, Schulze-Osthoff K, Belka C, Stuhler G, Wesselborg S (2003) Apoptotic cells induce migration of phagocytes via caspase-3-mediated release of a lipid attraction signal. Cell 113(6):717–730. https://doi.org/10.1016/s0092-8674(03)00422-7

Le Clorennec C, Youlyouz-Marfak I, Adriaenssens E, Coll J, Bornkamm GW, Feuillard J (2006) EBV latency III immortalization program sensitizes B cells to induction of CD95-mediated apoptosis via LMP1: role of NF-kappaB, STAT1, and p53. Blood 107(5):2070–2078. https://doi.org/10.1182/blood-2005-05-2053

Lee B (2013) HIV provides ample PAMPs for innate immune sensing. Proc Natl Acad Sci USA 110(48):19183–19184. https://doi.org/10.1073/pnas.1319118110

Li P, Nijhawan D, Budihardjo I, Srinivasula SM, Ahmad M, Alnemri ES, Wang X (1997) Cytochrome c and dATP-dependent formation of Apaf-1/caspase-9 complex initiates an apoptotic protease cascade. Cell 91(4):479–489. https://doi.org/10.1016/s0092-8674(00)80434-1

Li H, Zhu H, Xu CJ, Yuan J (1998) Cleavage of BID by caspase 8 mediates the mitochondrial damage in the Fas pathway of apoptosis. Cell 94(4):491–501. https://doi.org/10.1016/s0092-8674(00)81590-1

Li S-S, Yang S, Wang S, Yang X-M, Tang Q-L, Wang S-H (2011) Latent membrane protein 1 mediates the resistance of nasopharyngeal carcinoma cells to TRAIL-induced apoptosis by activation of the PI3K/Akt signalling pathway. Oncol Rep 26(6):1573–1579. https://doi.org/10.3892/or.2011.1423

Liang X, Liu Y, Zhang Q, Gao L, Han L, Ma C, Zhang L, Chen YH, Sun W (2007) Hepatitis B virus sensitizes hepatocytes to TRAIL-induced apoptosis through Bax. J Immunol 178(1):503–510. https://doi.org/10.4049/jimmunol.178.1.503

Liao H, Xu J, Huang J (2010) FasL/Fas pathway is involved in dengue virus induced apoptosis of the vascular endothelial cells. J Med Virol 82(8):1392–1399. https://doi.org/10.1002/jmv.21815

Liao Y, Wang H-X, Mao X, Fang H, Wang H, Li Y, Sun Y, Meng C, Tan L, Song C, Qiu X, Ding C (2017) RIP1 is a central signalling protein in regulation of TNF-α/TRAIL mediated apoptosis and necroptosis during Newcastle disease virus infection. Oncotarget 8(26):43201–43217. https://doi.org/10.18632/oncotarget.17970

Liu ZX, Govindarajan S, Okamoto S, Dennert G (2001) Fas-mediated apoptosis causes elimination of virus-specific cytotoxic T cells in the virus-infected liver. J Immunol 166(5):3035–3041. https://doi.org/10.4049/jimmunol.166.5.3035

Liu B, Behura SK, Clem RJ, Schneemann A, Becnel J, Severson DW, Zhou L (2013) P53-mediated rapid induction of apoptosis conveys resistance to viral infection in Drosophila melanogaster. PLoS Pathogens 9(2):e1003137. https://doi.org/10.1371/journal.ppat.1003137

Liu W, Lin Y-T, Yan X-L, Ding Y-L, Wu Y-L, Chen W-N, Lin X (2015) Hepatitis B virus core protein inhibits Fas-mediated apoptosis of hepatoma cells via regulation of mFas/FasL and sFas expression. FASEB J 29(3):1113–1123. https://doi.org/10.1096/fj.14-263822

Liu Y, Wang W, Zou Z, Fan Q, Hu Z, Feng Z, Zhu B, Xiong J (2017) Monocyte chemoattractant protein 1 released from macrophages induced by hepatitis C virus promotes monocytes migration. Virus Res 240:190–196. https://doi.org/10.1016/j.virusres.2017.08.013

Louten J (2016) Virus structure and classification. Essent Hum Virol 19–29. https://doi.org/10.1016/b978-0-12-800947-5.00002-8

Lum JJ, Pilon AA, Sanchez-Dardon J, Phenix BN, Kim JE, Mihowich J, Jamison K, Hawley-Foss N, Lynch DH, Badley AD (2001) Induction of cell death in human immunodeficiency virus-infected macrophages and resting memory CD4 T cells by TRAIL/Apo2l. J Virol 75(22):11128–11136. https://doi.org/10.1128/JVI.75.22.11128-11136.2001

Macias J, Japón MA, Sáez C, Palacios RB, Mira JA, García-García JA, Merchante N, Vergara S, Lozano F, Gómez-Mateos J, Pineda JA (2005) Increased hepatocyte fas expression and apoptosis in HIV and hepatitis C virus coinfection. J Infect Dis 192(9):1566–1576. https://doi.org/10.1086/491736

Mack C, Sickmann A, Lembo D, Brune W (2008) Inhibition of proinflammatory and innate immune signalling pathways by a cytomegalovirus RIP1-interacting protein. Proc Natl Acad Sci U S A 105(8):3094–3099. https://doi.org/10.1073/pnas.0800168105

Madigan MT, Bender KS, Buckley DH, Sattley WM, Stahl DA (2017) Brock biology of microorganisms, Global edition. Pearson, New York

Mahoney DJ, Cheung HH, Mrad RL, Plenchette S, Simard C, Enwere E, Arora V, Mak TW, Lacasse EC, Waring J, Korneluk RG (2008) Both cIAP1 and cIAP2 regulate TNFα-mediated NF-κB activation. Proc Natl Acad Sci USA 105(33):11778–11783. https://doi.org/10.1073/pnas.0711122105

Majchrzak M, Poręba M (2022) The roles of cellular protease interactions in viral infections and programmed cell death: a lesson learned from the SARS-CoV-2 outbreak and COVID-19 pandemic. Pharmacol Rep 74(6):1149–1165. https://doi.org/10.1007/s43440-022-00394-9

Majumder M, Ghosh AK, Steele R, Zhou XY, Phillips NJ, Ray R, Ray RB (2002) Hepatitis C virus NS5A protein impairs TNF-mediated hepatic apoptosis, but not by an anti-FAS antibody, in transgenic mice. Virology 294(1):94–105. https://doi.org/10.1006/viro.2001.1309

Manaka J, Kuraishi T, Shiratsuchi A, Nakai Y, Higashida H, Henson P, Nakanishi Y (2004) Draper-mediated and phosphatidylserine-independent phagocytosis of apoptotic cells by Drosophila hemocytes/macrophages. J Biol Chem 279(46):48466–48476. https://doi.org/10.1074/jbc.M408597200

Marshall B, Puthalakath H, Caria S, Chugh S, Doerflinger M, Colman PM, Kvansakul M (2015) Variola virus F1L is a Bcl-2-like protein that unlike its vaccinia virus counterpart inhibits apoptosis independent of Bim. Cell Death Dis 6(3):e1680. https://doi.org/10.1038/cddis.2015.52

Marshall JS, Warrington R, Watson W, Kim HL (2018) An introduction to immunology and immunopathology. Allergy Asthma Clin Immunol 14(2):49. https://doi.org/10.1186/s13223-018-0278-1

Mateu MG (2013) Introduction: the structural basis of virus function. Subcell Biochem 68:3–51. https://doi.org/10.1007/978-94-007-6552-8_1

McArthur K, Kile BT (2018) Apoptotic caspases: multiple or mistaken identities? Trends Cell Biol 28(6):475–493. https://doi.org/10.1016/j.tcb.2018.02.003

McCormick AL, Skaletskaya A, Barry PA, Mocarski ES, Goldmacher VS (2003) Differential function and expression of the viral inhibitor of caspase 8-induced apoptosis (vICA) and the viral mitochondria-localized inhibitor of apoptosis (vMIA) cell death suppressors conserved in primate and rodent cytomegaloviruses. Virology 316(2):221–233. https://doi.org/10.1016/j.virol.2003.07.003

Medana IM, Gallimore A, Oxenius A, Martinic MM, Wekerle H, Neumann H (2000) MHC class I-restricted killing of neurons by virus-specific CD8+ T lymphocytes is effected through the Fas/FasL, but not the perforin pathway. Eur J Immunol 30(12):3623–3633. https://doi.org/10.1002/1521-4141(200012)30:12<3623::AID-IMMU3623>3.0.CO;2-F

Medici MA, Sciortino MT, Perri D, Amici C, Avitabile E, Ciotti M, Balestrieri E, De Smaele E, Franzoso G, Mastino A (2003) Protection by herpes simplex virus glycoprotein D against Fas-mediated apoptosis: role of nuclear factor kappaB. J Biol Chem 278(38):36059–36067. https://doi.org/10.1074/jbc.M306198200

Melcarne C, Lemaitre B, Kurant E (2019) Phagocytosis in Drosophila: From molecules and cellular machinery to physiology. Insect Biochem Mol Biol 109:1–12. https://doi.org/10.1016/j.ibmb.2019.04.002

Migone TS, Zhang J, Luo X, Zhuang L, Chen C, Hu B, Hong JS, Perry JW, Chen SF, Zhou JXH, Cho YH, Ullrich S, Kanakaraj P, Carrell J, Boyd E, Olsen HS, Hu G, Pukac L, Liu D, Ni J, Kim S, Gentz R, Feng P, Moore PA, Ruben SM, Wei P (2002) TL1A is a TNF-like ligand for DR3 and TR6/DcR3 and functions as a T cell costimulator. Immunity 16(3):479–492. https://doi.org/10.1016/s1074-7613(02)00283-2

Mita E, Hayashi N, Iio S, Takehara T, Hijioka T, Kasahara A, Fusamoto H, Kamada T (1994) Role of Fas ligand in apoptosis induced by hepatitis C virus infection. Biochem Biophys Res Commun 204(2):468–474. https://doi.org/10.1006/bbrc.1994.2483

Miura Y, Koyanagi Y, Mizusawa H (2003) TNF-related apoptosis-inducing ligand (TRAIL) induces neuronal apoptosis in HIV-encephalopathy. J Med Dent Sci 50(1):17–25

Mogensen TH (2009) Pathogen recognition and inflammatory signalling in innate immune defenses. Clin Microbiol Rev 22(2):240–273. Table of Contents. https://doi.org/10.1128/cmr.00046-08

Molteni CG, Principi N, Esposito S (2014) Reactive oxygen and nitrogen species during viral infections. Free Radic Res 48(10):1163–1169. https://doi.org/10.3109/10715762.2014.945443

Moreira CGA, Jacinto A, Prag S (2013) Drosophila integrin adhesion complexes are essential for hemocyte migration in vivo. Biol Open 2(8):795–801. https://doi.org/10.1242/bio.20134564

Mori T, Ando K, Tanaka K, Ikeda Y, Koga Y (1997) Fas-mediated apoptosis of the hematopoietic progenitor cells in mice infected with murine cytomegalovirus. Blood 89(10):3565–3573. https://doi.org/10.1182/blood.V89.10.3565

Morrison TE, Mauser A, Klingelhutz A, Kenney SC (2004) Epstein-Barr virus immediate-early protein BZLF1 inhibits tumor necrosis factor alpha-induced signalling and apoptosis by downregulating tumor necrosis factor receptor 1. J Virol 78(1):544–549. https://doi.org/10.1128/jvi.78.1.544-549.2004

Mueller YM, De Rosa SC, Hutton JA, Witek J, Roederer M, Altman JD, Katsikis PD (2001) Increased CD95/Fas-induced apoptosis of HIV-specific CD8(+) T cells. Immunity 15(6):871–882. https://doi.org/10.1016/s1074-7613(01)00246-1

Mussabekova A, Daeffler L, Imler JL (2017) Innate and intrinsic antiviral immunity in Drosophila. Cell Mol Life Sci 74(11):2039–2054. https://doi.org/10.1007/s00018-017-2453-9

Nainu F, Tanaka Y, Shiratsuchi A, Nakanishi Y (2015) Protection of insects against viral infection by apoptosis-dependent phagocytosis. J Immunol 195(12):5696–5706. https://doi.org/10.4049/jimmunol.1500613

Nainu F, Shiratsuchi A, Nakanishi Y (2017) Induction of apoptosis and subsequent phagocytosis of virus-infected cells as an antiviral mechanism. Front Immunol 8:1220. https://doi.org/10.3389/fimmu.2017.01220

Natoli G, Costanzo A, Guido F, Moretti F, Bernardo A, Burgio VL, Agresti C, Levrero M (1998) Nuclear factor kB-independent cytoprotective pathways originating at tumor necrosis factor receptor-associated factor 2. J Biol Chem 273(47):31262–31272. https://doi.org/10.1074/jbc.273.47.31262

Nava VE, Cheng EH, Veliuona M, Zou S, Clem RJ, Mayer ML, Hardwick JM (1997) Herpesvirus saimiri encodes a functional homolog of the human bcl-2 oncogene. J Virol 71(5):4118–4122. https://doi.org/10.1128/JVI.71.5.4118-4122.1997

Neufeldt CJ, Cerikan B, Cortese M, Frankish J, Lee J-Y, Plociennikowska A, Heigwer F, Prasad V, Joecks S, Burkart SS, Zander DY, Subramanian B, Gimi R, Padmanabhan S, Iyer R, Gendarme M, El Debs B, Halama N, Merle U, Boutros M, Binder M, Bartenschlager R (2022) SARS-CoV-2 infection induces a pro-inflammatory cytokine response through cGAS-STING and NF-κB. Commun Biol 5(1):45. https://doi.org/10.1038/s42003-021-02983-5

Nidadavolu LS, Walston JD (2021) Underlying vulnerabilities to the cytokine storm and adverse COVID-19 outcomes in the aging immune system. J Gerontol A Biol Sci Med Sci 76(3):e13–e18. https://doi.org/10.1093/gerona/glaa209

O'Donnell DR, Milligan L, Stark JM (1999) Induction of CD95 (Fas) and apoptosis in respiratory epithelial cell cultures following respiratory syncytial virus infection. Virology 257(1):198–207. https://doi.org/10.1006/viro.1999.9650

Okamoto K, Fujisawa J-I, Reth M, Yonehara S (2006) Human T-cell leukemia virus type-I oncoprotein Tax inhibits Fas-mediated apoptosis by inducing cellular FLIP through activation of NF-kappaB. Genes Cells 11(2):177–191. https://doi.org/10.1111/j.1365-2443.2006.00927.x

Okamoto T, Campbell S, Mehta N, Thibault J, Colman PM, Barry M, Huang DCS, Kvansakul M (2012) Sheeppox virus SPPV14 encodes a Bcl-2-like cell death inhibitor that counters a distinct set of mammalian proapoptotic proteins. J Virol 86(21):11501–11511. https://doi.org/10.1128/JVI.01115-12

Omasta B, Tomaskova J (2022) Cellular lipids-hijacked victims of viruses. Viruses 14(9). https://doi.org/10.3390/v14091896

Orzalli MH, Kagan JC (2017) Apoptosis and necroptosis as host defense strategies to prevent viral infection. Trends Cell Biol 27(11):800–809. https://doi.org/10.1016/j.tcb.2017.05.007

Paiva CN, Bozza MT (2014) Are reactive oxygen species always detrimental to pathogens? Antioxid Redox Signal 20(6):1000–1037. https://doi.org/10.1089/ars.2013.5447

Pan G, Bauer JH, Haridas V, Wang S, Liu D, Yu G, Vincenz C, Aggarwal BB, Ni J, Dixit VM (1998) Identification and functional characterization of DR6, a novel death domain-containing TNF receptor. FEBS Lett 431(3):351–356. https://doi.org/10.1016/s0014-5793(98)00791-1

Parrish AB, Freel CD, Kornbluth S (2013) Cellular mechanisms controlling caspase activation and function. Cold Spring Harb Perspect Biol 5(6). https://doi.org/10.1101/cshperspect.a008672

Pasieka TJ, Baas T, Carter VS, Proll SC, Katze MG, Leib DA (2006) Functional genomic analysis of herpes simplex virus type 1 counteraction of the host innate response. J Virol 80(15): 7600–7612. https://doi.org/10.1128/jvi.00333-06

Pawlowski J, Kraft AS (2000) Bax-induced apoptotic cell death. Proc Natl Acad Sci U S A 97(2): 529–531. https://doi.org/10.1073/pnas.97.2.529

Pereira CF, Middel J, Jansen G, Verhoef J, Nottet HS (2001) Enhanced expression of fractalkine in HIV-1 associated dementia. J Neuroimmunol 115(1-2):168–175. https://doi.org/10.1016/s0165-5728(01)00262-4

Pianko S, Patella S, Ostapowicz G, Desmond P, Sievert W (2001) Fas-mediated hepatocyte apoptosis is increased by hepatitis C virus infection and alcohol consumption, and may be associated with hepatic fibrosis: mechanisms of liver cell injury in chronic hepatitis C virus infection. J Viral Hepat 8(6):406–413. https://doi.org/10.1046/j.1365-2893.2001.00316.x

Pobezinskaya YL, Kim Y-S, Choksi S, Morgan MJ, Li T, Liu C, Liu Z (2008) The function of TRADD in signalling through tumor necrosis factor receptor 1 and TRIF-dependent Toll-like receptors. Nat Immunol 9(9):1047–1054. https://doi.org/10.1038/ni.1639

Pollard AJ, Bijker EM (2021) A guide to vaccinology: from basic principles to new developments. Nat Rev Immunol 21(2):83–100. https://doi.org/10.1038/s41577-020-00479-7

Rampersad S, Tennant P (2018) Chapter 3—Replication and expression strategies of viruses. In: Tennant P, Fermin G, Foster JE (eds) Viruses. Academic, Cambridge, pp 55–82. https://doi.org/10.1016/B978-0-12-811257-1.00003-6

Ravichandran KS (2003) "Recruitment signals" from apoptotic cells: invitation to a quiet meal. Cell 113(7):817–820. https://doi.org/10.1016/s0092-8674(03)00471-9

Rawat S, Bouchard MJ (2015) The hepatitis B virus (HBV) HBx protein activates AKT to simultaneously regulate HBV replication and hepatocyte survival. J Virol 89(2):999–1012. https://doi.org/10.1128/jvi.02440-14

Ray RB, Meyer K, Steele R, Shrivastava A, Aggarwal BB, Ray R (1998) Inhibition of tumor necrosis factor (TNF-alpha)-mediated apoptosis by hepatitis C virus core protein. J Biol Chem 273(4):2256–2259. https://doi.org/10.1074/jbc.273.4.2256

Ren Y, Shu T, Wu D, Mu J, Wang C, Huang M, Han Y, Zhang X-Y, Zhou W, Qiu Y, Zhou X (2020) The ORF3a protein of SARS-CoV-2 induces apoptosis in cells. Cell Mol Immunol 17(8):881–883. https://doi.org/10.1038/s41423-020-0485-9

Rex DAB, Keshava Prasad TS, Kandasamy RK (2022) Revisiting Regulated Cell Death Responses in Viral Infections. International Journal of Molecular Sciences 23(13):7023. https://doi.org/10.3390/ijms23137023

Ring BA, Ferreira Lacerda A, Drummond DJ, Wangen C, Eaton HE, Brunetti CR (2013) Frog virus 3 open reading frame 97R localizes to the endoplasmic reticulum and induces nuclear invaginations. J Virol 87(16):9199–9207. https://doi.org/10.1128/JVI.00637-13

Rivera-Walsh I, Waterfield M, Xiao G, Fong A, Sun SC (2001) NF-kappaB signalling pathway governs TRAIL gene expression and human T-cell leukemia virus-I Tax-induced T-cell death. J Biol Chem 276(44):40385–40388. https://doi.org/10.1074/jbc.C100501200

Saheb Sharif-Askari N, Saheb Sharif-Askari F, Mdkhana B, Al Heialy S, Alsafar HS, Hamoudi R, Hamid Q, Halwani R (2021) Enhanced expression of immune checkpoint receptors during SARS-CoV-2 viral infection. Mol Ther Methods Clin Dev 20:109–121. https://doi.org/10.1016/j.omtm.2020.11.002

Saraste A, Pulkki K (2000) Morphologic and biochemical hallmarks of apoptosis. Cardiovasc Res 45(3):528–537. https://doi.org/10.1016/s0008-6363(99)00384-3

Schneider-Brachert W, Tchikov V, Merkel O, Jakob M, Hallas C, Kruse M-L, Groitl P, Lehn A, Hildt E, Held-Feindt J, Dobner T, Kabelitz D, Krönke M, Schütze S (2006) Inhibition of TNF receptor 1 internalization by adenovirus 14.7K as a novel immune escape mechanism. J Clin Invest 116(11):2901–2913. https://doi.org/10.1172/JCI23771

Sedger LM, Shows DM, Blanton RA, Peschon JJ, Goodwin RG, Cosman D, Wiley SR (1999) IFN-gamma mediates a novel antiviral activity through dynamic modulation of TRAIL and TRAIL receptor expression. J Immunol 163(2):920–926

Seirafian S, Prod'homme V, Sugrue D, Davies J, Fielding C, Tomasec P, Wilkinson GWG (2014) Human cytomegalovirus suppresses Fas expression and function. J Gen Virol 95(Pt 4):933–939. https://doi.org/10.1099/vir.0.058313-0

Selvarajan RS, Gopinath SCB, Zin NM, Hamzah AA (2021) Infection-mediated clinical biomarkers for a COVID-19 electrical biosensing platform. Sensors 21(11):3829

Shapiro NI, Filbin MR, Hou PC, Kurz MC, Han JH, Aufderheide TP, Ward MA, Pulia MS, Birkhahn RH, Diaz JL, Hughes TL, Harsch MR, Bell A, Suarez-Cuervo C, Sambursky R (2022) Diagnostic accuracy of a bacterial and viral biomarker point-of-care test in the outpatient setting. JAMA Network Open 5(10):e2234588. https://doi.org/10.1001/jamanetworkopen.2022.34588

Shi Y (2004) Caspase activation, inhibition, and reactivation: a mechanistic view. Protein Sci 13(8): 1979–1987. https://doi.org/10.1110/ps.04789804

Shimada M, Yamashita A, Saito M, Ichino M, Kinjo T, Mizuki N, Klinman DM, Okuda K (2020) The human papillomavirus E6 protein targets apoptosis-inducing factor (AIF) for degradation. Sci Rep 10(1):14195. https://doi.org/10.1038/s41598-020-71134-3

Shiratsuchi A, Kaido M, Takizawa T, Nakanishi Y (2000) Phosphatidylserine-mediated phagocytosis of influenza A virus-infected cells by mouse peritoneal macrophages. J Virol 74(19): 9240–9244. https://doi.org/10.1128/jvi.74.19.9240-9244.2000

Shisler J, Yang C, Walter B, Ware CF, Gooding LR (1997) The adenovirus E3-10.4K/14.5K complex mediates loss of cell surface Fas (CD95) and resistance to Fas-induced apoptosis. J Virol 71(11):8299–8306. https://doi.org/10.1128/JVI.71.11.8299-8306.1997

Silke J, Meier P (2013) Inhibitor of apoptosis (IAP) proteins-modulators of cell death and inflammation. Cold Spring Harb Perspect Biol 5(2). https://doi.org/10.1101/cshperspect.a008730

Simmons RA, Willberg CB, Paul K (2013) Immune evasion by viruses. eLS Wiley, Chichester. https://doi.org/10.1002/9780470015902.a0024790

Skaletskaya A, Bartle LM, Chittenden T, McCormick AL, Mocarski ES, Goldmacher VS (2001) A cytomegalovirus-encoded inhibitor of apoptosis that suppresses caspase-8 activation. Proc Natl Acad Sci U S A 98(14):7829–7834. https://doi.org/10.1073/pnas.141108798

Snow AL, Lambert SL, Natkunam Y, Esquivel CO, Krams SM, Martinez OM (2006) EBV can protect latently infected B cell lymphomas from death receptor-induced apoptosis. J Immunol 177(5):3283–3293. https://doi.org/10.4049/jimmunol.177.5.3283

Souza BSF, Sampaio GLA, Pereira CS, Campos GS, Sardi SI, Freitas LAR, Figueira CP, Paredes BD, Nonaka CKV, Azevedo CM, Rocha VPC, Bandeira AC, Mendez-Otero R, Dos Santos RR, Soares MBP (2016) Zika virus infection induces mitosis abnormalities and apoptotic cell death of human neural progenitor cells. Sci Rep 6:39775. https://doi.org/10.1038/srep39775

Speth C, Dierich MP (1999) Modulation of cell surface protein expression by infection with HIV-1. Leukemia 13(Suppl 1):S99–S105. https://doi.org/10.1038/sj.leu.2401322

Südhof TC, Rizo J (2011) Synaptic vesicle exocytosis. Cold Spring Harbor Perspect Biol 3(12). https://doi.org/10.1101/cshperspect.a005637

Sugden SM, Bego MG, Pham TNQ, Cohen ÉA (2016) Remodeling of the host cell plasma membrane by HIV-1 Nef and Vpu: a strategy to ensure viral fitness and persistence. Viruses 8(3):67

Sumbria D, Berber E, Mathayan M, Rouse BT (2020) Virus infections and host metabolism—can we manage the interactions? Front Immunol 11:594963. https://doi.org/10.3389/fimmu.2020.594963

Sun X, Whittaker GR (2003) Role for influenza virus envelope cholesterol in virus entry and infection. J Virol 77(23):12543–12551. https://doi.org/10.1128/jvi.77.23.12543-12551.2003

Tait SWG, Green DR (2010) Mitochondria and cell death: outer membrane permeabilization and beyond. Nat Rev Mol Cell Biol 11(9):621–632. https://doi.org/10.1038/nrm2952

Takizawa T, Matsukawa S, Higuchi Y, Nakamura S, Nakanishi Y, Fukuda R (1993) Induction of programmed cell death (apoptosis) by influenza virus infection in tissue culture cells. J Gen Virol 74(Pt 11):2347–2355. https://doi.org/10.1099/0022-1317-74-11-2347

Tanner JE, Alfieri C (1999) Epstein-Barr virus induces Fas (CD95) in T cells and Fas ligand in B cells leading to T-cell apoptosis. Blood 94(10):3439–3447

Tay MZ, Wiehe K, Pollara J (2019) Antibody-dependent cellular phagocytosis in antiviral immune responses. Front Immunol 10:332. https://doi.org/10.3389/fimmu.2019.00332

Taylor JM, Quilty D, Banadyga L, Barry M (2006) The vaccinia virus protein F1L interacts with Bim and inhibits activation of the pro-apoptotic protein Bax. J Biol Chem 281(51): 39728–39739. https://doi.org/10.1074/jbc.M607465200

Taylor RC, Cullen SP, Martin SJ (2008) Apoptosis: controlled demolition at the cellular level. Nat Rev Mol Cell Biol 9(3):231–241. https://doi.org/10.1038/nrm2312

Thai M, Graham NA, Braas D, Nehil M, Komisopoulou E, Kurdistani SK, McCormick F, Graeber TG, Christofk HR (2014) Adenovirus E4ORF1-induced MYC activation promotes host cell anabolic glucose metabolism and virus replication. Cell Metab 19(4):694–701. https://doi.org/10.1016/j.cmet.2014.03.009

Thaker SK, Ch'ng J, Christofk HR (2019) Viral hijacking of cellular metabolism. BMC Biol 17(1): 59. https://doi.org/10.1186/s12915-019-0678-9

Thome M, Schneider P, Hofmann K, Fickenscher H, Meinl E, Neipel F, Mattmann C, Burns K, Bodmer JL, Schröter M, Scaffidi C, Krammer PH, Peter ME, Tschopp J (1997) Viral FLICE-inhibitory proteins (FLIPs) prevent apoptosis induced by death receptors. Nature 386(6624): 517–521. https://doi.org/10.1038/386517a0

Thompson MR, Kaminski JJ, Kurt-Jones EA, Fitzgerald KA (2011) Pattern recognition receptors and the innate immune response to viral infection. Viruses 3(6):920–940. https://doi.org/10.3390/v3060920

Thomson BJ (2001) Viruses and apoptosis. Int J Exp Pathol 82(2):65–76. https://doi.org/10.1111/j.1365-2613.2001.iep0082-0065-x

Tisoncik JR, Korth MJ, Simmons CP, Farrar J, Martin TR, Katze MG (2012) Into the eye of the cytokine storm. Microbiol Mol Biol Rev 76(1):16–32. https://doi.org/10.1128/mmbr.05015-11

Tsao YT, Tsai YH, Liao WT, Shen CJ, Shen CF, Cheng CM (2020) Differential markers of bacterial and viral infections in children for point-of-care testing. Trends Mol Med 26(12): 1118–1132. https://doi.org/10.1016/j.molmed.2020.09.004

Turpin J, El Safadi D, Lebeau G, Krejbich M, Chatelain C, Desprès P, Viranaïcken W, Krejbich-Trotot P (2022) Apoptosis during ZIKA virus infection: too soon or too late? Int J Mol Sci 23(3). https://doi.org/10.3390/ijms23031287

Upton JW, Chan FK (2014) Staying alive: cell death in antiviral immunity. Mol Cell 54(2): 273–280. https://doi.org/10.1016/j.molcel.2014.01.027

Uribe-Querol E, Rosales C (2020) Phagocytosis: our current understanding of a universal biological process. Front Immunol 11:1066. https://doi.org/10.3389/fimmu.2020.01066

Vakil MK, Mansoori Y, Al-Awsi GRL, Hosseinipour A, Ahsant S, Ahmadi S, Ekrahi M, Montaseri Z, Pezeshki B, Mohaghegh P, Sohrabpour M, Bahmanyar M, Daraei A, Dadkhah Jouybari T, Tavassoli A, Ghasemian A (2022) Individual genetic variability mainly of proinflammatory cytokines, cytokine receptors, and toll-like receptors dictates pathophysiology of COVID-19 disease. J Med Virol 94(9):4088–4096. https://doi.org/10.1002/jmv.27849

van Grevenynghe J, Cubas RA, Noto A, DaFonseca S, He Z, Peretz Y, Filali-Mouhim A, Dupuy FP, Procopio FA, Chomont N, Balderas RS, Said EA, Boulassel M-R, Tremblay CL, Routy J-P, Sékaly R-P, Haddad EK (2011) Loss of memory B cells during chronic HIV infection is driven by Foxo3a- and TRAIL-mediated apoptosis. J Clin Invest 121(10):3877–3888. https://doi.org/10.1172/JCI59211

Verburg SG, Lelievre RM, Westerveld MJ, Inkol JM, Sun YL, Workenhe ST (2022) Viral-mediated activation and inhibition of programmed cell death. PLoS Pathog 18(8):e1010718. https://doi.org/10.1371/journal.ppat.1010718

Villanueva RA, Rouillé Y, Dubuisson J (2005) Interactions between virus proteins and host cell membranes during the viral life cycle. Int Rev Cytol 245:171–244. https://doi.org/10.1016/s0074-7696(05)45006-8

Vince JE, De Nardo D, Gao W, Vince AJ, Hall C, McArthur K, Simpson D, Vijayaraj S, Lindqvist LM, Bouillet P, Rizzacasa MA, Man SM, Silke J, Masters SL, Lessene G, Huang DCS, Gray DHD, Kile BT, Shao F, Lawlor KE (2018) The mitochondrial apoptotic effectors BAX/BAK activate caspase-3 and -7 to trigger NLRP3 inflammasome and caspase-8 driven IL-1β activation. Cell Rep 25(9):2339–2353.e2334. https://doi.org/10.1016/j.celrep.2018.10.103

Votteler J, Sundquist WI (2013) Virus budding and the ESCRT pathway. Cell Host Microbe 14(3): 232–241. https://doi.org/10.1016/j.chom.2013.08.012

Wada N, Matsumura M, Ohba Y, Kobayashi N, Takizawa T, Nakanishi Y (1995) Transcription stimulation of the Fas-encoding gene by nuclear factor for interleukin-6 expression upon influenza virus infection. J Biol Chem 270(30):18007–18012. https://doi.org/10.1074/jbc.270.30.18007

Walsh D, Mohr I (2011) Viral subversion of the host protein synthesis machinery. Nat Rev Microbiol 9(12):860–875. https://doi.org/10.1038/nrmicro2655

Walsh D, Mathews MB, Mohr I (2013) Tinkering with translation: protein synthesis in virus-infected cells. Cold Spring Harb Perspect Biol 5(1):a012351. https://doi.org/10.1101/cshperspect.a012351

Wang S, El-Deiry WS (2003) TRAIL and apoptosis induction by TNF-family death receptors. Oncogene 22(53):8628–8633. https://doi.org/10.1038/sj.onc.1207232

Wang G-H, Garvey TL, Cohen JI (1999) The murine gammaherpesvirus-68 M11 protein inhibits Fas- and TNF-induced apoptosis. J Gen Virol 80(Pt 10):2737–2740. https://doi.org/10.1099/0022-1317-80-10-2737

Wang G, Barrett JW, Nazarian SH, Everett H, Gao X, Bleackley C, Colwill K, Moran MF, McFadden G (2004) Myxoma virus M11L prevents apoptosis through constitutive interaction with Bak. J Virol 78(13):7097–7111. https://doi.org/10.1128/JVI.78.13.7097-7111.2004

Wang W, Fang Y, Sima N, Li Y, Li W, Li L, Han L, Liao S, Han Z, Gao Q, Li K, Deng D, Meng L, Zhou J, Wang S, Ma D (2011) Triggering of death receptor apoptotic signalling by human papillomavirus 16 E2 protein in cervical cancer cell lines is mediated by interaction with c-FLIP. Apoptosis 16(1):55–66. https://doi.org/10.1007/s10495-010-0543-3

Ward TH, Cummings J, Dean E, Greystoke A, Hou JM, Backen A, Ranson M, Dive C (2008) Biomarkers of apoptosis. Br J Cancer 99(6):841–846. https://doi.org/10.1038/sj.bjc.6604519

Watanabe Y, Shiratsuchi A, Shimizu K, Takizawa T, Nakanishi Y (2002) Role of phosphatidylserine exposure and sugar chain desialylation at the surface of influenza virus-infected cells in efficient phagocytosis by macrophages. J Biol Chem 277(20):18222–18228. https://doi.org/10.1074/jbc.M201074200

Westphal D, Ledgerwood EC, Tyndall JDA, Hibma MH, Ueda N, Fleming SB, Mercer AA (2009) The orf virus inhibitor of apoptosis functions in a Bcl-2-like manner, binding and neutralizing a set of BH3-only proteins and active Bax. Apoptosis 14(11):1317–1330. https://doi.org/10.1007/s10495-009-0403-1

Wheatley SP, Altieri DC (2019) Survivin at a glance. J Cell Sci 132(7). https://doi.org/10.1242/jcs.223826

White E, Blose SH, Stillman BW (1984) Nuclear envelope localization of an adenovirus tumor antigen maintains the integrity of cellular DNA. Mol Cell Biol 4(12):2865–2875. https://doi.org/10.1128/mcb.4.12.2865-2875.1984

Wilson AD, Redchenko I, Williams NA, Morgan AJ (1998) CD4+ T cells inhibit growth of Epstein-Barr virus-transformed B cells through CD95-CD95 ligand-mediated apoptosis. Int Immunol 10(8):1149–1157. https://doi.org/10.1093/intimm/10.8.1149

Xu J, Cherry S (2014) Viruses and antiviral immunity in Drosophila. Dev Comp Immunol 42(1): 67–84. https://doi.org/10.1016/j.dci.2013.05.002

Xu D, Woodfield SE, Lee TV, Fan Y, Antonio C, Bergmann A (2009) Genetic control of programmed cell death (apoptosis) in Drosophila. Fly (Austin) 3(1):78–90. https://doi.org/10.4161/fly.3.1.7800

Yang R, Zhao Q, Rao J, Zeng F, Yuan S, Ji M, Sun X, Li J, Yang J, Cui J, Jin Z, Liu L, Liu Z (2021) SARS-CoV-2 accessory protein ORF7b mediates tumor necrosis factor-α-induced apoptosis in cells. Front Microbiol 12:654709. https://doi.org/10.3389/fmicb.2021.654709

Yoon K, Jeong JG, Kim S (2001) Stable expression of human immunodeficiency virus type 1 Nef confers resistance against Fas-mediated apoptosis. AIDS Res Hum Retroviruses 17(2):99–104. https://doi.org/10.1089/08892220150217184

York IA, Stevens J, Alymova IV (2019) Influenza virus N-linked glycosylation and innate immunity. Biosci Rep 39(1). https://doi.org/10.1042/bsr20171505

Zajac AJ, Quinn DG, Cohen PL, Frelinger JA (1996) Fas-dependent CD4+ cytotoxic T-cell-mediated pathogenesis during virus infection. Proc Natl Acad Sci U S A 93(25): 14730–14735. https://doi.org/10.1073/pnas.93.25.14730

Zhang H, Huang C, Wang Y, Lu Z, Zhuang N, Zhao D, He J, Shi L (2015) Hepatitis B virus X protein sensitizes TRAIL-induced hepatocyte apoptosis by inhibiting the E3 ubiquitin ligase A20. PLoS One 10(5):e0127329. https://doi.org/10.1371/journal.pone.0127329

Zheng Q, Ma A, Yuan L, Gao N, Feng Q, Franc NC, Xiao H (2017) Apoptotic cell clearance in Drosophila melanogaster. Front Immunol 8:1881. https://doi.org/10.3389/fimmu.2017.01881

Zhu F, Zhang X (2013) The Wnt signalling pathway is involved in the regulation of phagocytosis of virus in Drosophila. Sci Rep 3(1):2069. https://doi.org/10.1038/srep02069

# Chapter 4
# The Art of Viral Membrane Fusion and Penetration

**Sophie L. Winter and Petr Chlanda**

**Abstract** As obligate pathogens, viruses have developed diverse mechanisms to deliver their genome across host cell membranes to sites of virus replication. While enveloped viruses utilize viral fusion proteins to accomplish fusion of their envelope with the cellular membrane, non-enveloped viruses rely on machinery that causes local membrane ruptures and creates an opening through which the capsid or viral genome is released. Both membrane fusion and membrane penetration take place at the plasma membrane or in intracellular compartments, often involving the engagement of the cellular machinery and antagonism of host restriction factors. Enveloped and non-enveloped viruses have evolved intricate mechanisms to enable virus uncoating and modulation of membrane fusion in a spatiotemporally controlled manner. This chapter summarizes and discusses the current state of understanding of the mechanisms of viral membrane fusion and penetration. The focus is on the role of lipids, viral scaffold uncoating, viral membrane fusion inhibitors, and host restriction factors as physicochemical modulators. In addition, recent advances in visualizing and detecting viral membrane fusion and penetration using cryo-electron microscopy methods are presented.

**Keywords** Enveloped viruses · Non-enveloped viruses · Membrane fusion · Fusion pore · Virus uncoating · Virus penetration · Virus scaffold · Cholesterol · Cryo-EM

## General Terminology and Mechanisms of How Viruses Cross Host Membranes

Membrane bilayers, comprising lipids and (glyco)proteins, constitute interacting barriers that effectively organize molecular processes in different cellular compartments. They provide a platform for protein and lipid interactions while spatially

S. L. Winter · P. Chlanda (✉)
Schaller Research Group, Department of Infectious Diseases, Virology, Heidelberg University Hospital, Heidelberg, Germany
e-mail: sophie.winter@bioquant.uni-heidelberg.de; petr.chlanda@bioquant.uni-heidelberg.de

 
S. Vijayakrishnan et al. (eds.), *Virus Infected Cells*, Subcellular Biochemistry 106,
https://doi.org/10.1007/978-3-031-40086-5_4

separating cellular processes from one another. Thus, it is not surprising that viruses have evolved to utilize membranes and membrane-bound compartments as sites to replicate their genomes and assemble their progeny. To initiate an infection, viruses need to enter host cells by crossing the host cell membrane and releasing their genome into the cytosol. To this end, they have evolved several mechanisms based on the architecture of the virus particles (***virion***) and particularly the organization of their outer protective coat: ***enveloped viruses*** are surrounded by a lipid bilayer acquired during the assembly and budding of viral progeny on the host cell membranes, whereas ***non-enveloped viruses***, sometimes referred to as naked virions, protect their genome with a proteinaceous capsid that confers stability and renders them resistant to harsh environments.

Enveloped viruses comprise particles of various shapes and sizes, in some cases even within the same viral population. Their large membrane surfaces typically harbor many viral glycoproteins whose main function is to attach to the host cell and carry out the fusion between the viral envelope and the host membrane. Viral fusion proteins are assembled and incorporated into viral particles in a metastable conformation and often require proteolytic cleavage of precursor fusion proteins referred to as ***priming***. Upon entry into host cells, the virions then encounter various molecular cues such as receptor binding or acidic pH, which trigger conformational rearrangements of the fusion protein into energetically more favorable states. The fusion peptides become exposed and inserted into the host membrane, thereby bridging the viral and host membrane. Subsequently, the fusion machinery pulls the two membranes together until they eventually mix, and the membranes fuse to form a fusion pore through which the viral genome is released.

As opposed to enveloped viruses, non-enveloped viruses do not have a lipid envelope and rely on intricate penetration machinery to cause local membrane ruptures, which allow either the entire capsid or the viral genome to enter the cytosol. Similar to the fusion machinery of enveloped viruses, proteolytical priming of the membrane penetration machinery and additional molecular cues (such as receptor binding and low pH) are often required to render non-enveloped virions infectious. Despite their less complex structural composition compared to enveloped virions, membrane penetration remains poorly characterized for many non-enveloped viruses and only a few structural studies on any penetration machinery exist.

Membrane fusion and penetration are followed by the release of the viral genome into the cytoplasm of infected cells. To this end, scaffold-forming proteins (in enveloped viruses) and capsids (in non-enveloped viruses) undergo conformational changes leading to global changes in the viral architecture and compromising the virions' integrity. This process is part of the ***virus uncoating***, which typically consists of several spatially and temporally controlled steps initiated prior to membrane fusion or penetration and completed once the free genome is available for transcription and replication. Virus entry into the host cell is a complex process encompassing virion binding, a crossing of the host membrane, and delivery of the viral genome to sites of virus replication. Some viruses are able to directly cross the plasma membrane, but the majority of viruses evolved to hitchhike the endocytic pathway and cross endosomal membranes to release their genome. This is sometimes referred to as ***viral endocytic escape*** and includes viral uncoating and membrane fusion or membrane penetration. Differences and similarities of enveloped versus non-enveloped endosomal escape including the terminology currently used in the literature are summarized in Fig. 4.1.

## Viral Membrane Fusion and Membrane Fusion Proteins

Membrane fusion is an essential cellular process, which likely evolved concomitantly with the development of the first membrane-surrounded cells, and greatly diversified to support compartmentalization in eukaryotic cells (reviewed by Martens and McMahon 2008). Governed by the hydrophobic effect, phospholipids assemble into bilayers to minimize the exposure of hydrophobic acyl chains to an aqueous solution. Lipid bilayers inherently form closed compartments that have negatively charged surfaces due to the phosphate groups present on the phospholipids. Hence, substantial energy is required to overcome the repulsion of the surfaces, bring two membranous compartments in proximity, and merge them without causing a rupture and leakage. Membrane fusion occurs through an intermediate called ***hemifusion*** at which the proximal monolayers are fused, but the distant monolayers remain separated. At this stage, lipids from the viral and target membrane can move from one to the other in a process called ***lipid demixing***. Hemifusion can assume different geometrical conformations, called either hemifusion stalk or hemifusion diaphragm (Chernomordik and Kozlov 2008). While the hemifusion stalk can proceed to fusion pore formation, an extended hemifusion diaphragm can represent a dead-end product which does not proceed to pore formation due to a high energy penalty (Chlanda et al. 2016).

Fusion proteins have evolved as molecular machines that upon refolding can bring two membranes in proximity and form a fusion pore that allows for lipid and content mixing. Perhaps, the most striking examples of cellular fusion proteins are Syncytin-2 implicated in placenta formation (Esnault et al. 2008) and the hapless 2 proteins (HAP2) (Fedry et al. 2017), which are responsible for membrane fusion of male and female gametes. Remarkably, both proteins are structural homologues to viral fusion proteins, raising the question of whether these proteins are ancestral to viral fusion proteins or vice versa (Doms 2017). Interestingly, membrane fusion proteins are not only limited to viruses infecting eukaryotes but are also found in bacteriophages such as the membrane-enveloped bacteriophage φ6 (Bamford et al. 1987) or haloarchaeal pleomorphic virus from the *Pleolipoviridae* family (El Omari et al. 2019). Viral membrane fusion proteins are in most cases initially synthesized in non-fusogenic forms and must be primed by proteolytic cleavage, which occurs either at or en route to the plasma membrane. Primed fusion proteins are metastable and undergo conformational changes upon triggering during host cell entry to assume a more stable, low-energy state. These conformational changes involve the formation of an extended intermediate structure that exposes a fusion peptide anchoring to the target membrane. The amphipathic fusion peptides form helices

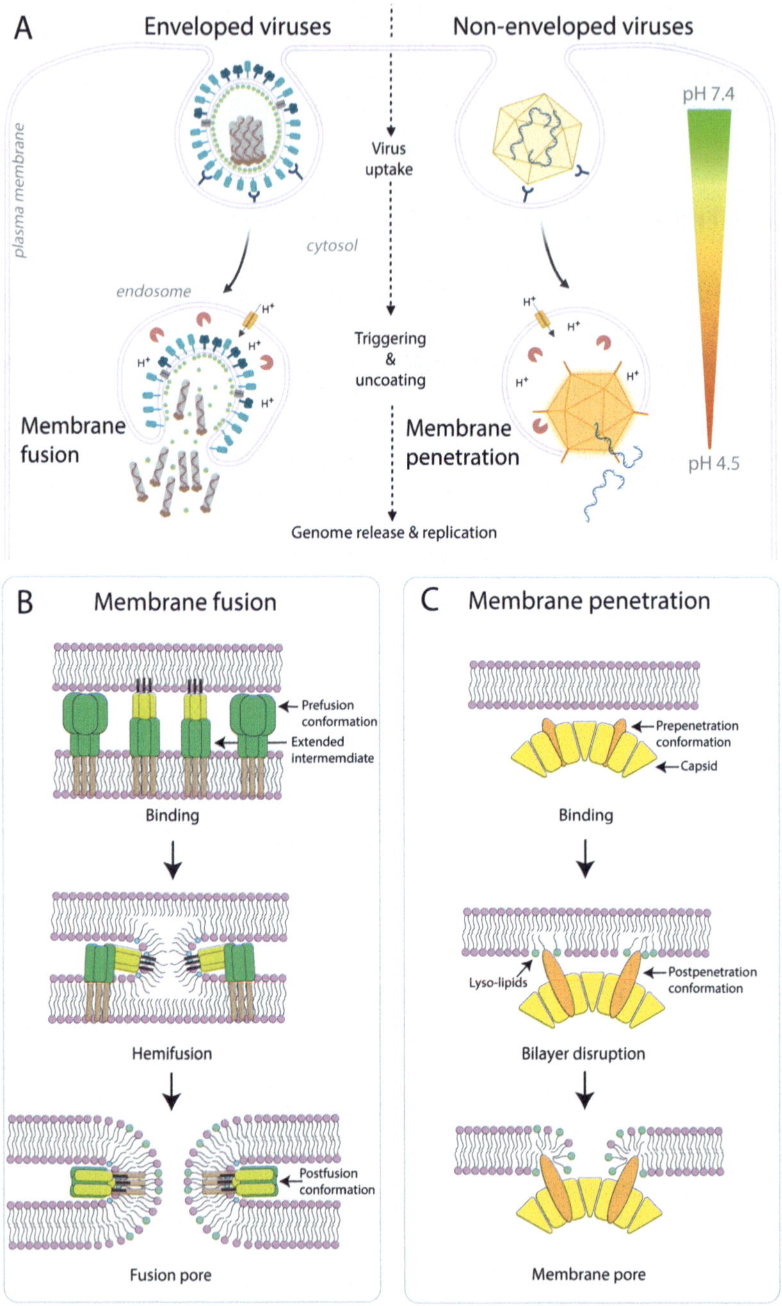


**Fig. 4.1** Membrane penetration mechanisms of enveloped and non-enveloped viruses. (**a**) Endosomal escape of enveloped and non-enveloped viruses. After cell attachment and entry by endocytosis, viruses are exposed to specific cues such as changes in pH, which (1) initiate a cascade of structural changes such as the weakening of capsid and nucleocapsid protomer interactions in a

or loops that upon insertion into the target membrane locally disrupt the bilayer architecture leading to the exposure of phospholipid acyl chains. The exerted work, which equals the released energy gained after assuming the low-energy postfusion conformational state, is invested into pulling the viral and host membrane together. Recent molecular dynamics simulations on the SARS-CoV-2 spike showed that the bound fusion peptide can withstand forces up to 250 pN before detaching from the target membrane (Schaefer et al. 2021). Subsequently, the extended intermediate folds back to bring the membranes to proximity below 1–2 nm (Harrison 2015). Based on the architecture and evolution, so far characterized viral fusion proteins have been classified into three classes (see Table 4.1), which mechanistically perform a similar task reviewed in detail by Rey and Lok (2018).

***Class I viral fusion proteins*** are non-covalently linked trimers in both prefusion conformation and postfusion conformation. They are synthesized as a single protein in the ER and trafficked to the plasma membrane via the Golgi, where they are cleaved into two subunits that remain linked in most cases by a disulfide bridge. One of the subunits entails a transmembrane domain and a fusion peptide, whereas the second subunit serves as the receptor binding domain and is often heavily glycosylated. In prefusion conformation, the amphipathic fusion peptides are localized in the trimeric interface in proximity to the viral membrane. After activation by low pH, the fusion peptides extend toward the target membrane and are inserted, thereby forming an extended intermediate. They then fold back by the formation of antiparallel α-helical coiled-coil assemblies composed of so-called heptad repeat domain 1 (HRD1) and HRD2 containing non-polar amino acids. Since the class I fusion proteins are trimeric, three pairs of antiparallel α-helices assemble into a so-called six-helix bundle in the postfusion conformation of the fusion protein (Kielian and Rey 2006).

**Fig. 4.1** (continued) process called viral uncoating and (2) activate the membrane fusion or penetration machinery (triggering). In both enveloped and non-enveloped viruses, these steps lead to the liberation of the viral genome into the cytosol followed by viral replication. (**b**) Membrane fusion is driven by enveloped viruses. A fusion protein in prefusion conformation binds to a receptor present on the host membrane. Upon triggering by low pH, proteolytic processing, or receptor binding, the prefusion spike protein rearranges its conformation to form an extended intermediate. The extended intermediate exposes an amphipathic fusion peptide that inserts into and disrupts the host membrane. Subsequently, the fusion protein folds back into a low-energy postfusion conformation that causes the formation of a hemifusion intermediate and ultimately leads to a fusion pore opening. (**c**) Membrane penetration by non-enveloped viruses. Non-enveloped viruses bind to a membrane using exposed binding domains on the surface of their proteinaceous capsid. Once internalized and upon triggering by low pH, proteolytic cleavage, or receptor binding, the viral capsid changes its conformation, leading to exposure of hydrophobic, amphipathic, or lipidated peptides (orange), which insert deeply into the host bilayer. Some non-enveloped viruses such as parvoviruses harbor phospholipases and thereby convert phospholipids to lysophospholipids (green lipids). Since lysolipids contain only one acyl chain and are able to form micelles, they play an important role in membrane lysis and increase the stability of open membrane pores. Image **a** was generated with BioRender under a personal license

**Table 4.1** List of enveloped viruses and properties of their membrane fusion machinery

| Virus family (example and reference) | Virus structure | Viral fusion machinery trigger | Membrane fusion site | Class | Prefusion conformation | Postfusion conformation |
|---|---|---|---|---|---|---|
| *Arenaviridae* (Lassa virus (York et al. 2008), lymphocytic choriomeningitis virus (Hastie et al. 2016)) | Pleomorphic | Low pH | Late endosome | I | Homotrimer | Homotrimer |
| Coronaviridae (SARS-CoV-2 (Tang et al. 2020)) | Spherical | Receptor binding | Plasma membrane, late endosome | | | |
| *Filoviridae* (Ebola virus (Das et al. 2020; Lee et al. 2008)) | Filamentous | Receptor binding, low pH, $Ca^{2+}$ | Late endosome-lysosome | | | |
| *Orthomyxoviridae* (Influenza A virus (Benton et al. 2020; Bullough et al. 1994)) | Pleomorphic-filamentous | Low pH | Late endosome | | | |
| *Paramyxoviridae* (Measles virus (Hashiguchi et al. 2018; Plattet et al. 2016)) | Pleomorphic-filamentous | Receptor binding<br>Low pH | Late endosome | | | |
| *Pneumoviridae* (Respiratory syncytial virus (Battles et al. 2016; Gilman et al. 2019; Krzyzaniak et al. 2013)) | Pleiomorphic | Receptor binding, pH-independent | Early endosome | | | |
| *Retroviridae* (Human immunodeficiency virus-1 (Chen 2019; Mangala Prasad et al. 2022b)) | Spherical | Receptor binding | Plasma membrane | | | |

| | | | | | | |
|---|---|---|---|---|---|---|
| *Flaviviridae*<br>(Dengue virus (Kaufmann et al. 2009), Zika virus (Sirohi et al. 2016), Binjari virus (Newton et al. 2021), hepatitis C virus (Torrents de la Pena et al. 2022)) | Icosahedral | Low pH | Late endosome | II | Homodimer | Homotrimer |
| *Togaviridae*<br>(Chikungunya virus (Mangala Prasad et al. 2022a)) | Icosahedral | Low pH | Late endosome | | Homodimer | |
| *Bunyaviridae*<br>(Hantavirus (Serris et al. 2020)) | Spherical, 4-fold symmetry-ordered glycoproteins | Low pH | Late endosomes | | Heterodimer | |
| *Bornaviridae*<br>(Borna disease virus (Bajramovic et al. 2003; Garry and Garry 2009)) | Spherical | Low pH | Late endosomes | III | Structure unknown | Structure unknown |
| *Herpesviridae*<br>(Herpes simplex virus-1<br>(Vollmer and Grunewald 2020; Vollmer et al. 2020)) | Spherical | Receptor binding | Plasma membrane, endosomes | III | Homotrimer | Homotrimer |
| *Rhabdoviridae*<br>(Vesicular stomatitis virus, rabies virus (Belot et al. 2019)) | Bullet shape | Low pH | Late endosomes | | Homotrimer | Homotrimer |
| *Baculoviridae*<br>(Baculovirus (Kadlec et al. 2008)) | Rod shape<br>Pleomorphic | Low pH | Late endosomes | | Homotrimer (structure unknown) | Homotrimer (gp64) |
| *Asfarviridae*<br>(African swine fever virus (Matamoros et al. 2020)) | Spherical, icosahedral | Low pH | Late endosomes | Unclassified | Structure unknown | Structure unknown |
| *Hepadnaviridae*<br>(Hepatitis B virus (Liu et al. 2022; Seitz et al. 2020)) | Spherical | Unknown, pH-independent | Endosomes | Unclassified | Structure unknown | Structure unknown |
| *Poxviridae*<br>(Vaccinia virus (Laliberte et al. 2011)) | Brick-shaped | Low pH | Late endosomes | Unclassified | Structure unknown | Structure unknown |

***Class II viral fusion proteins*** form homodimers that assemble icosahedral or highly ordered coats around the viral envelope. They function similarly to class I fusion proteins; however, in contrast to class I fusion proteins, they are mainly composed of β-sheets and do not contain long central α-helices. Upon activation, they first form a trimeric extended intermediate which folds back by domain rearrangement rather than by an extensive refolding of secondary structures (Modis 2014). In addition, class II proteins use fusion loops in contrast to α-helical fusion peptides typically utilized in class I fusion proteins.

***Class III viral fusion proteins*** share distinct structural characteristics with class I and class II proteins and utilize bipartite fusion loops (Albertini et al. 2012). While class III viral fusion proteins form trimers in prefusion and postfusion conformation as found in class I fusion proteins, they do not require proteolytic cleavage for activation. Members of the same viral family typically rely on the same class of fusion proteins, pointing toward a common ancestor of the families. An interesting exception is represented by Thogotoviruses, which belong to the family of *Orthomyxovididae*. Unlike influenza A viruses which harbor the class I fusion protein hemagglutinin (HA), Thogotoviruses express a class III viral fusion protein (Peng et al. 2017).

Overall, there is more information known about the viral fusion machinery of RNA viruses than of DNA viruses whose machinery is more complex. The fusion machinery of RNA viruses such as influenza A viruses, human coronaviruses, and filoviruses is composed of a single protein, "single-fusers." On the other hand, paramyxoviruses rely on more complex fusion machinery composed of two proteins, "tandem-fusers." DNA viruses typically rely on highly complex fusion machinery composed of multiple protein components, which impart a timely execution and regulation of membrane fusion. Perhaps the best characterized viral fusion machinery of DNA viruses is the fusion machinery of herpesviruses, which rely on glycoprotein B (gB) and the heterodimer of gH/gL constituting the core fusion machinery (Vollmer et al. 2020). Additional viral proteins specific to herpesvirus subfamilies are involved in regulation and binding (Vollmer and Grunewald 2020). Even more sophisticated machinery is utilized by poxviruses, which rely on a complex of 11 proteins (Laliberte et al. 2011). It remains to be seen how these proteins orchestrate membrane fusion. Vaccinia virus A26 is an important viral fusion protein controlling membrane fusion in the late endosome. However, in the absence of A26 membrane fusion occurs at the plasma membrane (Chang et al. 2019), indicating that A26 functions as a membrane fusion suppressor at the plasma membrane. Recent studies show that the African swine fever virus, which like the poxviruses belongs to the order of nucleocytoplasmic large DNA viruses (NCLDV), shares similarities with the poxviral fusion machinery (Matamoros et al. 2020).

## Conformational States of Viral Fusion Proteins

Influenza A virus HA has been used as a model system to understand the structure and dynamics of conformational changes of class I fusion proteins. HA is produced as a precursor (HA0), before being cleaved at or en route to the plasma membrane into the subunits HA1 and HA2, which remain linked by disulfide bonds. Upon entry, low pH encountered by the virions in late endosomes induces drastic conformational changes of the fusion protein leading to several intermediate structures. As these conformational changes occur rapidly, the existence of extended intermediates was experimentally confirmed by single-molecule Förster resonance energy transfer (FRET) (Das et al. 2018). To this end, click chemistry was employed to attach one fluorophore close to the fusion peptide and the second fluorophore within the D helix of HA2. By monitoring FRET, the dynamic landscape of HA conformational changes at low pH was determined. These data showed that initial conformational changes leading to the formation of the extended HA intermediate can be reversed by changing the pH back to neutral. This confirms earlier results from fusion inactivation assays, in which purified influenza A virions were subjected to low pH for a short time prior to re-neutralization and performing fusion assays (Leikina et al. 2002). The conformational change leading to the extended intermediate becomes irreversible after extended incubation in acidic pH and the presence of sialic acid receptors.

Recent advancements in fluorescence-based assays allow studying viral fusion on a single virion level and provide several advantages over assays assessing viral fusion in bulk (reviewed by Haldar 2022). Single virion fusion assays, which utilize supported lipid bilayers or liposomes of defined lipidic composition and virions labeled with fluorescent lipidic and content dyes, have proven to be an invaluable biophysical tool to dissect the kinetics of viral fusion (Otterstrom et al. 2014; Zawada et al. 2018). Such studies showed that 3–4 neighboring HAs, which undergo conformational changes, are required to form a membrane fusion pore. In addition, fusion kinetics measurements of HA with different mutations within the fusion peptide revealed that the withdrawal of the fusion peptide from a pocket in the prefusion trimer is a rate-limiting step of the conformational changes of HA (Ivanovic et al. 2013). Exciting structural data came from a recent single-particle cryo-EM study that uncovered the structure of the extended HA intermediate, which occurs within 10–20 s upon acidification and is short-lived (Benton et al. 2020). The HA trimer undergoes lateral dilation characterized by the separation of HA1 globular domains followed by the formation of the extended 150 Å-long triple-helical coils of HA2 (Fig. 4.2). Interestingly, an uncleaved HA0 precursor, which is unable to conduct membrane fusion, undergoes similar conformational changes leading to the formation of an extended intermediate; however, these conformational changes are reversible and HA0 returns into prefusion conformation at neutral pH (Garcia-Moro et al. 2022). Hence, it was proposed that the low pH structure of "unprimed" class I fusion proteins triggered by acidification may serve as a potential indicator of intermediates in the conformational changes of primed glycoproteins at low pH.

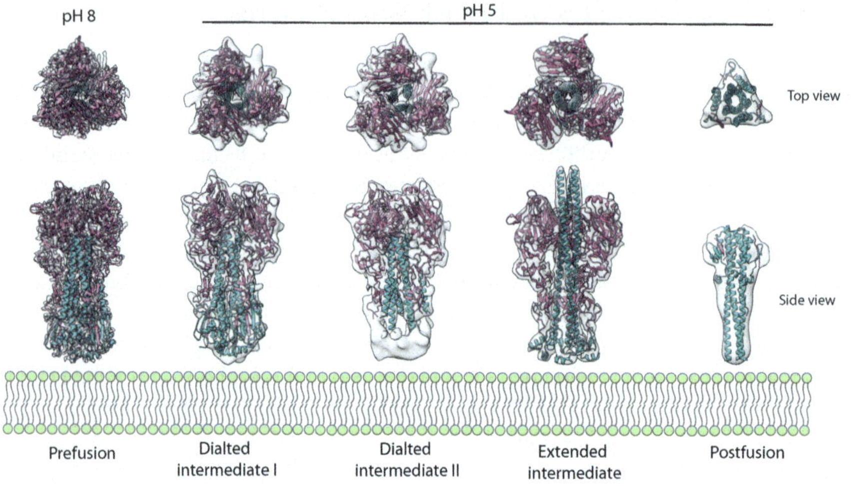


**Fig. 4.2** Single-particle reconstruction using cryo-EM of purified HA ectodomain from X31 influenza A virus subjected to low pH for 10, 20, 60 s, and 30 min. Top views and side views of electron microscopy maps (white isosurface) with a fitted atomic model showing HA structural rearrangement captured in five steps. The atomic structures of HA 2 fusogenic domains (HA2) are shown in magenta and binding domains HA2 are shown in cyan. The membrane bilayer is shown in green. HA conformational stages were visualized in ChimeraX (Pettersen et al. 2021) and correspond to PDB structures (from left to right): 6y5g, 6y5i, 6y5j, 6y5k from Benton et al. (2020) and PDB 1HTM from Bullough et al. (1994)

## Membrane Penetration by Non-enveloped Viruses

In contrast to enveloped virions, non-enveloped virions developed mechanisms to disrupt the bilayer architecture of the membrane via viral lytic peptides embedded into the viral capsid. Since most non-enveloped viruses require a trigger such as receptor binding (Hrebik et al. 2021) or low pH to initiate viral uncoating and exposure of the lytic peptide, they typically rely on the endosomal route for entry into the host (Kumar et al. 2018; Moyer and Nemerow 2011). The molecular mechanism and extent of membrane lysis caused during cytosolic entry (Fig. 4.1c) vary for different non-enveloped viruses. Based on the size and type of pore formation, non-enveloped viruses can be classified into (1) *proteinaceous pore-forming viruses*, which induce a protein-lining pore spanning the endosomal membrane to release the genome from the virion into the cytosol without completely disrupting the endosomal compartment, and (2) *membrane perforating viruses*, whose endosomal exit is characterized by the formation of a large pore which allows the capsid to reach the cytosol. Membrane perforating viruses either rely on the exposure of amphipathic peptides (*lytic peptide-driven membrane penetration*) that disrupt the membrane bilayer or on phospholipase activity (*lipid-modifying membrane rupturing viruses*). As a membranous compartment which undergoes a rupture tends to immediately reseal, bilayer edges of the open pore must be stabilized either by specialized lipids or by viral amphipathic and hydrophobic peptides.

## *Proteinaceous Pore-Forming Viruses*

*Picornaviridae* such as poliovirus, hepatitis A virus, foot-and-mouth disease virus, or rhinoviruses contain a genome of positive polarity in the form of single-stranded RNA. Upon uptake into late endosomes, the icosahedral capsids undergo low pH-induced conformational changes that lead to the exposure of the N-myristoylated peptide VP4 and the N-termini of VP1 proteins which together anchor to the endosomal membrane (Paul et al. 1987). Electron microscopy and biophysical assays revealed that purified rhinovirus VP4 proteins oligomerize and are able to form size-selective pores of approximately 12 nm in diameter (Panjwani et al. 2014) in supported membrane bilayers. Similar results were obtained with purified VP4 from the hepatitis A virus (Shukla et al. 2014). While a proposed model suggests that picornaviruses could release their genome via such a pore without disrupting the endosomal compartment, further structural studies using virions and liposome mixtures and cryo-electron tomography (cryo-ET) are needed to validate this model. It is possible that viral capsids form multiple pores across the endosomal membrane and thereby induce lysis of the entire endosomal compartment. Recent data show that in a cellular context endosomal escape is more complex and involves host proteins. For picornaviruses, for example, genome-wide haploid genetic screens identified $A_2$ (PLA2G16) as an essential host factor required for viral entry (Staring et al. 2017). While the authors show that in the absence of PLA2G16, galectin-8 targets endosomes harboring viruses for degradation by autophagy, it is currently not understood whether phospholipase A2 activity aids in the formation of the pore or increases its stability via generation of lysophospholipids (see below).

### Viruses Utilizing Lytic Peptide-Driven Membrane Penetration

The majority of non-enveloped viruses such as adenoviruses, reoviruses, and polyomaviruses enter host cells by endocytosis and undergo low pH-induced structural changes leading to the exposure of lipidated amphipathic peptides to induce membrane lysis directly in late endosomes. The molecular mechanism of how lytic peptides induce membrane rupture followed by stabilization and enlargement of a pore remains to be understood. Recently, all-atom molecular dynamics simulations of the γ lytic peptide from Flock House Virus, a member of the *Nodaviridae* family, interacting with two membrane bilayers showed oligomerization of the peptides followed by the formation of a water pore between the bilayers (Nangia et al. 2020). Interestingly, based on the α-helical secondary structure and physical behavior of γ lytic peptides it has been suggested that they may act similarly to antibacterial peptides, which represent a promising antibacterial therapy (Moretta et al. 2021).

It is plausible that upon insertion into the membrane bilayer, lytic peptides cause exposure of phospholipid acyl chains similar to that shown for fusion peptides of viral fusion peptides. Lytic peptide exposure and interaction with the membrane may subsequently cause adsorption of the endosomal membrane on the surface of the viral particle and thus induce a high positive curvature reminiscent of viral budding. In this model, lytic peptides anchored in the membrane would synergistically disrupt the membrane integrity to establish a pore. The lipidic edge of the pore would be stabilized by lytic peptides reducing the line tension as suggested for other model systems (Akimov et al. 2017) and was observed experimentally on giant unilamellar vesicles in the presence of an actin-binding protein called talin (Saitoh et al. 1998). This mode of action of lytic peptides would be further enhanced by lipidation such as myristylation which may stabilize the edge of the pore.

Interestingly, polyomaviruses, small icosahedral viruses with a DNA genome (e.g., simian virus 40 or BK virus), undertake a far more complex path on their way to establishing replication in the host cells (reviewed by Mayberry et al. 2021). Polyomaviruses enter through the caveolar endocytic pathway and undergo low pH-induced conformational changes leading to the exposure of a myristoylated amphipathic peptide in VP2. However, polyomaviruses are further trafficked into the endoplasmic reticulum (ER) starting 1–2 h post-infection (Kartenbeck et al. 1989). Although the mechanism of virus cytosolic escape from the ER is not yet fully understood, polyomavirus membrane rupture is dependent on the ER-resident protein folding machinery, which acts as a chaperone to destabilize the capsid (Schelhaas et al. 2007). It is not clear why polyomaviruses need to be trafficked into the ER via a pathway which bypasses the Golgi apparatus. Since the viral DNA genome replicates inside the nucleus, it is believed that ER exit brings capsids into the vicinity of the nucleus. Recent data show that polyomavirus particles residing in endosomes are directed for fusion with the ER via the Rab18 and SNARE machinery (Zhao and Imperiale 2017). To shed light on this fascinating process, early electron microscopy studies should be revisited with modern electron microscopy methods to provide a better understanding of this peculiar and complex entry pathway.

## Lipid-Modifying Membrane Rupturing Viruses

Endosomal rupture mediated by viruses of the family *Parvoviridae* relies on $A_2$ (PLA2) buried in the shell of the icosahedral capsid (Zadori et al. 2001). At neutral pH, PLA2 is buried in the capsid, and upon entry into the late endosome, acidification is required for structural rearrangements of the capsid and exposure of the N-terminal region of the minor capsid viral protein 1 (VP1) with PLA2 activity. PLA2 hydrolyzes the covalent bond between the glycerol backbone and the sn2 acyl chain, which leads to the formation of a lysophospholipid. Lysophospholipids have a conical shape and positive intrinsic curvature and thus promote pore formation and lysis of membrane bilayers (see Fig. 4.4). Thus, it is plausible that lysophospholipids could stabilize the rim of the membrane and prevent resealing of the pore. In vitro studies on capsids of the adenovirus-associated virus (AAV) showed that their

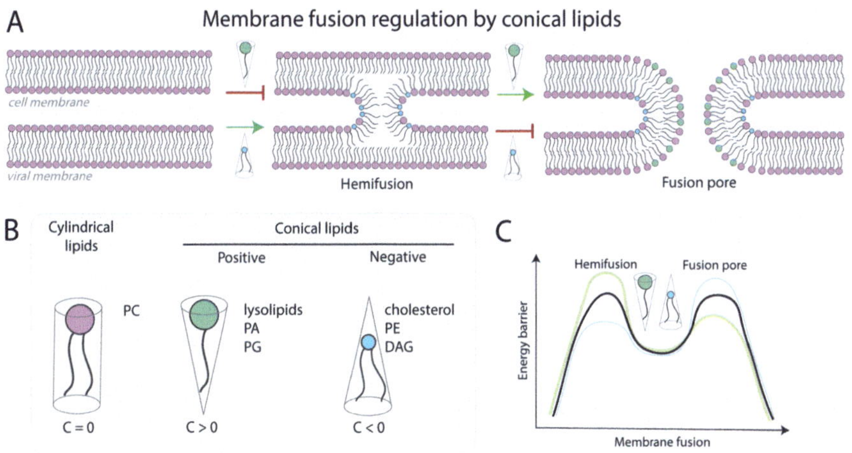


**Fig. 4.4** Regulation membrane fusion by lipids. (**a**) Role of lipids with positive (green) and negative spontaneous curvature (blue) in stabilizing or promoting hemifusion intermediate and pore formation. (**b**) Different shapes of lipids and their monolayer spontaneous curvature *C*. Monolayer spontaneous curvature is determined by the chemical structure of lipids. (**c**) Hypothetical energy profiles of membrane fusion showing two local energy barriers for hemifusion and fusion pore formation. Green and blue curves represent energy profiles of membrane fusion for conical and inverted conical lipids, respectively

stability is pH dependent, and capsids mixed with liposomes and analyzed by negative staining showed remodeling indicating that acidification is sufficient for the exposure of PLA2 (Lins-Austin et al. 2020). In the context of late endosomes, low pH-dependent proteases such as cathepsins may be critical to proteolytically activate capsids and render them pH sensitive or cathepsins contribute to capsid destabilization and exposure of the PLA2 domain (Akache et al. 2007) (Table 4.2).

## Visualization of Viral Membrane Fusion and Penetrations Using Cryo-electron Tomography

While fluorescence-based assays and microscopy have been instrumental to study membrane fusion at a high temporal resolution, cryo-ET continues to play a critical role in visualizing membrane fusion intermediates. Unlike single-particle cryo-electron microscopy, which is applied to study the structure of purified fusion proteins and their conformational changes at the subnanometer resolution, cryo-ET enables visualization of the steady state of the viral membrane fusion pathway in 3D (Fig. 4.3). This allows us not only to capture membrane fusion intermediates but also to shed light on structural changes of the virion scaffold and nucleocapsids, which are important viral constituents and implicated in the membrane fusion process. Cryo-ET entails sample preservation by plunge freezing into liquid ethane, which

**Table 4.2** List of non-enveloped viruses and properties of their penetration machinery

| Virus family (typical member) | Attachment protein | Penetration protein/ machinery | Priming of the penetration machinery | Cue to initiate membrane penetration |
|---|---|---|---|---|
| *Reoviridae* (Rhesus rotavirus (Gaudin 2021), bluetongue virus (Mohl and Roy 2014)) | VP8* (rhesus rotavirus) | VP5* (rhesus rotavirus) | VP4 cleavage by trypsin-like proteases into VP5* and VP8* fragments (rhesus rotavirus) | Decrease in calcium concentration (rhesus rotavirus) |
| | VP2 (bluetongue virus | VP5 (bluetongue virus) | No proteolytic priming of VP5 (bluetongue virus) | Low pH (bluetongue virus) |
| *Adenoviridae* (Human adenovirus (HAdV) (Nemerow et al. 2009)) | Fiber | pVI | – | Low pH-driven dissociation of vertex proteins |
| *Anelloviridae* (Torque teno virus (Desingu et al. 2022)) | Unknown | Unknown | Unknown | Unknown |
| *Astroviridae* (HAstV (Ykema and Tao 2021)) | V27/VP25 (probable) | Unknown | Unknown | Unknown |
| *Caliciviridae* (Norovirus (Campillay-Veliz et al. 2020; Graziano et al. 2019)) | VP1 (feline calicivirus, FCV, human Norovirus, HNoV) | Unknown | Unknown | Low pH (FCV) |
| *Hepeviridae* (Hepatitis E virus (Lin and Zhang 2021)) | P domain of the capsid protein | Unknown | Unknown | Unknown |
| *Nodaviridae* (Flock House virus, FHV (Nangia et al. 2020)) | Capsid protein | γ-peptide | – | Low pH |
| *Papillomaviridae* (Human papillomavirus (Nangia et al. 2020)) | L1 | L2 | KLK8 priming of L1 capsid to trigger conformational changes important for cytosolic entry | Unknown |
| *Parvoviridae* (Parvovirus (Tu et al. 2015)) | VP2 | VP1 | – | Low pH → conformational changes in VP1 |
| *Picobirnaviridae* (Human picobirnavirus) | CP | Unknown | Unknown | Unknown |

(continued)

**Table 4.2** (continued)

| Virus family (typical member) | Attachment protein | Penetration protein/ machinery | Priming of the penetration machinery | Cue to initiate membrane penetration |
|---|---|---|---|---|
| *Picornaviridae* (Poliovirus (Tuthill et al. 2006)) | VP1 | VP4 | – | Receptor-binding |
| *Polyomaviridae* (Simian virus 40 (Chen et al. 2019)) | VP1 | VP2 | Proteolytic priming of the virus particle by ER-resident redox proteins | Low pH |
| *Reoviridae* (Rotavirus (Arias and Lopez 2021)) | VP4 (subunit VP8) | VP4 (subunit VP5) | Trypsin cleavage of VP4 | Ca2+ decrease |

allows vitrifying a thin film of aqueous samples with cooling rates of $5 \times 10^5$ K/s. Rapid processes such as membrane fusion are thereby instantly immobilized. The vitrified sample is subsequently imaged in a transmission electron microscope at cryogenic temperatures to acquire datasets of 2D projection images while tilting the sample typically from 60° to –60°. After alignment of the projection images, they are reconstructed into a final tomogram, which captures the site of interest in 3D.

Cryo-ET has successfully been applied to study reconstituted viral membrane fusion using purified viruses and liposomes of defined lipid composition (Calder and Rosenthal 2016; Fontana and Steven 2015; Gui et al. 2016; Halldorsson et al. 2018; Lee 2010). Since many viruses engage either protein or lipid receptors to initiate fusion, liposomes must be equipped accordingly. The reconstitution of influenza A virus membrane fusion in vitro has for example been possible because the sialic acid receptor can be incorporated into liposomes in the form of gangliosides. A recent study using purified endosomes containing influenza A virions showed similar fusion kinetics as shown for viral membrane fusion with liposomes, confirming that liposomes are a good model system for studying membrane fusion (Haldar et al. 2020). Some viruses, however, require protein receptors and hence proteoliposomes must be used, which are produced by reconstituting the purified receptors into liposomes. Alternatively, giant plasma membrane vesicles (so-called blebs) or exosomes containing cellular receptors can be used to study viral membrane fusion. This has been demonstrated by a study showing human immunodeficiency virus (HIV) membrane fusion intermediates in the presence of membrane fusion restriction factors or fusion peptide inhibitors (Ward et al. 2020). Cryo-ET has also been applied to shed light on the function of fusion peptide inhibitors which are designed to block SARS-CoV-2 mediated membrane fusion (see below) (Marcink et al. 2022).

While most of the cryo-ET studies were performed to reconstitute membrane fusion of viruses harboring class I fusion proteins, cryo-ET was also applied to capture membrane fusion driven by class II viral proteins of the phlebovirus Rift Valley fever virus (Halldorsson et al. 2018) and the alphavirus chikungunya virus

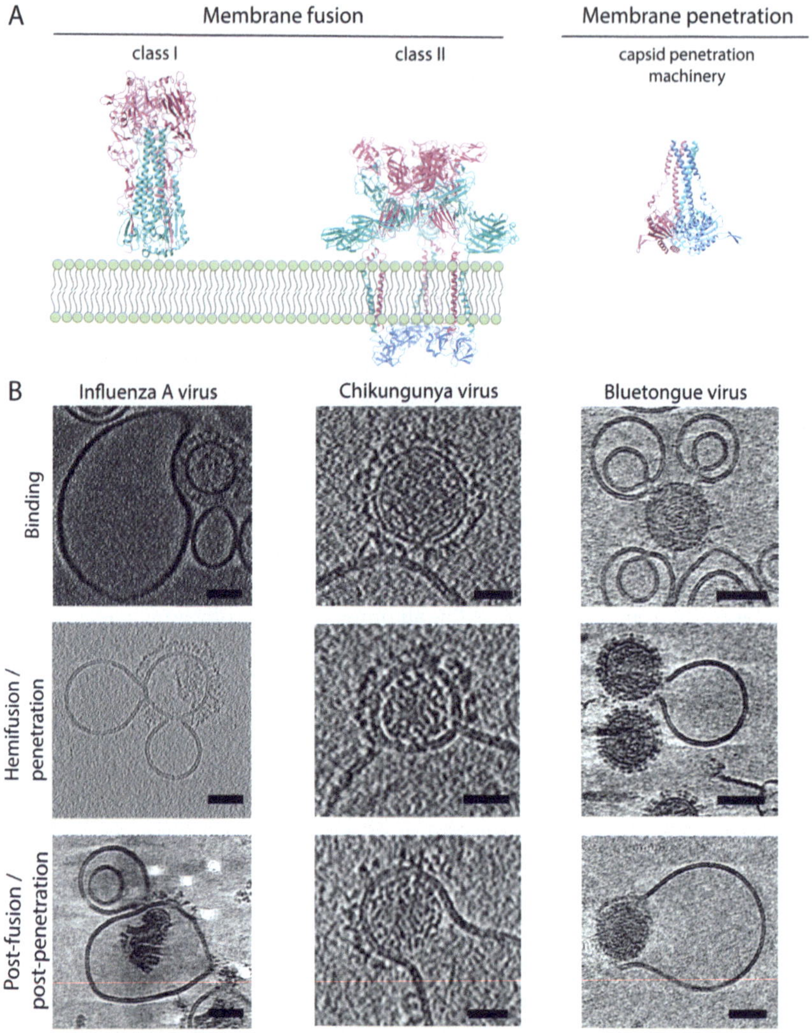


**Fig. 4.3** Cryo-ET of virus membrane fusion and penetration using liposomes. (**a**) Prefusion structures of influenza A virus HA (PDB:6Y5G, HA1 magenta, HA2 cyan) and chikungunya virus E1/E2 (PDB:3J2W, E1 cyan, E2 magenta, capsid blue) as examples of membrane fusion class I and class II machinery, respectively. The structure of bluetongue virus VP5 trimer at low pH (PDB:7RTN) is shown on the right (each monomer is shown in different colors). Images were prepared with ChimeraX (Pettersen et al. 2021). (**b**) Cryo-ET of virus–liposome mixtures undergoing membrane fusion (influenza A virus—left, chikungunya virus—middle) or membrane penetration by bluetongue virus (right). Three different subsequent states of membrane fusion or membrane penetration are shown in columns (binding, hemifusion/penetration, and postfusion/post-penetration events). Scale bars: influenza A virus and bluetongue virus: 50 nm; chikungunya virus: 20 nm. Images were used with the permission of corresponding authors of respective publications: influenza A virus (Chlanda et al. 2016); chikungunya virus (Mangala Prasad et al. 2022a); bluetongue virus (Xia et al. 2021). The image for the influenza A virus postfusion event was not previously published and was acquired for this chapter. White spots in the image are computationaly erased fiducial markers used for tilt series alignments

(Mangala Prasad et al. 2022a). For the chikungunya virus, the authors were able to directly visualize an array of trimeric fusion complexes and long glycoprotein connections binding to the liposomal membrane, consistent with chikungunya E1 homotrimers whose fusion peptides have been exposed and buried into the target membrane. Interestingly, large extended hemifusion intermediates and large fusion pores were revealed, which are not typically observed for class I fusion proteins. This difference may arise from the icosahedral structure of alphaviruses and may indicate that the formation of hemifusion intermediates and fusion pores requires the orchestrated action of more glycoproteins than in the case of class I fusion proteins. Intriguingly, after the completion of membrane fusion, which was reported to occur after 3 min at pH 5, the chikungunya nucleocapsid remains intact. This indicates that in the context of infection, additional factors are required to initiate nucleocapsid uncoating and liberation of viral RNA.

Finally, cryo-ET has also been applied to study membrane penetration by non-enveloped viruses. Xian Xia and colleagues thereby unraveled interactions of the non-enveloped icosahedral bluetongue virus with liposomes. Bluetongue virus is an endemic livestock pathogen, which belongs to the family of *Reoviridae* (Xia et al. 2021). It enters cells by clathrin-mediated endocytosis and utilizes the membrane penetration protein VP5, which forms 120 trimers on the viral surface. The VP5 trimer contains a histidine cluster and structurally resembles the prefusion conformation of the fusion domain of class I fusion proteins (Zhang et al. 2010, 2016). Single-particle cryo-EM of virions subjected to low pH revealed that VP5 undergoes structural rearrangements to form a central extended coiled-coil structure. The stalk formation appears to be irreversible as a change of pH back to neutral pH levels did not reverse the conformational changes. Cryo-ET of the virion and liposome mixtures showed that virions are decorated by rod-shaped stalks approximately 19.5 nm in length, which interact with the liposomal membrane (Xia et al. 2021). Interestingly, the interaction causes the formation of a single rupture site rather than multiple perforated areas. Hence, it is plausible that protruding stalks stabilize the ruptured edge of the liposomal membrane and successively enlarge the opening pore by engaging additional stalks all around the virion until the virion leaves the endosomal compartment. It remains to be determined whether, in the context of infection, the ruptured endosome spontaneously reseals after the completion of viral cytosolic entry.

Besides the many benefits of cryo-ET, as with any technique, several pitfalls should be considered during data interpretation. Membrane fusion is a rapid process, and while cryo-ET offers high-resolution imaging, it provides 3D snapshots of a steady state where long-lived intermediates are represented more frequently than short-lived membrane fusion intermediates. In addition, dead-end products or spurious by-products may be prevalent in certain buffer conditions or lipid compositions. Hence, it is highly advisable to complement cryo-ET studies with additional assays such as fluorescent dequenching assays to validate the membrane fusion system. Furthermore, misinterpretation of densities can arise in tomograms reconstructed from data acquired with an extensive defocus at thicker regions of samples due to decreased signal-to-noise ratio. While increasing the defocus leads to an increase in contrast, it also causes fringing artifacts manifested in a white halo around lipid monolayers and proteins. This can be mitigated for example by using the Volta phase plate and in-focus data collection (Danev et al. 2014), which was shown to improve the data interpretability of membrane fusion events (Chlanda et al. 2016). Additionally, the sample tilt range during acquisition is limited by the specimen holder, causing a so-called "missing wedge" of information and elongation of features in the Z-direction. This missing wedge information can be minimized by using dual-axis tomography which can also be combined with Volta phase plate imaging (Winter and Chlanda 2021).

Finally, to be able to localize rare fusion events and increase the throughput of cryo-ET, cryo-fluorescence microscopy can be applied to localize fluorescent dequenching events after vitrification. The cryo-light microscopy maps can be correlated with the cryo-EM map for targeted tomography. Metskas and Briggs used cryo-correlative light and electron tomography to correlate hemifusion events which are characterized by high fluorescent signal arising from lipid demixing and minimization of fluorophore self-quenching (Metskas and Briggs 2019).

## Endosomal Lipids as Regulators of Membrane Fusion

In endosomal compartments, membrane fusion and penetration are tightly controlled by lipids, ions, and host proteins determining the outcome of infection. Phospholipids are synthesized in the ER and are variously distributed to the different organelles, conferring specific properties to their membrane bilayers. It is well established that lipids are involved in signaling and give a specific physicochemical signature to membrane bilayers. They control the fluidity, rigidity, and membrane bending properties through their intrinsic shape, which can lower or increase the energy barrier of membrane fusion or rupture. Depending on the site of membrane interaction, lipids can inhibit or promote viral membrane fusion or penetration (reviewed here Sardar et al. 2022). Recent studies collectively indicate that membrane fusion is mainly controlled by lipid intrinsic curvature and lipid charge.

### *Effect of Lipid Curvature*

Based on effective molecular shape given by chemical structure, lipids are classified into cylindrical lipids (e.g., phosphatidyl choline, phosphatidyl serine); lipids with negative spontaneous curvature (e.g., phosphatidyl ethanolamine, diacylglycerol, cholesterol); and lipids with positive spontaneous curvature (e.g., lysolipids, phosphatidylinositol 4,5-bisphosphate) (Dymond 2021). Cylindrical lipids are able to form a membrane bilayer, whereas lipids with positive intrinsic curvature form a micelle, and lipids with negative intrinsic curvature form inverted micelles. The effect of lipid intrinsic curvature and its role in membrane fusion have been well characterized (Chernomordik and Kozlov 2003). Lipids with positive spontaneous curvature present in the membrane bilayer during membrane fusion will inhibit hemifusion but will facilitate the formation of a fusion pore. Conversely, lipids with negative intrinsic curvature will promote establishing hemifusion but will inhibit the progression from hemifusion to fusion pore opening (Fig. 4.4a–c). One of the most important lipidic regulators of membrane remodeling processes is cholesterol, which stands out from other lipids by its compact shape and the small hydroxy group as the only polar constituent of the molecule. Upon synthesis in the ER, cholesterol is trafficked toward the plasma membrane by cholesterol transporters that regulate the cholesterol homeostasis in the cell. Cholesterol is highly enriched in the plasma membrane and its concentration within the bilayer decreases along the retrograde trafficking pathway (Iaea and Maxfield 2015). Besides its high negative intrinsic curvature, cholesterol increases phospholipid packing and thereby increases bilayer thickness, reduces membrane permeability, and is critical for the formation of liquid-ordered domains in the membrane. Cholesterol also influences membrane bending rigidity, which represents the energy needed to change the membrane curvature (Karal et al. 2022). The impact of cholesterol in viral membrane fusion has extensively been studied using in vitro membrane fusion assays. In vitro experiments using the influenza A virus and liposomes with variable concentrations of cholesterol showed that at low cholesterol concentrations in liposomes, low pH-induced HA conformational changes lead to liposomal rupture rather than membrane fusion (Chlanda et al. 2016; Haldar et al. 2018). Volta phase plate cryo-ET revealed that membrane rupture is followed by spontaneous insertion of the ruptured membrane into the viral envelope. This data is supported by fluorescence microscopy of individual membrane fusion events of influenza A viruses showing that the efficiency of lipid mixing increases with increasing cholesterol concentration in the target membrane, whereas the rate of lipid mixing that occurs during HA-mediated membrane fusion is cholesterol independent (Liu and Boxer 2020). Recent studies showed that influenza A virions depleted of cholesterol or virions produced in the presence of lovastatin, which impairs cholesterol synthesis, exhibit reduced fusion efficiency (Zawada et al. 2016). Interestingly, similar results were obtained using Ebola virus-like particles produced in the presence of statins. Experiments employing photoactivatable cholesterol and NMR revealed that the transmembrane domain of the Ebola virus GP2 directly interacts with cholesterol via several glycine residues (Hacke et al. 2015; Lee et al. 2021), whose mutation increased the probability of hemifusion stalling and reduced GP-mediated membrane fusion (Lee et al. 2021). The importance of cholesterol as a mediator of membrane fusion has also been demonstrated for other viruses including SARS-CoV-2 (Sanders et al. 2021), and cholesterol within the HIV envelope influences the clustering of the envelope fusion glycoprotein (Env) to promote membrane fusion (Nieto-Garai et al. 2021). In conclusion, effective and non-leaky membrane fusion mediated by viral proteins requires cholesterol in both virion and the target membrane.

### *Effect of Lipid Charge*

Collective evidence indicates that negatively charged lipids play an important role in viral attachment and membrane fusion. It has been reported that filamentous viruses such as the Ebola virus and large viruses such as the vaccinia virus display phosphatidylserine on their envelope and trigger uptake into target cells by a process termed "apoptotic mimicry." The negatively charged phosphatidylserines are recognized by cellular receptors to trigger macropinocytosis (Acciani et al. 2021; Mercer and Helenius 2008; Nanbo et al. 2018).

Moreover, recent studies indicate that negatively charged lipids are implicated as cofactors of membrane fusion. For example, it was shown that anionic lipids are required for dengue virus class II fusion protein-mediated membrane fusion with liposomes and promote cell-to-cell fusion (Zaitseva et al. 2010). Similarly, HIV Env-mediated membrane fusion depends on the exposure of the negatively charged lipid phosphatidylserine. The Env trimer is composed of gp41 and gp120 that binds to CD4-positive cells and requires one of the two co-receptors CCR5 and CXCR4. Fluorescence microscopy-based assays revealed that the binding of HIV pseudoviruses induces phosphatidylserine flipping from the inner to the outer monolayer of the plasma membrane in a process driven by the scramblase TMEM16F. Interestingly, phosphatidylserine exposure and TMEM16F positively affect Env-mediated cell–cell and virion–cell membrane fusion, which indicates that the negative charge of serine might promote conformational changes of Env-promoting membrane fusion (Zaitseva et al. 2017).

Based on the morphological appearance obtained from EM studies, late endosomes are often referred to as multivesicular bodies due to a high number of intraluminal vesicles. The formation of intraluminal vesicles is driven by the ESCRT machinery and depends on bis-monoacylglycerol phosphate (BMP, known also as lysobisphosphatidic acid (LBPA)) (Fig. 4.5b), which is enriched in late endosomes (Hullin-Matsuda et al. 2007; Rabia et al. 2020). BMP is a negatively charged lipid which facilitates membrane fusion and the formation of intraluminal vesicles (Matsuo et al. 2004). Similar to cholesterol, BMP enhances lipid mixing in influenza A virus–liposome mixture (Gui et al. 2016). In addition, BMP facilitates the fusion of the phlebovirus Uukuniemi virus (UUKV) with liposomes and promotes UUKV-induced fusion with BHK-21 cells (Bitto et al. 2016). Bitto et al. investigated whether UUKV membrane fusion can be rescued by other negatively charged lipids in the absence of BMP, and fluorescence dequenching showed that only negatively charged lipids with negative intrinsic curvatures such as phosphatic acid and phosphatidylglycerol are able to support membrane fusion (Bitto et al. 2016). Another piece of evidence highlighting the role of BMP in the viral endosomal escape was recently reported for the Lassa virus and other related viruses from the family of *Arenaviridae*. Interestingly, the data of this study indicate that BMP does not influence the formation of the hemifusion intermediate but rather promotes the formation and enlargement of a fusion pore (Markosyan et al. 2021), which is unexpected considering the negative intrinsic curvature of BMP. Overall, recent

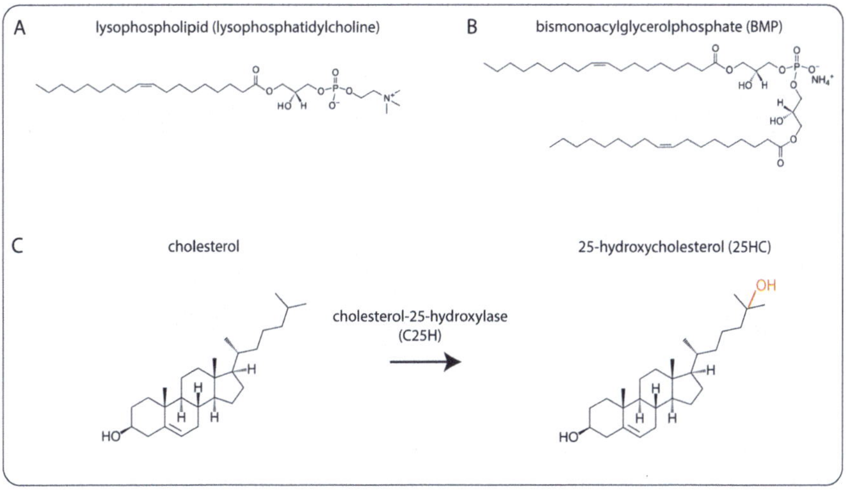


**Fig. 4.5** Chemical structures of lipids playing a role in viral membrane fusion. (**a**) Lysophospholipids such as lysophosphatidylcholine have positive spontaneous curvature due to the absence of one acyl chain. (**b**) Bismonoacylglycerolphosphate (BMP) is a lipid that regulates viral membrane fusion in endosomes. (**c**) An enzymatic reaction converting cholesterol to 25-hydroxycholesterol catalyzed by cholesterol-25-hydroxylase, which is a host restriction factor present in the ER. The hydroxy group at cholesterol carbon position 25 is highlighted in orange. We thank Avanti Polar Lipids for allowing us to use the lipid chemical structures from their webpage

data highlight BMP as a facilitator of class II fusion protein-mediated membrane fusion.

## Host Restriction Factors Controlling Viral Membrane Fusion and Penetration by Modulating the Endosomal Lipid Composition

While in vitro studies have revealed invaluable insights into the structure and mechanism of membrane fusion proteins and penetration machinery, the scenario of cellular endocytic escape is far more complex. By utilizing host restriction factors as part of the innate immune response, cells have evolved several defense mechanisms, which can delay the viral endocytic escape and thereby target incoming viruses for lysosomal degradation (Majdoul and Compton 2022). The expression of host restriction factors is stimulated by interferons, which are produced by cells sensing viral infection. Typically, this involves the recognition of viral material called pathogen-associated molecular patterns either by endosomal Toll-like receptors or by cytosolic viral sensors. Among those patterns are viral nucleic acids such as double-stranded RNA by-products, CpG DNA, or uncapped single-stranded RNA (Rojas et al. 2021), which are typically not found in the cellular cytoplasm. Perhaps, the most studied host restriction factors, which directly interfere with virus endosomal escape, are interferon-induced transmembrane proteins (IFITMs) (Brass et al. 2009). IFITM1, 2, and 3 have all been shown to block viral membrane fusion of several enveloped viruses (e.g., influenza A virus, Ebola virus, dengue virus, SARS coronaviruses, and HIV) but also some non-enveloped viruses such as reoviruses (Anafu et al. 2013; Perreira et al. 2013). IFITM3, the best-characterized IFITM ortholog with the highest antiviral potency of the three proteins, is a transmembrane protein with a single membrane-spanning α-helix and two amphipathic helices facing the cytosol. Post-translational modifications of IFITM3 such as phosphorylation, ubiquitination, and palmitoylation at Cys71, Cys72, and Cys105 are critical for IFITM3 endosomal targeting and its antiviral function in humans (Yount et al. 2010) and bats (Benfield et al. 2020). Despite its essential role in antagonizing the entry of several important viruses, the molecular mechanism of IFITM3-mediated inhibition of viral entry is not well understood, and several models have been proposed. Fluorescence dequenching of lipidic dyes showed that IFITM3 does not impair lipid mixing between viral and endosomal membranes (consistent with hemifusion intermediate states) but blocks the formation of a fusion pore (Desai et al. 2014). We have recently shown by in situ cryo-ET that IFITM3 does not modulate the number of intraluminal vesicles inside late endosomes as previously suggested (Klein et al. 2023) but arrests the influenza A virus at the hemifusion state. In addition, subtomogram averaging revealed that IFITM3-arrested hemifusion sites contain HA in postfusion conformation, indicating that IFITM3 stabilizes hemifusion intermediates without affecting the conformational transition of hemagglutinin from prefusion to postfusion state (Klein et al. 2023). How does IFITM3 then stabilize the hemifusion state? Although a recent study suggested that IFITM3 induces negative spontaneous curvature and thereby increases the energy barrier for pore formation (Guo et al. 2021), a growing body of evidence indicates that IFITM3 blocks viral membrane fusion by modulating the cholesterol concentration in late endosomes (Amini-Bavil-Olyaee et al. 2013; Das et al. 2021; Garst et al. 2021; Klein et al. 2023; Rahman et al. 2022). Recent data using photoactivable and clickable cholesterol and molecular dynamics simulations revealed that Cys72 S-palmitoylation is critical for IFITM3 interactions with cholesterol (Das et al. 2021; Klein et al. 2023). However, molecular dynamics simulations showed that despite the cholesterol interaction with palmitoyl, cholesterol is depleted from the immediate vicinity of IFITM3. This data led to a model supported by continuum membrane modeling, in which we proposed that IFITM3-induced lipid sorting leads to cholesterol accumulation in a hemifusion state and thereby stabilizing its lifetime (Klein et al. 2023).

Interestingly, IFITM3 seems to only block viruses which strongly depend on cholesterol during membrane fusion and enter via endosomes (influenza A virus, Ebola virus, SARS-CoV-2). Even though arenaviruses are also late-penetrating enveloped viruses, their membrane fusion is promoted by negatively charged lipids such as BMP (see above) and arenavirus entry is not restricted by IFITM3. This further highlights the important interplay between IFITM3 and cholesterol and shows that IFITM3 is functional only when cholesterol is required in viral membrane fusion.

Another interferon-inducible host restriction factor with broad antiviral function which directly affects cholesterol homeostasis is the cholesterol 25-hydroxylase (CH25H). This ER-associated enzyme catalyzes hydroxylation at the hydrophobic cholesterol acyl chain (Lund et al. 1998). This seemingly small chemical modification results in the formation of 25-hydroxycholesterol (25HC), a molecule whose physicochemical properties are diametrically distinct from those found for cholesterol (Fig. 4.5c). More specifically, 25HC has increased solubility in aqueous solutions and alters the mechanical properties of membranes, and it is unlikely to assume parallel packing with phospholipids in a membrane bilayer (Domingues et al. 2021; Gomes et al. 2019). Interestingly, 25HC inhibits the entry of several enveloped viruses including influenza A virus, Ebola virus, Nipah virus, HIV, herpes viruses (Blanc et al. 2013; Liu et al. 2013), and porcine coronaviruses (Zhang et al. 2019). Specifically, CH25H expression was shown to restrict the entry of pseudoviruses carrying SARS-CoV-2 S spikes to a similar extent as IFITM3, and 25HC accumulates in late endosomes where it blocks S-mediated membrane fusion (Zang et al. 2020). Biophysical approaches revealed that the membrane fusion efficiency of the HIV fusion peptide is reduced by 50% when cholesterol is replaced by 25HC in liposomes (Gomes et al. 2018). It remains to be investigated at which stage 25HC blocks membrane fusion and liposome–virus fusion assays have not been performed so far to further assess the membrane fusion efficiency in the presence of 25HC. Apart from inhibiting membrane fusion, 25HC was also shown to restrict the entry of the non-enveloped human rotaviruses by sequestering viral particles into late endosomes, demonstrating its broad effect on membrane processes in general. Whether 25HC stalls the entry of other non-enveloped viruses that exit from late endosomes remains to be elucidated.

Another well characterized viral restriction factor is serine incorporator 5 (Serinc5), which is incorporated into HIV virions during budding in a process antagonized by the viral protein Nef (Rosa et al. 2015). Serinc5 inhibits HIV entry by a so far not completely understood mechanism. A study employing Cryo-ET and TIRF microscopy investigated HIV-mediated membrane fusion between cell bleb-derived membranes, which harbor the HIV CD4 receptor and the CCR5 coreceptor in the absence or presence of Serinc isoforms 2, 3, and 5. While the presence of Serinc5 and Serinc3 delayed the membrane fusion and caused the accumulation of arrested membrane fusion intermediates including hemifusion, Serinc2 had no impact on Env-mediated membrane fusion. Fluorescence lifetime imaging microscopy using the fluorescent membrane dye FLIPPER-TR, which has been established as a reporter of membrane order (Colom et al. 2018), showed that Serinc5 but not Serinc2 directly modulates thickness and lipid order of lipid bilayers, which possibly leads to HIV membrane fusion inhibition (Ward et al. 2022). Interestingly, similar to IFITM3 restriction, Serinc5 membrane fusion inhibition can be reversed by the addition of the cholesterol-binding antimycotic amphotericin B. This suggests that the inhibitory effect of Serinc5 is cholesterol-dependent, a notion that is further promoted by the fact that HIV virions are enriched in cholesterol. Until recently, the inhibitory effect of Serinc5 on membrane fusion has only been reported for retroviruses; however, an important study showed that Serinc5 is also incorporated into SARS-CoV-2 virions in a process that is counteracted by the SARS-CoV-2 protein ORF7a (Timilsina et al. 2022). As well as in HIV, Serinc5 is able to inhibit membrane fusion of SARS-CoV-2 pseudoviruses, which indicates that Serinc5 may have a broader antiviral function than previously assumed. While envelope glycoproteins of vesicular stomatitis virus and Ebola virus are not susceptible to Serinc5 inhibition, Serinc5 is able to inhibit influenza A virus membrane fusion in a process dependent on HA glycosylation level (Zhao et al. 2022).

## Role of Viral Uncoating in Membrane Fusion

During virus entry, virions undergo structural changes to (at least partially) disintegrate and release the viral genome. These processes are referred to as *virus uncoating*, which typically begins prior to membrane fusion and finishes with the liberation of the viral genome into the cytosol. Different steps of virus uncoating and its spatiotemporal regulation depend highly on the viral architecture. Typically, viral genomes are protected by proteins in the form of nucleocapsids or viral ribonucleoprotein complexes. During the formation of viral progeny, they are recruited to assembly sites by interaction with viral membrane bending proteins forming either scaffolds (glycoprotein coats) or matrices (formed by matrix proteins). The majority of class I and class III fusion proteins are not able to form scaffolds that would induce membrane curvature and provide force during virion assembly. Accordingly, viruses harboring class I and class III fusion proteins rely on matrix proteins as major assembly drivers and determinants of viral shape. Exceptions are the influenza C virus and measles viruses, which have ordered viral fusion proteins assuming lattice-like arrangements but still rely on matrix proteins to drive virion assembly (Halldorsson et al. 2021). In contrast, class II fusion proteins are typically composed of two different proteins that interact and form a highly ordered coat on the virion envelope. For example, an icosahedral coat covers the envelope of flaviviruses and alphaviruses and highly ordered non-icosahedral coats are found on the surface of bunyaviruses. Class II proteins carry out two independent tasks: they drive viral assembly and membrane fusion during virus entry. In the context of membrane fusion, the transition of class II viral fusion machinery to postfusion conformation is responsible for virion softening and loss of integrity. In contrast, the transition of class I and class III viral proteins to postfusion conformation only has a minor influence on the overall integrity of viruses whose architecture is determined by matrix proteins. These viruses depend on the detachment of their matrix proteins from the viral envelope to allow fusion pore opening and genome release. A typical example of such a scenario is the influenza A virus, whose HA fusion protein depends on endosomal acidification to trigger its fusogenic activity. In addition, low pH is required to disassemble the viral matrix underneath the viral envelope and to detach the segmented genome from the viral particle. This is accompanied by

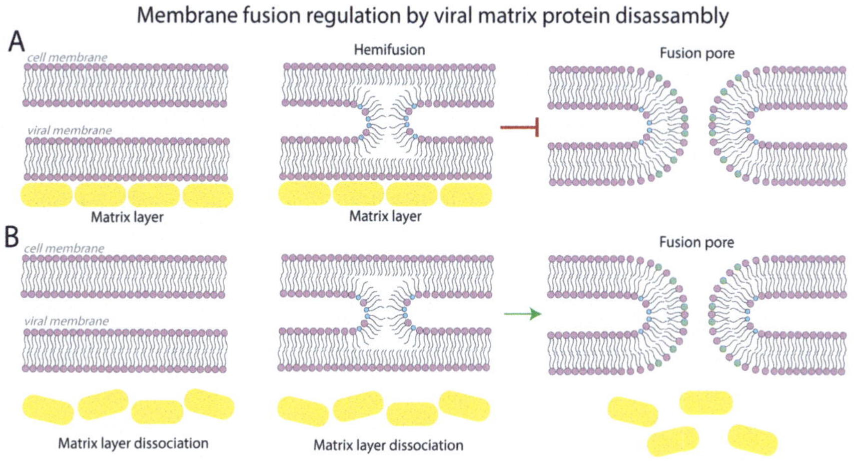


**Fig. 4.6** Regulation membrane fusion by matrix layer during virus uncoating. Membrane-enveloped viruses utilizing class I fusion proteins form a matrix layer (yellow) underneath the virion envelope, which is critical for viral assembly. The viral matrix layer must disassemble in order to release the viral genome into the cytosol. The model proposed by Winter et al. (2023) predicts that viral matrix layer disassembly plays a role in modulating the energy landscape of viral membrane fusion. (**a**) Matrix protein adhesion forces to the viral envelope do not prevent hemifusion formation, but the formation of a fusion pore is inhibited. (**b**) Matrix protein disassembly facilitates fusion pore formation

virion softening as determined by atomic force microscopy (Li et al. 2014). Recently, we showed that the influenza A virus might not be the only virus relying on low pH-mediated disassembly of matrix proteins. Similar to influenza A viruses, the Ebola virus shape is determined by a viral matrix protein, viral protein (VP)40, which detaches from the viral envelope upon acidification (Winter et al. 2023). In contrast to the influenza A virus, which relies on the activity of the viral ion channel M2 protein for luminal acidification, we show that Ebola virus protonation occurs by passive diffusion. In addition, the disassembly of the VP40 layer is critical for establishing a hemifusion stalk (Fig. 4.6). Due to the high functional similarity between matrix proteins of different enveloped viruses, it is plausible that pH-driven matrix disassembly in an ion channel-independent manner occurs in other late-penetrating enveloped viruses. We hypothesize that matrix proteins undergo disassembly during endosomal entry which facilitates subsequent membrane fusion driven by class I or class III membrane fusion proteins.

## Inhibitors of Viral Membrane Fusion

Viral membrane fusion represents an excellent target for antiviral therapy intervening early during viral infection. Numerous small molecules either blocking conformational changes of the viral membrane fusion protein or changing the fluidity of the membranes have been reported and some have been used therapeutically. Membrane fusion inhibitors can be used as direct-acting antivirals (DAA) or as host-targeting agents (HTA) (Ji and Li 2020). DAAs are used directly on the virus particles and thus can be employed as a means to inactivate viruses and produce virus-based inactivated vaccines. In contrast, HTA is used to target virus entry directly inside cells and thus is applied as a medication to minimize the disease outcome. Three types of inhibitors will be discussed in the following: *fusion peptide inhibitors, glycoprotein stabilizing inhibitors*, and recently developed *membrane ordering inhibitors*.

Among promising inhibitors targeting viral fusion proteins are fusion peptide inhibitors that prevent the formation of six-helix bundles in class I fusion proteins, thereby blocking conformational transitions to the postfusion conformation (reviewed in Behzadipour and Hemmati 2022). Fusion peptide inhibitors are analogues to the HRD 1 and 2 domains present in the fusion protein. They are designed to bind to one of the HRDs with high affinity and thereby sterically prevent the folding of the fusion protein and membrane fusion. To increase their potency, fusion peptide inhibitors are often lipidated, which targets them to the plasma membrane or endosomal membrane where viral fusion occurs. Currently, the only licensed fusion peptide inhibitor is enfuvirtide, commercially known as Fuzeon, which is applied in combination with antiviral therapy against HIV. Enfuvirtide targets the formation of the six-helix bundle by gp41, which constitutes the fusogenic protein domain of HIV Env. Although HRDs are highly conserved, enfuvirtide is not effective against all HIV-1 subtypes, and it was reported that mutations conferring resistance emerge in the first heptad repeat of gp41 during the treatment (Lu et al. 2006). Recently, fusion peptide inhibitors against SARS-CoV-2 were designed and tested, showing promising results in vitro and in vivo. Intranasal administration of lipidated fusion peptide inhibitors prevented SARS-CoV-2 transmission in ferrets (de Vries et al. 2021). In addition, the same lipidated fusion peptide inhibitors were applied to visualize arrested membrane fusion between ACE2-containing vesicles produced from 293T cells and SARS-CoV-2 virus-like particles by cryo-ET (Marcink et al. 2022). The data revealed extended and partially folded intermediate states of S2, which is the fusogenic protein of the S glycoprotein. Importantly, recent studies showed that fusion peptide inhibitors designed to target six-helix bundle formation of SARS-CoV-2 S confer broad-spectrum activity against divergent human coronaviruses (Zhu et al. 2021).

Conformational changes of the influenza A virus HA leading to the release of the fusion peptide represent the rate-limiting step during HA-mediated membrane fusion. Recently, small molecules stabilizing the prefusion conformation of glycoproteins emerged as promising inhibitors of viral entry (reviewed in Chen et al. 2021). Tertiary butylhydroquinone (TBHQ) and arbidol are small molecules that bind to the interface between central alpha helices on the trimeric interface of influenza A virus HA (Kadam and Wilson 2017; Russell et al. 2008). Interestingly, both molecules are more effective against HA phylogenetic group 2 and bind to a similar region and do not directly interact with the fusion peptide (Chen et al. 2021).

For example, arbidol binds with affinity KDs = 5.6–7.9 μM to group 2 HAs and with affinity KDs = 18.8–44.3 μM to group 1 HAs (Brancato et al. 2013). They stabilize the interaction between the alpha helices and serve as a molecular glue preventing conformational changes of HA and thereby contributing to its thermostability. A recent study showed that arbidol inhibits influenza A virus infectivity by 1000-fold at 19.5 μM concentration in vitro, and single virion membrane virus fusion assays using supported bilayers revealed that the hemifusion lag time is increased by a factor of 4 from 80 to around 320 s at a concentration of 40 μM and at pH 5.2 (Li et al. 2022). Arbidol is licensed in Russia and China where it is available over the counter as a medication against influenza A virus infection under the tradename Umifenovir. However, the compound was not submitted for FDA approval and further clinical studies are required to evaluate the side effects and efficacy of the drug. With the emergence of the SARS-CoV-2 pandemic, arbidol was tested as a potential inhibitor of SARS-CoV-2 S glycoprotein-mediated membrane fusion. Similar to the inhibition of influenza A virus entry, arbidol was shown to block SARS-CoV-2 entry in vitro by physically binding to the S glycoprotein (Shuster et al. 2021). Experiments testing arbidol as an inhibitor against other class I fusion proteins indicate that arbidol acts in vitro as a broad-spectrum inhibitor and blocks membrane fusion of other enveloped viruses such as SARS-CoV and hepatitis B virus (HBV), hepatitis C virus (HCV), chikungunya virus (CHIKV), and Hantan virus but also non-enveloped viruses such as reovirus and coxsackie virus B5 (Blaising et al. 2014). Since non-enveloped viruses do not rely on membrane fusion, arbidol may act by diverse synergistic mechanisms to increase the energy barrier of membrane fusion. To further shed light on the inhibitory effect of arbidol on virus entry, we performed cryo-ET of influenza A virus (A/WSN/33) with liposomes in the absence and presence of arbidol at concentration of 100 μM (Fig. 4.7). This unpublished data shows that HA is able to undergo conformational changes at low pH in the presence of arbidol and is able to execute membrane fusion with liposomes as indicated by the presence of postfusion events at pH5 plunge frozen after 10 minutes post-acidification. Overall, these preliminary results indicate that arbidol is not able to completely stabilize HA prefusion conformation but likely reduces the fusion kinetics and decreases the pH optimal threshold for HA triggering. Importantly, HA stabilizing drugs have great potential not only as potential medication but also for vaccine production. Compounds based on arbidol structure have shown greater inhibitory potential then arbidol (Brancato et al. 2013) and thus further research and novel structures might lead to even more promising spike-stabilizing drugs.

In contrast to arbidol, another small organic molecule toremifene binds between the Ebola virus glycoprotein domain 1 (GP1) and the GP2 domain to destabilize the prefusion structure of the glycoprotein. Fusion assays using HIV-1-derived pseudoviruses carrying Ebola virus GP showed that GP-mediated membrane fusion is completely inhibited at toremifene concentrations of 15 μM. Interestingly, data obtained from thermal shift assays revealed that toremifene increases the GP melting temperature at pH 5.2 (Zhao et al. 2016), which indicates that toremifene destabilizes

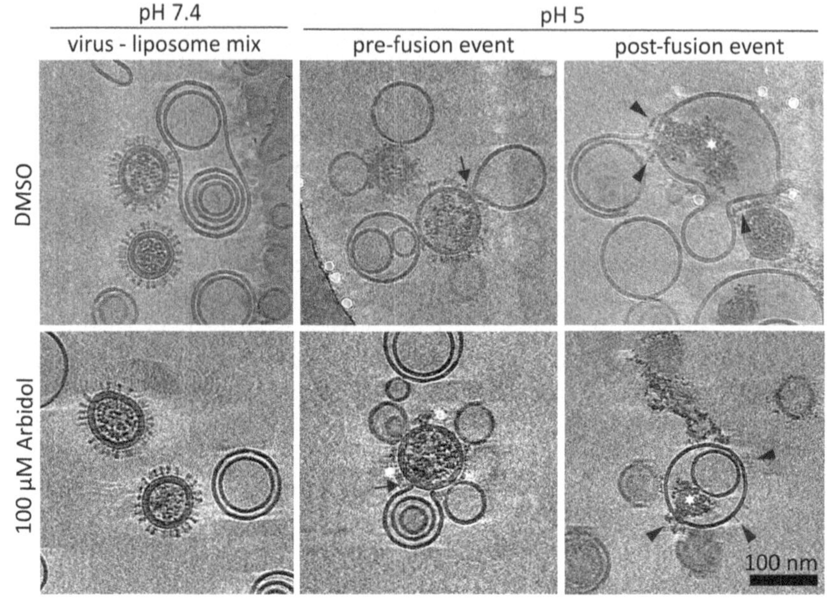


**Fig. 4.7** Cryo-ET of influenza A virus fusion with liposomes in the presence and absence of arbidol. At slices of cryo-electron tomograms showing influenza A virus–liposome interactions in the presence and absence of arbidol at pH 7.4 or 5. At neutral pH, influenza A viruses show hemagglutinin glycoproteins in the prefusion conformation (left column). At low pH, prefusion (middle column) and postfusion events (right column) can be observed. Liposome fusion sites are marked by black arrows; disassembled matrix layer and viral ribonucleoproteins are marked by a white star; hemagglutinin glycoproteins in postfusion conformation distributed on the surface of the fusion product are indicated by black arrowheads. Liposomes with a molar ratio of 55 mol% POPC, 40 mol% cholesterol, and 5 mol% total gangliosides were prepared by extrusion through a 200 nm filter. Sucrose-purified influenza A virus A/WSN/33 ($1 \times 10^6$ PFU/ml) was incubated either with 100 μM arbidol (Sigma Aldrich) or with DMSO (control) at room temperature for 4 h. Viruses were mixed with liposomes, and membrane fusion was triggered by the addition of citric acid to acidify the mixture to pH 5 and incubated for 10 min at 37 °C prior to addition plunge freezing and cryo-ET was performed using a Krios cryo-TEM (Thermo Fisher Scientific) equipped with K3 direct electron detector (Gatan). Tilt series were acquired using a dose-symmetric tilting scheme (Hagen et al. 2017) in SerialEM (Mastronarde 2005). Records were acquired with a target focus of −4 μm and 3 $e^{-}/Å^2$ electron dose at a magnification of 33,000× (pixel spacing of 2671 Å). Reconstruction was performed in the IMOD package (Kremer et al. 1996)

the GP structure, potentially leading to premature conformational changes and release of the fusion peptide. However, it is well established that the fusogenic activity of GP is dependent on cathepsin-mediated proteolytical processing in late endosomes, yielding GP2 connected to a 19 kDa GP1 fragment with an exposed fusion peptide. Hence, thermal shift assays using GP after cathepsin processing would impart further information on the stability of GP2 in the presence and absence of toremifene. In addition, further structural data in the presence of the Niemann-Pick C1 protein (NPC1), the endosomal Ebola virus receptor binding to GP1, are needed to obtain more information on the mode of action of toremifene. Toremifene interactions with GP2 may mimic the GP2 activation by NPC1 and could be used to structurally investigate the postfusion conformation of GP2, which remains to be determined.

Since membrane fusion is sensitive to the lipid composition of both viral and target membranes, small molecules that insert into membranes and alter membrane fluidity or lead to local changes in lipid composition in the vicinity of fusion sites would allow for the inhibition of a broad spectrum of viral fusion proteins. A small molecule that increases lipid order called XM-01 has been shown to potently inhibit membrane fusion of influenza A viruses (Pacheco et al. 2022). Interestingly, the authors proposed to use the compound for viral inactivation instead of formaldehyde which is required for the production of effective whole-inactivated enveloped virus vaccines. Influenza A viruses inactivated by formaldehyde or β-propiolactone are used as annual vaccines against the flu. However, formaldehyde inactivation reduces the immunogenicity of surface glycoproteins. The study shows that XM-01-inactivated influenza A viruses provide improved neutralizing antibody responses against both HA and neuraminidase. Therefore, inactivating viruses by altering viral envelope rigidity without compromising surface glycoprotein structure is a novel promising avenue in vaccine development.

## Conclusion

Entry of enveloped and non-enveloped viruses requires the viral fusion with or penetration of host cell membranes to deliver the viral genomes to the cytosol. While the entry of enveloped viruses has been studied in great mechanistic detail, the entry of the seemingly mechanistically simpler membrane penetration of non-enveloped viruses remains largely enigmatic. Tremendous advancements in microscopy techniques both fluorescence-based and in cryo-electron microscopy (briefly discussed above) at the single virion level have revealed important insights into fusion and penetration machinery. Despite major differences in lipid interactions and mechanisms to cross membrane barriers, these studies have revealed strategies shared between enveloped and non-enveloped viruses, including common triggers (such as low pH) and structural rearrangements that allow their interaction with host cell lipids. Since both types of viruses heavily rely on the host cell machinery and membrane properties at the site of entry, inhibitors targeting membrane fusion of enveloped viruses may be potent against non-enveloped viruses as well, as demonstrated for example by the inhibitory effect of 25-hydroxycholesterol. In perspective, integrative approaches along with more physiological models are needed to further understand viral membrane fusion and penetration. While supported bilayers and liposomes proved to be invaluable tools to study these processes, in situ imaging and assays capturing viral endosomal escape are needed to understand virus endosomal escape beyond the fusion and penetration machinery.

**Acknowledgments** We would like to thank Prof. Kelly Lee and Prof. Hong Zhou for providing images, and Dr. Sourav Haldar for critical reading and comments on the manuscript. Only a selection of recent primary literature has been cited and we apologize for not referencing all primary literature. The author acknowledges the funding provided by Chica and Heinz Schaller Foundation and by DFG—Deutsche Forschungsgemeinschaft (German Research Foundation), SFB1129 Project 19. We thank the Infectious Diseases Imaging Platform (IDIP) at the Center for Integrative Infectious Disease Research, Heidelberg, and the cryo-EM network at Heidelberg University (HD-cryoNET) for support and assistance. The authors gratefully acknowledge the data storage service SDS@hd supported by the Ministry of Science, Research, and the Arts Baden-Württemberg (MWK) and the German Research Foundation (DFG) through grant INST 35/1314-1 FUGG and INST 35/1503-1 FUGG.

## Glossary

**Endocytic escape** Exit of virions from endosomes by membrane fusion or penetration after internalization via the endocytic route typically leads to nucleocapsid or viral genome release into the cytosol

**Extended intermediate** Transient conformational state of fusion proteins with exposed fusion peptides

**Hemifusion diaphragm** Expanded hemifusion stalk, the potential dead-end product during the fusion of two outer membrane layers

**Hemifusion stalk** Fusion intermediate of two membranes, whose outer layers have fused and undergo lipid demixing

**Lipid demixing** Lipid exchange between two fusing membranes

**Membrane fusion protein** Transmembrane protein that undergoes conformational changes to drive a non-leaky fusion of viral and host membrane

**Non-enveloped virus** Virus lacking a membrane coat

**Penetration protein** Capsid-associated protein of non-enveloped viruses that interacts with lipids in the target membrane after activation by an external cue (such as low pH) and creates local ruptures or pores

**Postfusion conformation** Fusion and penetration protein machinery in a low-energy conformation after performing membrane crossing

**Prefusion conformation** Fusion and penetration protein machinery in metastable conformation able to facilitate membrane crossing

**Priming** Proteolytic processing of membrane fusion or perforation proteins

**Virion** Virus particle as a structural entity includes infectious and non-infectious particles

**Virus uncoating** Processes leading to the destabilization of virion structures that enable genome release from the viral particle

## References


Acciani MD, Lay Mendoza MF, Havranek KE, Duncan AM, Iyer H, Linn OL, Brindley MA (2021) Ebola virus requires phosphatidylserine scrambling activity for efficient budding and optimal infectivity. J Virol 95(20):e0116521. https://doi.org/10.1128/JVI.01165-21

Akache B, Grimm D, Shen X, Fuess S, Yant SR, Glazer DS, Park J, Kay MA (2007) A two-hybrid screen identifies cathepsins B and L as uncoating factors for adeno-associated virus 2 and 8. Mol Ther 15(2):330–339. https://doi.org/10.1038/sj.mt.6300053

Akimov SA, Volynsky PE, Galimzyanov TR, Kuzmin PI, Pavlov KV, Batishchev OV (2017) Pore formation in lipid membrane I: Continuous reversible trajectory from intact bilayer through hydrophobic defect to transversal pore. Sci Rep 7(1):12152. https://doi.org/10.1038/s41598-017-12127-7

Albertini AA, Baquero E, Ferlin A, Gaudin Y (2012) Molecular and cellular aspects of rhabdovirus entry. Viruses 4(1):117–139. https://doi.org/10.3390/v4010117

Amini-Bavil-Olyaee S, Choi YJ, Lee JH, Shi M, Huang IC, Farzan M, Jung JU (2013) The antiviral effector IFITM3 disrupts intracellular cholesterol homeostasis to block viral entry. Cell Host Microbe 13(4):452–464. https://doi.org/10.1016/j.chom.2013.03.006

Anafu AA, Bowen CH, Chin CR, Brass AL, Holm GH (2013) Interferon-inducible transmembrane protein 3 (IFITM3) restricts reovirus cell entry. J Biol Chem 288(24):17261–17271. https://doi.org/10.1074/jbc.M112.438515

Arias CF, Lopez S (2021) Rotavirus cell entry: not so simple after all. Curr Opin Virol 48:42–48. https://doi.org/10.1016/j.coviro.2021.03.011

Bajramovic JJ, Munter S, Syan S, Nehrbass U, Brahic M, Gonzalez-Dunia D (2003) Borna disease virus glycoprotein is required for viral dissemination in neurons. J Virol 77(22):12222–12231. https://doi.org/10.1128/jvi.77.22.12222-12231.2003

Bamford DH, Romantschuk M, Somerharju PJ (1987) Membrane fusion in prokaryotes: bacteriophage phi 6 membrane fuses with the Pseudomonas syringae outer membrane. EMBO J 6(5): 1467–1473. https://doi.org/10.1002/j.1460-2075.1987.tb02388.x

Battles MB, Langedijk JP, Furmanova-Hollenstein P, Chaiwatpongsakorn S, Costello HM, Kwanten L, Vranckx L, Vink P, Jaensch S, Jonckers TH, Koul A, Arnoult E, Peeples ME, Roymans D, McLellan JS (2016) Molecular mechanism of respiratory syncytial virus fusion inhibitors. Nat Chem Biol 12(2):87–93. https://doi.org/10.1038/nchembio.1982

Behzadipour Y, Hemmati S (2022) Viral prefusion targeting using entry inhibitor peptides: the case of SARS-CoV-2 and influenza A virus. Int J Pept Res Ther 28(1):42. https://doi.org/10.1007/s10989-021-10357-y

Belot L, Albertini A, Gaudin Y (2019) Structural and cellular biology of rhabdovirus entry. Adv Virus Res 104:147–183. https://doi.org/10.1016/bs.aivir.2019.05.003

Benfield CT, MacKenzie F, Ritzefeld M, Mazzon M, Weston S, Tate EW, Teo BH, Smith SE, Kellam P, Holmes EC, Marsh M (2020) Bat IFITM3 restriction depends on S-palmitoylation and a polymorphic site within the CD225 domain. Life Sci Alliance 3(1). https://doi.org/10.26508/lsa.201900542

Benton DJ, Gamblin SJ, Rosenthal PB, Skehel JJ (2020) Structural transitions in influenza haemagglutinin at membrane fusion pH. Nature 583(7814):150–153. https://doi.org/10.1038/s41586-020-2333-6

Bitto D, Halldorsson S, Caputo A, Huiskonen JT (2016) Low pH and anionic lipid-dependent fusion of uukuniemi phlebovirus to liposomes. J Biol Chem 291(12):6412–6422. https://doi.org/10.1074/jbc.M115.691113

Blaising J, Polyak SJ, Pecheur EI (2014) Arbidol as a broad-spectrum antiviral: an update. Antiviral Res 107:84–94. https://doi.org/10.1016/j.antiviral.2014.04.006

Blanc M, Hsieh WY, Robertson KA, Kropp KA, Forster T, Shui G, Lacaze P, Watterson S, Griffiths SJ, Spann NJ, Meljon A, Talbot S, Krishnan K, Covey DF, Wenk MR, Craigon M, Ruzsics Z, Haas J, Angulo A, Griffiths WJ, Glass CK, Wang Y, Ghazal P (2013) The transcription factor

STAT-1 couples macrophage synthesis of 25-hydroxycholesterol to the interferon antiviral response. Immunity 38(1):106–118. https://doi.org/10.1016/j.immuni.2012.11.004

Brancato V, Peduto A, Wharton S, Martin S, More V, Di Mola A, Massa A, Perfetto B, Donnarumma G, Schiraldi C, Tufano MA, de Rosa M, Filosa R, Hay A (2013) Design of inhibitors of influenza virus membrane fusion: synthesis, structure-activity relationship and in vitro antiviral activity of a novel indole series. Antiviral Res 99(2):125–135. https://doi.org/10.1016/j.antiviral.2013.05.005

Brass AL, Huang IC, Benita Y, John SP, Krishnan MN, Feeley EM, Ryan BJ, Weyer JL, van der Weyden L, Fikrig E, Adams DJ, Xavier RJ, Farzan M, Elledge SJ (2009) The IFITM proteins mediate cellular resistance to influenza A H1N1 virus, West Nile virus, and dengue virus. Cell 139(7):1243–1254. https://doi.org/10.1016/j.cell.2009.12.017

Bullough PA, Hughson FM, Skehel JJ, Wiley DC (1994) Structure of influenza haemagglutinin at the pH of membrane fusion. Nature 371(6492):37–43. https://doi.org/10.1038/371037a0

Calder LJ, Rosenthal PB (2016) Cryomicroscopy provides structural snapshots of influenza virus membrane fusion. Nat Struct Mol Biol 23(9):853–858. https://doi.org/10.1038/nsmb.3271

Campillay-Veliz CP, Carvajal JJ, Avellaneda AM, Escobar D, Covian C, Kalergis AM, Lay MK (2020) Human norovirus proteins: implications in the replicative cycle, pathogenesis, and the host immune response. Front Immunol 11:961. https://doi.org/10.3389/fimmu.2020.00961

Chang HW, Yang CH, Luo YC, Su BG, Cheng HY, Tung SY, Carillo KJD, Liao YT, Tzou DM, Wang HC, Chang W (2019) Vaccinia viral A26 protein is a fusion suppressor of mature virus and triggers membrane fusion through conformational change at low pH. PLoS Pathog 15(6): e1007826. https://doi.org/10.1371/journal.ppat.1007826

Chen B (2019) Molecular mechanism of HIV-1 entry. Trends Microbiol 27(10):878–891. https://doi.org/10.1016/j.tim.2019.06.002

Chen YJ, Liu X, Tsai B (2019) SV40 hijacks cellular transport, membrane penetration, and disassembly machineries to promote infection. Viruses 11(10). https://doi.org/10.3390/v11100917

Chen Z, Cui Q, Caffrey M, Rong L, Du R (2021) Small molecule inhibitors of influenza virus entry. Pharmaceuticals (Basel) 14(6). https://doi.org/10.3390/ph14060587

Chernomordik LV, Kozlov MM (2003) Protein-lipid interplay in fusion and fission of biological membranes. Annu Rev Biochem 72:175–207. https://doi.org/10.1146/annurev.biochem.72.121801.161504

Chernomordik LV, Kozlov MM (2008) Mechanics of membrane fusion. Nat Struct Mol Biol 15(7): 675–683. https://doi.org/10.1038/nsmb.1455

Chlanda P, Mekhedov E, Waters H, Schwartz CL, Fischer ER, Ryham RJ, Cohen FS, Blank PS, Zimmerberg J (2016) The hemifusion structure induced by influenza virus haemagglutinin is determined by physical properties of the target membranes. Nat Microbiol 1(6):16050. https://doi.org/10.1038/nmicrobiol.2016.50

Colom A, Derivery E, Soleimanpour S, Tomba C, Molin MD, Sakai N, Gonzalez-Gaitan M, Matile S, Roux A (2018) A fluorescent membrane tension probe. Nat Chem 10(11): 1118–1125. https://doi.org/10.1038/s41557-018-0127-3

Danev R, Buijsse B, Khoshouei M, Plitzko JM, Baumeister W (2014) Volta potential phase plate for in-focus phase contrast transmission electron microscopy. Proc Natl Acad Sci U S A 111(44):15635–15640. https://doi.org/10.1073/pnas.1418377111

Das DK, Govindan R, Nikic-Spiegel I, Krammer F, Lemke EA, Munro JB (2018) Direct visualization of the conformational dynamics of single influenza hemagglutinin trimers. Cell 174(4): 926–937 e912. https://doi.org/10.1016/j.cell.2018.05.050

Das DK, Bulow U, Diehl WE, Durham ND, Senjobe F, Chandran K, Luban J, Munro JB (2020) Conformational changes in the Ebola virus membrane fusion machine induced by pH, Ca2+, and receptor binding. PLoS Biol 18(2):e3000626. https://doi.org/10.1371/journal.pbio.3000626

Das T, Yang X, Lee H, Garst E, Valencia E, Chandran K, Im W, Hang H (2021) S-palmitoylation and sterol interactions mediate antiviral specificity of IFITM isoforms. Res Sq. https://doi.org/10.21203/rs.3.rs-1179000/v1

Desai TM, Marin M, Chin CR, Savidis G, Brass AL, Melikyan GB (2014) IFITM3 restricts influenza A virus entry by blocking the formation of fusion pores following virus-endosome hemifusion. PLoS Pathog 10(4):e1004048. https://doi.org/10.1371/journal.ppat.1004048

Desingu PA, Nagarajan K, Dhama K (2022) Can a torque teno virus (TTV) be a naked DNA particle without a virion structure? Front Virol 2. https://doi.org/10.3389/fviro.2022.821298

de Vries RD, Schmitz KS, Bovier FT, Predella C, Khao J, Noack D, Haagmans BL, Herfst S, Stearns KN, Drew-Bear J, Biswas S, Rockx B, McGill G, Dorrello NV, Gellman SH, Alabi CA, de Swart RL, Moscona A, Porotto M (2021) Intranasal fusion inhibitory lipopeptide prevents direct-contact SARS-CoV-2 transmission in ferrets. Science 371(6536):1379–1382. https://doi.org/10.1126/science.abf4896

Domingues MM, Gomes B, Hollmann A, Santos NC (2021) 25-hydroxycholesterol effect on membrane structure and mechanical properties. Int J Mol Sci 22(5). https://doi.org/10.3390/ijms22052574

Doms RW (2017) What came first—the virus or the egg? Cell 168(5):755–757. https://doi.org/10.1016/j.cell.2017.02.012

Dymond MK (2021) Lipid monolayer spontaneous curvatures: A collection of published values. Chem Phys Lipids 239:105117. https://doi.org/10.1016/j.chemphyslip.2021.105117

El Omari K, Li S, Kotecha A, Walter TS, Bignon EA, Harlos K, Somerharju P, De Haas F, Clare DK, Molin M, Hurtado F, Li M, Grimes JM, Bamford DH, Tischler ND, Huiskonen JT, Stuart DI, Roine E (2019) The structure of a prokaryotic viral envelope protein expands the landscape of membrane fusion proteins. Nat Commun 10(1):846. https://doi.org/10.1038/s41467-019-08728-7

Esnault C, Priet S, Ribet D, Vernochet C, Bruls T, Lavialle C, Weissenbach J, Heidmann T (2008) A placenta-specific receptor for the fusogenic, endogenous retrovirus-derived, human syncytin-2. Proc Natl Acad Sci U S A 105(45):17532–17537. https://doi.org/10.1073/pnas.0807413105

Fedry J, Liu Y, Pehau-Arnaudet G, Pei J, Li W, Tortorici MA, Traincard F, Meola A, Bricogne G, Grishin NV, Snell WJ, Rey FA, Krey T (2017) The ancient gamete fusogen HAP2 is a eukaryotic class II fusion protein. Cell 168(5):904–915 e910. https://doi.org/10.1016/j.cell.2017.01.024

Fontana J, Steven AC (2015) Influenza virus-mediated membrane fusion: Structural insights from electron microscopy. Arch Biochem Biophys 581:86–97. https://doi.org/10.1016/j.abb.2015.04.011

Garcia-Moro E, Zhang J, Calder LJ, Brown NR, Gamblin SJ, Skehel JJ, Rosenthal PB (2022) Reversible structural changes in the influenza hemagglutinin precursor at membrane fusion pH. Proc Natl Acad Sci U S A 119(33):e2208011119. https://doi.org/10.1073/pnas.2208011119

Garry CE, Garry RF (2009) Proteomics computational analyses suggest that the bornavirus glycoprotein is a class III viral fusion protein (gamma penetrene). Virol J 6:145. https://doi.org/10.1186/1743-422X-6-145

Garst EH, Lee H, Das T, Bhattacharya S, Percher A, Wiewiora R, Witte IP, Li Y, Peng T, Im W, Hang HC (2021) Site-specific lipidation enhances IFITM3 membrane interactions and antiviral activity. ACS Chem Biol 16(5):844–856. https://doi.org/10.1021/acschembio.1c00013

Gaudin Y (2021) Penetration of non-enveloped viruses. Nat Microbiol 6(11):1343–1344. https://doi.org/10.1038/s41564-021-00991-z

Gilman MSA, Furmanova-Hollenstein P, Pascual G, BvtW A, Langedijk JPM, McLellan JS (2019) Transient opening of trimeric prefusion RSV F proteins. Nat Commun 10(1):2105. https://doi.org/10.1038/s41467-019-09807-5

Gomes B, Goncalves S, Disalvo A, Hollmann A, Santos NC (2018) Effect of 25-hydroxycholesterol in viral membrane fusion: Insights on HIV inhibition. Biochim Biophys Acta Biomembr 1860(5):1171–1178. https://doi.org/10.1016/j.bbamem.2018.02.001

Gomes B, Sanna G, Madeddu S, Hollmann A, Santos NC (2019) Combining 25-hydroxycholesterol with an HIV fusion inhibitor peptide: interaction with biomembrane model systems and human blood cells. ACS Infect Dis 5(4):582–591. https://doi.org/10.1021/acsinfecdis.8b00321

Graziano VR, Wei J, Wilen CB (2019) Norovirus attachment and entry. Viruses 11(6). https://doi.org/10.3390/v11060495

Gui L, Ebner JL, Mileant A, Williams JA, Lee KK (2016) Visualization and sequencing of membrane remodeling leading to influenza virus fusion. J Virol 90(15):6948–6962. https://doi.org/10.1128/JVI.00240-16

Guo X, Steinkuhler J, Marin M, Li X, Lu W, Dimova R, Melikyan GB (2021) Interferon-induced transmembrane protein 3 blocks fusion of diverse enveloped viruses by altering mechanical properties of cell membranes. ACS Nano 15(5):8155–8170. https://doi.org/10.1021/acsnano.0c10567

Hacke M, Bjorkholm P, Hellwig A, Himmels P, Ruiz de Almodovar C, Brugger B, Wieland F, Ernst AM (2015) Inhibition of Ebola virus glycoprotein-mediated cytotoxicity by targeting its transmembrane domain and cholesterol. Nat Commun 6:7688. https://doi.org/10.1038/ncomms8688

Hagen WJH, Wan W, Briggs JAG (2017) Implementation of a cryo-electron tomography tilt-scheme optimized for high resolution subtomogram averaging. J Struct Biol 197(2):191–198. https://doi.org/10.1016/j.jsb.2016.06.007

Haldar S (2022) Recent developments in single-virus fusion assay. J Membr Biol. https://doi.org/10.1007/s00232-022-00270-w

Haldar S, Mekhedov E, McCormick CD, Blank PS, Zimmerberg J (2018) Lipid-dependence of target membrane stability during influenza viral fusion. J Cell Sci 132(4). https://doi.org/10.1242/jcs.218321

Haldar S, Okamoto K, Dunning RA, Kasson PM (2020) Precise triggering and chemical control of single-virus fusion within endosomes. J Virol 95(1). https://doi.org/10.1128/JVI.01982-20

Halldorsson S, Li S, Li M, Harlos K, Bowden TA, Huiskonen JT (2018) Shielding and activation of a viral membrane fusion protein. Nat Commun 9(1):349. https://doi.org/10.1038/s41467-017-02789-2

Halldorsson S, Sader K, Turner J, Calder LJ, Rosenthal PB (2021) In situ structure and organization of the influenza C virus surface glycoprotein. Nat Commun 12(1):1694. https://doi.org/10.1038/s41467-021-21818-9

Harrison SC (2015) Viral membrane fusion. Virology 479–480:498–507. https://doi.org/10.1016/j.virol.2015.03.043

Hashiguchi T, Fukuda Y, Matsuoka R, Kuroda D, Kubota M, Shirogane Y, Watanabe S, Tsumoto K, Kohda D, Plemper RK, Yanagi Y (2018) Structures of the prefusion form of measles virus fusion protein in complex with inhibitors. Proc Natl Acad Sci U S A 115(10):2496–2501. https://doi.org/10.1073/pnas.1718957115

Hastie KM, Igonet S, Sullivan BM, Legrand P, Zandonatti MA, Robinson JE, Garry RF, Rey FA, Oldstone MB, Saphire EO (2016) Crystal structure of the prefusion surface glycoprotein of the prototypic arenavirus LCMV. Nat Struct Mol Biol 23(6):513–521. https://doi.org/10.1038/nsmb.3210

Hrebik D, Fuzik T, Gondova M, Smerdova L, Adamopoulos A, Sedo O, Zdrahal Z, Plevka P (2021) ICAM-1 induced rearrangements of capsid and genome prime rhinovirus 14 for activation and uncoating. Proc Natl Acad Sci U S A 118(19). https://doi.org/10.1073/pnas.2024251118

Hullin-Matsuda F, Kawasaki K, Delton-Vandenbroucke I, Xu Y, Nishijima M, Lagarde M, Schlame M, Kobayashi T (2007) De novo biosynthesis of the late endosome lipid, bis(monoacylglycero)phosphate. J Lipid Res 48(9):1997–2008. https://doi.org/10.1194/jlr.M700154-JLR200

Iaea DB, Maxfield FR (2015) Cholesterol trafficking and distribution. Essays Biochem 57:43–55. https://doi.org/10.1042/bse0570043

Ivanovic T, Choi JL, Whelan SP, van Oijen AM, Harrison SC (2013) Influenza-virus membrane fusion by cooperative fold-back of stochastically induced hemagglutinin intermediates. Elife 2:e00333. https://doi.org/10.7554/eLife.00333

Ji X, Li Z (2020) Medicinal chemistry strategies toward host targeting antiviral agents. Med Res Rev 40(5):1519–1557. https://doi.org/10.1002/med.21664

Kadam RU, Wilson IA (2017) Structural basis of influenza virus fusion inhibition by the antiviral drug Arbidol. Proc Natl Acad Sci U S A 114(2):206–214. https://doi.org/10.1073/pnas.1617020114

Kadlec J, Loureiro S, Abrescia NG, Stuart DI, Jones IM (2008) The postfusion structure of baculovirus gp64 supports a unified view of viral fusion machines. Nat Struct Mol Biol 15(10):1024–1030. https://doi.org/10.1038/nsmb.1484

Karal MAS, Mokta NA, Levadny V, Belaya M, Ahmed M, Ahamed MK, Ahammed S (2022) Effects of cholesterol on the size distribution and bending modulus of lipid vesicles. PLoS One 17(1):e0263119. https://doi.org/10.1371/journal.pone.0263119

Kartenbeck J, Stukenbrok H, Helenius A (1989) Endocytosis of simian virus 40 into the endoplasmic reticulum. J Cell Biol 109(6 Pt 1):2721–2729. https://doi.org/10.1083/jcb.109.6.2721

Kaufmann B, Chipman PR, Holdaway HA, Johnson S, Fremont DH, Kuhn RJ, Diamond MS, Rossmann MG (2009) Capturing a flavivirus pre-fusion intermediate. PLoS Pathog 5(11): e1000672. https://doi.org/10.1371/journal.ppat.1000672

Kielian M, Rey FA (2006) Virus membrane-fusion proteins: more than one way to make a hairpin. Nat Rev Microbiol 4(1):67–76. https://doi.org/10.1038/nrmicro1326

Klein S, Golani G, Lolicato F, Beyer D, Herrmann A, Wachsmuth-Melm M, Reddmann N, Brecht R, Lahr C, Hosseinzadeh M, Kolovou A, Schorb M, Schwab Y, Brügger B, Nickel W, Schwarz US, Chlanda P (2023) IFITM3 blocks viral entry by sorting lipids and stabilizing hemifusion. Cell Host Microbe. https://doi.org/10.1016/j.chom.2023.03.005

Kremer JR, Mastronarde DN, McIntosh JR (1996) Computer visualization of three-dimensional image data using IMOD. J Struct Biol 116(1):71–76. https://doi.org/10.1006/jsbi.1996.0013

Krzyzaniak MA, Zumstein MT, Gerez JA, Picotti P, Helenius A (2013) Host cell entry of respiratory syncytial virus involves macropinocytosis followed by proteolytic activation of the F protein. PLoS Pathog 9(4):e1003309. https://doi.org/10.1371/journal.ppat.1003309

Kumar CS, Dey D, Ghosh S, Banerjee M (2018) Breach: host membrane penetration and entry by nonenveloped viruses. Trends Microbiol 26(6):525–537. https://doi.org/10.1016/j.tim.2017.09.010

Laliberte JP, Weisberg AS, Moss B (2011) The membrane fusion step of vaccinia virus entry is cooperatively mediated by multiple viral proteins and host cell components. PLoS Pathog 7(12): e1002446. https://doi.org/10.1371/journal.ppat.1002446

Lee KK (2010) Architecture of a nascent viral fusion pore. EMBO J 29(7):1299–1311. https://doi.org/10.1038/emboj.2010.13

Lee JE, Fusco ML, Hessell AJ, Oswald WB, Burton DR, Saphire EO (2008) Structure of the Ebola virus glycoprotein bound to an antibody from a human survivor. Nature 454(7201):177–182. https://doi.org/10.1038/nature07082

Lee J, Kreutzberger AJB, Odongo L, Nelson EA, Nyenhuis DA, Kiessling V, Liang B, Cafiso DS, White JM, Tamm LK (2021) Ebola virus glycoprotein interacts with cholesterol to enhance membrane fusion and cell entry. Nat Struct Mol Biol 28(2):181–189. https://doi.org/10.1038/s41594-020-00548-4

Leikina E, Ramos C, Markovic I, Zimmerberg J, Chernomordik LV (2002) Reversible stages of the low-pH-triggered conformational change in influenza virus hemagglutinin. EMBO J 21(21): 5701–5710. https://doi.org/10.1093/emboj/cdf559

Li S, Sieben C, Ludwig K, Hofer CT, Chiantia S, Herrmann A, Eghiaian F, Schaap IA (2014) pH-Controlled two-step uncoating of influenza virus. Biophys J 106(7):1447–1456. https://doi.org/10.1016/j.bpj.2014.02.018

Li Z, Li T, Liu M, Ivanovic T (2022) Hemagglutinin stability determines influenza A virus susceptibility to a broad-spectrum fusion inhibitor arbidol. ACS Infect Dis 8(8):1543–1552. https://doi.org/10.1021/acsinfecdis.2c00178

Lin S, Zhang YJ (2021) Advances in hepatitis E virus biology and pathogenesis. Viruses 13(2). https://doi.org/10.3390/v13020267

Lins-Austin B, Patel S, Mietzsch M, Brooke D, Bennett A, Venkatakrishnan B, Van Vliet K, Smith AN, Long JR, McKenna R, Potter M, Byrne B, Boye SL, Bothner B, Heilbronn R, Agbandje-

McKenna M (2020) Adeno-associated virus (AAV) capsid stability and liposome remodeling during endo/lysosomal pH trafficking. Viruses 12(6). https://doi.org/10.3390/v12060668

Liu KN, Boxer SG (2020) Target membrane cholesterol modulates single influenza virus membrane fusion efficiency but not rate. Biophys J 118(10):2426–2433. https://doi.org/10.1016/j.bpj.2020.03.021

Liu SY, Aliyari R, Chikere K, Li G, Marsden MD, Smith JK, Pernet O, Guo H, Nusbaum R, Zack JA, Freiberg AN, Su L, Lee B, Cheng G (2013) Interferon-inducible cholesterol-25-hydroxylase broadly inhibits viral entry by production of 25-hydroxycholesterol. Immunity 38(1):92–105. https://doi.org/10.1016/j.immuni.2012.11.005

Liu H, Hong X, Xi J, Menne S, Hu J, Wang JC (2022) Cryo-EM structures of human hepatitis B and woodchuck hepatitis virus small spherical subviral particles. Sci Adv 8(31):eabo4184. https://doi.org/10.1126/sciadv.abo4184

Lu J, Deeks SG, Hoh R, Beatty G, Kuritzkes BA, Martin JN, Kuritzkes DR (2006) Rapid emergence of enfuvirtide resistance in HIV-1-infected patients: results of a clonal analysis. J Acquir Immune Defic Syndr 43(1):60–64. https://doi.org/10.1097/01.qai.0000234083.34161.55

Lund EG, Kerr TA, Sakai J, Li WP, Russell DW (1998) cDNA cloning of mouse and human cholesterol 25-hydroxylases, polytopic membrane proteins that synthesize a potent oxysterol regulator of lipid metabolism. J Biol Chem 273(51):34316–34327. https://doi.org/10.1074/jbc.273.51.34316

Majdoul S, Compton AA (2022) Lessons in self-defence: inhibition of virus entry by intrinsic immunity. Nat Rev Immunol 22(6):339–352. https://doi.org/10.1038/s41577-021-00626-8

Mangala Prasad V, Blijleven JS, Smit JM, Lee KK (2022a) Visualization of conformational changes and membrane remodeling leading to genome delivery by viral class-II fusion machinery. Nat Commun 13(1):4772. https://doi.org/10.1038/s41467-022-32431-9

Mangala Prasad V, Leaman DP, Lovendahl KN, Croft JT, Benhaim MA, Hodge EA, Zwick MB, Lee KK (2022b) Cryo-ET of Env on intact HIV virions reveals structural variation and positioning on the Gag lattice. Cell 185(4):641–653 e617. https://doi.org/10.1016/j.cell.2022.01.013

Marcink TC, Kicmal T, Armbruster E, Zhang Z, Zipursky G, Golub KL, Idris M, Khao J, Drew-Bear J, McGill G, Gallagher T, Porotto M, des Georges A, Moscona A (2022) Intermediates in SARS-CoV-2 spike-mediated cell entry. Sci Adv 8(33):eabo3153. https://doi.org/10.1126/sciadv.abo3153

Markosyan RM, Marin M, Zhang Y, Cohen FS, Melikyan GB (2021) The late endosome-resident lipid bis(monoacylglycero)phosphate is a cofactor for Lassa virus fusion. PLoS Pathog 17(9):e1009488. https://doi.org/10.1371/journal.ppat.1009488

Martens S, McMahon HT (2008) Mechanisms of membrane fusion: disparate players and common principles. Nat Rev Mol Cell Biol 9(7):543–556. https://doi.org/10.1038/nrm2417

Mastronarde DN (2005) Automated electron microscope tomography using robust prediction of specimen movements. J Struct Biol 152(1):36–51. https://doi.org/10.1016/j.jsb.2005.07.007

Matamoros T, Alejo A, Rodriguez JM, Hernaez B, Guerra M, Fraile-Ramos A, Andres G (2020) African swine fever virus protein pE199L mediates virus entry by enabling membrane fusion and core penetration. mBio 11(4). https://doi.org/10.1128/mBio.00789-20

Matsuo H, Chevallier J, Mayran N, Le Blanc I, Ferguson C, Faure J, Blanc NS, Matile S, Dubochet J, Sadoul R, Parton RG, Vilbois F, Gruenberg J (2004) Role of LBPA and Alix in multivesicular liposome formation and endosome organization. Science 303(5657):531–534. https://doi.org/10.1126/science.1092425

Mayberry CL, Bond AC, Wilczek MP, Mehmood K, Maginnis MS (2021) Sending mixed signals: polyomavirus entry and trafficking. Curr Opin Virol 47:95–105. https://doi.org/10.1016/j.coviro.2021.02.004

Mercer J, Helenius A (2008) Vaccinia virus uses macropinocytosis and apoptotic mimicry to enter host cells. Science 320(5875):531–535. https://doi.org/10.1126/science.1155164

Metskas LA, Briggs JAG (2019) Fluorescence-based detection of membrane fusion state on a cryo-EM grid using correlated cryo-fluorescence and cryo-electron microscopy. Microsc Microanal 25(4):942–949. https://doi.org/10.1017/S1431927619000606

Modis Y (2014) Relating structure to evolution in class II viral membrane fusion proteins. Curr Opin Virol 5:34–41. https://doi.org/10.1016/j.coviro.2014.01.009

Mohl BP, Roy P (2014) Bluetongue virus capsid assembly and maturation. Viruses 6(8): 3250–3270. https://doi.org/10.3390/v6083250

Moretta A, Scieuzo C, Petrone AM, Salvia R, Manniello MD, Franco A, Lucchetti D, Vassallo A, Vogel H, Sgambato A, Falabella P (2021) Antimicrobial peptides: a new hope in biomedical and pharmaceutical fields. Front Cell Infect Microbiol 11:668632. https://doi.org/10.3389/fcimb.2021.668632

Moyer CL, Nemerow GR (2011) Viral weapons of membrane destruction: variable modes of membrane penetration by non-enveloped viruses. Curr Opin Virol 1(1):44–49. https://doi.org/10.1016/j.coviro.2011.05.002

Nanbo A, Maruyama J, Imai M, Ujie M, Fujioka Y, Nishide S, Takada A, Ohba Y, Kawaoka Y (2018) Ebola virus requires a host scramblase for externalization of phosphatidylserine on the surface of viral particles. PLoS Pathog 14(1):e1006848. https://doi.org/10.1371/journal.ppat.1006848

Nangia S, Boyd KJ, May ER (2020) Molecular dynamics study of membrane permeabilization by wild-type and mutant lytic peptides from the non-enveloped Flock House virus. Biochim Biophys Acta Biomembr 1862(2):183102. https://doi.org/10.1016/j.bbamem.2019.183102

Nemerow GR, Pache L, Reddy V, Stewart PL (2009) Insights into adenovirus host cell interactions from structural studies. Virology 384(2):380–388. https://doi.org/10.1016/j.virol.2008.10.016

Newton ND, Hardy JM, Modhiran N, Hugo LE, Amarilla AA, Bibby S, Venugopal H, Harrison JJ, Traves RJ, Hall RA, Hobson-Peters J, Coulibaly F, Watterson D (2021) The structure of an infectious immature flavivirus redefines viral architecture and maturation. Sci Adv 7(20). https://doi.org/10.1126/sciadv.abe4507

Nieto-Garai JA, Arboleya A, Otaegi S, Chojnacki J, Casas J, Fabrias G, Contreras FX, Krausslich HG, Lorizate M (2021) Cholesterol in the viral membrane is a molecular switch governing HIV-1 Env clustering. Adv Sci (Weinh) 8(3):2003468. https://doi.org/10.1002/advs.202003468

Otterstrom JJ, Brandenburg B, Koldijk MH, Juraszek J, Tang C, Mashaghi S, Kwaks T, Goudsmit J, Vogels R, Friesen RH, van Oijen AM (2014) Relating influenza virus membrane fusion kinetics to stoichiometry of neutralizing antibodies at the single-particle level. Proc Natl Acad Sci U S A 111(48):E5143–E5148. https://doi.org/10.1073/pnas.1411755111

Pacheco A, Buchholz D, Monreal I, Xu S, Imbiakha B, Sahler J, Jager M, Lai A, Contreras E, Cook B, Ralalage E, Liu Q, Yeo YY, Ma A, Zamora J, Shil N, Jones-Burrage S, Pritchard S, Yang C, Zhao Y, Mohamed Z, Xu C, Daniel S, Jung M, de Walle GV, Mukhopadhyay S, Shimamura M, Goodman A, Hardy M, Bose S, Nicola A, Freed J, August A, Jones J, Xian M, Aguilar H (2022) New class of broad-spectrum antivirals improves influenza virus vaccine development. Res Sq. https://doi.org/10.21203/rs.3.rs-1193746/v1

Panjwani A, Strauss M, Gold S, Wenham H, Jackson T, Chou JJ, Rowlands DJ, Stonehouse NJ, Hogle JM, Tuthill TJ (2014) Capsid protein VP4 of human rhinovirus induces membrane permeability by the formation of a size-selective multimeric pore. PLoS Pathog 10(8): e1004294. https://doi.org/10.1371/journal.ppat.1004294

Paul AV, Schultz A, Pincus SE, Oroszlan S, Wimmer E (1987) Capsid protein VP4 of poliovirus is N-myristoylated. Proc Natl Acad Sci U S A 84(22):7827–7831. https://doi.org/10.1073/pnas.84.22.7827

Peng R, Zhang S, Cui Y, Shi Y, Gao GF, Qi J (2017) Structures of human-infecting Thogotovirus fusogens support a common ancestor with insect baculovirus. Proc Natl Acad Sci U S A 114(42):E8905–E8912. https://doi.org/10.1073/pnas.1706125114

Perreira JM, Chin CR, Feeley EM, Brass AL (2013) IFITMs restrict the replication of multiple pathogenic viruses. J Mol Biol 425(24):4937–4955. https://doi.org/10.1016/j.jmb.2013.09.024

Pettersen EF, Goddard TD, Huang CC, Meng EC, Couch GS, Croll TI, Morris JH, Ferrin TE (2021) UCSF ChimeraX: Structure visualization for researchers, educators, and developers. Protein Sci 30(1):70–82. https://doi.org/10.1002/pro.3943

Plattet P, Alves L, Herren M, Aguilar HC (2016) Measles virus fusion protein: structure, function and inhibition. Viruses 8(4):112. https://doi.org/10.3390/v8040112

Rabia M, Leuzy V, Soulage C, Durand A, Fourmaux B, Errazuriz-Cerda E, Koffel R, Draeger A, Colosetti P, Jalabert A, Di Filippo M, Villard-Garon A, Bergerot C, Luquain-Costaz C, Moulin P, Rome S, Delton I, Hullin-Matsuda F (2020) Bis(monoacylglycero)phosphate, a new lipid signature of endosome-derived extracellular vesicles. Biochimie 178:26–38. https://doi.org/10.1016/j.biochi.2020.07.005

Rahman K, Datta SAK, Beaven AH, Jolley AA, Sodt AJ, Compton AA (2022) Cholesterol binds the amphipathic helix of IFITM3 and regulates antiviral activity. J Mol Biol 434(19):167759. https://doi.org/10.1016/j.jmb.2022.167759

Rey FA, Lok SM (2018) Common features of enveloped viruses and implications for immunogen design for next-generation vaccines. Cell 172(6):1319–1334. https://doi.org/10.1016/j.cell.2018.02.054

Rojas JM, Alejo A, Martin V, Sevilla N (2021) Viral pathogen-induced mechanisms to antagonize mammalian interferon (IFN) signaling pathway. Cell Mol Life Sci 78(4):1423–1444. https://doi.org/10.1007/s00018-020-03671-z

Rosa A, Chande A, Ziglio S, De Sanctis V, Bertorelli R, Goh SL, McCauley SM, Nowosielska A, Antonarakis SE, Luban J, Santoni FA, Pizzato M (2015) HIV-1 Nef promotes infection by excluding SERINC5 from virion incorporation. Nature 526(7572):212–217. https://doi.org/10.1038/nature15399

Russell RJ, Kerry PS, Stevens DJ, Steinhauer DA, Martin SR, Gamblin SJ, Skehel JJ (2008) Structure of influenza hemagglutinin in complex with an inhibitor of membrane fusion. Proc Natl Acad Sci U S A 105(46):17736–17741. https://doi.org/10.1073/pnas.0807142105

Saitoh A, Takiguchi K, Tanaka Y, Hotani H (1998) Opening-up of liposomal membranes by talin. Proc Natl Acad Sci U S A 95(3):1026–1031. https://doi.org/10.1073/pnas.95.3.1026

Sanders DW, Jumper CC, Ackerman PJ, Bracha D, Donlic A, Kim H, Kenney D, Castello-Serrano I, Suzuki S, Tamura T, Tavares AH, Saeed M, Holehouse AS, Ploss A, Levental I, Douam F, Padera RF, Levy BD, Brangwynne CP (2021) SARS-CoV-2 requires cholesterol for viral entry and pathological syncytia formation. Elife 10. https://doi.org/10.7554/eLife.65962

Sardar A, Dewangan N, Panda B, Bhowmick D, Tarafdar PK (2022) Lipid and lipidation in membrane fusion. J Membr Biol. https://doi.org/10.1007/s00232-022-00267-5

Schaefer SL, Jung H, Hummer G (2021) Binding of SARS-CoV-2 fusion peptide to host endosome and plasma membrane. J Phys Chem B 125(28):7732–7741. https://doi.org/10.1021/acs.jpcb.1c04176

Schelhaas M, Malmstrom J, Pelkmans L, Haugstetter J, Ellgaard L, Grunewald K, Helenius A (2007) Simian Virus 40 depends on ER protein folding and quality control factors for entry into host cells. Cell 131(3):516–529. https://doi.org/10.1016/j.cell.2007.09.038

Seitz S, Habjanic J, Schutz AK, Bartenschlager R (2020) The hepatitis B virus envelope proteins: molecular gymnastics throughout the viral life cycle. Annu Rev Virol 7(1):263–288. https://doi.org/10.1146/annurev-virology-092818-015508

Serris A, Stass R, Bignon EA, Muena NA, Manuguerra JC, Jangra RK, Li S, Chandran K, Tischler ND, Huiskonen JT, Rey FA, Guardado-Calvo P (2020) The hantavirus surface glycoprotein lattice and its fusion control mechanism. Cell 183(2):442–456 e416. https://doi.org/10.1016/j.cell.2020.08.023

Shukla A, Padhi AK, Gomes J, Banerjee M (2014) The VP4 peptide of hepatitis A virus ruptures membranes through formation of discrete pores. J Virol 88(21):12409–12421. https://doi.org/10.1128/JVI.01896-14

Shuster A, Pechalrieu D, Jackson CB, Abegg D, Choe H, Adibekian A (2021) Clinical antiviral drug arbidol inhibits infection by SARS-CoV-2 and variants through direct binding to the spike protein. ACS Chem Biol 16(12):2845–2851. https://doi.org/10.1021/acschembio.1c00756

Sirohi D, Chen Z, Sun L, Klose T, Pierson TC, Rossmann MG, Kuhn RJ (2016) The 3.8 A resolution cryo-EM structure of Zika virus. Science 352(6284):467–470. https://doi.org/10.1126/science.aaf5316

Staring J, von Castelmur E, Blomen VA, van den Hengel LG, Brockmann M, Baggen J, Thibaut HJ, Nieuwenhuis J, Janssen H, van Kuppeveld FJ, Perrakis A, Carette JE, Brummelkamp TR (2017) PLA2G16 represents a switch between entry and clearance of Picornaviridae. Nature 541(7637): 412–416. https://doi.org/10.1038/nature21032

Tang T, Bidon M, Jaimes JA, Whittaker GR, Daniel S (2020) Coronavirus membrane fusion mechanism offers a potential target for antiviral development. Antiviral Res 178:104792. https://doi.org/10.1016/j.antiviral.2020.104792

Timilsina U, Umthong S, Ivey EB, Waxman B, Stavrou S (2022) SARS-CoV-2 ORF7a potently inhibits the antiviral effect of the host factor SERINC5. Nat Commun 13(1):2935. https://doi.org/10.1038/s41467-022-30609-9

Torrents de la Pena A, Sliepen K, Eshun-Wilson L, Newby ML, Allen JD, Zon I, Koekkoek S, Chumbe A, Crispin M, Schinkel J, Lander GC, Sanders RW, Ward AB (2022) Structure of the hepatitis C virus E1E2 glycoprotein complex. Science 378(6617):263–269. https://doi.org/10.1126/science.abn9884

Tu M, Liu F, Chen S, Wang M, Cheng A (2015) Role of capsid proteins in parvoviruses infection. Virol J 12:114. https://doi.org/10.1186/s12985-015-0344-y

Tuthill TJ, Bubeck D, Rowlands DJ, Hogle JM (2006) Characterization of early steps in the poliovirus infection process: receptor-decorated liposomes induce conversion of the virus to membrane-anchored entry-intermediate particles. J Virol 80(1):172–180. https://doi.org/10.1128/JVI.80.1.172-180.2006

Vollmer B, Grunewald K (2020) Herpesvirus membrane fusion—a team effort. Curr Opin Struct Biol 62:112–120. https://doi.org/10.1016/j.sbi.2019.12.004

Vollmer B, Prazak V, Vasishtan D, Jefferys EE, Hernandez-Duran A, Vallbracht M, Klupp BG, Mettenleiter TC, Backovic M, Rey FA, Topf M, Grunewald K (2020) The prefusion structure of herpes simplex virus glycoprotein B. Sci Adv 6(39). https://doi.org/10.1126/sciadv.abc1726

Ward AE, Kiessling V, Pornillos O, White JM, Ganser-Pornillos BK, Tamm LK (2020) HIV-cell membrane fusion intermediates are restricted by Serincs as revealed by cryo-electron and TIRF microscopy. J Biol Chem 295(45):15183–15195. https://doi.org/10.1074/jbc.RA120.014466

Ward AE, Sokovikova D, Waxham MN, Heberle FA, Levental I, Levental KR, Kiessling V, White JM, Tamm LK (2022) Serinc5 restricts HIV membrane fusion by altering lipid order and heterogeneity in the viral membrane. bioRxiv. https://doi.org/10.1101/2022.08.25.505328

Winter SL, Chlanda P (2021) Dual-axis Volta phase plate cryo-electron tomography of Ebola virus-like particles reveals actin-VP40 interactions. J Struct Biol 213(2):107742. https://doi.org/10.1016/j.jsb.2021.107742

Winter SL, Golani G, Lolicato F, Vallbracht M, Thiyagarajah K, Ahmed SS, Lüchtenborg C, Fackler OT, Brügger B, Hoenen T, Nickel W, Schwarz US, Chlanda P (2023) The Ebola virus VP40 matrix undergoes endosomal disassembly essential for membrane fusion. EMBO J. https://doi.org/10.15252/embj.2023113578

Xia X, Wu W, Cui Y, Roy P, Zhou ZH (2021) Bluetongue virus capsid protein VP5 perforates membranes at low endosomal pH during viral entry. Nat Microbiol 6(11):1424–1432. https://doi.org/10.1038/s41564-021-00988-8

Ykema M, Tao YJ (2021) Structural insights into the human astrovirus capsid. Viruses 13(5). https://doi.org/10.3390/v13050821

York J, Dai D, Amberg SM, Nunberg JH (2008) pH-induced activation of arenavirus membrane fusion is antagonized by small-molecule inhibitors. J Virol 82(21):10932–10939. https://doi.org/10.1128/JVI.01140-08

Yount JS, Moltedo B, Yang YY, Charron G, Moran TM, Lopez CB, Hang HC (2010) Palmitoylome profiling reveals S-palmitoylation-dependent antiviral activity of IFITM3. Nat Chem Biol 6(8):610–614. https://doi.org/10.1038/nchembio.405

Zadori Z, Szelei J, Lacoste MC, Li Y, Gariepy S, Raymond P, Allaire M, Nabi IR, Tijssen P (2001) A viral phospholipase A2 is required for parvovirus infectivity. Dev Cell 1(2):291–302. https://doi.org/10.1016/s1534-5807(01)00031-4

Zaitseva E, Yang ST, Melikov K, Pourmal S, Chernomordik LV (2010) Dengue virus ensures its fusion in late endosomes using compartment-specific lipids. PLoS Pathog 6(10):e1001131. https://doi.org/10.1371/journal.ppat.1001131

Zaitseva E, Zaitsev E, Melikov K, Arakelyan A, Marin M, Villasmil R, Margolis LB, Melikyan GB, Chernomordik LV (2017) Fusion stage of HIV-1 entry depends on virus-induced cell surface exposure of phosphatidylserine. Cell Host Microbe 22(1):99–110 e117. https://doi.org/10.1016/j.chom.2017.06.012

Zang R, Case JB, Yutuc E, Ma X, Shen S, Gomez Castro MF, Liu Z, Zeng Q, Zhao H, Son J, Rothlauf PW, Kreutzberger AJB, Hou G, Zhang H, Bose S, Wang X, Vahey MD, Mani K, Griffiths WJ, Kirchhausen T, Fremont DH, Guo H, Diwan A, Wang Y, Diamond MS, Whelan SPJ, Ding S (2020) Cholesterol 25-hydroxylase suppresses SARS-CoV-2 replication by blocking membrane fusion. Proc Natl Acad Sci U S A 117(50):32105–32113. https://doi.org/10.1073/pnas.2012197117

Zawada KE, Wrona D, Rawle RJ, Kasson PM (2016) Influenza viral membrane fusion is sensitive to sterol concentration but surprisingly robust to sterol chemical identity. Sci Rep 6:29842. https://doi.org/10.1038/srep29842

Zawada KE, Okamoto K, Kasson PM (2018) Influenza hemifusion phenotype depends on membrane context: differences in cell-cell and virus-cell fusion. J Mol Biol 430(5):594–601. https://doi.org/10.1016/j.jmb.2018.01.006

Zhang X, Boyce M, Bhattacharya B, Zhang X, Schein S, Roy P, Zhou ZH (2010) Bluetongue virus coat protein VP2 contains sialic acid-binding domains, and VP5 resembles enveloped virus fusion proteins. Proc Natl Acad Sci U S A 107(14):6292–6297. https://doi.org/10.1073/pnas.0913403107

Zhang X, Patel A, Celma CC, Yu X, Roy P, Zhou ZH (2016) Atomic model of a nonenveloped virus reveals pH sensors for a coordinated process of cell entry. Nat Struct Mol Biol 23(1):74–80. https://doi.org/10.1038/nsmb.3134

Zhang Y, Song Z, Wang M, Lan M, Zhang K, Jiang P, Li Y, Bai J, Wang X (2019) Cholesterol 25-hydroxylase negatively regulates porcine intestinal coronavirus replication by the production of 25-hydroxycholesterol. Vet Microbiol 231:129–138. https://doi.org/10.1016/j.vetmic.2019.03.004

Zhao L, Imperiale MJ (2017) Identification of Rab18 as an essential host factor for BK polyomavirus infection using a whole-genome RNA interference screen. mSphere 2(4). https://doi.org/10.1128/mSphereDirect.00291-17

Zhao Y, Ren J, Harlos K, Jones DM, Zeltina A, Bowden TA, Padilla-Parra S, Fry EE, Stuart DI (2016) Toremifene interacts with and destabilizes the Ebola virus glycoprotein. Nature 535(7610):169–172. https://doi.org/10.1038/nature18615

Zhao F, Xu F, Liu X, Hu Y, Wei L, Fan Z, Wang L, Huang Y, Mei S, Guo L, Yang L, Cen S, Wang J, Liang C, Guo F (2022) SERINC5 restricts influenza virus infectivity. PLoS Pathog 18(10):e1010907. https://doi.org/10.1371/journal.ppat.1010907

Zhu Y, Yu D, Hu Y, Wu T, Chong H, He Y (2021) SARS-CoV-2-derived fusion inhibitor lipopeptides exhibit highly potent and broad-spectrum activity against divergent human coronaviruses. Signal Transduct Target Ther 6(1):294. https://doi.org/10.1038/s41392-021-00698-x

# Chapter 5
# Single-Particle Tracking of Virus Entry in Live Cells

**Xiaowei Zhang, Wei Li, and Zongqiang Cui**

**Abstract** Novel imaging technologies such as single-particle tracking provide tools to study the intricate process of virus infection in host cells. In this chapter, we provide an overview of studies in which single-particle tracking technologies were applied for the analysis of the viral entry pathways in the context of the live host cell. Single-particle tracking techniques have been dependent on advances in the fluorescent labeling microscopy method and image analysis. The mechanistic and kinetic insights offered by this technique will provide a better understanding of virus entry and may lead to a rational design of antiviral interventions.

**Keywords** Single-particle tracking · Virus dynamic entry · Fluorescent labeling · Microscopy configurations · Image processing

## Introduction

Viruses are obligate intracellular parasites, which contain either an RNA or DNA genome surrounded by the protective virus-encoded protein coat or a membrane of cellular origin. Virus entry typically occurs in several tightly controlled, consecutive steps. These steps start with binding at the cell surface and end with the release of the genome at the site of replication. As the virus progresses in its entry program, it undergoes changes that result in events such as internalization, membrane fusion, and uncoating of the genome (Fig. 5.1). Viruses exploit cellular signals not only to induce changes in the particle but also to coordinate intracellular transport via virus–cell interactions, ensuring that each step in the entry program occurs at the right point in the sequence, at the right time, and in the right place (Marsh and Helenius 2006).

X. Zhang · W. Li · Z. Cui (✉)
State Key Laboratory of Virology, Wuhan Institute of Virology, Center for Biosafety Mega-Science, Chinese Academy of Sciences, Wuhan, China

University of Chinese Academy of Sciences, Beijing, China
e-mail: zhangxw@wh.iov.cn; liwei@wh.iov.cn; czq@wh.iov.cn

 
S. Vijayakrishnan et al. (eds.), *Virus Infected Cells*, Subcellular Biochemistry 106,
https://doi.org/10.1007/978-3-031-40086-5_5

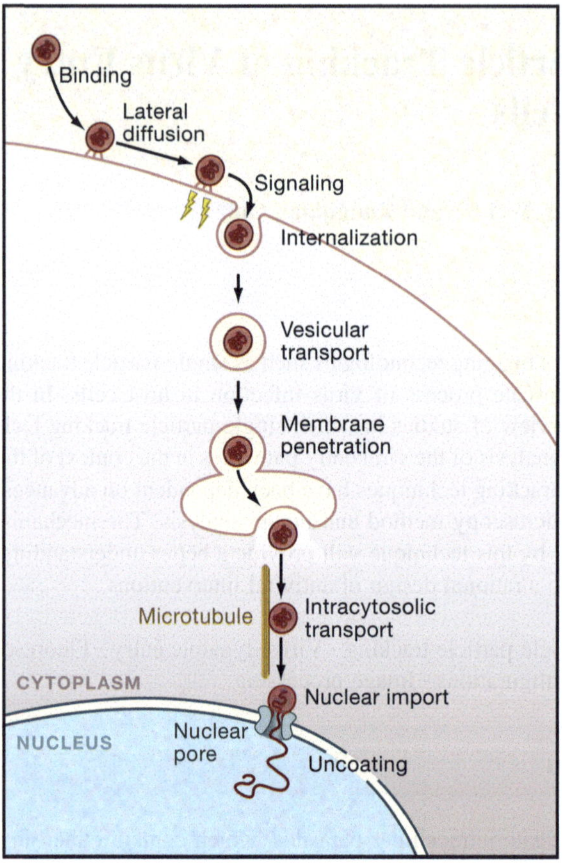


**Fig. 5.1** Steps in the entry program of viruses. Whether enveloped or non-enveloped, many viruses depend on the host cell's endocytic pathways for entry. They follow a multistep entry and uncoating program that allows them to move from the cell periphery to the perinuclear space. In this example, the virus proceeds to deliver its uncoated genome into the nucleoplasm. The interaction between the virus and the host cell starts with virus binding to receptors on the cell surface, followed by lateral movement of the virus–receptor complexes and the induction of signals that result in the endocytic internalization of the virus particle. After trafficking and delivery into the lumen of endosomes, caveosomes, or the ER, a change in the virus conformation is induced by cellular cues. This alteration results in the penetration of the virus or its capsids through the vacuole membrane into the cytosolic compartment. Enveloped viruses use membrane fusion for penetration, whereas non-enveloped viruses induce lysis or pore formation. After targeting and transport along microtubules, the virus or the capsid binds, as in this example, to the nuclear pore complex, undergoes a final conversion, and releases the viral genome into the nucleus. The details in the entry program vary for different viruses and cell types, but many of the key steps shown here are general. Adapted with permission from reference Marsh and Helenius (2006). Copyright 2006 Elsevier (order number 5393950747869)

Single-particle tracking of virus entry in live cells captures the dynamic, transient, and multistep processes unresolved through traditional ensemble and fixed cell assays. It allows researchers to follow the fate of individual virus particles, probe dynamic interactions between viruses and cellular components, and dissect the steps of the entry program (Brandenburg and Zhuang 2007). In this chapter, we will highlight the dynamic entry of viruses and describe how to use single-particle tracking methods for elucidating individual steps of entry programs. Furthermore, in the following section, we will give an insight on how to use fluorophores for virus labeling, imaging instruments, and data analysis methods.

## Dynamic Entry of Viruses into Host Cells

Viruses are obligate intracellular parasites that must gain entry into the host cells and utilize cellular machinery to survive. Virus entry is a highly dynamic motion involving multiple transient steps including binding, plasma membrane surfing, internalization/endocytosis, intracellular trafficking, and uncoating/endosomal fusion. Viruses are relatively simple in composition and lack any locomotive capacity; thus viruses evolved to exploit host cell proteins during entry. The interactions of the virus with different cellular structures during the various entry steps are dynamic, and so are the cellular structures themselves.

### *Virus–Receptor Interactions*

Viruses first exploit cell surface receptors to initiate infection of target cells. Specific virus–receptor interactions will activate the downstream signaling cascades, and trigger the internalization of viruses or conformational alterations in viral structural proteins for genome release. The detailed processes of how viruses navigate along the cell surface to bind specific receptors are difficult to decipher in traditional ensemble measurements. Single-particle tracking provides intricate information on the transient interaction between the virus and receptor molecules, as exemplified by the different patterns of surface movement, spatio-temporal clustering of receptors, and sequential exploitation of multiple receptors.

For example, different pathways are used by Avian sarcoma and leukosis virus (ASLV) depend upon usage of two natural isoforms of its receptor. The entry of labeled ASLV into cells has been imaged using fluorescence microscopy in live cells in real time. Important differences in ASLV entry have been found, in that virus was attached to a transmembrane receptor or a lipid-anchored receptor. It revealed functional implications of ASLV entry at different rates via alternative receptors (Jha et al. 2011).

## *Virus Internalization*

Virus binding to receptors and ensuing cellular signaling promote endocytosis or induce fusion at the plasma membrane. Most viruses hijack different cellular endocytotic pathways to enter the host cells. Endocytosis is a sequential multistep process mainly involving activation, cargo capture and sorting, endosomal organelle targeting, and specific fusion of cargoes with a target membrane (Doherty and McMahon 2009). Clathrin-mediated endocytosis is the most commonly used pathway for receptor-mediated endocytosis of viruses. In addition, the internalization of viruses may also occur via several other endocytotic pathways, such as caveolae-dependent, clathrin- and caveolae-independent endocytosis, and macropinocytosis (Mercer et al. 2010; Schelhaas 2010). On the other hand, several enveloped viruses such as human immunodeficiency virus (HIV), herpes simplex virus, and Sendai virus and a number of non-enveloped viruses including picornaviruses and polyomaviruses have been reported to release the capsids or genome directly into the cytoplasm. However, many of these viruses are also capable of utilizing endocytosis for penetration.

Single-virus tracking offers an ideal approach to discriminate the internalization pathways and pinpoint the accurate release site of viral nucleocapsids or genome. Single-virus tracking can also be combined with the co-tracking of cellular components. For example, the entry mechanism of the influenza virus has been revealed by tracking the interaction of single viruses with cellular endocytic components using fluorescence microscopy. Thus, it has been found that the influenza virus can simultaneously exploit clathrin-mediated and clathrin- and caveolin-independent endocytic pathways with similar efficiency (Rust et al. 2004).

## *Virus Transport*

Once in the cytoplasm, viruses need to be transported to a particular subcellular location to release the genome and initiate replication. Due to the size of viruses ranging from about 20 to 400 nm and the high viscosity of the cytoplasm, polarized or directional movements of viruses toward concrete regions of cells by free diffusion are difficult (Greber and Way 2006). Therefore, viruses have evolved efficient strategies to hijack the host intracellular trafficking machinery components such as cytoskeletal tracks to transport to sites of replication within the cytosol (most RNA viruses) or the nucleus (most DNA viruses) (Hernandez-Gonzalez et al. 2021). The dynamic and constantly changing cytoskeleton may provide the force necessary to move viruses within cells. On the other hand, cytoskeleton tracks exemplified by microtubules and microfilaments may serve as a "highway" along which viruses can efficiently traffic. Transport of viruses along the cytoskeleton is mediated by microtubule motor proteins (dynein/dynactin and kinesin) or microfilament motor proteins (myosins) (Radtke et al. 2006). As a result, the transport of viral cargo molecules is a complex and multistep process that requires the orchestration of multiple host and viral components. As shown by single-virus tracking and live cell imaging, HIV-1 was endocytosed and transported to endosomes in a clathrin- and actin-dependent manner in primary macrophages. Furthermore, the authors showed that a dynamic actin cytoskeleton plays a pivotal role in HIV-1 entry and intracellular migration in macrophages (Li et al. 2017). By using single-particle tracking, trajectories of individual, fluorescence-labeled murine polyoma virus on the surface of tissue culture cells were recorded. It has been shown that virus particles underwent free, cholesterol-dependent, lateral diffusion, rapidly followed by a period of confinement through an actin cytoskeleton-dependent mechanism (Ewers et al. 2005).

### *Virus Uncoating*

Following internalization, viruses undergo an uncoating process to release the genome into a particular cellular compartment capable of supporting virus replication. For enveloped viruses, uncoating follows the fusion of their lipid membranes with a host membrane, in most cases of endosomal origin (Mercer et al. 2010). For non-enveloped viruses, it involves direct pore formation or disassembly to deliver the viral genome (Suomalainen and Greber 2013). Membrane fusion plays a pivotal role in the entry of enveloped viruses into cells. The merging of viral and cellular membranes is mediated by viral fusion proteins, which involves multiple sequential steps in the fusion process (Joo et al. 2010). Non-enveloped viruses undergo conformational changes upon entry into cells leading to the release of the genome, which is a stepwise process induced by cellular cues. The dynamics of these fusion and uncoating steps remain unresolved, and the trigger mechanism of some viruses is not fully understood.

Measurement of fusion/uncoating by tracking single-virus particles will allow more accurate characterization of the pathways and mechanisms by which viral entry occurs (Haldar et al. 2020). For example, many non-enveloped viruses such as poliovirus (PV) have been reported to release their genome at the plasma membrane. However, endocytosis of these viruses has also been observed, making it difficult to identify the productive entry pathway. By using the approach of single-virus tracking, it was found that PV uncoating occurs only after the internalization of virus particles (Brandenburg et al. 2007).

## Fluorescent Labeling of Viruses and Cellular Components

Virus entry encompasses the initial steps of infection, from virion attachment to genome release. Fluorescent labeling of viral and cellular components and imaging in live cells enable intricate studies on this process. To perform single-virus tracking, both virions and relevant cellular components need to be labeled to visualize them in live cells. Virions are commonly labeled by fusing fluorescent proteins to viral proteins, by incorporating organic dyes, or by incorporating fluorescent nanoparticles into the virion. Meanwhile, relevant cellular components are also often labeled with fluorescent probes, without impairing their physiological localization and cellular functions.

## *Fluorescent Proteins*

Green fluorescent protein (GFP) from the jellyfish *Aequorea victoria* and its homologs from diverse marine animals and its mutant variants are widely used as genetically encoded fluorescent probes (Chudakov et al. 2005). Identification and development of fluorescent proteins offer a powerful toolkit for the visualization of structural organization and dynamic processes in live cells. Up to now, fluorescent proteins of different colors and properties have been used in a variety of applications to visualize morphology, location, and movement of proteins of interest, and the emission wavelength of fluorescent proteins spans from blue to near-infrared spectrum (Chudakov et al. 2010).

Fluorescent proteins can be used as a convenient fluorescent probe to label viruses and relevant cellular components for virus tracking. Recombinant viruses encoding fluorescent fusion proteins facilitate real-time analysis of the entry motility of viruses. For example, EGFP fusions to tegument proteins, which are located between the lipid envelope and the capsid shell, have been used to follow the complex anterograde and retrograde directional movements of herpesviruses in synaptically connected neurons (Greber 2005; Luxton et al. 2005). The pseudorabies virus (PRV) protein Us9 is a small membrane protein that is highly conserved among alpha herpesviruses and is essential for the anterograde axonal spread in neurons. PRV strain that expresses functional green fluorescent protein (GFP)-Us9 fusion proteins has been constructed to visualize its anterograde direction transportation in living neurons during infection (Taylor et al. 2012; Lyman et al. 2008).

Some viruses can enter cells via the endocytosis pathway, along which the pH drops from moderately acidic early and recycling endosomes maturing into more acidic late endosomes to most acidic in the lysosomes (Sorkin and Von Zastrow 2002). Viral structural proteins will pass through acidic compartments at some point in the virus entry phase. Therefore, when using fluorescent fusion proteins to monitor viral entry in live cells, it is important to consider how fluorescent proteins will behave in these acidic compartments. The current generations of blue, cyan, and red fluorescent proteins (e.g., ECFP and mKate2) are relatively pH insensitive and will maintain relatively high fluorescence in even the most acidic intracellular compartments, like lysosomes (pH 4.5–5.0). These kinds of fluorescent proteins are suitable for imaging viral particles in acidic organelles (Shaner et al. 2005). Other fluorescent proteins such as EYFP, pHluorin, and pHuji are highly pH sensitive. These proteins present high fluorescence under neutral pH conditions, while the fluorescence signal decreases markedly at acid pH. The pH sensitivity of such fluorescent proteins can be exploited as a pH sensor to study virus fusion and entry (Melikyan et al. 2005; Miyauchi et al. 2011). By co-labeling an avian retrovirus with pH-sensitive pHluorin and pH-resistant mKate2 simultaneously, single-virus imaging showed that the virus co-trafficked and fused with intracellular compartments tagged with fluorescent Rab5 and Rab7 proteins in live cells (Padilla-Parra et al. 2014).

## *Organic Dyes*

Organic dyes are important fluorophores used in fluorescence imaging in live cells (Gonçalves 2009). Due to their relatively small size of 1–2 nm, organic dyes are suitable for fluorescence labeling of viruses, which have a size range of 20–400 nm in diameter. Besides, organic dyes have several outstanding advantages, such as excellent optical properties, good biocompatibility, commercial availability, and a wide range of color palettes, which confer their broad usage for single-virus tracking (Miyawaki et al. 2003; Resch-Genger et al. 2008).

Viral glycoproteins (envelope proteins) or capsid proteins for non-enveloped viruses can be readily labeled with amine-reactive dyes, as exemplified by cyanine dyes and Alexa Fluor dyes. Cy3 and Cy5 are the most commonly used cyanine dyes for single virus tracking. Cy3 has been used to label the envelope protein of the influenza virus (Lakadamyali et al. 2003, 2006). As a long-wavelength fluorophore, Cy5 has the advantage of reduced interference from autofluorescence in live cells. Therefore, Cy5 has been much more popularly applied to label viruses, including the envelope protein of Semliki forest virus (SFV) (Vonderheit and Helenius 2005) and rabies virus (Xu et al. 2015), and the capsid protein of PV (Brandenburg et al. 2007) and adeno-associated virus (Seisenberger et al. 2001). It is worthy of note that researchers have viewed live scenes of viral infection by tagging individual viruses with one or two Cy5 molecules (Wang et al. 2014). Besides, virions and cellular components could be labeled by cyanine dyes or fluorescent proteins, respectively, for imaging in live cells. For instance, the intracellular pathway taken by individual Cy5-labeled SFV particles was followed using triple-color fluorescence microscopy in live cells transfected with GFP- and RFP-tagged Rab5, Rab7, Rab4, and Arf1, the cellular components that mediate cargo transfer (Vonderheit and Helenius 2005).

Alexa Fluor dyes are another group of amine-reactive dyes that are frequently used for single-virus tracking alone or in combination with other fluorophores. Alexa Fluor dyes cover an extremely broad range of excitation/emission wavelengths spanning from ultraviolet (UV) to visible and near-infrared regions. They are derivatives of certain well-known organic dyes such as rhodamine, fluorescein, and cyanine dyes. Alexa Fluor dyes possess highly demanded photophysical properties including high quantum yields, large molar extinction coefficients, and high photostabilities. Moreover, Alexa Fluor dyes tend to be more resistant to fluorescence quenching and retain their intense fluorescence even at high degrees of labeling compared to the cyanine dyes (Berlier et al. 2003; Gebhardt et al. 2021).

Thus, Alexa Fluor dyes have been widely used for investigating the cellular uptake and trafficking of viruses. For example, tracking of Alexa Fluor 594-labeled simian virus 40 revealed that individual viruses moved laterally in a random fashion until trapped in stationary spots positive for caveolin-1 tagged with a green fluorescent protein (Pelkmans et al. 2002). Alexa Fluor derivatives have also been applied to label other non-enveloped viruses, including foot-and-mouth disease virus (Martín-Acebes et al. 2011) and human papillomavirus (Schelhaas et al. 2008), or enveloped viruses such as vesicular stomatitis virus (VSV) (Cureton et al. 2010) and dengue virus (Dejarnac et al. 2018).

Enveloped viruses are surrounded by a continuous lipid bilayer derived from the host cell plasma membrane or intracellular membrane (Lenard 2008). Fluorescent lipophilic dyes are popular for membrane labeling and particle tracing applications and thus can be utilized to label viruses by targeting the lipid envelope. It has been known that hydrophobic–lipophilic interactions facilitate the incorporation of lipophilic dyes into the outer membrane of enveloped viruses (McClelland et al. 2021). Up to now, the most commonly used fluorescent lipophilic dyes are rhodamine derivatives and long-chain dialkylcarbocyanines such as DiO, DiI, and DiD. Lipophilic dyes at high concentrations tend to fluorescence self-quench due to close proximity constraints (Klymchenko et al. 2012). Upon fusion where viral membranes merge with the membranes of host cells, the lipophilic dyes will diffuse resulting in dequenching and an increase in fluorescence indicating the occurrence of membrane fusion. For instance, the self-quenching rhodamine derivatives R18 and R110 have been applied to study influenza virus envelope fusion with cellular membranes (Lowy et al. 1990; Floyd et al. 2008). By labeling respiratory syncytial virus (RSV) particles with DiD, the fluorescent signal was used for real-time monitoring of the virus fusion and entry upon clustering with surface trafficked NCL, the coreceptor for RSV fusion (Griffiths et al. 2020). It is noteworthy that this approach using fluorescent lipophilic dyes detects fusion based on the lipid mixing activity, but not the actual release of the interior core/content of virus, which is a prerequisite for infection. More accurate approaches involve co-labeling the viral core and the viral membrane with different fluorescent probes are needed.

## *Quantum Dots*

Quantum dots (QDs), a kind of semiconductor nanocrystals, are among the most investigated fluorescent nanoparticles in bioimaging applications (Michalet et al. 2005; Wegner and Hildebrandt 2015). QDs offer a wide range of useful optical properties for bioimaging assays with single-molecule readouts, including high photostability, high quantum yield, large absorption coefficients, and narrow emission spectra (Zhou et al. 2015). The emission wavelengths of QDs range from UV to infrared and can be tuned by controlling their size and composition (Wegner and Hildebrandt 2015). Their excellent brightness (10−100 times higher than organic dyes or fluorescent proteins) and resistance to photobleaching (100−1000 times

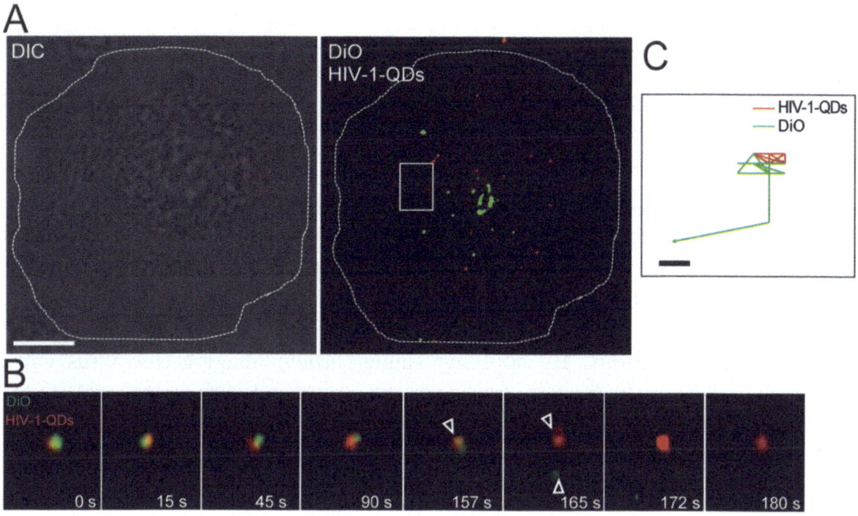


**Fig. 5.2** Real-time imaging of the HIV-1-QD viral core release in primary macrophages. The cellular boundary of the macrophage is highlighted by a dashed line. (**a**) Dual-labeled HIV-1-QD-DiO particle in a macrophage (shown in the rectangular region) was tracked. Scale bar: 10 μm. (**b**) Sequential snapshots are shown for the separation of QD-Vpr and the DiO-labeled membrane of the viral particle that is shown in (**a**) in the macrophage. Arrowhead indicates the separation of QD and DiO signals. (**c**) Trajectories of HIV-1-QD (red) and DiO-labeled membrane (green) for the viral particle that is shown in rectangular region of (**a**). Scale bar: 0.2 μm. Adapted with permission from reference Li et al. (2017). Copyright 2017 American Chemical Society

higher than organic dyes or fluorescent proteins) make them especially useful for live-cell imaging and single-particle tracking (Alivisatos et al. 2005; Liu et al. 2020). The wide emission wavelength and distinct emission spectra of QDs also facilitate multicolor imaging where various components of a virion can be labeled and tracked simultaneously.

In a noteworthy study, QDs were successfully exploited to site specifically label enveloped viruses, leading to the real-time monitoring of the intracellular movement of single viruses in live cells (Joo et al. 2008). Specific internal viral components can also be labeled with QDs during viral replication and prior to assembly. The use of this approach could obviate incorporation of QDs on the viral surface which may affect the virus–host interaction and thus the efficiency of viral entry. For example, QDs conjugated with modified genomic RNAs, which contain a packaging signal sequence, can be successfully incorporated into intact virions in live cells without modification of the viral surface. This QD-containing VSV-G pseudotyped lentivirus demonstrated a Rab5+ endosome- and microtubule-dependent entry route (Zhang et al. 2013). In the context of QDs-based labeling of viral interior proteins, our group demonstrated that QDs could be encapsulated in HIV-1 virions by incorporating viral accessory protein Vpr-conjugated QDs during virus assembly. With the HIV-1 particles encapsulating QDs, we not only monitor the entry route of the virus at a single-particle level in live human primary macrophages (Li et al. 2017)

(Fig. 5.2), but also explored detailed scenarios of HIV dynamic entry and crossing the cortical actin barrier in resting CD4+ T lymphocytes (Yin et al. 2020). In addition, we have constructed a type of infectious HIV-1 virus encapsulating QDs through the site-specific decoration of the viral matrix protein and used it to visualize virus entry in human primary macrophages by single-particle imaging (Li et al. 2018).

To track the intricate sequential entry process of a single-virus particle in live cells, labeling of both the interior and exterior components of viruses is needed without an impact on the infectivity of viruses. Our group constructed dual-color IAV particles by combining internal encapsulation and surface decoration using QDs with different colors or by incorporating differently colored QD-vRNP segments into single virions. By applying single-particle imaging, the virus entry process including internalization, transport, uncoating, and intranuclear vRNP trafficking of individual IAV virions was visualized in real time in live cells (Qin et al. 2019; Yamauchi 2019).

# Microscopy Methods and Image Processing

## *Microscopy Configurations*

To visualize and track the dynamic entry processes of viruses, a number of different microscopy configurations need to be employed.

Wide-field epifluorescence microscopy is the simplest optical implementation for single virus tracking experiments among the available microscopy configurations. The epifluorescence configuration illuminates a relatively large area (approximately 100 μm × 100 μm) of the sample via the excitation light and the fluorescence signal can be captured quickly and easily by a highly sensitive camera (Rust et al. 2011). This method has large excitation depth and low signal loss, allowing the study of the long-range movement of individual viruses. However, owing to a large amount of background autofluorescence of cells, epifluorescence detection often shows weak signals originating from individual virions that contain only a few fluorescent molecules (Liu et al. 2020). This method needs to extract an in-focus signal from a high-noise background.

The confocal microscopy modality has a thinner depth of field and higher resolution compared to wide-field epifluorescence microscopy. It uses point illumination and pinhole optics to eliminate the background noise but at the cost of signal reduction. In order to track the dynamic behavior of viruses in live cells, different sites of volume must be scanned simultaneously. The usage of spinning-disk confocal microscopy (SDCM) meets the demand. Cross sections of a sample with a fast piezo z-scan device can be combined with SDCM to reconstruct all three-dimensional images (X, Y, and Z), which has been used for single-virus tracking in live cells (Han et al. 2012). Besides, SDCM can markedly improve imaging speed and allows the elucidation of dynamic events of viruses in live cells. The main disadvantage of SDCM is the cross-talk between pinholes, which can be circumvented by increasing the inter-pinhole distance or by decreasing the out-of-focus light. Besides, exposure to strong light can be damaging to live cells (Shimozawa et al. 2013).

Total internal reflection fluorescence microscopy (TIRFM) is useful to observe fluorescence signals occurring at interfaces. The TIRFM makes good use of total internal reflection to produce an evanescent field. Typically, the evanescent wave extends only a few hundred nanometers into the cell. Therefore, TIRFM facilitates the visualization of biological events occurring on the plasma membrane of live cells, such as clathrin-based entry kinetics of VSV (Johannsdottir et al. 2009).

## *Image Processing*

After choosing and utilization of a virus labeling method and microscopy configuration, the dynamic information of virus entry can be extracted from a time series of images. Several challenges have to be overcome with the analysis of viral entry in live cells. It is important to eliminate background noise and thereby enhance the signal-to-noise ratio. To combat the noise of concern, several particle detection and image restoration techniques have been developed specifically for single-virus tracking. For example, in each frame of images containing fluorescent particles, the location of the individual particle with high precision will be determined. To locate the center position, the point spread-function (PSF) is regarded as a suitable parameter to measure the shape of individual particles, as well as the 3D diffraction pattern (Liu et al. 2020). Then the isolated fluorescent peaks that correspond to virus particles will be automatically identified by PSF-fitting methods. The Gaussian nonlinear least-squares fitting method is the most common way for single-virus tracking. It possesses very high localization accuracy and is compatible with localization-based super-resolution techniques such as stochastical optical reconstruction microscopy (STORM) and photoactivated-localization microscopy (PALM). The radial symmetry approach is another type of the geometric-based method which can be utilized for localizing particle centers. In terms of computation speed, it is significantly faster than that of the Gaussian fitting method (Liu et al. 2013).

Trajectories of virus particles can be reconstructed by connecting particles in each frame by using nearest-neighbor association and the motion history of individual virus particles (Brandenburg and Zhuang 2007; Shen et al. 2017; Sbalzarini and Koumoutsakos 2005). Then mean squared displacements are applied to determine whether the virus particles display different types of movement models, such as directed, normal, or anomalous diffusion. Viruses tend to change their motion behavior from 2D diffusion on the cell membrane to directed transport and then to 3D anomalous diffusion within the cytosol (Brandenburg and Zhuang 2007). The diffusion type can reveal the type of interactions between different viral structures or the interactions between viral and cellular structures. By measuring the

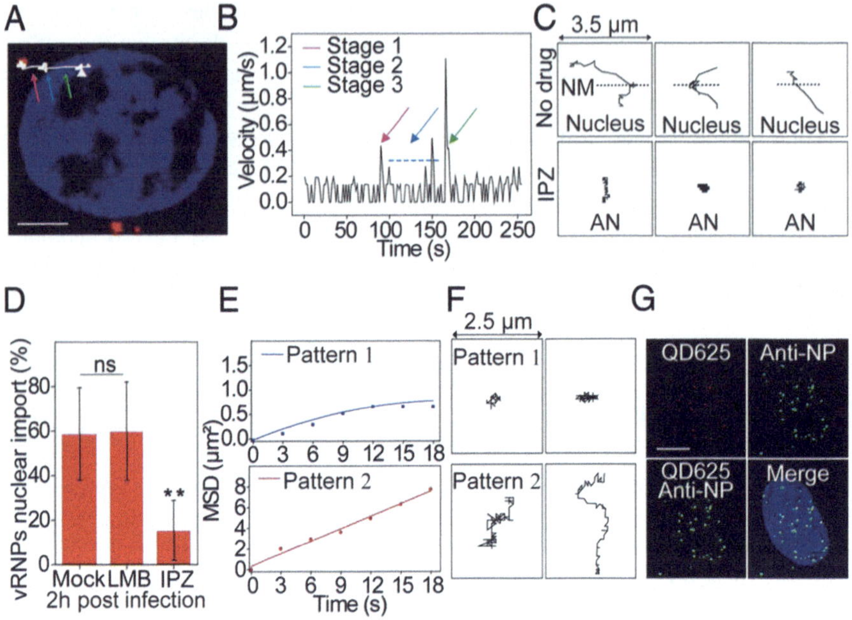


**Fig. 5.3** Characterization of vRNP nuclear import and intranuclear movement. (**a**) Trajectory of a representative QD625-vRNP transport. The colored arrows indicate stage 1 (pink), stage 2 (blue), and stage 3 (green) (scale bar: 5 μm). (**b**) Analysis of the mean velocities of QD625-labeled vRNPs shown in **a**. (**c**) Trajectories of the QD signals in untreated and importazole (IPZ)-treated cells. NM, nuclear membrane; AN, adjacent nuclear. (**d**) Statistical analysis of /IAV vRNP nuclear import with drug treatment. ns, no significant difference; $n = 3$; $^{**}P < 0.01$. (**e**) Analysis of the MSD plots of two vRNP diffusive patterns in the nucleus. (**f**) Trajectories of two vRNP diffusive patterns. (**g**) Immunofluorescence assay of QD625-vRNPs in the nucleus at 2 hpi (scale bar: 5 μm). Adapted with permission from reference Qin et al. (2019). Copyright © 2019 the Author(s). Published by PNAS. This open access article is distributed under Creative Commons Attribution-NonCommercial-NoDerivatives License 4.0 (CC BY-NC-ND)

instantaneous speed versus time, the different motions from a trajectory can be distinguished and analyzed separately. For instance, it has been shown that individual QD-labeled vRNPs of influenza viruses display a three-stage movement to the cell nucleus and show two diffusion patterns when inside the nucleus during the entry process (Qin et al. 2019) (Fig. 5.3).

## Concluding Remarks

Single-particle tracking techniques are powerful tools for investigating the dynamic entry processes of viruses into cells with high spatio-temporal resolution. They can help elucidate the distinct entry pathway of the same virus or similar entry routes of different viruses. With the development of new fluorescent probes, better labeling strategies, and novel microscope configurations in the upcoming decade, we will witness an ongoing and broader application of single particle tracking in the field of virology. Particularly, over recent years, functional genetic screens have been carried out using genome-wide siRNA libraries, insertional mutagenesis of haploid cell lines, and CRISPR-Cas9 knockout screens. A subset of host factors required to support virus entry has been identified (Hirsch 2010; Puschnik et al. 2017; Carette et al. 2011). Given that viruses exploit a variety of cellular factors to get entry into host cells, single-particle tracking of viruses in live cells will undoubtedly provide an important tool for dissecting the dynamic roles of the identified host proteins in viral entry. To reduce the influence of fluorescent labeling on viral infectivity, only a few fluorescent molecules can be used to label viral structures. Fluorescent nanoparticles such as QDs are useful labeling schemes and have the potential to track the whole entry process of individual viruses in live cells. Lastly, most live cell imaging studies on virus entry are currently performed with 2D culture cells. Viral tracking in conditions that more accurately mimic in vivo circumstances such as 3D organoids or physiological barriers (Strange et al. 2019; Premeaux et al. 2021) is likely to give a better understanding of the virus entry process.

## References


Alivisatos AP, Gu W, Larabell C (2005) Quantum dots as cellular probes. Annu Rev Biomed Eng 7: 55–76

Berlier JE, Rothe A, Buller G, Bradford J, Gray DR, Filanoski BJ, Telford WG, Yue S, Liu J, Cheung CY, Chang W, Hirsch JD, Beechem JM, Haugland RP (2003) Quantitative comparison of long-wavelength Alexa Fluor dyes to Cy dyes: fluorescence of the dyes and their bioconjugates. J Histochem Cytochem 51(12):1699–1712

Brandenburg B, Zhuang X (2007) Virus trafficking—learning from single-virus tracking. Nat Rev Microbiol 5(3):197–208

Brandenburg B, Lee LY, Lakadamyali M, Rust MJ, Zhuang X, Hogle JM (2007) Imaging poliovirus entry in live cells. PLoS Biol 5(7):10

Carette JE, Guimaraes CP, Wuethrich I, Blomen VA, Varadarajan M, Sun C, Bell G, Yuan B, Muellner MK, Nijman SM, Ploegh HL, Brummelkamp TR (2011) Global gene disruption in human cells to assign genes to phenotypes by deep sequencing. Nat Biotechnol 29 (6):542–546

Chudakov DM, Lukyanov S, Lukyanov KA (2005) Fluorescent proteins as a toolkit for in vivo imaging. Trends Biotechnol 23(12):605–613

Chudakov DM, Matz MV, Lukyanov S, Lukyanov KA (2010) Fluorescent proteins and their applications in imaging live cells and tissues. Physiol Rev 90(3):1103–1163

Cureton DK, Massol RH, Whelan SP, Kirchhausen T (2010) The length of vesicular stomatitis virus particles dictates a need for actin assembly during clathrin-dependent endocytosis. PLoS Pathog 6(9):1001127

Dejarnac O, Hafirassou ML, Chazal M, Versapuech M, Gaillard J, Perera-Lecoin M, Umana-Diaz C, Bonnet-Madin L, Carnec X, Tinevez JY, Delaugerre C, Schwartz O, Roingeard P, Jouvenet N, Berlioz-Torrent C, Meertens L, Amara A (2018) TIM-1 ubiquitination mediates dengue virus entry. Cell Rep 23(6):1779–1793

Doherty GJ, McMahon HT (2009) Mechanisms of endocytosis. Annu Rev Biochem 78:857–902

Ewers H, Smith AE, Sbalzarini IF, Lilie H, Koumoutsakos P, Helenius A (2005) Single-particle tracking of murine polyoma virus-like particles on live cells and artificial membranes. Proc Natl Acad Sci U S A 102(42):15110–15115

Floyd DL, Ragains JR, Skehel JJ, Harrison SC, van Oijen AM (2008) Single-particle kinetics of influenza virus membrane fusion. Proc Natl Acad Sci U S A 105(40):15382–15387

Gebhardt C, Lehmann M, Reif MM, Zacharias M, Gemmecker G, Cordes T (2021) Molecular and spectroscopic characterization of green and red cyanine fluorophores from the alexa fluor and AF series*. Chemphyschem 22(15):1566–1583

Gonçalves MS (2009) Fluorescent labeling of biomolecules with organic probes. Chem Rev 109(1): 190–212

Greber UF (2005) Viral trafficking violations in axons: the herpesvirus case. Proc Natl Acad Sci U S A 102(16):5639–5640

Greber UF, Way M (2006) A superhighway to virus infection. Cell 124(4):741–754

Griffiths CD, Bilawchuk LM, McDonough JE, Jamieson KC, Elawar F, Cen Y, Duan W, Lin C, Song H, Casanova JL, Ogg S, Jensen LD, Thienpont B, Kumar A, Hobman TC, Proud D, Moraes TJ, Marchant DJ (2020) IGF1R is an entry receptor for respiratory syncytial virus. Nature 583(7817):615–619

Haldar S, Okamoto K, Dunning RA, Kasson PM (2020) Precise triggering and chemical control of single-virus fusion within endosomes. J Virol 95(1):01982-20

Han JJ, Kiss C, Bradbury AR, Werner JH (2012) Time-resolved, confocal single-molecule tracking of individual organic dyes and fluorescent proteins in three dimensions. ACS Nano 6(10): 8922–8932

Hernandez-Gonzalez M, Larocque G, Way M (2021) Viral use and subversion of membrane organization and trafficking. J Cell Sci 134(5):252676

Hirsch AJ (2010) The use of RNAi-based screens to identify host proteins involved in viral replication. Future Microbiol 5 (2):303–311

Jha NK, Latinovic O, Martin E, Novitskiy G, Marin M, Miyauchi K, Naughton J, Young JA, Melikyan GB (2011) Imaging single retrovirus entry through alternative receptor isoforms and intermediates of virus-endosome fusion. PLoS Pathog 7(1):1001260

Johannsdottir HK, Mancini R, Kartenbeck J, Amato L, Helenius A (2009) Host cell factors and functions involved in vesicular stomatitis virus entry. J Virol 83(1):440–453

Joo KI, Lei Y, Lee CL, Lo J, Xie J, Hamm-Alvarez SF, Wang P (2008) Site-specific labeling of enveloped viruses with quantum dots for single virus tracking. ACS Nano 2(8):1553–1562

Joo KI, Tai A, Lee CL, Wong C, Wang P (2010) Imaging multiple intermediates of single-virus membrane fusion mediated by distinct fusion proteins. Microsc Res Tech 73(9):886–900

Klymchenko AS, Roger E, Anton N, Anton H, Shulov I, Vermot J, Mely Y, Vandamme TF (2012) Highly lipophilic fluorescent dyes in nano-emulsions: towards bright non-leaking nano-droplets. RSC Adv 2(31):11876–11886

Lakadamyali M, Rust MJ, Babcock HP, Zhuang X (2003) Visualizing infection of individual influenza viruses. Proc Natl Acad Sci U S A 100(16):9280–9285

Lakadamyali M, Rust MJ, Zhuang X (2006) Ligands for clathrin-mediated endocytosis are differentially sorted into distinct populations of early endosomes. Cell 124(5):997–1009

Lenard J (2008) Viral membranes. Encycl Virol 2008:308–314. https://doi.org/10.1016/B978-012374410-4.00530-6. Epub 2008 Jul 30

Li Q, Li W, Yin W, Guo J, Zhang ZP, Zeng D, Zhang X, Wu Y, Zhang XE, Cui Z (2017) Single-particle tracking of human immunodeficiency virus type 1 productive entry into human primary macrophages. ACS Nano 11(4):3890–3903

Li Q, Yin W, Li W, Zhang Z, Zhang X, Zhang XE, Cui Z (2018) Encapsulating quantum dots within HIV-1 virions through site-specific decoration of the matrix protein enables single virus tracking in live primary macrophages. Nano Lett 18(12):7457–7468

Liu SL, Li J, Zhang ZL, Wang ZG, Tian ZQ, Wang GP, Pang DW (2013) Fast and high-accuracy localization for three-dimensional single-particle tracking. Sci Rep 3(2462)

Liu SL, Wang ZG, Xie HY, Liu AA, Lamb DC, Pang DW (2020) Single-virus tracking: from imaging methodologies to virological applications. Chem Rev 120(3):1936–1979

Lowy RJ, Sarkar DP, Chen Y, Blumenthal R (1990) Observation of single influenza virus-cell fusion and measurement by fluorescence video microscopy. Proc Natl Acad Sci U S A 87(5): 1850–1854

Luxton GW, Haverlock S, Coller KE, Antinone SE, Pincetic A, Smith GA (2005) Targeting of herpesvirus capsid transport in axons is coupled to association with specific sets of tegument proteins. Proc Natl Acad Sci U S A 102(16):5832–5837

Lyman MG, Curanovic D, Brideau AD, Enquist LW (2008) Fusion of enhanced green fluorescent protein to the pseudorabies virus axonal sorting protein Us9 blocks anterograde spread of infection in mammalian neurons. J Virol 82(20):10308–10311

Marsh M, Helenius A (2006) Virus entry: open sesame. Cell 124(4):729–740

Martín-Acebes MA, Vázquez-Calvo A, González-Magaldi M, Sobrino F (2011) Foot-and-mouth disease virus particles inactivated with binary ethylenimine are efficiently internalized into cultured cells. Vaccine 29(52):9655–9662

McClelland RD, Culp TN, Marchant DJ (2021) Imaging flow cytometry and confocal immunofluorescence microscopy of virus-host cell interactions. Front Cell Infect Microbiol 11:749039

Melikyan GB, Barnard RJ, Abrahamyan LG, Mothes W, Young JA (2005) Imaging individual retroviral fusion events: from hemifusion to pore formation and growth. Proc Natl Acad Sci U S A 102(24):8728–8733

Mercer J, Schelhaas M, Helenius A (2010) Virus entry by endocytosis. Annu Rev Biochem 79:803–833

Michalet X, Pinaud FF, Bentolila LA, Tsay JM, Doose S, Li JJ, Sundaresan G, Wu AM, Gambhir SS, Weiss S (2005) Quantum dots for live cells, in vivo imaging, and diagnostics. Science 307(5709):538–544

Miyauchi K, Marin M, Melikyan GB (2011) Visualization of retrovirus uptake and delivery into acidic endosomes. Biochem J 434(3):559–569

Miyawaki A, Sawano A, Kogure T (2003) Lighting up cells: labelling proteins with fluorophores. Nat Cell Biol 7:S1–S7

Padilla-Parra S, Marin M, Kondo N, Melikyan GB (2014) Pinpointing retrovirus entry sites in cells expressing alternatively spliced receptor isoforms by single virus imaging. Retrovirology 11(47):1742–4690

Pelkmans L, Püntener D, Helenius A (2002) Local actin polymerization and dynamin recruitment in SV40-induced internalization of caveolae. Science 296(5567):535–539

Premeaux TA, Mediouni S, Leda A, Furler RL, Valente ST, Fine HA, Nixon DF, Ndhlovu LC (2021) Next-generation human cerebral organoids as powerful tools to advance neuroHIV research. mBio 12(4):00680-00621

Puschnik AS, Majzoub K, Ooi YS, Carette JE (2017) A CRISPR toolbox to study virus-host interactions. Nat Rev Microbiol 15 (6):351–364

Qin C, Li W, Li Q, Yin W, Zhang X, Zhang Z, Zhang XE, Cui Z (2019) Real-time dissection of dynamic uncoating of individual influenza viruses. Proc Natl Acad Sci U S A 116(7):2577–2582

Radtke K, Döhner K, Sodeik B (2006) Viral interactions with the cytoskeleton: a Hitchhiker's guide to the cell. Cell Microbiol 8(3):387–400

Resch-Genger U, Grabolle M, Cavaliere-Jaricot S, Nitschke R, Nann T (2008) Quantum dots versus organic dyes as fluorescent labels. Nat Methods 5(9):763–775

Rust MJ, Lakadamyali M, Zhang F, Zhuang X (2004) Assembly of endocytic machinery around individual influenza viruses during viral entry. Nat Struct Mol Biol 11(6):567–573

Rust MJ, Lakadamyali M, Brandenburg B, Zhuang X (2011) Single-virus tracking in live cells. Cold Spring Harb Protoc 1(9)

Sbalzarini IF, Koumoutsakos P (2005) Feature point tracking and trajectory analysis for video imaging in cell biology. J Struct Biol 151 (2):182–195

Schelhaas M (2010) Come in and take your coat off - how host cells provide endocytosis for virus entry. Cell Microbiol 12 (10):1378–1388

Schelhaas M, Ewers H, Rajamäki ML, Day PM, Schiller JT, Helenius A (2008) Human papillomavirus type 16 entry: retrograde cell surface transport along actin-rich protrusions. PLoS Pathog 4(9):1000148

Seisenberger G, Ried MU, Endress T, Büning H, Hallek M, Bräuchle C (2001) Real-time single-molecule imaging of the infection pathway of an adeno-associated virus. Science 294(5548): 1929–1932

Shaner NC, Steinbach PA, Tsien RY (2005) A guide to choosing fluorescent proteins. Nat Methods 2(12):905–909

Shen H, Tauzin LJ, Baiyasi R, Wang W, Moringo N, Shuang B, Landes CF (2017) Single Particle Tracking: From Theory to Biophysical Applications. Chem Rev 117 (11):7331–7376

Shimozawa T, Yamagata K, Kondo T, Hayashi S, Shitamukai A, Konno D, Matsuzaki F, Takayama J, Onami S, Nakayama H, Kosugi Y, Watanabe TM, Fujita K, Mimori-Kiyosue Y (2013) Improving spinning disk confocal microscopy by preventing pinhole cross-talk for intravital imaging. Proc Natl Acad Sci U S A 110(9):3399–3404

Sorkin A, Von Zastrow M (2002) Signal transduction and endocytosis: close encounters of many kinds. Nat Rev Mol Cell Biol 3(8):600–614

Strange DP, Jiyarom B, Pourhabibi Zarandi N, Xie X, Baker C, Sadri-Ardekani H, Shi PY, Verma S (2019) Axl promotes zika virus entry and modulates the antiviral state of human sertoli cells. mBio 10(4):01372-19

Suomalainen M, Greber UF (2013) Uncoating of non-enveloped viruses. Curr Opin Virol 3(1): 27–33

Taylor MP, Kramer T, Lyman MG, Kratchmarov R, Enquist LW (2012) Visualization of an alphaherpesvirus membrane protein that is essential for anterograde axonal spread of infection in neurons. mBio 3(2):00063-12

Vonderheit A, Helenius A (2005) Rab7 associates with early endosomes to mediate sorting and transport of Semliki forest virus to late endosomes. PLoS Biol 3(7):21

Wang S, Huang X, Huang Y, Hao X, Xu H, Cai M, Wang H, Qin Q (2014) Entry of a novel marine DNA virus, Singapore grouper iridovirus, into host cells occurs via clathrin-mediated endocytosis and macropinocytosis in a pH-dependent manner. J Virol 88(22):13047–13063

Wegner KD, Hildebrandt N (2015) Quantum dots: bright and versatile in vitro and in vivo fluorescence imaging biosensors. Chem Soc Rev 44(14):4792–4834

Xu H, Hao X, Wang S, Wang Z, Cai M, Jiang J, Qin Q, Zhang M, Wang H (2015) Real-time imaging of rabies virus entry into live vero cells. Sci Rep 5:11753

Yamauchi Y (2019) Quantum dots crack the influenza uncoating puzzle. Proc Natl Acad Sci U S A 116(7):2404–2406

Yin W, Li W, Li Q, Liu Y, Liu J, Ren M, Ma Y, Zhang Z, Zhang X, Wu Y, Jiang S, Zhang XE, Cui Z (2020) Real-time imaging of individual virion-triggered cortical actin dynamics for human immunodeficiency virus entry into resting CD4 T cells. Nanoscale 12(1):115–129

Zhang Y, Ke X, Zheng Z, Zhang C, Zhang Z, Zhang F, Hu Q, He Z, Wang H (2013) Encapsulating quantum dots into enveloped virus in live cells for tracking virus infection. ACS Nano 7(5): 3896–3904

Zhou J, Yang Y, Zhang CY (2015) Toward biocompatible semiconductor quantum dots: from biosynthesis and bioconjugation to biomedical application. Chem Rev 115(21):11669–11717

# Chapter 6
# Correlative Cryo-imaging Using Soft X-Ray Tomography for the Study of Virus Biology in Cells and Tissues

**Archana C. Jadhav and Ilias Kounatidis**

**Abstract** Viruses are obligate intracellular pathogens that depend on their host cell machinery and metabolism for their replicative life cycle. Virus entry, replication, and assembly are dynamic processes that lead to the reorganisation of host cell components. Therefore, a complete understanding of the viral processes requires their study in the cellular context where advanced imaging has been proven valuable in providing the necessary information. Among the available imaging techniques, soft X-ray tomography (SXT) at cryogenic temperatures can provide three-dimensional mapping to 25 nm resolution and is ideally suited to visualise the internal organisation of virus-infected cells. In this chapter, the principles and practices of synchrotron-based cryo-soft X-ray tomography (cryo-SXT) in virus research are presented. The potential of the cryo-SXT in correlative microscopy platforms is also demonstrated through working examples of reovirus and hepatitis research at Beamline B24 (Diamond Light Source Synchrotron, UK) and BL09-Mistral beamline (ALBA Synchrotron, Spain), respectively.

**Keywords** Soft X-ray tomography · Super-resolution microscopy · Structured illumination microscopy · Virus · Reovirus · Herpesvirus · SARS-CoV-2 · Beamline · Cryo-SXT · CryoSIM · Synchrotron

## Abbreviations

| | |
|---|---|
| ACE2 | Angiotensin-converting enzyme 2 |
| CCD | Charge-coupled device |

A. C. Jadhav
Beamline B24, Diamond Light Source, Harwell Science and Innovation Campus, Didcot, UK
e-mail: archana.jadhav@diamond.ac.uk

I. Kounatidis (✉)
Beamline B24, Diamond Light Source, Harwell Science and Innovation Campus, Didcot, UK

School of Life, Health and Chemical Sciences, The Open University, Milton Keynes, UK
e-mail: ilias.kounatidis@open.ac.uk



S. Vijayakrishnan et al. (eds.), *Virus Infected Cells*, Subcellular Biochemistry 106,
https://doi.org/10.1007/978-3-031-40086-5_6

| | |
|---|---|
| CLXT | Correlative light X-ray tomography |
| Cryo-ET | Cryo-electron tomography |
| CryoFIB-SEM | Cryo-focused ion beam milling scanning electron microscopy |
| CryoSIM | Cryo-structured illumination microscope |
| Cryo-SIM | Cryo-structured illumination microscopy |
| Cryo-SXT | Cryo-soft X-ray tomography |
| DMV | Double-membrane vesicle |
| eBIC | Electron Bio-Imaging Centre |
| EM | Electron microscopy |
| ER | Endoplasmic reticulum |
| EV | Enveloped virus |
| FZP | Fresnel zone plate |
| HCV | Hepatitis C virus |
| HFF | Human foreskin fibroblasts |
| HFF-hTERT cells | Human foreskin fibroblast cells immortalised with human telomerase reverse transcriptase |
| HSV-1 | Herpes simplex virus-1 (HSV-1) |
| LAC | Linear absorption coefficient |
| LD | Lipid droplet |
| MRV | Mammalian reovirus |
| MV | Mature virus |
| ROI | Regions of interest |
| SARS-CoV-2 | Severe acute respiratory syndrome coronavirus-2 |
| SIM | Structured illumination microscopy |
| SMV | Single membrane vesicle |
| SNR | Signal-to-noise ratio |
| SRM | Super-resolution microscopy |
| SR-$\mu$FTIR | Fourier transform infrared microspectroscopy |
| ssRNAs | Single-stranded RNAs |
| SXT | Soft X-ray tomography |
| TEM | Transmission electron microscopy |
| TXRM | Transmission X-ray microscope |
| U2OS cells | Human bone osteosarcoma epithelial cells |
| VACV | Vaccinia virus |
| VF | Viral factories |
| XANES | X-ray absorption near edge spectroscopy |
| XRF | X-ray fluorescence |
| ZIKV | Zika virus |

## Soft X-Ray Technique

X-rays were first discovered by Wilhelm Konrad Röntgen in November 1895, and they had immediate applications in medicine with many hospitals using them for assessing fractures (Cheng and Jan 1987; Jacobsen 2020). However, the lack of X-ray sources in combination with the need for suitable optics was an obstacle to the early development of X-ray microscopy for cellular imaging.

During the 1970s, Fresnel zone plates became available and along with advanced condenser objectives allowed improved image resolution, improved contrast, and precise handling of photon density and radiation dose. These technical advances in combination with the introduction of cryopreservation as a sample preparation method (previously validated in the field of cryo-electron microscopy: Cryo-EM) paved the way for the optimisation of soft X-ray microscopy for imaging biological samples (Young 1972; Brüggeller and Mayer 1980; Dubochet and McDowall 1981).

The first demonstration of the use of X-ray microscopy on cryogenically preserved samples (Ptk2 cells—kidney cancer cells derived from a rat kangaroo and *Chlamydomonas reinhardtii* algae) took place in 1998 by the Göttingen beamline team at the BESSY storage ring in Berlin (Germany) (Schneider 1998). Until that moment the radiation damage due to the exposure of the samples to X-rays was the main limitation in imaging biological samples. The cryo-preparation method, which was originally introduced by the same group in 1994, significantly increased the levels of the radiation that a biological sample could withstand, allowing thus longer exposure and serial imaging, leading to the current configurations of cryo-SXT (Schneider et al. 1995).

### *SXT Principles and Advantages*

Cryo-SXT delivers images by absorption contrast at defined spectral areas (Harkiolaki et al. 2018; Okolo 2022), taking advantage of the absorption characteristics of biological matter at the "water window" spectral area, as defined by the X-ray absorption edges of carbon and oxygen (284 and 583 eV photon energy, respectively), and it is the current standard. Soft X-rays in this region are attenuated by bioorganic structures as they are preferentially absorbed by carbon- and nitrogen-rich biological structures in comparison to the surrounding oxygen-rich medium. The above relative difference regarding the attenuation of X-rays generates the absorption contrast across each projection (Spiller et al. 1976; Beese et al. 1986). Across a sample each biochemical component shows a specific X-ray linear absorption coefficient (LAC) that can be determined from the reconstructed slices. Different LAC values allow the discrimination of different organelles, such as lipid bodies, mitochondria, nuclei, or nucleoli (Uchida et al. 2011).

Cryo-SXT cannot compete in resolution with EM techniques (tens of nanometres compared to a few nanometres); however, it offers access to much larger imaging

**Fig. 6.1** TXRM end station at Beamline B24 in Diamond Light Source (**a**). Side and top view of the sample environment inside the TXRM chamber (**b**, **c**)

volumes, facilitating observational studies within the context of whole cells. This is especially important for viral infection studies at the cellular level where the host cellular organisation and the ultrastructural changes brought about by infection are both primary areas of focus (Mazzon and Mercer 2014; de Armas-Rillo et al. 2016; Hernandez-Gonzalez et al. 2021; Garriga et al. 2021).

Advantages of cryo-SXT compared to other high-resolution imaging techniques include the imaging of relatively thick samples (up to 10 μm) with a lateral resolution of up to 25 nm (depending on the objective set-up used) without the need for cell sectioning or milling. Samples are cryopreserved without the need for chemical fixative agents; therefore, they retain their native distribution and architecture through faithful preservation of all cellular ultrastructural features (Kirz et al. 1995; Harkiolaki et al. 2018; Okolo 2022).

Presently, cryo-SXT is available at six international facilities across the world including Beamline B24 at Diamond Light Source (Didcot, UK) (Harkiolaki et al. 2018), HZB X-Ray microscopy beamline U41-PGM1-XM at BESSY II (Berlin, Germany) (Weiss et al. 2000), BE2.1 beamline (XM-2) at the Advanced Light Source (Berkeley, CA, USA) (Le Gros et al. 2014), BL09-MISTRAL beamline at ALBA (Barcelona, Spain) (Pereiro et al. 2009), the soft X-ray microscopy beamline at the National Synchrotron Radiation Laboratory (Hefei, China) (Bai et al. 2020), and the 24A soft X-ray tomography beamline at the National Synchrotron Radiation Research Center (Hsinchu, Taiwan) (Lee et al. 2016).

At Beamline B24 at the Diamond Light Source (the UK's national synchrotron), the source of the X-rays is a bending magnet in the synchrotron ring. X-rays pass through a series of focusing and dispersive optics, and they reach the vessel of the transmission X-ray microscope (TXRM) (UltraXRM-S220c Zeiss) (Fig. 6.1). X-rays are focused and delivered to the sample plane through a capillary condenser. The

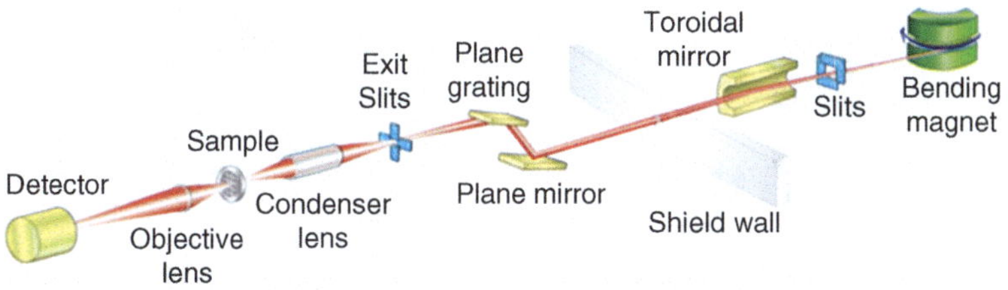


**Fig. 6.2** A schematic of Beamline B24 from source to detector in the synchrotron (reprinted from Harkiolaki et al. 2008)

**Table 6.1** Time required for SXT data acquisition and processing

| Experiment design | Time required on the user-specific projects |
|---|---|
| *Sample preparation* | |
| • Cell culture/maintenance | 0–6 days |
| • Cell seeding on TEM grids (when needed) | 6–24 h |
| • Cell treatment (dependent on user requirements) and labelling | 0.5–1 h |
| • Cryopreservation (plunge freezing) | 0.5–1 h |
| *Data collection* | |
| • Samples loaded to TXM (air to vacuum; sample remains vitrified) | 1–2 h |
| • Sample (TEM grid with multiple imaging areas) loaded and mapped | 15–30 min |
| • Single tilt series acquisition | 5–25 min |
| • Complete grid data collection (several tilt series; 20–50 Gigabytes) | 3–6 h |
| • Data processing and reconstruction (concurrent to data acquisition) | 3–6 h |
| *Post processing/analyses* | |
| • Data mining | Dependent on user needs |

resulting projections are directed to a charge-coupled device (CCD) detector through a Fresnel zone plate (FZP) (Fig. 6.2) (Harkiolaki et al. 2018).

## *SXT Data Collection*

Prior to cryopreservation and data acquisition, cells are directly loaded from their growth media either on holey carbon TEM grids or into full-rotational thin-walled capillaries tubes (this is the case for the Advanced Light Source, CA, USA). As soon as a sample is loaded, a full tomographic image series can be acquired within a few minutes (for a description of the duration of each step of the cell preparation and data collection processes, see Table 6.1). In the case of the glass capillary, a full-rotation

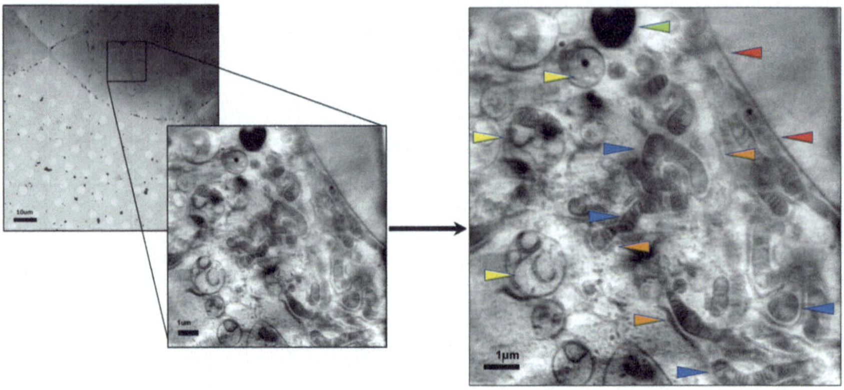


**Fig. 6.3** Projection through a human bone osteosarcoma epithelial (U2OS) cell; field of view (10 × 10 μm) captures a cytoplasmic perinuclear area full of organelles and a fraction of the nucleus (top right). Projection collected on B24 at 500 eV using a 25 nm zone plate objective. Scale bars represent 1 μm. Arrows point to selected organelles; blue—mitochondria; green—lipid droplets (LDs); orange—endoplasmic reticulum; yellow—multivesicular bodies; red—nuclear membrane

series can be collected, while for the TEM grids the maximum rotation is limited by the beam attenuation at high tilts due to the sample thickness which results from missing wedge artefacts. TEM grids, however, have the advantage of accommodating both cell suspensions and adherent cells, while capillary tubes are more suited to imaging cell suspensions (Harkiolaki et al. 2018; Larabell and Nugent 2010).

In limited angle SXT (which is the case for Beamline B24, Diamond Light Source, UK) SXT data are collected in the form of tilt series. Each consists of a 2D X-ray projection of the sample at a different rotation around the central Y axis. Tilt series are then reconstructed to a 3D image volume (tomogram) after alignment correction. The field of view of each captured image can extend to 20 × 20 μm$^2$ and cumulative volumes typically have heights up to 10 μm with a maximum resolution depth of focus of 1–3 μm or more, depending on the optical performance. A typical acquisition of a tomogram lasts between 5 and 25 min depending on the exposure time for each projection (Table 6.1) and the tilt angle value (Harkiolaki et al. 2018; Okolo 2022). Raw SXT data are subsequently reconstructed into 3D tomograms with the use of dedicated open-source algorithms such as IMOD (Kremer et al. 1996). The tomograms are fundamentally Z-stacks of images that capture the cellular ultrastructure at near-physiological conditions under high-resolution examination (Fig. 6.3).

## Use of SXT in Viral Studies

Viruses are omnipresent in nature and are obligatory intracellular parasites, where they infect living cells of all kinds for their survival within both the prokaryotic and eukaryotic domains of life (Taylor 2014; Harris and Hill 2021). As pathogens, viruses are responsible for mild to severe diseases in almost all living forms including humans (Harris and Hill 2021), but can also develop beneficial relationships with their hosts through symbiosis (Roossinck 2011). The steps involved in the virus life cycle are virus entry, translation, replication, assembly, and egress of viral offspring from the infected host cells (Mercer et al. 2010). The process of virus infection in the eukaryotic cell is complex and dynamic, and it begins with the initial virus–host cell interaction followed by hijacking of the host cell machinery to produce progeny virions which brings about the spatiotemporal reprogramming and/or reorganisation of the cellular genome, biochemistry, and structures (Glingston et al. 2019). The life cycle of viruses in infected host cells requires the recruitment of numerous cellular pathways and brings the offender in direct contact with different cellular structures (Sanchez and Lagunoff 2015). To develop effective antiviral therapies, it is vital to understand virus structure and its interactions within host cells during viral pathogenesis; thus far, different conventional and advanced light and electron microscopic techniques have been successfully utilised to date (Risco 2021).

Here we focus on how the state-of-the-art imaging technique of cryo-SXT has been successfully applied globally at different SXT facilities to obtain detailed three-dimensional ultrastructural information at the cellular level, at the multiple steps involved in virus-infected host cells in their near-native physiological state, at a few tens of nanometres spatial resolution. The potential of the cryo-SXT platform for virological research was first demonstrated in the study of vitrified, unstained, purified particles of vaccinia virus (VACV) from monkey kidney BSC40 cell cultures at the HZB X-ray microscopy beamline at BESSY II (Berlin, Germany) (Carrascosa et al. 2009). VACV is a member of enveloped, large DNA poxviruses and exists in two infectious forms called mature virus (MV) and enveloped virus (EV). The two forms differ in the number of the membranes and exert a different role during the VACV life cycle with MV being the most abundant (Roberts and Smith 2008). X-ray data analysis reconfirmed the complexity regarding the structure of the MV form as previously shown by the use of cryo-electron tomography, including a well-defined external envelope surrounding the inner core. The inner region comprises a lighter core surrounded by a denser envelope (Fig. 6.4A). The same group in 2012 conducted an in-depth comparative study in VACV-infected PtK2 cells combining cryo-SXT and transmission electron microscopy (TEM) (Chichón et al. 2012). For this study, PtK2 cells were grown on the flat surfaces of TEM Au grids and infected with a vaccinia virus strain expressing a green fluorescent protein (GFP). Infected cells on sample holders were cryopreserved without any staining or sectioning of the sample using a rapid plunge freezing method. The infection status of the host cells was confirmed by the expression of GFP using a light

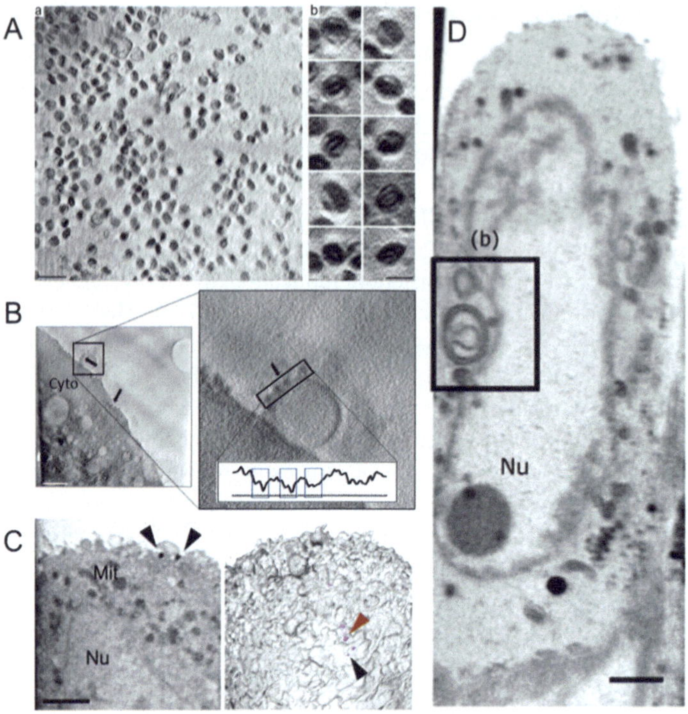


**Fig. 6.4** Collation of SXT applications to investigate virus biology at different SXT facilities. (**A**) Structural analysis of vaccinia virus particles using 25 nm zone plate at HZB X-ray microscopy beamline at BESSY II (Berlin, Germany). X-ray analysis confirms the brick shape form of mature virus. Scale bar represents 1 μm (**a**) and subtomographic analysis shows different types of virions with the structural details of the virus. Scale bar represents 300 nm (**b**) (reprinted from Carrascosa et al. 2009). (**B**) Investigation of SARS-CoV-2 infection at Beamline B24 at Diamond Light Source (Didcot, UK) using 40 nm zone plate. Tomographic slice representative of SARS-CoV-2-infected Vero cell indicating the exocytosis of virions (black arrows) at cell surface and infected cell cytoplasm is filled with DMVs. Scale bar represents 1 μm (reprinted from Mendonça et al. 2021). (**C**) Study conducted on HEK cells stably expressing ACE2 receptors to investigate the mechanism of SARS-CoV-2 infection at BE2.1 beamline (XM-2) at the Advanced Light Source (Berkeley, CA, USA) using a 50 nm zone plate. The X-ray orthoslice indicates the virion entry (black arrows) and the aggregates of virions (orange arrows) identified in 3D volume rendering at the cell surface. Scale bar represents 2 μm (reprinted from Loconte et al. 2021). (**D**) Investigation of HSV-1 infection in human B cells using 50 nm zone plate at BE2.1 beamline (XM-2). HSV-1 infection after 20 h indicating the membrane alteration to form DMV structure with the concave invagination in B-cells (rectangle). Scale bar represents 2 μm (reprinted from Chen et al. 2022a). Nu, nucleus; Mit, mitochondria; Cyto, cytoplasm

fluorescence microscope prior to X-ray data collection. The cell thickness for X-ray imaging was in the range of 3–10 μm. For TEM data collection, plunge frozen virus-infected cells were processed using a freeze-substitution method, stained with uranyl acetate and lead citrate followed by epoxy resin embedding prior to the ultrathin sectioning to create 70 nm thick slices. The cartographic analysis of unstained and non-sectioned tomograms obtained by limited angle tomography from VACV-infected PtK2 cells revealed structures resembling vaccinia virus factories related to different maturation steps of VACV (immature and mature particles) in the cell cytoplasm, which was confirmed in the heavily processed ultrathin TEM images at higher resolution. The analysis also revealed the reorganisation of various host cell organelles (nucleus, mitochondria, ER) due to the presence of the virus albeit at slightly lower resolution compared to the TEM data.

At the BL09-MISTRAL beamline at ALBA (Barcelona, Spain) the host cellular modifications of Zika virus (ZIKV)-infected human glioblastoma U251 cells prepared on TEM grids were recently studied (Garriga et al. 2021). ZIKV belongs to the *Flaviviridae* family and has an enveloped icosahedral capsid harbouring a linear positive ssRNA genome. Previous studies using electron tomography in hepatoma and neuronal progenitor ZIKV-infected cells showed extensive host structural modifications (Cortese et al. 2017). X-ray data analysis of the ZIKV-infected glioblastoma cells elucidated the ZIKV-mediated cytopathic effects including cell rounding, deformation of the nucleus into a kidney-shaped structure, fragmentation of ER resulting in the formation of Zika virus replication vesicles, and the formation of convolution membranes with rare membrane tubules in the cytosol of infected cells.

Mendonça and her colleagues described the use of a multi-scale cryo-imaging platform for the investigation of the replication cycle of the severe acute respiratory syndrome coronavirus-2 (SARS-CoV-2) in Vero cells at the Beamline B24 and the Electron Bio-Imaging Centre (eBIC) at the Diamond Light Source (Didcot, UK) (Mendonça et al. 2021). SARS-CoV-2 is a member of the coronavirus RNA family and induces the development of a range of membrane compartments during its life cycle that contribute to viral replication efficiency and host recognition escape via potection of viral components (Ertel et al. 2017; Klein et al. 2020). The focus of the study was the investigation of the SARS-CoV-2 replication cycle aiming to understand the viral assembly and egress processes, combining cryo serial focused ion beam scanning electron microscopy (cryoFIB-SEM), cryo-SXT, and high-resolution cryo-electron tomography (cryo-ET). Serial cryoFIB/SEM image analysis of SARS-CoV-2-infected Vero cells revealed cytopathological alterations on a large scale showing a large number of vesicles in the perinuclear space and membrane tunnel-like structures coated with the electron-dense particles consistent with the size of SARS-CoV-2. SXT data analyses of infected Vero cells further elucidated cytological alterations at the whole cell level showing substantial changes in the mitochondrial network (a reduced number and altered shape of mitochondria in infected cells was recorded) accompanied by the formation of extensive cytoplasmic invaginations and nuclear blebbing. The final step of the virus life cycle including the viral release via an exocytosis-like mechanism was also recorded (Fig. 6.4B).

The mechanism of SARS-CoV-2 infection and pathogenesis was studied at the same period at the BE2.1 beamline (XM-2), the Advanced Light Source (Berkeley, CA, USA) using full-itation cryo-SXT (Loconte et al. 2021). For this study, angiotensin-converting enzyme 2 (ACE2) receptor stably expressing HEK293-ACE2 cells were infected with SARS-CoV-2 at 0.5 MOI for 6 and 24 h and fixed cells were added to thin-walled glass capillaries and rapidly plunge frozen to achieve a full-rotation tomography. The tomographic analysis of SARS-CoV-2-infected HEK293-ACE2 cells captured virion entry and aggregation (Fig. 6.4C) and confirmed the presence of characteristic structures associated with SARS-CoV-2 replication, such as the formation of double-membrane vesicles (DMVs) and convoluted membranes in the infected cell cytoplasm. The DMV structures were found to be in physical contact with mitochondria in the infected cell cytoplasm suggesting the facilitation of a crosstalk between mitochondria and ER. Ptogression through infection from 6 to 24 h in HEK293-ACE2 cells revealed prominent cellular remodelling where a significant increase in the volume of lipid droplets and, an increase in the number and volume of cytoplasmic vesicles were observed (Loconte et al. 2021).

The host cell remodelling in the case of the herpes simplex virus-1 (HSV-1) infection was the focus of a series of studies conducted at the BE2.1 beamline (XM-2) at the Advanced Light Source (Berkeley, CA, USA). HSV-1 is a large, enveloped DNA virus and a member of the Herpesviridae family capable of establishing a persistent long-lasting infection in sensory and sympathetic neurons (Nicoll et al. 2012). The production of viral particles during the lytic replication is a complex process and involves multiple cellular compartments (Johnson and Baines 2011; Arii 2021 and Brown & Newcomb (2011)). Cryo-SXT analysis of HSV-1-infected human B-cells revealed changes in cellular structures including the presence of cytoplasmic stress granules and multivesicular structures, the formation of nuclear dense bodies, aggregation of capsids, elongation of mitochondria, and increased abundance endosomal and lysosomal vesicles (Fig. 6.4D) (Chen et al. 2022a). Previous studies at the same beamline have focused on the molecular organisation of chromatin in HSV-1 infection and for that reason a combinatorial approach of cryo-SXT, cryo-fluorescence, confocal, and TEM imaging was then followed (Myllys et al. 2016; Aho et al. 2017, 2019). SXT data analyses showed that during the late stages of the infection the formation of the viral replication compartment results in the enrichment of heterochromatin in the nuclear periphery accompanied by the compaction of chromatin. While this peripheral compacted chromatin restricts the viral capsid diffusion, capsids are eventually able to reach the nuclear envelope moving into the site of their nuclear egress through interchromatin channels.

The cryo-SXT technique aiming to elucidate the complexity of the HSV-1 life cycle has also been applied at Beamline B24 at Diamond Light Source. Nahas and colleagues used a consolidated multimodal imaging platform, combining cryo-widefield fluorescence microscopy to identify stages of infection and cryo-SXT technique to investigate morphological changes induced by HSV-1 infection at the subcellular level (Nahas et al. 2022a). Various stages of HSV-1 infections were identified in fully-hydrated human foreskin fibroblast cells immortalised with human telomerase reverse transcriptase (HFF-hTERT) using cryo-SXT (Fig. 6.5A). Spatio-temporal changes were traced in U2OS cells when infected with a recombinant "timestamp" strain that expressed fluorescent chimaeras of the early protein ICP0 and the late protein gC to distinguish early and late stages of the infection cycle (Scherer et al. 2021). Discriminating between early and late stages of HSV-1 infection made it possible to investigate temporal changes to organelles during HSV-1 infection. This investigation confirmed substantial remodelling of the

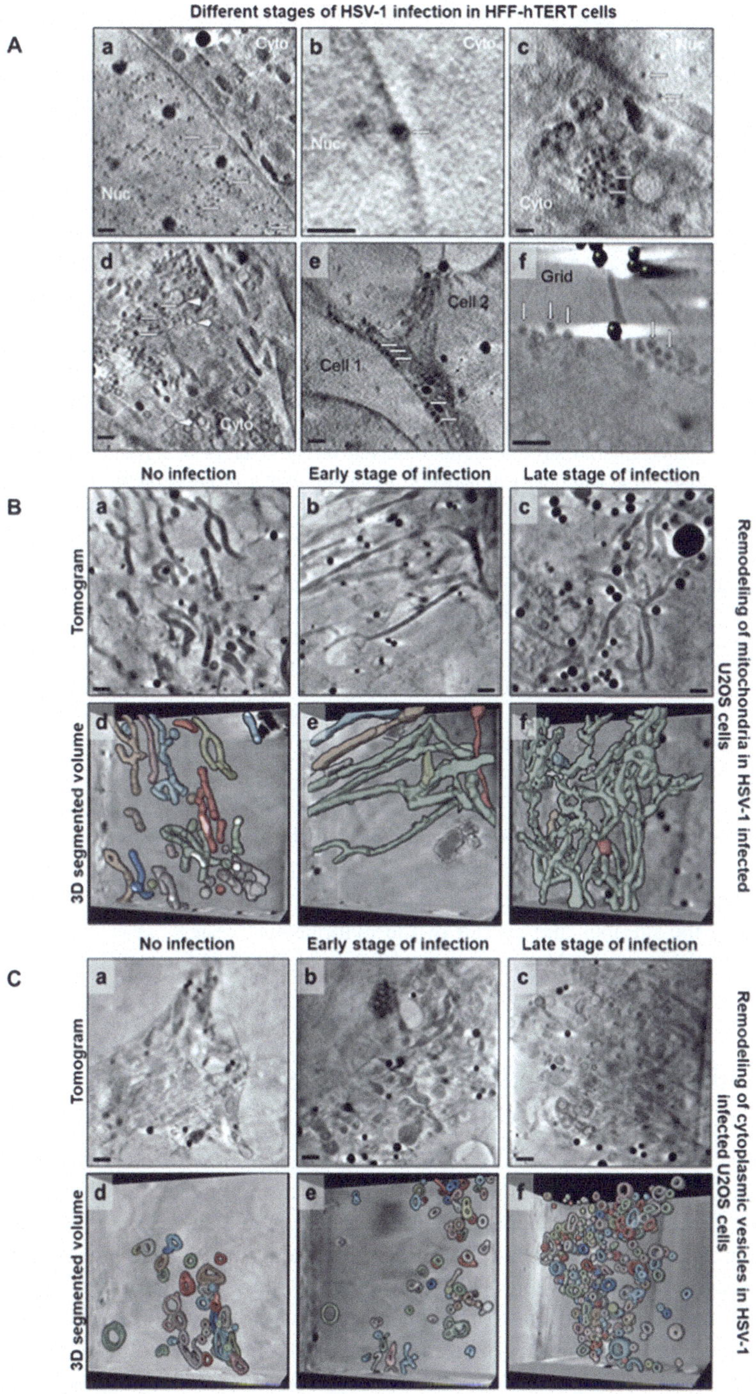


**Fig. 6.5** (**A**) Soft X-ray imaging of HSV-1-infected HFF-hTERT cells exhibiting virus particles at different stages of HSV-1 life cycle: HFF-hTERT cells were infected with HSV-1 at MOI of 2 for

**Fig. 6.5** (continued) 16 h before cryofixation. All tomograms were reconstructed from X-ray projections acquired using a 40 nm zone plate objective; scale bars represent 1 μm. HSV-1-infected cells show multiple dark puncta in the nucleus (arrows) indicating newly assembled naked HSV-1 capsids (**a**). Trapped HSV-1 capsids were observed inside the inner and outer nuclear envelope, suggesting the transport of naked capsids from nucleus to the cytoplasm by nuclear membrane modifications (**b**). HSV-1 capsids (arrows) were found in the infected cell cytoplasm surrounded by cytoplasmic vesicles (arrowheads) (**c**, **d**). Transmission of HSV-1 virions (arrows) is observed at the intersection of two neighbouring cells (**e**). Exocytosis of mature HSV-1 virions was observed at the infected cell surface. Gold nanoparticles used for image registration are indicated with stars (**f**). (**B**) Remodelling of mitochondria in HSV-1-infected U2OS cells: Changes in mitochondrial morphology were assessed from cryo-SXT tomograms collected using a 25 nm zone plate objective from non-infected cells and cells infected at early or late stages of infection with recombinant timestamp strains of HSV-1. Data are representative of three independent experiments. Scale bars represent 1 μm. Non-infected U2OS cells had mitochondria with different morphologies indicating spherical, short, and long structures in tomograms (**a**) and segmented 3D volumes of tomograms (**d**). The alteration in mitochondrial structures was observed in early stages of infection represented as elongated bodies (**b**, **e**) and branched networks in the late stages of infection (**c**, **f**). Mitochondrial structures were segmented and differentiated using *Contour* (Nahas et al. 2022b). (**C**) Remodelling of cytoplasmic vesicles in HSV-1-infected U2OS cells: Changes in the number and spread of cytoplasmic vesicles were documented in cryo-SXT tomograms and widefield fluorescence cryo-microscopy images acquired from uninfected cells, or cells at an early or late stage of infection with the recombinant timestamp HSV-1 strains. Data are representative of three independent experiments. Scale bars represent 1 μm. A significant increase in the accumulation of vesicles was observed at perinuclear spaces of infected cells from early stages of infection (**b**, **e**) to the late stages of infection (**c**, **f**) in comparison with the uninfected U2OS cells (**a**, **d**). Vesicles representing spherical, ellipsoidal, or dumbbell shapes were considered in the *Contour* analysis, but vesicles different to these shapes were omitted. All images were kindly provided by Dr Kamal Nahas, Diamond Light Source, UK

mitochondria, the cytoplasmic vesicles, and the lipid droplets. At early stages of infection, mitochondria were heterogenous in morphology. They became elongated at late stages of infection, often forming branched networks. Similarly, a significant change in the number and spread of the cytoplasmic vesicles was observed with progression through HSV-1 infection (Nahas et al. 2022a) (Fig. 6.5B, C).

## Extending the Information Content of SXT Data in the Context of Correlative Microscopy Schemes

Cryo-SXT data provide valuable structural information of intact, fully hydrated cells without the need for sectioning techniques or contrast-enhancing methods. The utility of SXT techniques has been further enhanced by the development of correlative imaging schemes where SXT is combined with other microscopy techniques for the localisation of events at the subcellular level. Such a correlative scheme is exemplified by combining synchrotron-based techniques (cryo-SXT, XRF, and XANES) to investigate the mechanism of photoactivated anticancer agents in human cancer cells in their near-native state (Bolitho et al. 2021). Through cryo-

SXT, morphological changes to organelles within cancer cells were analysed in 3D after exposure to photoactivable Pt anticancer complexes with or without activating irradiation. 2D cellular distribution and quantification of Pt and endogenous elements were investigated using nano focused X-ray fluorescence (XRF). Co-localisation of Pt and endogenous Zn was detected in the nucleus of all treated cells, which confirmed the formation of DNA-Pt lesions in the nucleus. Subsequently, the changes were correlated to the cellular distribution of Pt identified using XRF, with an established anticancer drug cisplatin, using XRF technique. The intracellular photoreduction of the Pt anticancer agents was detected using X-ray absorption near edge structure (XANES) spectroscopy. Therefore, the complementary approach of cryo-SXT, XRF, and XANES techniques provided detailed information on biological responses and oxidation state of the photoactive complexes and their photoproducts in treated cancer cells when subjected to the photoactivated chemotherapy (Bolitho et al. 2021).

SXT has benefitted from the development of cryogenic light microscopes for correlative imaging. This combination of light and X-ray imaging techniques (CLXT) allows tracing of tagged molecules—localisation in the context of a high-resolution three-dimensional structural information obtained from the X-ray image of a cell. This corresponds to a long-standing need for chemical localisation in cell research and as such has enormous potential in both clinical and industrial research. The first approaches for establishing SXT-based correlative microscopy schemes were employed using either cryogenically adapted fluorescence microscopes prior to cryo-SXT imaging (McDermott et al. 2009; Chichón et al. 2012; Le Gros et al. 2009; McDermott et al. 2012; Duke et al. 2014; Smith et al. 2014; Elgass et al. 2015; Myllys et al. 2016; Bai et al. 2020) or fluorescence optics located inside the X-ray microscope (Schneider et al. 2012; Hagen et al. 2012).

The above-mentioned instances were based on the use of conventional optical fluorescence light microscopy, which has been widely used in the field of modern biological research for the tracking of fluorescently tagged biomolecules in the specimen which deliver insights on cellular physiology at both functional and spatial levels. With the new advances in the development of optics and the availability of various fluorescent markers for biological research, several fluorescence microscopy variants have been developed ranging from traditional epifluorescence microscopy to confocal and multi-photon microscopy (Combs and Shroff 2017). Under optimal conditions, such as homogeneous refractive index of diffraction barrier, high signal intensity, and minimal noise from the background, the optical resolution of light microscopic techniques can be achieved to the theoretical limit as defined by the Rayleigh criterion (Rayleigh 1879).

As an alternative to conventional microscopy, research over the past few decades has been developing optical imaging techniques aiming to provide resolution beyond the diffraction limit; these techniques are collectively known as super-resolution microscopy (SRM) techniques, inclusive of both near-field and far-field methods (Schermelleh et al. 2019). There are currently numerous outstanding custom-built instruments or commercially available microscopes being developed for super-resolution imaging of biological samples (Schermelleh et al. 2010). To

**Fig. 6.6** Diffraction limited (panel **a**, bottom half) and reconstructed SIM (panel **a**, top half) image of human foreskin fibroblast cells (HFF) expressing vimentin-mIFP, a marker of intermediate filaments, which shows the improvement in resolution and background light suppression in cryo-SIM imaging. Scale bar is 10 μm (image courtesy by Dr Kamal Nahas, Diamond Light Source, UK). The cryo-SIM set-up at the Beamline B24 (**b**) with a closeup of the Linkam cryo-stage (**c**) that holds the samples on a motorised z-stage. Top view of the four excitation wavelengths travelling paths through cryoSIM optics (**d**)

achieve CLXT at cryogenic temperature taking advantage of the resolution levels of the SRM technique, a bespoke custom-built SRM platform that offers three-dimensional cryo-structured illumination microscopy (cryo-SIM) has been developed at Beamline B24 in collaboration with the Micron imaging facility at the University of Oxford (Fig. 6.6) (Phillips et al. 2020). The platform is called cryo-structured illumination microscope (cryoSIM) and can serve as a complementary tool to cryo-SXT techniques for imaging whole cells under cryogenic conditions. The fundamental principle of SIM involves the use of a patterned light interaction with the fluorescently tagged biomolecules in the specimen, resulting in a low-frequency moire interference fringes corresponding to fine structural details in the sample (which would not be detected normally) to be shifted into lower frequencies which are acquired in the image. The patterned light is created using a nematic liquid-crystal-on-silicon spatial light modulator and multiple images are acquired for sample thickness over 10 μm at five phases and three different angles. All the information acquired contains a set of equations of separate high-frequency and low-frequency contributions, which can then be computationally extracted and shifted back to its originating place in frequency space (Phillips et al. 2020). The information is then mapped during a reconstruction procedure to create a super-resolution image which can lead to an eightfold increment in volumetric resolution. Raw SIM data can be assessed for reconstruction artefacts through the *SIM Check* plugin in Fiji (Phillips et al. 2020; Vyas et al. 2021).

The cryo-SIM technique offers distinctive advantages for cryo-imaging including the following aspects: (1) easy application to multi-channel image registration by using a range of common fluorophores or fluorescently tagged biomolecules achieving a resolution up to 200 nm (depending on the selected fluorescence wavelength), (2) effective thermal sample insulation as the air objective used has a long working distance (2 mm), so that the gap between the objective and frozen sample prevents the objective from heating the specimen, (3) reduced heat damage and lowered ice crystal formation as cryo-SIM uses relatively lower illumination intensities (10–100 W/cm$^2$ laser power) and shorter image acquisition times (20–100 ms per single exposure), and (4) acquisition of high-contrast images in samples of over 10 μm in thickness without the need of sectioning as the techniques show effective out of focus light suppression (Demmerle et al. 2017). The use of SIM has been previously documented in the same cell correlative scheme along with single molecule localisation microscopy to visualise cryopreserved cells and then imaged using scanning EM via FIB milling, after they are processed with freeze-substitution staining (Hoffman et al. 2020).

## Sample Preparation and B24 Experimental Workflow

Beamline B24 has established a correlative imaging scheme (CLXT) for biological samples at cryogenic temperatures that uses the purpose-built cryoSIM and the synchrotron-based X-ray microscope that is located in the B24 cabin. Among the available cryo-SXT beamlines worldwide, Beamline B24 at Diamond Light Source is unique in offering X-ray imaging data correlated with super-resolution data at cryogenic conditions. The B24 correlative scheme is based on an experimental workflow that includes the sample preparation, the sample imaging, and the data analysis (Fig. 6.7).

The first step is the sample preparation for limited angle tomography in which cells grow onto flat surfaces of Quantifoil TEM grids (3 mm in diameter) coated with holey carbon support film where positions are marked with numbered letters. Due to the hydrophobic chemical nature of the grids, they need pre-treatment to submerge and make them cell attachment friendly. There are three ways to pre-treat TEM grids (Okolo et al. 2021): (1) hydrophilisation of grids by coating them with filter-sterile foetal bovine serum (FBS) for 6–24 h before the addition of biological samples,

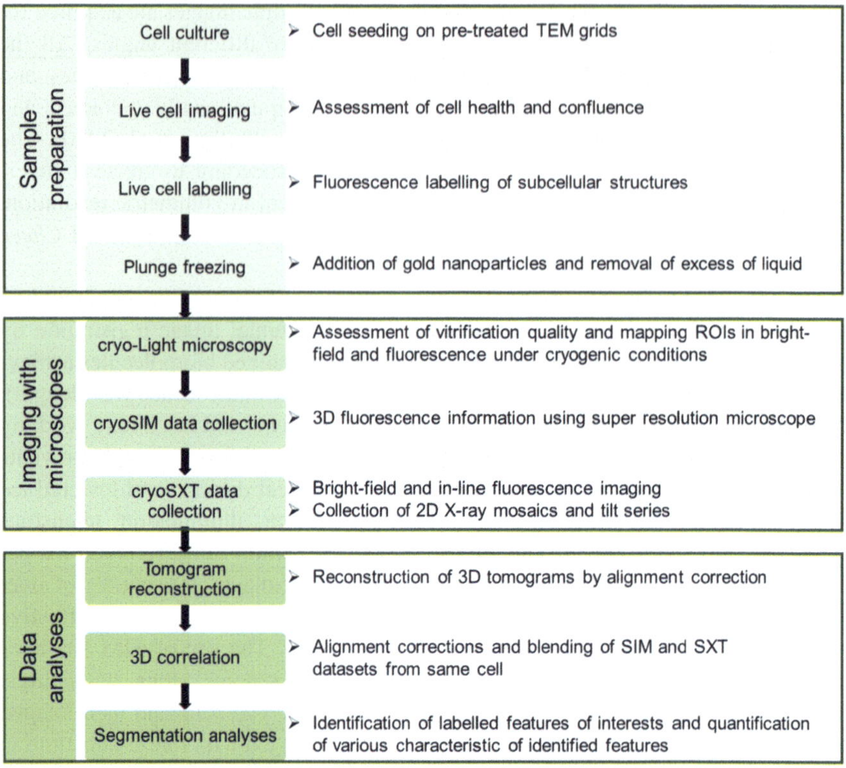


**Fig. 6.7** A schematic of Beamline B24 step-by-step CLXT workflow explaining the sample preparation, imaging with different microscopes, and data analyses stages

(2) glow/plasma discharging of grids using an easiGlow unit, and (3) pre-coating grids with cell attachment facilitating polymers such as poly-L-lysine, vitronectin, and laminin. The choice of pre-treatment method is dependent on the type of cells being used in the experiment. After pre-treatment of TEM grids, cells are seeded onto the grids at optimum cell density in order to achieve up to 70% confluence followed by a 37 °C/5% $CO_2$/24 h incubation. Next day, cell health and confluence are assessed under a conventional bright-field microscope before proceeding with the virus infection. Cells could endogenously express fluorescently tagged protein/s of interest or alternatively cells are labelled with fluorescent markers upon infection. Fluorescence labelling of cell organelles such as mitochondria, lipid droplets, and lysosomes not only provides information of chemical localisation but also facilitates data registration with labelled organelles considered as correlation markers (Okolo et al. 2021). Plunge freezing (Dobro et al. 2010) is the method of choice for sample preservation in Beamline B24, while for thicker samples (>10 μm) high-pressure freezing (HPF) (Moor 1987) is the recommended fixation method. In plunge freezing, vitrification is achieved via rapid cooling to temperatures below –170 °C by plunging the sample grid into liquid nitrogen-cooled liquid ethane or propane. Prior

to plunge freezing, gold nanoparticles of 100–250 nm in diameter are added for later uses as image registration markers for accurate alignment of X-ray projections to process images into reconstructed 3D volumes.

The next step involves the mapping of samples using conventional widefield fluorescence microscopy to assess the vitrification quality and identify regions of interest (ROI) before collecting the super-resolution 3D cryo-SIM data followed by cryo-SXT data acquisition. The last step refers to the necessary raw SXT data reconstruction into 3D tomograms (Kremer et al. 1996) and the following image registration with the use of the appropriate software (eC-CLEM) (Paul-Gilloteaux et al. 2017). The CLXT experimental workflow is adaptable for a wide range of biological samples such as multi-species adherent or suspension cell cultures (primary cells and immortalised cell lines), bacteria, parasites, algae, fungi, and viruses (Harkiolaki et al. 2018).

## SXT Correlative Schemes in the Service of Virus Research

### *Reovirus*

Reoviruses are non-enveloped double-stranded RNA viruses that result in either an asymptomatic, non-pathogenic infection or mild gastroenteritis in humans (Coombs 2006). The reovirus replication cycle is entirely cytoplasmic with the virus entering the cell through clathrin-mediated endocytosis and subsequently trafficked through the endosomal pathway from the early to the late endosomes. Transcriptionally active particles are released into the cytoplasm and synthesise positive-sense, single-stranded RNAs (ssRNAs). Reoviral ssRNA copies are translated to new viral proteins and work as templates for negative-sense ssRNAs in order to produce genomic dsRNA. Genome replication and assembly takes place in specialised compartments known as viral factories (VFs) or inclusions. In the final stages, outer-capsid proteins are added onto the newly formed viral cores and egress occurs either through nonlytic egress, common with polarised cell types, or though lytic egress mediated by an NF-κB-dependent apoptotic pathway (Miller and Krijnse-Locker 2008; Tenorio et al. 2019; Roth et al. 2021).

Kounatidis and his colleagues demonstrated for the first time the use of the CLXT workflow developed at Beamline B24 and applied the method for the case study of reovirus infection in human cancer cells for the first time (Kounatidis et al. 2020). The applied correlation protocol is based on the use of open-source software (eC-CLEM plugin in Icy) (Paul-Gilloteaux et al. 2017) and involves the 2D positioning of bright-field, fluorescence, X-ray mosaic, and X-ray tomogram data followed by 3D image registration to position the cryo-SIM data on the X-ray tomogram. Features such as gold nanoparticles or labelled cell organelles like lipid droplets, mitochondria, and lysosomes are used as registration markers for alignment of the datasets. The successful application of this protocol in endogenously

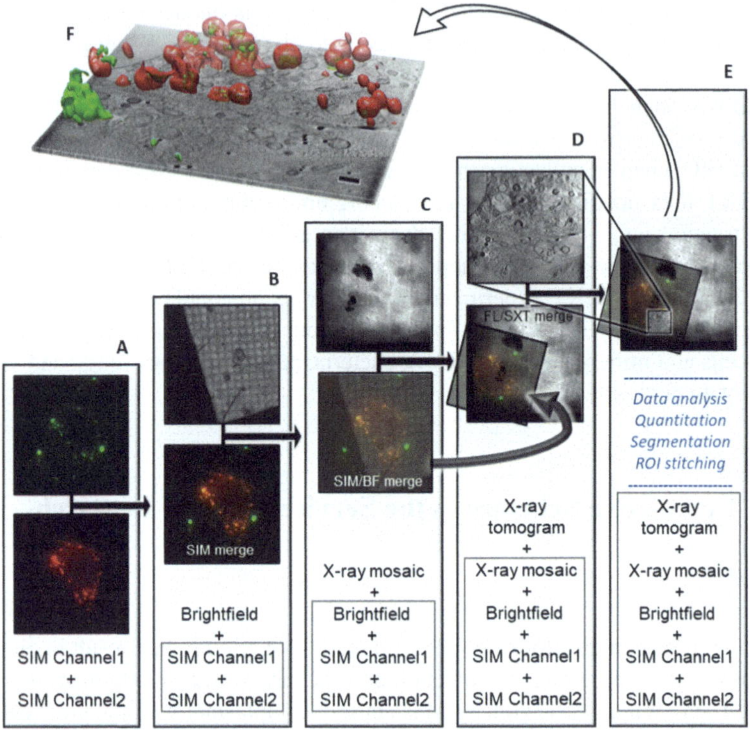


**Fig. 6.8** A stepwise correlation protocol illustrating the image registration workflow using eC-CLEM plugin in the open-source ICY application (**a**–**e**), and visualisation of correlated 3D rendering of fluorescence data in Z-slice using Chimera application (**f**) for fluorescently Alexa488-labelled (green fluorescence) reovirus 4 h post-infected U2OS cells endogenously expressing mCherry (red fluorescence)-labelled Galectin-3 protein using cryoSIM and TXRM imaging platforms (reprinted from Kounatidis et al. 2020)

mCherry-labelled Galectin-3 protein expressing U2OS cells infected with fluorescently Alexa488-labelled reovirus 4 h post-infection is shown in Fig. 6.8.

For the purposes of the experimental workflow, human U2OS cells expressing mCherry-Gal3 (serves as an indicator of endosomal membrane disruption) were seeded onto pre-treated TEM gold grids and infected with mammalian reovirus (MRV) for 16 h. MRV particles were labelled with Alexa488 NHS ester antibody and used at MOI of 100. The infection status of cells was monitored via confocal microscopy confirming the presence of green fluorescence of labelled virus components intracellularly. Infected cells were vitrified after applying 250 nm gold fiducials in cooled liquid ethane using a Leica EM GP2 plunge freezer at 1 h intervals. Vitrified samples on grids were transferred to the cryoSIM and imaged in green and red fluorescence to identify cells infected with MRV and expressing Gal3. 3D-SIM data were collected on representative cells at 1, 2, 3, and 5 h after infection. X-ray data were collected with an UltraXRM-S/L220x X-ray microscope (Carl Zeiss) at

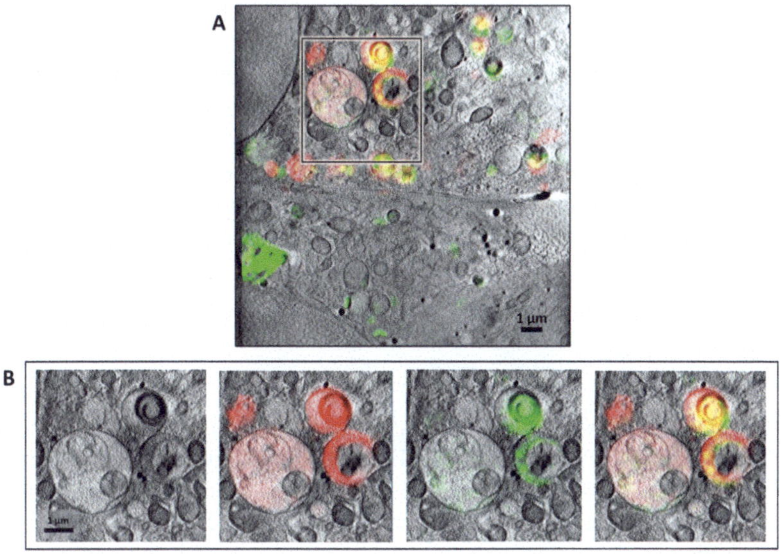


**Fig. 6.9** Endosomal trafficking and escape of reoviruses in U2OS cells. (**a**) Single Z-slice from a reconstructed tomogram with Gal3 (red) and MRV (green) localisation from 2 h post-infected cell (**b**). ROI was identified with either X-ray grayscale imaging alone or superimposed with fluorescence signal. Multivesicular bodies (MVBs) are easily distinguishable with the recognisable sub-compartments where Gal3 is accumulated (suggesting the fusion of virus-loaded vesicles with virus free late endosomes) (reprinted from Kounatidis et al. 2020)

Beamline B24 with a 40 nm zone plate objective set-up. Tilt series were collected from –65° to +65° at increments of 0.5°. Data were aligned and reconstructed automatically using the in-house pipeline which employs Batchruntomo as part of the IMOD package (Kremer et al. 1996, Mastronarde 2005). Same cell cryo-SIM and cryo-SXT datasets were overlaid with the registration software eC-CLEM (Paul-Gilloteaux et al. 2017) (Fig. 6.9).

The CLXT data analyses elucidated the early stages of the reovirus infection revealing a series of events: (1) by 1 h post-infection (PI) the virus is contained within a distinct endosomal population, (2) from 1 h PI localised Gal3 within virus-bearing vesicles is detected suggesting the permissibility of cytoplasmic proteins or other small cellular constituents inside vesicles (the fact that the integrity of the virus-loaded endosomes is not compromised suggests a mechanism of selective shuttling of core particles with pores closing or reducing to size when not in use), and (3) by 2 h PI endosomes were loaded with virus and endosome structures changed to larger, more complex, and multivesicular bodies, and were observed in close proximity to the nucleus, indicating that they are trafficked for either recycling within or evicted from the host cell (Fig. 6.9).

With the above study, a detailed and clear timeline for the transport of viral material through the conventional endosomal pathway and the viral escape is now available. The study was the first validation of the cryo-SXT platform developed at

Beamline B24, which is now fully commissioned and freely accessible to the global scientific community.

Beyond CLXT, cryo-SXT can be used in other imaging correlative schemes bringing together the power of different microscopy techniques for the study of viral infections. During recent years, cryo-SXT has been combined with complementary imaging methods such as electron microscopy, electron tomography, and infrared spectroscopy for the purposes of scouting virus-induced changes in infected cells and understanding the orchestration of virus–host remodelling quantitatively.

## *Hepatitis C Virus*

Hepatitis C virus (HCV) is an enveloped, positive-sense single-stranded RNA virus member of the *Flaviviridae* family. The HCV polyprotein is processed by viral and cellular proteases to produce ten major viral proteins both structural and non-structural with the latter ones sufficient to support HCV RNA replication. HCV infection cycle begins with the host cell receptor-mediated endocytosis of HCV, where the viral genome is released and processed in the host cell cytoplasm to produce progeny virions through a complex and dynamic replication and assembly process. During HCV life cycle, a host cell membrane rearrangement is well documented and known to form a membranous web to harbour HCV replication complexes and viral proteins. This rearrangement includes the formation of double-membrane vesicles where the virus replication likely takes place (Dubuisson and Cosset 2014). Based on EM studies, membranous web derives primarily from the ER and contains markers of rough ER, early and late endosomes, mitochondria, and LDs (Romero-Brey et al. 2012).

Perez-Berna et al. (2021) have employed a consolidated approach involving two microscopy methods based on the different beamlines in the ALBA Synchrotron facility. Cryo-SXT has been used to study the structure of the membranous web structure under cryogenic conditions at the whole cell level. Subsequently, synchrotron-based Fourier transform infrared microspectroscopy (SR-$\mu$FTIR) was used for the spatial correlation between the membranous web in the cells and the specific chemical footprint of lipids and proteins.

For the needs of the experimental workflow, a surrogate model of infection was used (in accordance with the biosafety protocols of ALBA Synchrotron facility) that did not involve the handling of infectious material. Therefore, the adherent human hepatoma cell line (HuH-7) was electroporated with subgenomic bicistronic HCV RNA. Effective replication of the RNA genome was confirmed by the resistance to certain antibiotic treatment (G418). The viral load was quantified with the application of RT-qPCR. Hepatoma cells were seeded onto gold TEM grids and 24 h later were cryopreserved by plunge freezing. Prior to cryopreservation 100 nm gold nanoparticles were added for tomographic alignment purposes. Soft X-ray data were acquired with the use of a 40 nm Fresnel zone plate objective. IMOD (Kremer et al. 1996) was used for the alignment of the projections and SIRT reconstruction option in TOMO3D (Agulleiro and Fernandez 2011) for the final reconstructions. Visualisation and segmentation of reconstructed data were carried out with the Amira-Avizo Software and UCSF Chimera (Pettersen et al. 2004).

After cryo-SXT data collection, the TEM grids were dipped from liquid nitrogen to 3.7% paraformaldehyde (PFA) at room temperature. The cells were dried in a desiccator and then were used for SR-$\mu$FTIR spectroscopic analysis at the MIRAS beamline at the ALBA Synchrotron (Yousef et al. 2017) using a Hyperion 3000 Microscope equipped with a 36-magnification objective coupled to a Vertex 70 spectrometer (Bruker). Spectroscopic data were acquired in two different ways: (1) minimum 50 spectra from single cells for each sample in each region and (2) maps with minimum dimensions of 50 × 50 μm and a step size of 6 × 6 μm. SR-$\mu$FTIR spectra of single independent cells and various cell maps were analysed using Opus 7.5 (Bruker).

By correlating 2D SR-$\mu$FTIR data, they have demonstrated a link of the overall increase in the oxidation of the cytoplasm of the HCV-replicating cell with its structural alteration. Cryo-SXT data analysis allowed the 3D characterisation of the membranous web. Interestingly, the recorded HCV-induced alterations were reversed using different antiviral drugs, thus providing a mechanism of action of the used drugs. This reversing of virus-induced cytological alterations came at a cost for the host cell as it was accompanied by certain side effects including the formation of large self-treated aggresome organelles due to the cellular stress induced by the drugs provided. Data correlation for the case of infected cells revealed (1) a lipid peak in the perinuclear region, (2) a maximum protein distribution across the region where abnormal mitochondria were present, and (3) high oxidation levels in areas with enhanced membrane alterations. Therefore, for the first time the chemical nature of the virus-induced ultrastructural change in the case of HCV was understood (Fig. 6.10) (Perez-Berna et al. 2021).

## Challenges and Prospects

Cryo-SXT has been proven to be a powerful approach for the study of the host–virus interactions as it facilitates capturing of dynamic processes throughout the virus life cycle at different time points and in different intracellular locations. Several studies related to viral infections have been performed in soft X-ray microscopes worldwide demonstrating the potential of the technique in the field of virological research (Hagen et al. 2012; Myllys et al. 2016; Aho et al. 2017; Aho et al. 2019; Kounatidis et al. 2020; Loconte et al. 2021; Mendonça et al. 2021; Chen et al. 2022a; Nahas et al. 2022a). Cryo-SXT allows a spatiotemporal analysis of the host cellular remodelling during the infection cycle at the ultrastructural level by elucidating the role of specific cellular compartments in a near-native state, as the samples are cryopreserved. Therefore, all the ultrastructural alterations recorded are not an artefact introduced by disruptive alternative techniques such as chemical fixation,

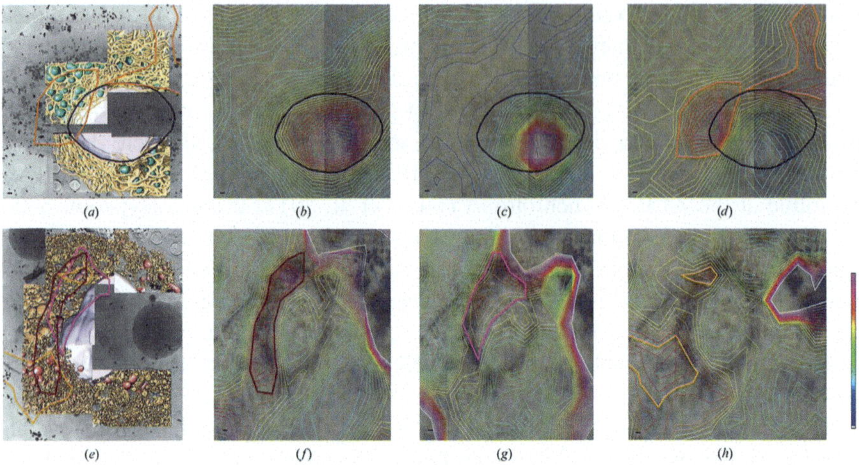


**Fig. 6.10** Correlative SXT and infrared and visible maps of control and HCV-replicating cells. (**a**, **e**) Three-dimensional reconstruction of whole-cell volumes of control and HCV replicon-bearing cell lines by cryo-SXT. Tile-scanned projections showing the area selected for SXT as well as volume tomograms are shown. A colour-coded manual segmentation of the surface boundaries identifying different organelles is presented, with normal mitochondria in green, abnormal mitochondria type I in red, abnormal mitochondria type II in purple, ER in yellow, modified ER in brown, and nuclear envelope in pink. (**b**, **f**) Lipid area distribution, (**c**. **g**) Amide I area distribution. (**d**, **h**) Absorbance ratios corresponding to oxidation ($A_{1740}/A_{2960+2925}$). Blue for low intensity, red for high intensity. 1 μm scale bar (reprinted from Perez-Berna et al. 2021)

dehydration, resin embedding, or sectioning used in routine EM-based techniques (Harkiolaki et al. 2018; Okolo 2022).

The main drawback of the cryo-SXT technique is the limited resolution especially when compared to other available techniques such as TEM, cryo-EM, or cryo-ET. The maximum effective resolution by using the 25 nm zone plate is approximately 30 nm, and while cellular compartments and virus particles could be imaged at these levels, structures such as cytoskeleton components and macromolecular machines such as ribosomes require higher resolution levels. The limitations in the production of more advanced X-ray diffractive lenses impair further progress in the spatial resolution provided by the X-ray microscopes. Research teams have demonstrated that by the use of higher-order imaging (Rehbein et al. 2009) or by applying new alternative technologies into the construction of new zone plates (combining electron-beam lithography with atomic layer deposition and focused ion beam induced deposition (Vila-Comamala et al. 2009), or by introducing new zone plate designs with reduced zone widths (Chao et al. 2012), a 10 nm spatial resolution is achievable. However, a 25 nm resolution is currently the ideal applied image resolution as the smaller the width of the ZP's outermost zone the higher the achieved resolution but also the smaller the associated field of view and depth of view (Okolo 2022).

The other main limitation of the cryo-SXT technique is the difficulty to distinguish between different intracellular compartments resembling similar shapes and X-ray attenuation levels (resulting in similar density projections). However, when cryo-SXT is coupled with a SRM technique such as the cryo-SIM, the above obstacles can be overcome with either the application of external dyes for the unambiguous recognition of specific cellular organelles or with the endogenous overexpression of fluorescently tagged biomolecules (Kounatidis et al. 2020; Nahas et al. 2022a).

In the future, further development of the CLXT technique (higher imaging resolution levels, broader fluorophore repertoire) coupled with advanced quantitative analysis software tools will allow an in-depth analysis of statistically significant numbers of virus-infected cells. The development of this kind of high statistical power imaging platforms will be beneficial for the evaluation of potential antiviral agents at the preclinical stage, through the evaluation of their structural consequences at the subcellular level.

**Acknowledgements** The authors would like to thank Dr Harkiolaki Maria and Dr Kamal Nahas for insightful discussions, critical reading of the chapter and the provision of illustration material.

## References

Agulleiro JI, Fernandez JJ (2011) Fast tomographic reconstruction on multicore computers. Bioinformatics 27:582–583. https://doi.org/10.1093/bioinformatics/btq692

Aho V, Myllys M, Ruokolainen V, Hakanen S, Mäntylä E, Virtanen J, Hukkanen V, Kühn T, Timonen J, Mattila K, Larabell CA, Vihinen-Ranta M (2017) Chromatin organization regulates viral egress dynamics. Sci Rep 7(1):3692. https://doi.org/10.1038/s41598-017-03630-y

Aho V, Mäntylä E, Ekman A, Hakanen S, Mattola S, Chen JH, Weinhardt V, Ruokolainen V, Sodeik B, Larabell C, Vihinen-Ranta M (2019) Quantitative microscopy reveals stepwise alteration of chromatin structure during herpesvirus infection. Viruses 11(10):935. https://doi.org/10.3390/v11100935

Arii J (2021) Host and viral factors involved in nuclear egress of herpes simplex virus 1. Viruses 13:754. https://doi.org/10.3390/v13050754

Bai H, Guan Y, Liu J, Chen L, Wei W, Liu G, Tian Y (2020) Precise correlative method of Cryo-SXT and Cryo-FM for organelle identification. J Synchrotron Radiat 1:176–184. https://doi.org/10.1107/S1600577519015194

Beese L, Feder R, Sayre D (1986) Contact x-ray microscopy. A new technique for imaging cellular fine structure. Biophys J 49:259–268. https://doi.org/10.1016/S0006-3495(86)83639-6

Bolitho EM, Sanchez-Cano C, Shi H, Quinn PD, Harkiolaki M, Imberti C, Sadler PJ (2021) Single-cell chemistry of photoactivatable platinum anticancer complexes. J Am Chem Soc 143:20224–20240. https://doi.org/10.1021/jacs.1c08630

Brown JC, Newcomb WW (2011) Herpesvirus capsid assembly: insights from structural analysis. Curr Opin Virol 1:142–149. https://doi.org/10.1016/j.coviro.2011.06.003

Brüggeller P, Mayer E (1980) Complete vitrification in pure liquid water and dilute aqueous solutions. Nature 288:569–571. https://doi.org/10.1038/288569a0

Carrascosa JL, Chichón FJ, Pereiro E, Rodríguez MJ, Fernández JJ, Esteban M, Heim S, Guttmann P, Schneider G (2009) Cryo-X-ray tomography of vaccinia virus membranes and inner compartments. J Struct Biol 168:234–239. https://doi.org/10.1016/j.jsb.2009.07.009

Chao W, Fischer P, Tyliszczak T, Rekawa S, Anderson E, Naulleau P (2012) Real space soft x-ray imaging at 10 nm spatial resolution. Opt Express 20(9):9777–9783. https://doi.org/10.1364/OE.20.009777

Chen JH, Vanslembrouck B, Ekman A, Aho V, Larabell CA, Le Gros MA, Vihinen-Ranta M, Weinhardt V (2022a) Soft X-ray tomography reveals HSV-1-induced remodelling of human B cells. Viruses 14(12):2651. https://doi.org/10.3390/v14122651

Cheng PC, Jan GJ (1987) X-ray Microscopy: Instrumentation and Biological Applications. Springer-Verlag, Berlin. https://doi.org/10.1007/978-3-642-72881-5

Chichón FJ, Rodríguez MJ, Pereiro E, Chiappi M, Perdiguero B, Guttmann P, Werner S, Rehbein S, Schneider G, Esteban M, Carrascosa JL (2012) Cryo X-ray nano-tomography of vaccinia virus infected cells. J Struct Biol 177:202–211. https://doi.org/10.1016/j.jsb.2011.12.001

Combs CA, Shroff H (2017) Fluorescence microscopy: a concise guide to current imaging methods. Curr Protoc Neurosci 10:2.1.1–2.1.25. https://doi.org/10.1002/cpns.29

Coombs KM (2006) Reovirus structure and morphogenesis. Curr Top Microbiol Immunol 309: 117–167. https://doi.org/10.1007/3-540-30773-7_5

Cortese M, Goellner S, Acosta EG, Neufeldt CJ, Oleksiuk O, Lampe M, Haselmann U, Funaya C, Schieber N, Ronchi P, Schorb M, Pruunsild P, Schwab Y, Chatel-Chaix L, Ruggieri A, Bartenschlager R (2017) Ultrastructural characterization of zika virus replication factories. Cell Rep 18:2113–2123. https://doi.org/10.1016/j.celrep.2017.02.014

de Armas-Rillo L, Valera MS, Marrero-Hernández S, Valenzuela-Fernández A (2016) Membrane dynamics associated with viral infection. Rev Med Virol 26:146–160. https://doi.org/10.1002/rmv.1872

Demmerle J, Innocent C, North AJ, Ball G, Müller M, Miron E, Matsuda A, Dobbie IM, Markaki Y, Schermelleh L (2017) Strategic and practical guidelines for successful structured illumination microscopy. Nat Protoc 12(5):988. https://doi.org/10.1038/nprot.2017.019

Dobro MJ, Melanson LA, Jensen GJ, McDowall AW (2010) Plunge freezing for electron cryomicroscopy. Methods Enzymol 481:63–82. https://doi.org/10.1016/S0076-6879(10)81003-1

Dubochet J, McDowall AW (1981) Vitrification of pure water for electron microscopy. J Microsc 124:3–4. https://doi.org/10.1111/j.1365-2818.1981.tb02483.x

Dubuisson J, Cosset FL (2014) Virology and cell biology of the hepatitis C virus life cycle: an update. J Hepatol 61:S3–S13. https://doi.org/10.1016/j.jhep.2014.06.031

Duke EM, Razi M, Weston A, Guttmann P, Werner S, Henzler K, Schneider G, Tooze SA, Collinson LM (2014) Imaging endosomes and autophagosomes in whole mammalian cells using correlative cryo-fluorescence and cryo-soft X-ray microscopy (cryo-CLXM). Ultramicroscopy 143:77–87. https://doi.org/10.1016/j.ultramic.2013.10.006

Elgass KD, Smith EA, LeGros MA, Larabell CA, Ryan MT (2015) Analysis of ER-mitochondria contacts using correlative fluorescence microscopy and soft X-ray tomography of mammalian cells. J Cell Sci 128:2795–2804. https://doi.org/10.1242/jcs.169136

Ertel KJ, Benefield D, Castaño-Diez D, Pennington JG, Horswill M, den Boon JA, Otegui MS, Ahlquist P (2017) Cryo-electron tomography reveals novel features of a viral RNA replication compartment. Elife 6:e25940. https://doi.org/10.7554/eLife.25940

Garriga D, Chichón FJ, Calisto BM, Ferrero DS, Gastaminza P, Pereiro E, Pérez-Berna AJ (2021) Imaging of virus-infected cells with soft X-ray tomography. Viruses 13(11):2109. https://doi.org/10.3390/v13112109

Glingston RS, Deb R, Kumar S, Nagotu S (2019) Organelle dynamics and viral infections: at cross roads. Microbes Infect 21:20–32. https://doi.org/10.1016/j.micinf.2018.06.002

Hagen C, Guttmann P, Klupp B, Werner S, Rehbein S, Mettenleiter TC, Schneider G, Grünewald K (2012) Correlative VIS-fluorescence and soft X-ray cryo-microscopy/tomography of adherent cells. J Struct Biol 177:193–201. https://doi.org/10.1016/j.jsb.2011.12.012

Harkiolaki M, Darrow MC, Spink MC, Kosior E, Dent K, Duke E (2018) Cryo-soft X-ray tomography: using soft X-rays to explore the ultrastructure of whole cells. Emerg Top Life Sci 20:81–92. https://doi.org/10.1042/ETLS20170086

Harris HMB, Hill C (2021) A place for viruses on the tree of life. Front Microbiol 14(11):604048. https://doi.org/10.3389/fmicb.2020.604048

Hernandez-Gonzalez M, Larocque G, Way M (2021) Viral use and subversion of membrane organisation and trafficking. J Cell Sci 134(5):jcs252676. https://doi.org/10.1242/jcs.252676

Hoffman DP, Shtengel G, Xu CS, Campbell KR, Freeman M, Wang L, Milkie DE, Pasolli HA, Iyer N, Bogovic JA, Stabley DR, Shirinifard A, Pang S, Peale D, Schaefer K, Pomp W, Chang CL, Lippincott-Schwartz J, Kirchhausen T, Solecki DJ, Betzig E, Hess HF (2020) Correlative three-dimensional super-resolution and block-face electron microscopy of whole vitreously frozen cells. Science 17(367(6475)):eaaz5357. https://doi.org/10.1126/science.aaz5357. PMID: 31949053; PMCID: PMC7339343

Jacobsen C (2020) X-ray microscopy, 1st edn. Cambridge University Press, Cambridge

Johnson DC, Baines JD (2011) Herpesviruses remodel host membranes for virus egress. Nat Rev Microbiol M9(5):382–394. https://doi.org/10.1038/nrmicro2559. PMID: 21494278

Kirz J, Jacobsen C, Howells M (1995) Soft X-ray microscopes and their biological applications. Q Rev Biophys 28:33–130. https://doi.org/10.1017/s0033583500003139

Klein S, Cortese M, Winter SL, Wachsmuth-Melm M, Neufeldt CJ, Cerikan B, Stanifer ML, Boulant S, Bartenschlager R, Chlanda P (2020) SARS-CoV-2 structure and replication characterized by in situ cryo-electron tomography. Nat Commun 11:5885. https://doi.org/10.1038/s41467-020-19619-7

Kounatidis I, Stanifer ML, Phillips MA, Paul-Gilloteaux P, Heiligenstein X, Wang H, Okolo CA, Fish TM, Spink MC, Stuart DI, Davis I, Boulant S, Grimes JM, Dobbie IM, Harkiolaki M (2020) 3D correlative cryo-structured illumination fluorescence and soft X-ray microscopy elucidates reovirus intracellular release pathway. Cell 23:515–530.e17. https://doi.org/10.1016/j.cell.2020.05.051

Kremer JR, Mastronarde DN, McIntosh JR (1996) Computer visualization of three-dimensional image data using IMOD. J Struct Biol 116:71–76. https://doi.org/10.1006/jsbi.1996.0013

Larabell CA, Nugent KA (2010) Imaging cellular architecture with X-rays. Curr Opin Struct Biol 20:623–631. https://doi.org/10.1016/j.sbi.2010.08.008

Lee MH, Chang CY, Chang CH, Chang SH, Chen C, Chiu CC, Huang L, Lai L, Lee L, Liu DG, Su Y, Yan HY (2016) Developing white beamline components of TPS Beamline 24A. In: Mechanical engineering design of synchrotron radiation equipment and instrumentation conference (9th). https://doi.org/10.18429/JACoW-MEDSI2016-TUPE12

Le Gros MA, McDermott G, Uchida M, Knoechel CG, Larabell CA (2009) High-aperture cryogenic light microscopy. J Microsc 235:1–8. https://doi.org/10.1111/j.1365-2818.2009.03184.x

Le Gros MA, McDermott G, Cinquin BP, Smith EA, Do M, Chao WL, Naulleau PP, Larabell CA (2014) Biological soft X-ray tomography on beamline 2.1 at the advanced light source. J Synchrotron Radiat 21:1370–1377. https://doi.org/10.1107/S1600577514015033

Loconte V, Chen JH, Cortese M, Ekman A, Le Gros MA, Larabell C, Bartenschlager R, Weinhardt V (2021) Using soft X-ray tomography for rapid whole-cell quantitative imaging of SARS-CoV-2-infected cells. Cell Rep Methods 1(7):100117. https://doi.org/10.1016/j.crmeth.2021.100117

Mastronarde DN (2005) Automated electron microscope tomography using robust prediction of specimen movements. J Struct Biol 152:36–51. https://doi.org/10.1016/j.jsb.2005.07.007

Mazzon M, Mercer J (2014) Lipid interactions during virus entry and infection. Cell Microbiol 16:1493–1502. https://doi.org/10.1111/cmi.12340

McDermott G, Le Gros MA, Knoechel CG, Uchida M, Larabell CA (2009) Soft X-ray tomography and cryogenic light microscopy: the cool combination in cellular imaging. Trends Cell Biol 19:587–595. https://doi.org/10.1016/j.tcb.2009.08.005

McDermott G, Le Gros MA, Larabell CA (2012) Visualizing cell architecture and molecular location using soft x-ray tomography and correlated cryo-light microscopy. Annu Rev Phys Chem 63:225–239. https://doi.org/10.1146/annurev-physchem-032511-143818

Mendonça L, Howe A, Gilchrist JB, Sheng Y, Sun D, Knight ML, Zanetti-Domingues LC, Bateman B, Krebs AS, Chen L, Radecke J, Li VD, Ni T, Kounatidis I, Koronfel MA,

Szynkiewicz M, Harkiolaki M, Martin-Fernandez ML, James W, Zhang P (2021) Correlative multi-scale cryo-imaging unveils SARS-CoV-2 assembly and egress. Nat Commun 12:4629. https://doi.org/10.1038/s41467-021-24887-y

Mercer J, Schelhaas M, Helenius A (2010) Virus entry by endocytosis. Annu Rev Biochem 79:803–833. https://doi.org/10.1146/annurev-biochem-060208-104626

Miller S, Krijnse-Locker J (2008) Modification of intracellular membrane structures for virus replication. Nat Rev Microbiol 6:363–374. https://doi.org/10.1038/nrmicro1890

Moor H (1987) Theory and practice of high pressure freezing. In: Steinbrecht RA, Zierold K (eds) Cryotechniques in biological electron microscopy. Springer, Berlin, pp 175–191. https://doi.org/10.1007/978-3-642-72815-0_8

Myllys M, Ruokolainen V, Aho V, Smith EA, Hakanen S, Peri P, Salvetti A, Timonen J, Hukkanen V, Larabell CA, Vihinen-Ranta M (2016) Herpes simplex virus 1 induces egress channels through marginalized host chromatin. Sci Rep 6:28844. https://doi.org/10.1038/srep28844

Nahas KL, Connor V, Scherer KM, Kaminski CF, Harkiolaki M, Crump CM, Graham SC (2022a) Near-native state imaging by cryo-soft-X-ray tomography reveals remodelling of multiple cellular organelles during HSV-1 infection. PLoS Pathog 18:e1010629. https://doi.org/10.1371/journal.ppat.1010629

Nahas KL, Fernandes JF, Vyas N, Crump C, Graham S, Harkiolaki M (2022b) Contour, a semi-automated segmentation and quantitation tool for cryo-soft-X-ray tomography. Biol Imaging 2: e3. https://doi.org/10.1017/S2633903X22000046

Nicoll MP, Proença JT, Efstathiou S (2012) The molecular basis of herpes simplex virus latency. FEMS Microbiol Rev 36:684–705. https://doi.org/10.1111/j.1574-6976.2011.00320.x

Okolo CA (2022) A guide into the world of high-resolution 3D imaging: the case of soft X-ray tomography for the life sciences. Biochem Soc Trans 50:649–663. https://doi.org/10.1042/BST20210886

Okolo CA, Kounatidis I, Groen J, Nahas KL, Balint S, Fish TM, Koronfel MA, Cortajarena AL, Dobbie IM, Pereiro E, Harkiolaki M (2021) Sample preparation strategies for efficient correlation of 3D SIM and soft X-ray tomography data at cryogenic temperatures. Nat Protoc 16:2851–2885. https://doi.org/10.1038/s41596-021-00522-4

Paul-Gilloteaux P, Heiligenstein X, Belle M, Domart MC, Larijani B, Collinson L, Raposo G, Salamero J (2017) eC-CLEM: flexible multidimensional registration software for correlative microscopies. Nat Methods 14:102–103. https://doi.org/10.1038/nmeth.4170

Pereiro E, Nicolás J, Ferrer S, Howells MR (2009) A soft X-ray beamline for transmission X-ray microscopy at ALBA. J Synchrotron Radiat 16:505–512. https://doi.org/10.1107/S0909049509019396

Perez-Berna AJ, Benseny-Cases N, Rodríguez MJ, Valcarcel R, Carrascosa JL, Gastaminza P, Pereiro E (2021) Monitoring reversion of hepatitis C virus-induced cellular alterations by direct-acting antivirals using cryo soft X-ray tomography and infrared microscopy. Acta Crystallogr D Struct Biol 77:1365–1377. https://doi.org/10.1107/S2059798321009955

Pettersen EF, Goddard TD, Huang CC, Couch GS, Greenblatt DM, Meng EC, Ferrin TE (2004) UCSF chimera—a visualization system for exploratory research and analysis. J Comput Chem 25:1605–1612. https://doi.org/10.1002/jcc.20084

Phillips MA, Harkiolaki M, Susano Pinto DM, Parton RM, Palanca A, Garcia-Moreno M, Kounatidis I, Sedat JW, Stuart DI, Castello A, Booth MJ, Davis I, Dobbie IM (2020) CryoSIM: super-resolution 3D structured illumination cryogenic fluorescence microscopy for correlated ultrastructural imaging. Optica 13:802–812. https://doi.org/10.1364/OPTICA.393203

Rayleigh L (1879) XXXI. Investigations in optics, with special reference to the spectroscope. Lond Edinb Dublin Philos Mag J Sci 8:261–274

Rehbein S, Heim S, Guttmann P, Werner S, Schneider G (2009) Ultrahigh-resolution soft-x-ray microscopy with zone plates in high orders of diffraction. Phys Rev Lett 103(11):110801. https://doi.org/10.1103/PhysRevLett.103.110801

Risco C (2021) Application of advanced imaging to the study of virus-host interactions. Viruses 13(10):1958. https://doi.org/10.3390/v13101958

Roberts KL, Smith GL (2008) Vaccinia virus morphogenesis and dissemination. Trends Microbiol 16:472–479. https://doi.org/10.1016/j.tim.2008.07.009

Romero-Brey I, Merz A, Chiramel A, Lee JY, Chlanda P, Haselman U, Santarella-Mellwig R, Habermann A, Hoppe S, Kallis S, Walther P, Antony C, Krijnse-Locker J, Bartenschlager R (2012) Three-dimensional architecture and biogenesis of membrane structures associated with hepatitis C virus replication. PLoS Pathog 8(12):e1003056. https://doi.org/10.1371/journal.ppat.1003056

Roossinck MJ (2011) The good viruses: viral mutualistic symbioses. Nat Rev Microbiol 9:99–108. https://doi.org/10.1038/nrmicro2491

Roth AN, Aravamudhan P, Fernández de Castro I, Tenorio R, Risco C, Dermody TS (2021) Ins and outs of reovirus: vesicular trafficking in viral entry and egress. Trends Microbiol 29:363–375. https://doi.org/10.1016/j.tim.2020.09.004

Sanchez EL, Lagunoff M (2015) Viral activation of cellular metabolism. Virology 479-480:609–618. https://doi.org/10.1016/j.virol.2015.02.038

Scherer KM, Manton JD, Soh TK, Mascheroni L, Connor V, Crump CM, Kaminski CF (2021) A fluorescent reporter system enables spatiotemporal analysis of host cell modification during herpes simplex virus-1 replication. J Biol Chem 296:100236. https://doi.org/10.1074/jbc.RA120.016571

Schermelleh L, Heintzmann R, Leonhardt H (2010) A guide to super-resolution fluorescence microscopy. J Cell Biol 190:165–175. https://doi.org/10.1083/jcb.201002018

Schermelleh L, Ferrand A, Huser T, Eggeling C, Sauer M, Biehlmaier O, Drummen GPC (2019) Super-resolution microscopy demystified. Nat Cell Biol 21:72–84. https://doi.org/10.1038/s41556-018-0251-8

Schermelleh L, Heintzmann R, Leonhardt H (2010) A guide to super-resolution fluorescence microscopy. J Cell Biol 190:165–175. https://doi.org/10.1083/jcb.201002018

Schneider G (1998) Cryo X-ray microscopy with high spatial resolution in amplitude and phase contrast. Ultramicroscopy 75:85–104. https://doi.org/10.1016/s0304-3991(98)00054-0

Schneider G, Niemann B, Guttmann P, Rudolph D, Schmahl G (1995) Cryo X-ray microscopy. Synchrotron Radiat News 8:19–28. https://doi.org/10.1080/08940889508602810

Schneider G, Guttmann P, Rehbein S, Werner S, Follath R (2012) Cryo X-ray microscope with flat sample geometry for correlative fluorescence and nanoscale tomographic imaging. J Struct Biol 177:212–223. https://doi.org/10.1016/j.jsb.2011.12.023

Smith EA, McDermott G, Do M, Leung K, Panning B, Le Gros MA, Larabell CA (2014) Quantitatively imaging chromosomes by correlated cryo-fluorescence and soft x-ray tomographies. Biophys J 107:1988–1996. https://doi.org/10.1016/j.bpj.2014.09.011

Spiller E, Feder R, Topalian J, Eastman D, Gudat W, Sayre D (1976) X-ray microscopy of biological objects with carbon kappa and with synchrotron radiation. Science 191:1172–1174. https://doi.org/10.1126/science.1257741

Taylor MW (2014) What is a virus? Viruses Man Hist Interact 22:23–40. https://doi.org/10.1007/978-3-319-07758-1_2

Tenorio R, Fernández de Castro I, Knowlton JJ, Zamora PF, Sutherland DM, Risco C, Dermody TS (2019) Function, architecture, and biogenesis of reovirus replication neoorganelles. Viruses 11:288. https://doi.org/10.3390/v11030288

Uchida M, Sun Y, McDermott G, Knoechel C, Le Gros MA, Parkinson D, Drubin DG, Larabell CA (2011) Quantitative analysis of yeast internal architecture using soft X-ray tomography. Yeast 28:227–236. https://doi.org/10.1002/yea.1834

Vila-Comamala J, Jefimovs K, Raabe J, Pilvi T, Fink RH, Senoner M, Maassdorf A, Ritala M, David C (2009) Advanced thin film technology for ultrahigh resolution X-ray microscopy. Ultramicroscopy 109:1360–1364. https://doi.org/10.1016/j.ultramic

Vyas N, Perry N, Okolo CA, Kounatidis I, Fish TM, Nahas KL, Jadhav A, Koronfel MA, Groen J, Pereiro E, Dobbie IM, Harkiolaki M (2021) Cryo-structured illumination microscopic data collection from cryogenically preserved cells. J Vis Exp 28(171). https://doi.org/10.3791/62274

Weiss D, Schneider G, Niemann B, Guttmann P, Rudolph D, Schmahl G (2000) Computed tomography of cryogenic biological specimens based on X-ray microscopic images. Ultramicroscopy 84:185–197. https://doi.org/10.1016/s0304-3991(00)00034-6

Young M (1972) Zone plates and their aberrations. J Opt Soc Am 197262:972–976. https://doi.org/10.1364/JOSA.62.000972

Yousef I, Ribó L, Crisol A, Šics I, Ellis G, Ducic T, Kreuzer M, Benseny-Cases N, Quispe M, Dumas P, Lefrançois S, Moreno T, García G, Ferrer S, Nicolas J, MAG A (2017) MIRAS: the infrared synchrotron radiation beamline at ALBA. Synchrotron Radiat News 30:4–6. https://doi.org/10.1080/08940886.2017.1338410

# Chapter 7
# The Virus-Induced Cytopathic Effect

**Daniel Céspedes-Tenorio and Jorge L. Arias-Arias**

**Abstract** The cytopathic effect comprises the set of cellular alterations produced by a viral infection. It is of great relevance since it constitutes a direct marker of infection. Likewise, these alterations are often virus-specific which makes them a phenotypic marker for many viral species. All these characteristics have been used to complement the study of the dynamics of virus-cell interactions through the kinetic study of the progression of damage produced by the infection. Various approaches have been used to monitor the cytopathic effect, ranging from light microscopy, immunofluorescence assays, and direct labeling with fluorescent dyes, to plaque assay for the characterization of the infection over time. Here we address the relevance of the study of cytopathic effect and describe different experimental alternatives for its application.

**Keywords** Viruses · Cytopathic Effect (CPE) · Immunofluorescence Assay (IFA) · Plaque assay · Live-cell imaging · Fluorescence microscopy

## Introduction

Viruses are considered cell parasites. When infecting a cell, they hijack the cellular machinery to ensure the production of viral progeny. That is, they need the cell to replicate and perpetuate (Mandell et al. 2010). This concept is closely linked to the interactions that are generated when the cell comes into direct contact with the virus. Different viral species rely on multiple signals and receptors that determine whether

---

D. Céspedes-Tenorio
Centro de Investigación en Enfermedades Tropicales, Facultad de Microbiología, Universidad de Costa Rica, San José, Costa Rica
e-mail: daniel.cespedestenorio@ucr.ac.cr

J. L. Arias-Arias (✉)
Centro de Investigación en Enfermedades Tropicales, Facultad de Microbiología, Universidad de Costa Rica, San José, Costa Rica

Dulbecco Lab Studio, Residencial Lisboa 2G, Alajuela, Costa Rica
e-mail: jorgeluis.arias@ucr.ac.cr



S. Vijayakrishnan et al. (eds.), *Virus Infected Cells*, Subcellular Biochemistry 106,
https://doi.org/10.1007/978-3-031-40086-5_7

they can infect a particular cell type. In most species, there are well-characterized receptors and co-receptors required for viral entry into the host cell. Likewise, the production of viral progeny depends on the cell in question, allowing virus replication (Flint et al. 2009).

In consequence, different concepts are derived to describe whether a cell can be infected and whether it allows viral replication. Susceptibility is linked to the presence of specific receptors and co-receptors in the cell that allow viral recognition, adsorption, and entry. On the other hand, permissiveness refers to the possibility that viral replication can occur in the cell. Finally, these two concepts are linked to tropism, which determines the cell lineages in which the virus finds the best conditions to establish the infection and replicate (Mandell et al. 2010; Flint et al. 2009).

Once the infection has occurred, the hijacking of the cellular machinery and the structural alteration of the cell results in important changes. The set of cellular alterations produced by a viral infection is known as the cytopathic effect (CPE). This can include changes in metabolic pathways, alteration of the cell morphology, perturbations in growth kinetics and even cell death. Over time, these effects result in ultrastructural manifestations of the infection that can be monitored with standard light and fluorescence microscopes. Since the direct observation of viral particles requires the use of electron microscopy techniques, detection of the CPE is considered an alternative way to recognize a viral infection with conventional equipment. In addition, it can serve as a phenotypic recognition marker of the viral species that is causing the infection (Doms 2016; Mandell et al. 2010; Flint et al. 2009).

The effects exerted by a virus on the cell can be direct or indirect. In the former case, the virus directly causes damage to the cell. Many viruses, such as Ebola, are lytic viruses able to activate the apoptosis cascade and produce specific cellular changes that characterize this type of cell death (Baseler et al. 2017; Doms 2016). Cytopathic changes related to cell morphology include fragmentation, cell rounding, and loss of cell–cell adhesion, among others. In the latter case, cells remain metabolically active, but their cytoskeleton is compromised.

On the other hand, indirect changes are those in which the virus does not cause damage to the cell, but the pathology is secondary to some products generated during the infection. For example, enveloped viruses produce glycoproteins that are released from the cell and can bind to membrane receptors of adjacent cells. This stimulates the fusion of cell membranes causing the formation of syncytia (Doms 2016). Likewise, some viruses trigger the pathology from an exacerbated immune response generated by the expression of the different viral proteins in the infected cells. Particularly, in the case of hepatitis B virus (HBV), the integration of the viral DNA into the host genome generates the expression of the surface antigen (HBsAg) that, when recognized by CD8+ T cells (TCD8+), causes most of the death observed in infected cells (Iannacone and Guidotti 2021).

However, not all viruses cause cell death. Cytopathic changes may comprise alterations in different metabolic and molecular pathways involved in cell replication and survival. This could be translated as decreased cell death and cell transformation involving changes both in morphology and in the expression of different cellular markers that can lead to the development of cancer. Oncogenic viruses such as the human papilloma virus (HPV) are examples of this type of infection (Mandell et al. 2010; Mighty and Laimins 2014).

## Cytopathic Effects across Viral Species

The observation of cytopathic effects in cell culture is of great diagnostic utility since it has been implemented as a confirmation of infection and as a hallmark of many viral species with very distinctive effects that allow them to be identified and distinguished directly from others (Doms 2016; Flint and American Society for Microbiology 2009).

Lysis and cell death are the most evident cellular effects observed in the context of a viral infection. These mechanisms can be direct, when the virus activates death signals in the cell, or as is the case of most viruses like the human immunodeficiency virus (HIV), death is secondary to disturbances that generate significant damage. In vitro, HIV triggers two cytopathic effects: syncytia formation and direct cell lysis. Each of these two alterations depends on the type of co-receptor employed by the virus. Viruses using CXCR4 induce syncytia formation and the appearance of multinucleated giant cells, while viruses using CCR5 do not cause this alteration. This virus also induces in the cells it directly infects, mainly TCD4+ cells, a phenomenon known as cell engulfment degeneration, where the infected cells enlarge until they lose their membrane integrity and are consequently lysed (Costin 2007; Lang et al. 2011).

On the other hand, viruses of the genus *Flavivirus* cause gradual cell death in different cell types. In general, the alteration of cell death pathways induced by these viruses primarily causes apoptosis, necrosis, and autophagy in the infected cells. Dengue virus (DENV) induces apoptosis in different cell lineages including cells of the immune system, nervous system, and liver. Apoptosis is triggered by both the virion and specific viral proteins, activating both the extrinsic and the intrinsic pathways. This phenomenon is also typical of the West Nile virus (WNV) infection. Likewise, DENV can also induce autophagy, which results in a lower immune response and the enhancement of viral replication (John et al. 2015; Roy et al. 2014). These phenomena can be monitored in cell culture by an increase in cell death or in the expression of specific markers using fluorescence microscopy.

The formation of syncytia is a characteristic present in infections by paramyxoviruses and coronaviruses, as a viral strategy for dissemination that allows minimal exposure to the extracellular medium and its components, including the immune system. Although this effect is not exclusive of one viral species, the formation of syncytia in lung cells is pathognomonic of the respiratory syncytial virus (RSV) (Fig. 7.1). Among the different proteins involved, the F protein of RSV plays the most important role in breaking the hydrophobic bonds of the membrane in infected cells and inducing fusion of membranes with neighbor cells (Gower et al. 2005). Following membrane fusion and viral replication, cell death is triggered by necrosis,

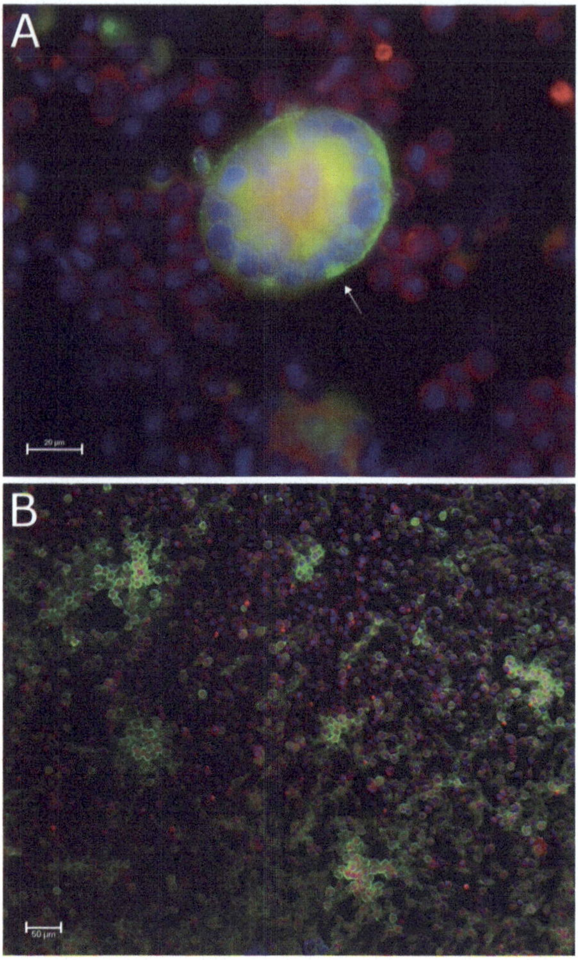


**Fig. 7.1** Immunofluorescence analysis of virus-induced CPE. (**a**) Syncytia formation (arrow) in HEp-2 cells infected with RSV (strain A2) and immunostained with an anti-RSV F protein monoclonal antibody (green) at 72 hours postinfection. (**b**) Clustering CPE in Vero cells infected with HSV-1 (strain F) and immunolabeled with an anti-HSV-1 D glycoprotein monoclonal antibody (green) at 48 hours postinfection. Total magnifications of 400 X and 200 X, respectively. In both cases cell cytoplasms and nuclei were counterstained with Evans blue (red) and Hoechst 33342 (blue), respectively. Images by Jorge L. Arias-Arias, Universidad de Costa Rica

leading to further tissue destruction and viral spread (Gagliardi et al. 2017; Valdovinos and Gómez 2003).

More severe systemic infections such as those produced by the Ebola virus (EBOV) are more difficult to fit into a classification due to the differential affectations in several tissues. The cell lineages disturbed are many; however, hepatocytes, lymphocytes, and epithelial cells of the vessels are the most affected. In the hepatic tissue, intracytoplasmic lipid vacuoles engulfed by the cells and inclusion bodies in which the virus replicates and assembles can be observed until hepatocytes rupture and die by necrosis. Likewise, lymphocytes and cells of the vascular epithelium activate death pathways of both necrosis and apoptosis. The vascular epithelial cells suffer alterations such as cell rounding, loss of adhesion, and destruction of the intercellular junctions, resulting in loss of integrity of the vascular membrane, which leads to the severe hemorrhages observed in the Ebola Virus Disease (EVD) (Barrientos and Rollin 2007; Baseler et al. 2017; Francica et al. 2009).

However, EBOV is not the only virus with multiple and complex cellular disturbances. For example, the cytopathic effect caused by Zika virus (ZIKV) infection is not fully defined. Nevertheless, important findings have been documented that characterize the damage to the infected cells. One of the most predominant is rapid and widespread vacuolization, particularly in astrocytes and skin fibroblasts. This vacuolization starts from the outer zone of the cell and culminates in a process that resembles paraptosis, a caspase-independent cell death mechanism that results in the formation of large cytoplasmic vacuoles (Monel et al. 2017). It has also been observed that these vesicles are not only intracellular, but they can behave as exosomes that are released into the extracellular space and promote the dissemination and immune evasion of the viral infection (Huang et al. 2018). Moreover, as a part of the induction of cell death by apoptotic and necrotic mechanisms and the reduction in cell proliferation, ZIKV infection causes various mitotic abnormalities easily observed over time by methods such as fluorescence microscopy. These include chromosomal aberrations, formation of micronuclei, multipolarity, and multiplication of the number of centrosomes. This results in chromosomal instability that promotes death in these abnormal cells or, when they survive, aneuploidy in the different tissues (Souza et al. 2016).

## Approaches to Studying the Cytopathic Effect

### *Bright-Field Microscopy*

The analysis of virus-infected cell cultures using a light microscope permits the direct observation of cytopathic changes produced by viral infections over time and constitutes the more economical approach for this purpose (Fig. 7.2). However, it requires expertise and well-trained analysts as these alterations could be virus-specific, allowing a preliminary diagnosis (Chong et al. 2014; Doms 2016). This is characteristic of pathological studies to observe the dysplastic changes produced by the Human Papillomavirus (HPV) in genital tissue. In this case, cervical cytology samples are stained with hematoxylin/eosin to detect the alterations produced by this virus, which could be associated with cancer (Mighty and Laimins 2014; Schiffman and Wentzensen 2013).

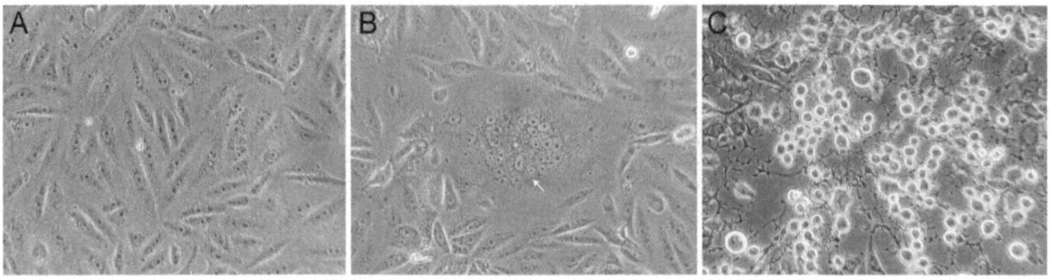


**Fig. 7.2** Bright-field microscopy analysis of virus-induced CPE in Vero cells. (**a**) Noninfected control cells at 48 hours postseeding. (**b**) Syncytia formation (arrow) at 48 hours postinfection with measles virus (MV, strain Moraten). (**c**) Ballooning and clustering CPE at 48 hours postinfection with HSV-1 (strain F). Total magnification of 200 X. Images by Francisco Vega-Aguilar, Universidad de Costa Rica

## *Fluorescence Microscopy*

Fluorescence microscopy enables the specific identification in a single experiment of several viral and cellular markers of damage in the infected cells using molecular sensors, such as labeled antibodies or chemical dyes, in a variety of colors. It also allows the analysis over time of the dynamics and kinetics of viral infection (Parveen et al. 2018). Among the experimental approaches assisted by fluorescence microscopy for the study of virus-induced cytopathic effect, immunofluorescence assay (IFA) and direct labeling with fluorescent dyes are described herein.

### Immunofluorescence Assay (IFA)

Immunofluorescence assays allow the detection of viral particles as well as their components inside cells by means of labeled antibodies directed against one or more of the viral structural or nonstructural proteins (Fig. 7.1). This is important for the understanding of the pathological process behind the cytopathic effect, as co-localization of specific viral proteins with counterstained cellular structures allows the elucidation of the viral components involved in the development of the cell damage observed in particular organelles (Arias-Arias and Mora-Rodríguez 2021). This method also allows easy identification of cells that are more susceptible to infection and supports viral replication and dissemination, even before the appearance of clear cell damage, which permits the corroboration of the preliminary bright-field diagnosis.

### Labeling with Fluorescent Dyes

Based on the principle that the cytopathic effect relies on the interaction between the virus and cellular components, we envisaged that the use of fluorescent dyes directed against such cellular elements could be applied to obtain information about the

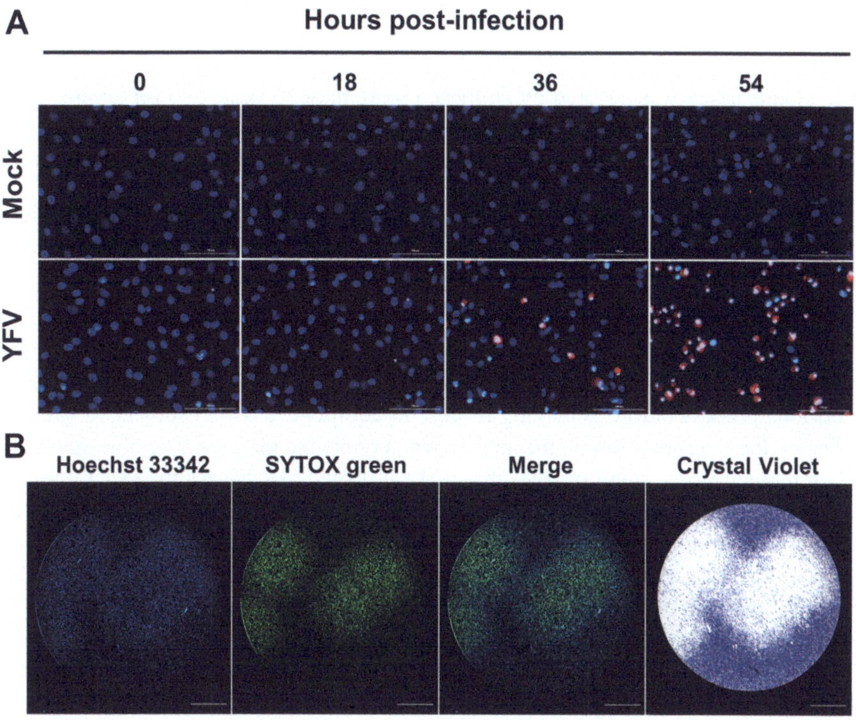


**Fig. 7.3** Labeling with fluorescent dyes of CPE in flavivirus-infected cell cultures. (**a**) CPE induction kinetics in YFV-infected BHK-21 cells with the nucleic acid dyes Hoechst 33342 (cells with condensed chromatin, saturated blue) and TO-PRO-3 iodide (dead cells, red) at a total magnification of 200 X (scale bar = 100 μm). (**b**) Plaque assay on live BHK-21 cells by CPE labeling with the DNA stainings Hoechst 33342 (chromatin condensation, blue) and SYTOX green (cell death, green) at 120 hours postinfection with ZIKV (total magnification of 40 X; scale bar = 1000 μm). Images by Jorge L. Arias-Arias, Universidad de Costa Rica

damage produced to the infected cell. Taking this into consideration, we developed a method to monitor the kinetics of CPE formation by the use of chemical probes commonly used in cell biology and cancer research to label DNA (Arias-Arias et al. 2020).

In this respect, nuclear probes such as Hoechst 33342 or cell death markers such as propidium iodide (PI), SYTOX green, or TO-PRO-3 iodide allow the monitoring of damage at different stages of the infectious process in terms of chromatin condensation and plasma membrane permeabilization, respectively (Fig. 7.3a). This permits the real-time monitoring of viral infection with single-cell resolution and the observation by live-cell imaging of events corresponding to early (chromatin condensation) and late (membrane permeabilization) stages of virus-induced CPE (Arias-Arias et al. 2020). For further details about this method, please refer to the cited reference.

## *Plaque Assay*

The plaque assay was adapted by Dulbecco more than 70 years ago for applications with animal viruses (Dulbecco 1952) and remains the method of choice for the quantification of lytic viruses. This method is based on the infection of a cell monolayer embedded in a semi-solid matrix, such as agarose, to restrain viral dissemination. Viral infection and multiplication generate foci of lysis which, after staining with dyes such as crystal violet, can be observed as plaques that can be counted by the analyst. The number of plaques correlates with the viral titers present in the original viral seed (Ryu 2017).

Despite the advantages of this method for viral quantification, it is an endpoint approach that does not collect data about the kinetics of infection and requires a laborious process before achieving results. In addition, it is also subjective since it relies on the interpretation by the analyst who counts the plaques. The bias increases when there are very small plaques that make it difficult to discern the exact number of foci of infection (Wen et al. 2019). Due to these limitations, we decided to apply our above-mentioned technique of labeling with fluorescent dyes for the improvement of the conventional plaque assay, since virus-induced CPE is essential for the formation of viral plaques and our technique is suitable for live imaging (Arias-Arias et al. 2021).

The infection kinetics and CPE development involve many events prior to cell death that cannot be monitored by other approaches but live-cell imaging. Our method with fluorescent DNA dyes allowed the real-time monitoring of the plaque formation process and the characterization of early (chromatin condensation) and late (cell death) cytopathic effects with single-plaque resolution. In addition, we developed an image analysis pipeline for the characterization of the plaques at a single-cell level (Arias-Arias et al. 2021).

After validation in Vero cells of this fluorescent plaque assay, we concluded that our method was not only an improvement over the conventional plaque assay for viral quantification but also a tool for the study and characterization of the CPE generated by both RNA and DNA viruses, since in the same experimental condition there are viral plaques surrounded by a monolayer of uninfected cells. This constitutes the perfect internal control for the comparison of the structural differences between infected cells that form the plaques and their noninfected counterparts on the periphery.

In our real-time fluorescent plaque assay, a first labeling is performed with Hoechst 33342 as a marker of chromatin condensation (blue fluorescence) and a second staining with SYTOX green as a marker of cell death upon membrane permeabilization (green fluorescence). The results showed a green fluorescence focus of dead cells in the inner zone of the plaques and a concentration of blue fluorescence in the outer region corresponding to cells with condensed chromatin (Fig. 7.3b). This indicated that chromatin condensation processes occur prior to cell death, a conclusion that cannot be obtained from a conventional plaque assay (Arias-Arias et al. 2021).

Finally, a deeper single-cell analysis by the quantification of the percentage of cells showing chromatin condensation and cell death within individual viral plaques allowed us to identify interesting variations in the CPE generated by three different viral species (VSV, YFV, and HSV-1). The results for each case suggested differential rates of infection and virus interaction with the cell death programs. This also supported the utility of our approach for the identification of mixed infections based on viral-specific CPE fingerprints at the level of the plaques generated (Arias-Arias et al. 2021).

## Monitoring of the Cytopathic Effect and Its Relevance

The importance of in vitro measurement of viral-induced cytopathic effect is not only related to its application in clinical virology for the diagnosis of a particular infection, but also it is extensively used for research in this field. This has been widely applied in three aspects: characterization of the cellular tropism to different viral species, evaluation of antiviral drugs in the pharmaceutical industry, and the study of neutralizing antibodies for the development of vaccines. These three applications are discussed below using the recently discovered SARS-CoV-2 virus as an example.

### *Cellular Tropism Evaluation*

By monitoring the cytopathic effect in various cell lines, it is possible to characterize the permissiveness of these cells to infection by a given viral species and thus elucidate the cell tropism of the virus of interest (Arias-Arias and Mora-Rodríguez 2021). In the case of SARS-CoV-2, as the number of studies on this virus increases due to the rise in the number of cases worldwide and the need to understand the physiopathology of this infection, one of the first issues to be considered is the cellular tropism of the virus and the extent of the infection in the body. Manuscripts such as those published by Zhu et al. (2020) and Hao et al. (2020) focused on describing the viral replication profile and cellular damage in airway epithelial cells. The former describes distinctive cytopathic effects such as cell fusion, apoptosis, and destruction of intercellular junctions in in vitro cultures of human ciliated and secretory airway epithelial cells (Zhu et al. 2020). This is a feature that differentiates the tropism of this virus from other coronaviruses. Likewise, the latter study broadens these findings by evaluating long-term cytopathic changes to describe the kinetics and dynamics of this viral infection (Hao et al. 2020).

As with this virus, the study of the cytopathic effect of different viral species has been helpful in correlating the pathology of infection with the cell tropism of the virus. This is the case with ZIKV. Infections in children are associated with cases of microcephaly, and this can be explained because the virus has tropism for neural cell progenitors and astrocytes. In the latter, especially, persistent infection produces significant chromosomal abnormalities that contribute to cell death and neurological damage (Souza et al. 2016).

The above-mentioned examples highlight tropism for single-cell lineages. However, for most viruses, this is not the case. Viruses possess the ability to infect diverse cell lineages and thus favor the spread of infection to different tissues. As an example, we demonstrated by plaque assay and IFA the permissiveness of primary human umbilical artery smooth muscle cells (HUASMC) to the four DENV serotypes in comparison with the infection produced in both primary human umbilical vein endothelial cells (HUVEC) and rhesus macaque kidney cells (LLC-MK2). Our results showed that HUASMC is susceptible and productive to infection with all DENV serotypes, although to a lesser degree when compared with the other two cell lines evaluated (Arias-Arias et al. 2018).

## *Antiviral Drug Screening*

The development of antiviral drugs to treat SARS-CoV-2 infections has been a major focus of the biomedical industry since the virus began its global spread. Such studies mainly focus on assays measuring cytopathic/cytotoxic effects or variations of the well-known plaque assay. Many different chemical compounds, ranging from already approved drugs to compounds under evaluation, have been tested by cytotoxicity assays. These involve culturing virus-infected cells and then adding the chemical compound in evaluation to assess whether it decreases the percentage of cytotoxicity resulting from the viral infection (Arias-Arias et al. 2018).

Chen and collaborators evaluated the cytopathic effect inhibition for the antiviral screening of 8810 compounds against SARS-CoV-2, comprising FDA-approved and experimental drugs. Of these, 319 were found to have antiviral activity against the virus. Among the approved drugs, three compounds (chlorprothixene, methotrimeprazine, and piperacetazine) showed a prominent antiviral effect that was previously unknown (Chen et al. 2021). Similarly, Taylor and collaborators applied a plaque reduction assay to demonstrate the antiviral activity of the approved drug galidesivir against SARS-CoV-2 (Taylor et al. 2021).

Other studies focused more on experimental compounds that have not been previously studied, such as plant extracts. That is the case for the research by Ogbole and collaborators, in which the antiviral effect of 27 crude plant extracts was tested against three echovirus serotypes (E7, E13, and E19) using a neutralization assay to measure the inhibition of cytopathic effect in cell culture. The results showed that ten out of all compounds tested presented a significant therapeutic effect against these serotypes, opening the possibility of its use as an alternative treatment for these infections (Ogbole et al. 2018).

The above-discussed experimental approach developed by our research group combining a fluorescence-assisted real-time plaque assay with automated image analysis allowed the application of a plaque reduction assay for the simultaneous screening of cytotoxic and antiviral effects of drugs in cell culture. For this, HSV-1 virus-infected Vero cells were stained with Hoechst 33342 and incubated with a medium containing PI and increasing concentrations of acyclovir (0–3000 ng/ml). Image analysis of live-cell photomicrographs taken over 96 hours allowed us to demonstrate the well-known antiviral activity of acyclovir in inhibiting the cytopathic effect and reducing the titers of HSV-1 by several logarithms. In addition, it was evidenced that concentrations equal to or greater than 3000 ng/ml of acyclovir were highly cytotoxic. The latter was determined by analyzing the uninfected cells at the periphery of viral plaques. With these data, it was possible to determine in a single experiment the 50% inhibitory concentration (IC50), 50% cytotoxic concentration (CC50), and selectivity index (SI: CC50/IC50) for in vitro acyclovir treatment of HSV-1 infection in Vero cells (Arias-Arias et al. 2021).

## *Neutralizing Antibodies Assessment*

One of the most significant applications of the virus-induced cytopathic effect measurement is the evaluation of the efficacy of neutralizing antibodies. This is of relevance in the development of vaccines, humoral immunotherapy, and serological assays for viral diagnosis. The principle of these assays is mainly focused on the evaluation of the inhibition of the cytopathic effect upon exposure to antibodies. A mixture of the virus with the serum containing the antibodies is inoculated into cell cultures and the cytotoxicity is measured after incubation to evaluate the neutralizing effect of the antibodies (Fenwick et al. 2021).

A current example of the neutralizing capacity of an antibody against SARS-CoV-2 is described by Fenwick and collaborators. This research team isolated the fragment antigen-binding region (Fab) of a monoclonal antibody (P5C3 Fab) that presents picomolar neutralizing capacity upon infection with any of the SARS-CoV-2 variants described to date. The neutralizing ability of the antibody was validated by experiments based on the in vitro inhibition of the cytopathic effect and the in vivo measurement of viral loads produced in animal models. Spike protein variants corresponding to each of the viral strains included were generated by recombinant expression technology and the neutralization potential of the antibodies was evaluated for each strain, demonstrating their effectiveness in all cases (Fenwick et al. 2021).

Other authors, such as Manenti and collaborators, used these experimental approaches to develop immunological techniques for rapid and accurate diagnosis of infection. The authors developed a colorimetric microneutralization assay that can be used in clinical practice and allows the quantification of the concentration of neutralizing antibodies in serum in a more specific way. The results obtained by the neutralization assay with measurement of the cytopathic effect were compared with those obtained by ELISA and a correlation of more than 50% was determined between both methods (Manenti et al. 2020).

## Conclusions

Advances in microscopy technologies as well as the development of computer platforms for image analysis, have enabled the enhancement of methods for the study of viral infections at the cellular level. Nevertheless, the determination of the virus-induced cytopathic effect in cell culture has still been one of the preferred strategies for the study of virus-cell interactions and the screening of antivirals and vaccines. The emergence of new viruses, such as SARS-CoV-2, increasingly merits the need to fully decipher those interactions and the importance of the herein-discussed approaches.

**Acknowledgments** We want to thank Facultad de Microbiología and Centro de Investigación en Enfermedades Tropicales, Universidad de Costa Rica, for supporting our research. This work is dedicated to our students.

## References

Arias-Arias, J. L., and Mora-Rodríguez, R. (2021). Fluorescence imaging approaches in Flavivirus research. En S. I. Ahmad (Ed.), Human viruses: diseases, treatments and vaccines (pp. 713–729). Springer International Publishing, Cham. https://doi.org/10.1007/978-3-030-71165-8_34

Arias-Arias JL, Vega-Aguilar F, Corrales-Aguilar E, Hun L, Loría GD, Mora-Rodríguez R (2018) Dengue virus infection of primary human smooth muscle cells. Am J Trop Med Hygiene 99(6): 1451–1457. https://doi.org/10.4269/ajtmh.18-0175

Arias-Arias JL, MacPherson DJ, Hill ME, Hardy JA, Mora-Rodríguez R (2020) A fluorescence-activatable reporter of flavivirus NS2B–NS3 protease activity enables live imaging of infection in single cells and viral plaques. J Biol Chem 295(8):2212–2226. https://doi.org/10.1074/jbc.RA119.011319

Arias-Arias JL, Corrales-Aguilar E, Mora-Rodríguez RA (2021) A fluorescent real-time plaque assay enables single-cell analysis of virus-induced cytopathic effect by live-cell imaging. Viruses 13(7):1193. https://doi.org/10.3390/v13071193

Barrientos LG, Rollin PE (2007) Release of cellular proteases into the acidic extracellular milieu exacerbates Ebola virus-induced cell damage. Virology 358(1):1–9. https://doi.org/10.1016/j.virol.2006.08.018

Baseler L, Chertow DS, Johnson KM, Feldmann H, Morens DM (2017) The pathogenesis of Ebola virus disease. Annu Rev Pathol: Mechanisms of Disease 12(1):387–418. https://doi.org/10.1146/annurev-pathol-052016-100506

Chen CZ, Shinn P, Itkin Z, Eastman RT, Bostwick R, Rasmussen L, Huang R, Shen M, Hu X, Wilson KM, Brooks BM, Guo H, Zhao T, Klump-Thomas C, Simeonov A, Michael SG, Lo DC, Hall MD, Zheng W (2021) Drug repurposing screen for compounds inhibiting the cytopathic effect of SARS-CoV-2. Front Pharmacol 11:592737. https://doi.org/10.3389/fphar.2020.592737

Chong MK, Chua AJS, Tan TTT, Tan SH, Ng ML (2014) Microscopy techniques in flavivirus research. Micron 59:33–43. https://doi.org/10.1016/j.micron.2013.12.006

Costin JM (2007) Cytopathic mechanisms of HIV-1. Virol J 4(1):100. https://doi.org/10.1186/1743-422X-4-100

Doms RW (2016) Basic concepts. In: *Viral pathogenesis*. Elsevier, pp 29–40. https://doi.org/10.1016/B978-0-12-800964-2.00003-3

Dulbecco R (1952) Production of plaques in monolayer tissue cultures by single particles of an animal virus. PNAS 38:747–752

Fenwick C, Turelli P, Perez L, Pellaton C, Esteves-Leuenberger L, Farina A, Campos J, Lana E, Fiscalini F, Raclot C, Pojer F, Lau K, Demurtas D, Descatoire M, Joo VS, Foglierini M, Noto A, Abdelnabi R, Foo CS et al (2021) A highly potent antibody effective against SARS-CoV-2 variants of concern. Cell Rep 37(2):109814. https://doi.org/10.1016/j.celrep.2021.109814

Flint SJ, American Society for Microbiology (eds) (2009) Principles of virology, 3rd edn. ASM Press

Flint SJ, Enquist LW, Racaniello VR, Shaika AM (2009) Principles of virology, 3rd edn. ASM Press

Francica JR, Matukonis MK, Bates P (2009) Requirements for cell rounding and surface protein down-regulation by Ebola virus glycoprotein. Virology 383(2):237–247. https://doi.org/10.1016/j.virol.2008.10.029

Gagliardi TB, Criado MF, Proença-Módena JL, Saranzo AM, Iwamoto MA, de Paula FE, Cardoso RS, Delcaro LS, Silva ML, Câmara AA, Arruda E (2017) Syncytia induction by clinical isolates of human respiratory syncytial virus a. Intervirology 60(1–2):56–60. https://doi.org/10.1159/000480014

Gower TL, Pastey MK, Peeples ME, Collins PL, McCurdy LH, Hart TK, Guth A, Johnson TR, Graham BS (2005) RhoA signaling is required for respiratory syncytial virus-induced syncytium formation and filamentous Virion morphology. J Virol 79(9):5326–5336. https://doi.org/10.1128/JVI.79.9.5326-5336.2005

Hao S, Ning K, Kuz CA, Vorhies K, Yan Z, Qiu J (2020) Long-term modeling of SARS-CoV-2 infection of *in vitro* cultured polarized human airway epithelium. MBio 11(6):e02852-20. https://doi.org/10.1128/mBio.02852-20

Huang Y, Li Y, Zhang H, Zhao R, Jing R, Xu Y, He M, Peer J, Kim YC, Luo J, Tong Z, Zheng J (2018) Zika virus propagation and release in human fetal astrocytes can be suppressed by neutral sphingomyelinase-2 inhibitor GW4869. Cell Discovery 4(1):19. https://doi.org/10.1038/s41421-018-0017-2

Iannacone M, Guidotti LG (2021) Immunobiology and pathogenesis of hepatitis B virus infection. Nat Rev Immunol 22:19. https://doi.org/10.1038/s41577-021-00549-4

John DV, Lin Y-S, Perng GC (2015) Biomarkers of severe dengue disease – a review. J Biomed Sci 22(1):83. https://doi.org/10.1186/s12929-015-0191-6

Lang TU, Khalbuss WE, Monaco SE, Michelow P, Pantanowitz L (2011) Review of HIV-related cytopathology. Pathol Res Int 2011:1–12. https://doi.org/10.4061/2011/256083

Mandell GL, Bennett JE, Dolin R (eds) (2010) Mandell, Douglas, and Bennett's principles and practice of infectious diseases, 7th edn. Churchill Livingstone/Elsevier

Manenti A, Maggetti M, Casa E, Martinuzzi D, Torelli A, Trombetta CM, Marchi S, Montomoli E (2020) Evaluation of SARS-CoV-2 neutralizing antibodies using a CPE-based colorimetric live virus micro-neutralization assay in human serum samples. J Med Virol 92(10):2096–2104. https://doi.org/10.1002/jmv.25986

Mighty KK, Laimins LA (2014) The role of human papillomaviruses in oncogenesis. In: Chang MH, Jeang K-T (eds) Viruses and human cancer, vol 193. Springer, Berlin, Heidelberg, pp 135–148. https://doi.org/10.1007/978-3-642-38965-8_8

Monel B, Compton AA, Bruel T, Amraoui S, Burlaud-Gaillard J, Roy N, Guivel-Benhassine F, Porrot F, Génin P, Meertens L, Sinigaglia L, Jouvenet N, Weil R, Casartelli N, Demangel C, Simon-Lorière E, Moris A, Roingeard P, Amara A, Schwartz O (2017) Zika virus induces massive cytoplasmic vacuolization and paraptosis-like death in infected cells. EMBO J 36(12): 1653–1668. https://doi.org/10.15252/embj.201695597

Ogbole OO, Akinleye TE, Segun PA, Faleye TC, Adeniji AJ (2018) In vitro antiviral activity of twenty-seven medicinal plant extracts from Southwest Nigeria against three serotypes of echoviruses. Virol J 15(1):110. https://doi.org/10.1186/s12985-018-1022-7

Parveen N, Borrenberghs D, Rocha S, Hendrix J (2018) Single viruses on the fluorescence microscope: imaging molecular mobility, interactions and structure sheds new light on viral replication. Viruses 10(5):250. https://doi.org/10.3390/v10050250

Roy SG, Sadigh B, Datan E, Lockshin RA, Zakeri Z (2014) Regulation of cell survival and death during Flavivirus infections. World J Biol Chem 5(2):14

Ryu W-S (2017) Diagnosis and methods. In: *Molecular virology of human pathogenic viruses*. Elsevier, pp 47–62. https://doi.org/10.1016/B978-0-12-800838-6.00004-7

Schiffman M, Wentzensen N (2013) Human papillomavirus infection and the multistage carcinogenesis of cervical cancer. Cancer Epidemiol Biomark Prev 22(4):553–560. https://doi.org/10.1158/1055-9965.EPI-12-1406

Souza BSF, Sampaio GLA, Pereira CS, Campos GS, Sardi SI, Freitas LAR, Figueira CP, Paredes BD, Nonaka CKV, Azevedo CM, Rocha VPC, Bandeira AC, Mendez-Otero R, dos Santos RR, Soares MBP (2016) Zika virus infection induces mitosis abnormalities and apoptotic cell death of human neural progenitor cells. Sci Rep 6(1):39775. https://doi.org/10.1038/srep39775

Taylor R, Bowen R, Demarest JF, DeSpirito M, Hartwig A, Bielefeldt-Ohmann H, Walling DM, Mathis A, Babu YS (2021) Activity of Galidesivir in a hamster model of SARS-CoV-2. Viruses 14(1):8. https://doi.org/10.3390/v14010008

Valdovinos MR, Gómez B (2003) Establishment of respiratory syncytial virus persistence in cell lines: association with defective interfering particles. Intervirology 46(3):190–198. https://doi.org/10.1159/000071461

Wen Z, Citron M, Bett AJ, Espeseth AS, Vora KA, Zhang L, DiStefano DJ (2019) Development and application of a higher throughput RSV plaque assay by immunofluorescent imaging. J Virol Methods 263:88–95. https://doi.org/10.1016/j.jviromet.2018.10.022

Zhu N, Wang W, Liu Z, Liang C, Wang W, Ye F, Huang B, Zhao L, Wang H, Zhou W, Deng Y, Mao L, Su C, Qiang G, Jiang T, Zhao J, Wu G, Song J, Tan W (2020) Morphogenesis and cytopathic effect of SARS-CoV-2 infection in human airway epithelial cells. Nat Commun 11(1):3910. https://doi.org/10.1038/s41467-020-17796-z

# Part II
# Specific Viruses

# Chapter 8
# Human Papilloma Virus-Infected Cells

**Alfredo Cruz-Gregorio and Ana Karina Aranda-Rivera**

**Abstract** Human papillomavirus (HPV) is associated with infection of different tissues, such as the cervix, anus, vagina, penis, vulva, oropharynx, throat, tonsils, back of the tongue, skin, the lungs, among other tissues. HPV infection may or may not be associated with the development of cancer, where HPVs not related to cancer are defined as low-risk HPVs and are associated with papillomatosis disease. In contrast, high-risk HPVs (HR-HPVs) are associated with developing cancers in areas that HR-HPV infects, such as the cervix. In general, infection of HPV target cells is regulated by specific molecules and receptors that induce various conformational changes of HPV capsid proteins, allowing activation of HPV endocytosis mechanisms and the arrival of the HPV genome to the human cell nucleus. After the transcription of the HPV genome, the HPV genome duplicates exponentially to lodge in a new HPV capsid, inducing the process of exocytosis of HPV virions and thus releasing a new HPV viral particle with a high potential of infection. This infection process allows the HPV viral life cycle to conclude and enables the growth of HPV virions. Understanding the entire infection process has been a topic that researchers have studied and developed for decades; however, there are many things to still understand about HPV infection. A thorough understanding of these HPV infection processes will allow new potential treatments for HPV-associated cancer and papillomatosis. This chapter focuses on HPV infection, the process that will enable HPV to complete its HPV life cycle, emphasizing the critical role of different molecules in allowing this infection and its completion during the HPV viral life cycle.

**Keywords** HPV infection · L1 protein · L2 protein · Furin · Dynein · Endocytosis

A. Cruz-Gregorio (✉)
Departamento de Fisiología, Instituto Nacional de Cardiología Ignacio Chávez, Mexico City, Mexico
e-mail: alfredo.cruz@cardiologia.org.mx

A. K. Aranda-Rivera
Laboratorio F-315, Departamento de Biología, Facultad de Química, Universidad Nacional Autónoma de México, Mexico City, Mexico
e-mail: anitaaranda023@comunidad.unam.mx

 
S. Vijayakrishnan et al. (eds.), *Virus Infected Cells*, Subcellular Biochemistry 106,
https://doi.org/10.1007/978-3-031-40086-5_8

## Introduction

Human papillomaviruses (HPV) infections can be transmitted by skin-to-skin contact or by sexual intercourse through vaginal, anal, and oral contact. This virus infects different tissues, such as the cervix, anus, vagina, penis, vulva, oropharynx, throat, tonsils, back of the tongue, skin, and lungs, among other tissues (Rubin et al. 2001; Daling et al. 2002; Clark et al. 2004; D'Souza and Dempsey 2011). HPV infection may or may not be associated with the development of cancer. HPVs that are not associated with cancer are defined as low-risk HPVs (LR-HPV) and are associated with papillomatosis disease. In contrast, high-risk HPVs (HR-HPVs) are associated with the development of cancers in areas that HR-HPV infects, such as the cervix. The latter was proved by molecular, clinical, virological, and epidemiological evidence, showing that HPV is the etiological agent that develops cancer (zur Hausen 2002; zur Hausen 2009).

HPV is a nonenveloped virus with a double-stranded DNA genome belonging to the papillomaviridae family. For its study, the HPV genome is divided into three regions: the region that corresponds to the early (E) proteins (E1, E2, E4, E5, E6, and E7), necessary for viral replication, regulation of transcription and immortalization, and cell transformation; the region that encodes the late (L) proteins (L1 and L2) that build the structure of the capsid inducing the release of the virion and the long region (LCR), which contains the origin of replication and the early promoter that allow replication and expression of the viral genome (McMurray et al. 2001). The HPV genome is surrounded by a 55 nm icosahedral capsid composed of two proteins, L1 and L2. The L1 and L2 proteins form 72 capsomeres that build the capsid structure, where L2 is located internally surrounded by the L1 proteins in the capsomeres. L1 is involved in capsid stabilization, endosomal escape of HPV virions, and nuclear transport of the HPV genome (Doorbar et al. 2012). Meanwhile, L2 is associated with the interaction with different substrates to allow HPV to enter the cell and is also necessary for the encapsidation of the viral DNA, in addition to facilitating the endosomal escape of the viral genome after infection, possibly due to the interaction with sorting nexin 17 (Bergant Marušič et al. 2012). The infection of HPV target cells is regulated by L1 and L2 associated with different molecules and receptors that induce conformational changes of the HPV capsid proteins, allowing the activation of HPV endocytosis mechanisms and the arrival of the HPV genome to the cell nucleus. In the nucleus, the replication and transcription of the HPV genome are carried out, increasing its genetic material exponentially. Then the genome is housed in a new HPV capsid, inducing the process of exocytosis of the HPV virions, releasing new HPV viral particles with the potential for a new infection (Doorbar et al. 2012) (Fig. 8.1). Although the HPV infection process has been a topic that researchers have studied and developed, there are many issues to understand about HPV infection. The entire understanding of these HPV infection processes will allow to development of new potential treatments for HPV-associated cancer and papillomatosis. This chapter focuses on HPV infection, the process that allows HPV

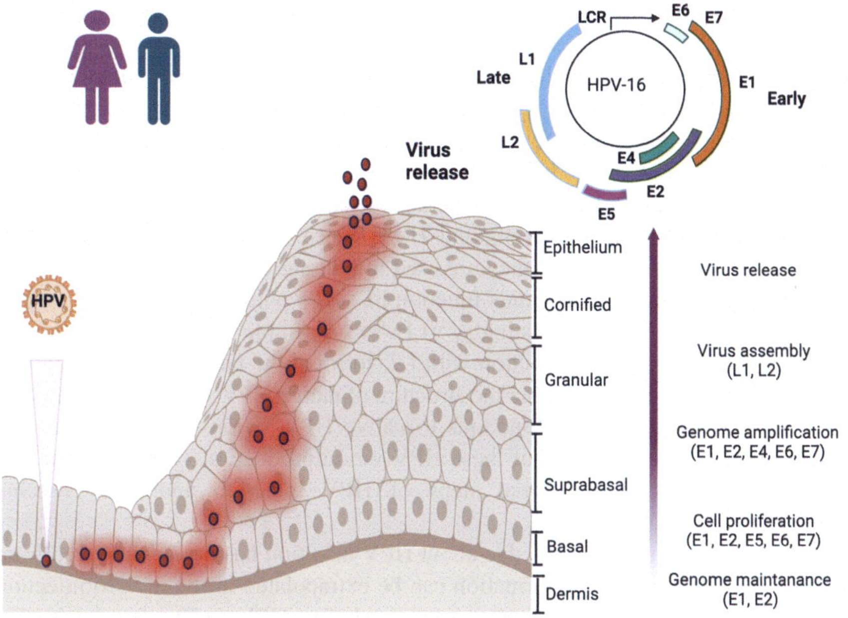


**Fig. 8.1** Human Papillomavirus (HPV) life cycle. HPV infects basal cells to induce the transcription of E1 and E2, which trigger the HPV genome replication. Likewise, the E5, E6, and E7 oncoproteins are transcribed, preventing cell death and entry into the cell cycle. In the intermediate layers of the epithelium, the HPV genome is transcribed exponentially, and in the upper layers of the epithelium, E4 is transcribed for keratin breakdown, allowing the virion formed by the encapsidation of the HPV genome by L1 and L2 to exit the cell, and start a new infection

to complete its HPV life cycle. It emphasizes the important role of different molecules in allowing this infection and its completion during the HPV viral life cycle.

## HPV Entry into Its Target Cells

Replication and transcription of HPV require basal epithelial cell differentiation since undifferentiated basal epithelial cells do not express key cellular transcription factors (TFs) like transcription factor IID (TFIID), which induce the expression of the early protein of HPV (Moody 2017; Ribeiro et al. 2018). Thus, the viral transcription of E1, E2, E5, E6, and E7 and genome amplification are maintained at low levels in the basal layer of the squamous epithelium. As the epithelium differentiates, the genes such as E1 and E2 are expressed, inducing amplification of the viral genome and increasing the replication of the HPV genome in high quantities. Because of this dependence, the generation of papillomavirus in vitro has been difficult, so the production of infectious viral particles or virions for its study in vitro was only possible following the development of keratinocyte-based organotypic cultures and mice xerographs (Dollard et al. 1992; Bonnez et al. 1998). These limitations were partially overcome using free virus particles (noninfectious virus-like particles or VLPs) with reporter plasmids (Meyers et al. 1992) for the development of packaging cell lines and codon optimization. Therefore, these technologies permitted the high expression of capsid proteins that occurred in conjunction with the large-scale production of VLPs, leading to a better study of interactions between infected cells and HPV. Since VLPs do not have an HPV genome and are noninfectious, they are used in binding studies. On the other hand, VLPs that harbor reporter plasmids function like HPV viral "genomes" are used to quantify infection levels. Interestingly, VLPs self-assemble into L1-VLPs when L1 is expressed alone or in conjunction with L2, forming the basis of current prophylactic vaccines since VLPs mimic the structure of native HPV virions, and therefore share similar immunological functions to native HPV virions (Wang and Roden 2013). Thus, the development of these HPV virion production systems has helped the understanding of the initial steps of HPV infection and HPV vaccine development.

Although most of the knowledge about HPV infection has been discovered from in vitro models, the major information can be extrapolated to the in vivo infection process. For instance, it has been suggested that before HPV reaches basal epithelial cells, mainly through microlesions, HPV binds to the heparan sulfate (HS) and laminin 332 of the extracellular matrix (ECM). This suggestion was made after researchers reported that in vitro essays, VLPs bind to HS and laminin 332 (Shafti-Keramat et al. 2003; Culp et al. 2006a, b). Therefore, it has been suggested that in vivo, the virus first uses these molecules to bind to the basement membrane (BM) (which plays the same role as the ECM in vitro models), and then, during wound healing, the HPV is transferred to basal keratinocytes as they migrate to the wounded area (Roberts et al. 2007). Different groups of researchers have discovered that HS modifications, such as O-linked sulfation and N-linked sulfation, induce better binding between viral particles with their target cell and that even the level of HSPG on the cell surface correlates with the number of bound HPVs (Selinka et al. 2003). However, whether these interactions occur in vivo remains to be investigated. Knowing if these molecules are key to HPV infection in vivo could be key to considering these biomolecules as possible targets to reduce HPV infection. Next, in basal keratinocytes virus binds to the heparan sulfate proteoglycans (HSPG) (Sapp and Day 2009). HSPGs are founded on the membrane surface of the basal cells of the epithelium and in the extracellular matrix. If HSPGs are attached to cell membranes they are called syndecans or if they are attached to the extracellular matrix they are called perlecans. HPGS-syndecans 1 and 4, which are overexpressed during wound healing, serve as the first anchorage receptors on basal cells (Sapp and Day 2009). After that virus is anchored to the cells other bindings are needed, such as $\alpha6\beta4$ integrins plus the tetraspanin CD151 (Abban and Meneses 2010; Scheffer et al. 2014). Moreover, this complex can bind and activate epidermal growth factor receptor (EGFR) signaling, leading to Src kinase phosphorylation of annexin A2, which induces the extracellular translocation of the annexin A2/S100A10 heterotetramer (A2t) (Woodham et al. 2012; Dziduszko and Ozbun 2013; Scheffer et al. 2014). Interestingly, EGFR signaling activation deactivates autophagy through the activation of PI3K signaling. Autophagy is an intrinsic cellular sensor that inhibits HPV infection, permitting an efficient HPV infection process (Surviladze et al. 2013; Griffin et al. 2013). These interactions and activating different molecules allow HPV uptake and successful HPV infection.

The HPV binding with HS also induces a conformational change in the HPV capsid. This conformational change of the capsid weakens the link of the virus with the HPGS, leading to the exposure of the N-terminus of L2. Thus, this conformational change induces L2 to unmask since most of L2 is hidden on the surface of the capsid. When L2 is exposed, it is cleaved by furin (Richards et al. 2006). Cleavage of the N-terminus of L2 by furin exposes the binding site for cell membrane receptors involved in infectious internalization, allowing HPV to enter by endocytosis. Notably, L2 cleavage is essential for the insertion and protrusion of L2 across the vesicular membrane. However, other proteins are needed during L2 insertion; for instance, it is known that the multisubunit intramembrane protease γ-secretase acts as a chaperone that inserts L2 into the cell membrane (Harwood et al. 2020). Only when L2 inserts into the cell membrane is proper downward trafficking of L2/vDNA allowed, as L2 in the cytosol allows for the recruitment of sorting nexins and the retromeric complex. Furthermore, L2 is critical for the retrograde trafficking of L2/vDNA complexes from endosomes to the trans-Golgi network (TGN), which is a required step of the initial infection (Fig. 8.2). Thus, L2 acts as a transmembrane protein that directs HPV DNA trafficking into the TGN and into the cell nucleus for productive infection away from lysosomal compartments that might degrade it.

## *HPV Endocytosis as a Journey to the Nucleus*

HPV in the cytosol recruits sorting nexin 17 (SNX17), which prevents HPV following a rapid lysosomal sorting and its degradation. This late endosome escape of the viral genome is mediated by L2 (Bergant Marušič et al. 2012). L2 also interacts with the complexes Vps26, Vps29, and Vps35, a trimeric retromer that fulfills endosome tubulation and vesicle formation that induces the retrograde transport of HPV from the endosome to TGN. This showed that different molecules such as Rab-GTPases SNX proteins, TBC1 domain family member 5 (TBC1D5), and the endoplasmic reticulum (ER)-anchored protein vesicle-associated membrane protein (VAMP)-associated protein (VAP) assist this process (Day et al. 2013; Siddiqa et al. 2018; Pim et al. 2021). Where TBC1D5 is recruited to the L2/retromer at the endosomal membrane for stimulating hydrolysis of Rab7-GTP, it induces the disassembly of the retromer from the HPV (Xie et al. 2020). After the retromer is dissociated, vesicles that contain the virus traffic to the TGN, where the cargo is delivered by membrane fusion. This cytoplasmic traffic uses the cellular microtubules as highways mediated by L2 through its binding to the dynein light chain. Dynein is a kinesin and the most crucial motor protein associated with microtubules. Its function is critical in the

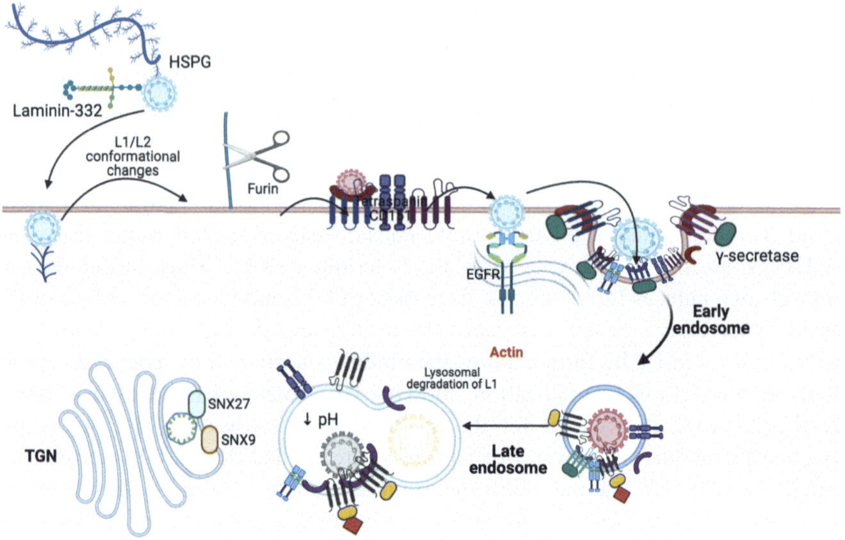


**Fig. 8.2** Entry and endocytosis of Human Papilloma Virus. HPV viral particles binds to different receptor and molecules such as heparan sulfate, which induce capsid conformational changes, inducing that furin cleavages L2, this cut is essential for the insertion and protrusion of L2 across the vesicular membrane. However, other proteins are needed during L2 insertion such as γ-secretase, which acts as a chaperone that inserts L2 into the cell membrane. Only when L2 inserts into the cell membrane is proper downward trafficking of L2/vDNA complex from endosomes to the trans-Golgi network (TGN)

retrograde transport of substances within the cell and in inducing chromosome movement during cell mitosis. Thus, this interaction drives the transportation of HPV to the nucleus.

It is important to mention that the study of VLP-associated endocytosis has yielded different endocytosis pathways such as clathrin, caveolin, lipid vesicles, clathrin-independent, and cholesterol-mediated endocytosis mechanisms and, therefore, it is still a matter of scientific debate. This could be because there are different "maturity" states of VLPs. That is when HPV capsids are extracted from replicating cultured cells, capsids can be larger, less regular, and less protease resistant or "immature" in comparison with "mature" capsids; this indicates that immature capsids must suffer a substantial change in conformation during the maturation process (Buck et al. 2005). Thus, this high variability between "immature" and "mature" capsids could direct one process of endocytosis or another. The latter is supported by the fact that HPV that are phylogenetically closely related, such as HPV-16 and -31, follow different endocytosis routes, where HPV-16 activates clathrin-mediated endocytosis while HPV31 activates caveolin-mediated endocytosis (Bousarghin et al. 2003). Furthermore, HPV16 PsV uptake has been reported to be clathrin-independent, and HPV16 endocytosis occurs via tetraspanin-rich microdomains (Spoden et al. 2008). Thus, it seems that there is not only one endocytosis pathway used by HPV; it will depend both on the microenvironment and the receptors associated with the viral particle. The latter is because the virus enters the cells asynchronously. This means that while some virions enter in minutes, others take hours. It has been proposed that HSPG-binding virions have a slower entry rate than receptor-binding virions, which mediate a high entry rate (Williams and Fuki 1997). We should not forget to mention that these interactions could also be associated with the endocytosis pathway that is activated. Therefore, the entry of the virus will be faster or slower.

## Replication and Transcription of the HPV Genome

The entry of the HPV genome into the cell nucleus is more associated with the nuclear membrane breakdown during mitosis than with the transporters via karyopherin (Pyeon et al. 2009). In this process, L2 remains in a membrane-spanning conformation, interacting with the proteins of the mitotic spindle motor to transport the vesicular HPV DNA genome from the remnants of the TGN to the pericentriolar space in prometaphase, waiting for the chromosomes breakdown in the metaphase (DiGiuseppe et al. 2016). L2 also binds to chromosomes in the delivery of vesicle-bound episomal HPV DNA into the nuclei of daughter cells (Aydin et al. 2017). This vesicle that surrounds L2/vDNA is maintained through mitosis and into the next G1 in the cell cycle (DiGiuseppe et al. 2016). Then, it has been observed that upon G1 re-entry into HPV-infected cells, nuclear entry and nuclear domain 10 (ND10)/promyelocytic leukemia (PML) components bind around L2/vesicular vDNA within nuclei (Guion et al. 2019). Thus, the HPV genome can arrive at the cell nucleus to start replication and transcription of the virus genome. Once the HPV genome is transported to the nucleus, the transcription and replication of the virus begin. The LCR contains several regulatory sites for both viral replication and transcription (Bernard 2013). These cellular processes are induced by the direct interaction between different cellular and viral components with the LCR. For example, the LCR has specific sequences for the association of E1 and E2 that promote virus replication, as well as its transcription. In addition, it contains binding sequences to components of the transcription machinery such as specificity protein 1 (SP1) and TATA-binding protein (TBP), as well as glucocorticoid receptors that induce transcription of the viral genome (Bernard 2013). When these transcription factors associated with HPV genome induce the transcription of its different proteins with the different functions that permit the HPV life cycle to be concluded. For instance, the HPV E1 protein is a helicase that participates in viral DNA replication, it is composed of around 681 amino acids (aa) and is divided into three regions: the carboxyl-terminal region related to helicase activity, the region where the domain is located, Binding to DNA or the DNA Binding Domain (DBD) and an amino-terminal region that is the target of phosphorylation, sumoylation or acetylation that influence E1 activity and therefore transcription and viral replication. In addition to containing signals that allow the entry and exit from the cell nucleus (Bergvall et al. 2013), E1 associates with the origin of replication on the LCR, inducing the formation of the initiation complex and the unwinding of the DNA to promote replication of the viral genome. It is also known that E1 directly interacts with DNA polymerase, which in turn associates with the replication complex and thus initiates the replication process (Bergvall et al. 2013). Also, E2 is a regulatory protein of viral replication and transcription. This 300–500 amino acid protein consists of two conserved domains: the amino-terminal region where the transactivation domain (TAD) is responsible for transcriptional regulation and viral replication, interacting with different cellular proteins. The other E2 domain is known as DBD. This domain recognizes and binds to specific sequences of ACCGN4CGGT found in the viral genome (Gillitzer et al. 2000). Between the TAD and DBD domains is a hinge region. This region does not participate in the basic functions of virus replication and transcription; however, it provides stability to E2, participates in cytoplasm-nucleus localization, and functions as a spacer between the two domains, avoiding steric hindrance (McBride 2013). When E2 associates with viral DNA, it recruits different cellular proteins involved in viral transcription and replication. E2 participates both in the segregation of the viral genome, anchoring it to the cell chromosomes during cell mitosis, and in the packaging for the formation of virions (McBride 2013). Among other cellular targets of E2 is p53, this interaction induces cell arrest and apoptosis (Desaintes et al. 1999). An important factor in cancer development lies in the antiproliferative role exerted by E2, since it can repress cell growth and induce apoptosis in HPV-positive cells. Partly to the repression of the transcription of E6 and E7, with the consequent increase of p53 and pRb (Demeret et al. 1997). The HPV oncoproteins, E5, E6 and E7, are transcribed to avoid to cell differentiation, cell senescence or cell death, even avoid immune recognition. That is, although E5 is the smallest protein encoded by HPV, with only 83 amino acids and a molecular weight of 9 kDa, it can induce cell proliferation associated with overexposure of the epidermal growth factor receptor (EGFR) receptor (Crusius et al. 1998). It is located mainly in the endoplasmic reticulum (ER) and the Golgi apparatus (GA), inducing stress of these organelles that help viral replication and persistence (Venuti et al. 2011). Likewise, it has been shown that E5 binds to the 16 K subunit of the vacuolar V-ATPase of the endosomes, decreasing its activity and inhibiting its acidification, intervening in vesicular traffic (Di Domenico et al. 2009). Regarding E6 and E7, the most oncoproteins studied so far, they have the ability to induce immortalization, inhibit cell apoptosis, or prevent cell-cell interactions (Mantovani and Banks 2001). For example, it has been found that E6 can associate with human telomerase reverse transcriptase (hTERT), a polymerase that lengthens the telomeric ends of cell chromosomes, inducing their activation and allowing cell immortalization (Oh et al. 2001). Another important characteristic of the E6 proteins of high-risk HPVs is their association with the E6 associated protein (E6AP) ubiquitin ligase, through which it can associate with p53, promoting its degradation via the proteasome and thus blocking cell apoptosis (Thomas et al. 1999). HR-HPV E6 can also bind to proteins with PDZ domains such as human disc large (hDLG), multi-PDZ domain protein 1 (MUPP) and hSCRIBb for their degradation using the same mechanism as with p53. These proteins are related to cell transformation due to the ability to regulate cell-cell interactions and cell polarity (Mantovani and Banks 2001).

The E7 oncoprotein is a protein with an enormous capacity for transformation due to its ability to induce the degradation of the tumor suppressor protein pRb, via the proteasome (Münger et al. 2001). The degradation of pRb induces the release of E2F, inducing the cells to enter the synthesis phase (S phase) of the cell cycle (Boyer et al. 1996). Thus, these oncoproteins permits that HPV genome is synthetized in huge quantities, thereby avoiding cell death, activation of cell cycle and proliferation.

## Exocytosis and Exit toward a New Infection

After a large amount of the HPV genome has accumulated, it is time to package it; however, HPV virions need make their way back through the intracellular network of cytoskeletal proteins. This process is induced by HPV E4 protein, a protein of approximately 17 kDa. Interestingly, although the E4 open reading frame (ORF) is contained within the E2 ORF, it is expressed late, just before L1 and L2. The late expression of E4 is due to the fact that it is regulated by a specific differentiation promoter (p670 for HR-HPV16), and therefore it only accumulates in differentiated cells of the superficial layers of the epithelium (Doorbar 2013). Thus, the main function of E4 has been associated with the collapse of keratin filaments located in the cell cytoplasm, a collapse necessary for the release of virions. However, E4 has also been associated with the inhibition of Cdk1 activity, inducing cell cycle arrest in the G2 phase (Davy et al. 2002). The L1 and L2 proteins are proteins expressed in the late phase of the viral life cycle. These proteins assemble to build the HPV capsid, a process that requires oxidative stress to produce disulfide bonds between the L1 proteins, permitting the maturation of the capsid (Buck et al. 2005). It is important to mention that HPV proteins differentially regulate the redox state (Cruz-Gregorio et al. 2018a, b), regulating both organelles that produce reactive oxygen species and antioxidants (Cruz-Gregorio et al. 2019, 2020, 2023) of cells and that this provides the appropriate conditions for the virus to infect and conclude its viral life cycle. Thus, five L1 units are associated with each other forming an L1 pentamer. In the gap of each L1 pentamer, an L2 protein associates, this complex is called a capsomer, so the HPV capsid is made up of 72 capsomeres in an icosahedral network. Once the virions are assembled, they are released by shedding, and the viral cycle can be started again through L1, which binds to the heparan sulfate (HS) receptor contained in the extracellular matrix and the HPV target cells (Fig. 8.3).

There remain many questions to be answered to fully resolve HPV infection, and the investigation of the process is likely to provide information to avoid infection that is associated with papillomatosis and cancer development. Moreover, the

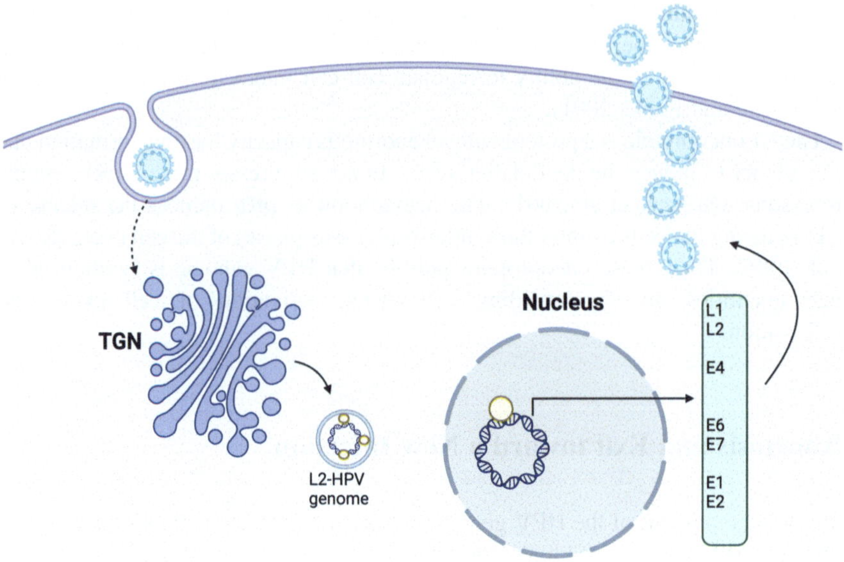


**Fig. 8.3** Replication, transcription, and exocytosis of the HPV. As soon as the HPV genome is carried from the TGN to the nucleus, it is transcribed and replicated by the cellular transcription and replication machinery aided by the E1 and E2 proteins. The E5, E6, and E7 proteins are also expressed, which prevent the cells from dying by apoptosis and activate the cell cycle, allowing an exponential increase of the viral genome. With high amounts of the viral genome, it is time to package it in the capsid formed by L1 and L2, which will make their way through the cytoskeleton thanks to the keratin degradation induced by E4 so that the virions can leave the infected cell and start another cycle of infection

majority of the knowledge of papillomavirus infection has developed in HR-HPV, which can be extrapolated to LR-HPV or other papillomaviruses such as canine or feline, as papillomaviruses share characteristics (Cruz-Gregorio et al. 2022); however, it is necessary to study infection in these types of virus and different species.

## Conclusions

Understanding the entire HPV infection process has been a topic that researchers have studied and developed for decades; however, there is still much to understand about papillomavirus infection. A deep understanding of these infection processes by these viruses will allow potential new treatments for papillomavirus-associated cancer and papillomatosis in humans and other species.

## References


Abban CY, Meneses PI (2010) Usage of heparan sulfate, integrins, and FAK in HPV16 infection. Virology 403:1–16. https://doi.org/10.1016/j.virol.2010.04.007

Aydin I, Villalonga-Planells R, Greune L et al (2017) A central region in the minor capsid protein of papillomaviruses facilitates viral genome tethering and membrane penetration for mitotic nuclear entry. PLoS Pathog 13:e1006308. https://doi.org/10.1371/journal.ppat.1006308

Bergant Marušič M, Ozbun MA, Campos SK et al (2012) Human papillomavirus L2 facilitates viral escape from late endosomes via sorting nexin 17. Traffic 13:455–467. https://doi.org/10.1111/j.1600-0854.2011.01320.x

Bergvall M, Melendy T, Archambault J (2013) The E1 proteins. Virology 445:35–56. https://doi.org/10.1016/j.virol.2013.07.020

Bernard H-U (2013) Regulatory elements in the viral genome. Virology 445:197–204. https://doi.org/10.1016/j.virol.2013.04.035

Bonnez W, DaRin C, Borkhuis C et al (1998) Isolation and propagation of human papillomavirus type 16 in human xenografts implanted in the severe combined immunodeficiency mouse. J Virol 72:5256–5261. https://doi.org/10.1128/JVI.72.6.5256-5261.1998

Bousarghin L, Touzé A, Sizaret P-Y, Coursaget P (2003) Human papillomavirus types 16, 31, and 58 use different endocytosis pathways to enter cells. J Virol 77:3846–3850. https://doi.org/10.1128/jvi.77.6.3846-3850.2003

Boyer SN, Wazer DE, Band V (1996) E7 protein of human papilloma virus-16 induces degradation of retinoblastoma protein through the ubiquitin-proteasome pathway. Cancer Res 56:4620–4624

Buck CB, Thompson CD, Pang Y-YS et al (2005) Maturation of papillomavirus capsids. J Virol 79: 2839–2846. https://doi.org/10.1128/JVI.79.5.2839-2846.2005

Clark MA, Hartley A, Geh JI (2004) Cancer of the anal canal. Lancet Oncol 5:149–157. https://doi.org/10.1016/S1470-2045(04)01410-X

Crusius K, Auvinen E, Steuer B et al (1998) The human papillomavirus type 16 E5-protein modulates ligand-dependent activation of the EGF receptor family in the human epithelial cell line HaCaT. Exp Cell Res 241:76–83. https://doi.org/10.1006/excr.1998.4024

Cruz-Gregorio A, Manzo-Merino J, Gonzaléz-García MC et al (2018a) Human papillomavirus types 16 and 18 early-expressed proteins differentially modulate the cellular redox state and DNA damage. Int J Biol Sci 14:21–35. https://doi.org/10.7150/ijbs.21547

Cruz-Gregorio A, Manzo-Merino J, Lizano M (2018b) Cellular redox, cancer and human papillomavirus. Virus Res 246:35–45. https://doi.org/10.1016/j.virusres.2018.01.003

Cruz-Gregorio A, Aranda-Rivera AK, Aparicio-Trejo OE et al (2019) E6 Oncoproteins from high-risk human papillomavirus induce mitochondrial metabolism in a head and neck squamous cell carcinoma model. Biomol Ther 9:351. https://doi.org/10.3390/biom9080351

Cruz-Gregorio A, Aranda-Rivera AK, Pedraza-Chaverri J (2020) Human papillomavirus-related cancers and mitochondria. Virus Res 286:198016. https://doi.org/10.1016/j.virusres.2020.198016

Cruz-Gregorio A, Aranda-Rivera AK, Pedraza-Chaverri J (2022) Pathological similarities in the development of papillomavirus-associated cancer in humans, dogs, and cats. Animals (Basel) 12:2390. https://doi.org/10.3390/ani12182390

Cruz-Gregorio A, Aranda-Rivera AK, Roviello GN, Pedraza-Chaverri J (2023) Targeting mitochondrial therapy in the regulation of HPV infection and HPV-related cancers. Pathogens 12: 402. https://doi.org/10.3390/pathogens12030402

Culp TD, Budgeon LR, Christensen ND (2006a) Human papillomaviruses bind a basal extracellular matrix component secreted by keratinocytes which is distinct from a membrane-associated receptor. Virology 347:147–159. https://doi.org/10.1016/j.virol.2005.11.025

Culp TD, Budgeon LR, Marinkovich MP et al (2006b) Keratinocyte-secreted laminin 5 can function as a transient receptor for human papillomaviruses by binding virions and transferring them to adjacent cells. J Virol 80:8940–8950. https://doi.org/10.1128/JVI.00724-06

D'Souza G, Dempsey A (2011) The role of HPV in head and neck cancer and review of the HPV vaccine. Prev Med 53(Suppl 1):S5–S11. https://doi.org/10.1016/j.ypmed.2011.08.001

Daling JR, Madeleine MM, Schwartz SM et al (2002) A population-based study of squamous cell vaginal cancer: HPV and cofactors. Gynecol Oncol 84:263–270. https://doi.org/10.1006/gyno.2001.6502

Davy CE, Jackson DJ, Wang Q et al (2002) Identification of a G(2) arrest domain in the E1 wedge E4 protein of human papillomavirus type 16. J Virol 76:9806–9818. https://doi.org/10.1128/jvi.76.19.9806-9818.2002

Day PM, Thompson CD, Schowalter RM et al (2013) Identification of a role for the trans-Golgi network in human papillomavirus 16 pseudovirus infection. J Virol 87:3862–3870. https://doi.org/10.1128/JVI.03222-12

Demeret C, Desaintes C, Yaniv M, Thierry F (1997) Different mechanisms contribute to the E2-mediated transcriptional repression of human papillomavirus type 18 viral oncogenes. J Virol 71:9343–9349. https://doi.org/10.1128/JVI.71.12.9343-9349.1997

Desaintes C, Goyat S, Garbay S et al (1999) Papillomavirus E2 induces p53-independent apoptosis in HeLa cells. Oncogene 18:4538–4545. https://doi.org/10.1038/sj.onc.1202818

Di Domenico F, Foppoli C, Blarzino C et al (2009) Expression of human papilloma virus type 16 E5 protein in amelanotic melanoma cells regulates endo-cellular pH and restores tyrosinase activity. J Exp Clin Cancer Res 28:4. https://doi.org/10.1186/1756-9966-28-4

DiGiuseppe S, Luszczek W, Keiffer TR et al (2016) Incoming human papillomavirus type 16 genome resides in a vesicular compartment throughout mitosis. Proc Natl Acad Sci 113: 6289–6294. https://doi.org/10.1073/pnas.1600638113

Dollard SC, Wilson JL, Demeter LM et al (1992) Production of human papillomavirus and modulation of the infectious program in epithelial raft cultures. OFF Genes Dev 6:1131–1142. https://doi.org/10.1101/gad.6.7.1131

Doorbar J (2013) The E4 protein; structure, function and patterns of expression. Virology 445:80–98. https://doi.org/10.1016/j.virol.2013.07.008

Doorbar J, Quint W, Banks L et al (2012) The biology and life-cycle of human papillomaviruses. Vaccine 30:F55–F70. https://doi.org/10.1016/j.vaccine.2012.06.083

Dziduszko A, Ozbun MA (2013) Annexin A2 and S100A10 regulate human papillomavirus type 16 entry and intracellular trafficking in human keratinocytes. J Virol 87:7502–7515. https://doi.org/10.1128/JVI.00519-13

Gillitzer E, Chen G, Stenlund A (2000) Separate domains in E1 and E2 proteins serve architectural and productive roles for cooperative DNA binding. EMBO J 19:3069–3079. https://doi.org/10.1093/emboj/19.12.3069

Griffin LM, Cicchini L, Pyeon D (2013) Human papillomavirus infection is inhibited by host autophagy in primary human keratinocytes. Virology 437:12–19. https://doi.org/10.1016/j.virol.2012.12.004

Guion L, Bienkowska-Haba M, DiGiuseppe S et al (2019) PML nuclear body-residing proteins sequentially associate with HPV genome after infectious nuclear delivery. PLoS Pathog 15: e1007590. https://doi.org/10.1371/journal.ppat.1007590

Harwood MC, Dupzyk AJ, Inoue T et al (2020) p120 catenin recruits HPV to γ-secretase to promote virus infection. PLoS Pathog 16:e1008946. https://doi.org/10.1371/journal.ppat.1008946

Mantovani F, Banks L (2001) The human papillomavirus E6 protein and its contribution to malignant progression. Oncogene 20:7874–7887. https://doi.org/10.1038/sj.onc.1204869

McBride AA (2013) The papillomavirus E2 proteins. Virology 445:57–79. https://doi.org/10.1016/j.virol.2013.06.006

McMurray HR, Nguyen D, Westbrook TF, McAnce DJ (2001) Biology of human papillomaviruses. Int J Exp Pathol 82:15–33. https://doi.org/10.1046/j.1365-2613.2001.00177.x

Meyers C, Frattini MG, Hudson JB, Laimins LA (1992) Biosynthesis of human papillomavirus from a continuous cell line upon epithelial differentiation. Science 257:971–973. https://doi.org/10.1126/science.1323879

Moody C (2017) Mechanisms by which HPV induces a replication competent environment in differentiating keratinocytes. Viruses 9:261. https://doi.org/10.3390/v9090261

Münger K, Basile JR, Duensing S et al (2001) Biological activities and molecular targets of the human papillomavirus E7 oncoprotein. Oncogene 20:7888–7898. https://doi.org/10.1038/sj.onc.1204860

Oh ST, Kyo S, Laimins LA (2001) Telomerase activation by human papillomavirus type 16 E6 protein: induction of human telomerase reverse transcriptase expression through Myc and GC-rich Sp1 binding sites. J Virol 75:5559–5566. https://doi.org/10.1128/JVI.75.12.5559-5566.2001

Pim D, Broniarczyk J, Siddiqa A et al (2021) Human papillomavirus 16 L2 recruits both Retromer and retriever complexes during retrograde trafficking of the viral genome to the cell nucleus. J Virol 95:e02068–e02020. https://doi.org/10.1128/JVI.02068-20

Pyeon D, Pearce SM, Lank SM et al (2009) Establishment of human papillomavirus infection requires cell cycle progression. PLoS Pathog 5:e1000318. https://doi.org/10.1371/journal.ppat.1000318

Ribeiro AL, Caodaglio AS, Sichero L (2018) Regulation of HPV transcription. Clinics (Sao Paulo) 73:e486s. https://doi.org/10.6061/clinics/2018/e486s

Richards RM, Lowy DR, Schiller JT, Day PM (2006) Cleavage of the papillomavirus minor capsid protein, L2, at a furin consensus site is necessary for infection. Proc Natl Acad Sci U S A 103:1522–1527. https://doi.org/10.1073/pnas.0508815103

Roberts JN, Buck CB, Thompson CD et al (2007) Genital transmission of HPV in a mouse model is potentiated by nonoxynol-9 and inhibited by carrageenan. Nat Med 13:857–861. https://doi.org/10.1038/nm1598

Rubin MA, Kleter B, Zhou M et al (2001) Detection and typing of human papillomavirus DNA in penile carcinoma: evidence for multiple independent pathways of penile carcinogenesis. Am J Pathol 159:1211–1218. https://doi.org/10.1016/S0002-9440(10)62506-0

Sapp M, Day PM (2009) Structure, attachment and entry of polyoma- and papillomaviruses. Virology 384:400–409. https://doi.org/10.1016/j.virol.2008.12.022

Scheffer KD, Berditchevski F, Florin L (2014) The tetraspanin CD151 in papillomavirus infection. Viruses 6:893–908. https://doi.org/10.3390/v6020893

Selinka H-C, Giroglou T, Nowak T et al (2003) Further evidence that papillomavirus capsids exist in two distinct conformations. J Virol 77:12961–12967. https://doi.org/10.1128/JVI.77.24.12961-12967.2003

Shafti-Keramat S, Handisurya A, Kriehuber E et al (2003) Different heparan sulfate proteoglycans serve as cellular receptors for human papillomaviruses. J Virol 77:13125–13135. https://doi.org/10.1128/jvi.77.24.13125-13135.2003

Siddiqa A, Massimi P, Pim D et al (2018) Human papillomavirus 16 infection induces VAP-dependent endosomal Tubulation. J Virol 92:e01514–e01517. https://doi.org/10.1128/JVI.01514-17

Spoden G, Freitag K, Husmann M et al (2008) Clathrin- and caveolin-independent entry of human papillomavirus type 16--involvement of tetraspanin-enriched microdomains (TEMs). PLoS One 3:e3313. https://doi.org/10.1371/journal.pone.0003313

Surviladze Z, Sterk RT, DeHaro SA, Ozbun MA (2013) Cellular entry of human papillomavirus type 16 involves activation of the phosphatidylinositol 3-kinase/Akt/mTOR pathway and inhibition of autophagy. J Virol 87:2508–2517. https://doi.org/10.1128/JVI.02319-12

Thomas M, Pim D, Banks L (1999) The role of the E6-p53 interaction in the molecular pathogenesis of HPV. Oncogene 18:7690–7700. https://doi.org/10.1038/sj.onc.1202953

Venuti A, Paolini F, Nasir L et al (2011) Papillomavirus E5: the smallest oncoprotein with many functions. Mol Cancer 10:140. https://doi.org/10.1186/1476-4598-10-140

Wang JW, Roden RBS (2013) Virus-like particles for the prevention of human papillomavirus-associated malignancies. Expert Rev Vaccines 12:129–141. https://doi.org/10.1586/erv.12.151

Williams KJ, Fuki IV (1997) Cell-surface heparan sulfate proteoglycans: dynamic molecules mediating ligand catabolism. Curr Opin Lipidol 8:253–262. https://doi.org/10.1097/00041433-199710000-00003

Woodham AW, Silva DMD, Skeate JG et al (2012) The S100A10 subunit of the Annexin A2 Heterotetramer facilitates L2-mediated human papillomavirus infection. PLoS One 7:e43519. https://doi.org/10.1371/journal.pone.0043519

Xie J, Heim EN, Crite M, DiMaio D (2020) TBC1D5-catalyzed cycling of Rab7 is required for Retromer-mediated human papillomavirus trafficking during virus entry. Cell Rep 31:107750. https://doi.org/10.1016/j.celrep.2020.107750

zur Hausen H (2002) Papillomaviruses and cancer: from basic studies to clinical application. Nat Rev Cancer 2:342–350. https://doi.org/10.1038/nrc798

zur Hausen H (2009) Papillomaviruses in the causation of human cancers - a brief historical account. Virology 384:260–265. https://doi.org/10.1016/j.virol.2008.11.046

# Chapter 9
# Defining the Assembleome of the Respiratory Syncytial Virus

**Richard J. Sugrue and Boon Huan Tan**

**Abstract** During respiratory syncytial virus (RSV) particle assembly, the mature RSV particles form as filamentous projections on the surface of RSV-infected cells. The RSV assembly process occurs at the / on the cell surface that is modified by a virus infection, involving a combination of several different host cell factors and cellular processes. This induces changes in the lipid composition and properties of these lipid microdomains, and the virus-induced activation of associated Rho GTPase signaling networks drives the remodeling of the underlying filamentous actin (F-actin) cytoskeleton network. The modified sites that form on the surface of the infected cells form the nexus point for RSV assembly, and in this review chapter, they are referred to as the RSV assembleome. This is to distinguish these unique membrane microdomains that are formed during virus infection from the corresponding membrane microdomains that are present at the cell surface prior to infection. In this article, an overview of the current understanding of the processes that drive the formation of the assembleome during RSV particle assembly is given.

**Keywords** Respiratory syncytial virus · Virus assembly · Rac-1 protein · Lovastatin · Lipid raft · F-actin · Rho GTPase

---

R. J. Sugrue (✉)
School of Biological Sciences, Nanyang Technological University, Singapore, Republic of Singapore
e-mail: rjsugrue@ntu.edu.sg

B. H. Tan
LKC School of Medicine, Nanyang Technological University, Singapore, Republic of Singapore
e-mail: tboonhuan@ntu.edu.sg



S. Vijayakrishnan et al. (eds.), *Virus Infected Cells*, Subcellular Biochemistry 106,
https://doi.org/10.1007/978-3-031-40086-5_9

## Respiratory Syncytial Virus Infection Is a Global Health Problem

Respiratory syncytial virus (RSV) infections produce a range of clinical manifestations in humans, from relatively mild disease in the upper airway, to more severe infections that are associated with lower airway infection that can lead to viral pneumonia. RSV is a leading cause of viral pneumonia in young children worldwide, and lower respiratory tract infections in neonates and young children less than 5 years of age can be fatal. The virus is responsible for high levels of global infant morbidity and mortality, with many infant deaths occurring in mid and low-income counties (Nair et al. 2010; Li et al. 2022; Shi et al. 2017). Although young children are the traditional high-risk group for severe RSV infection, the elderly are also increasingly being recognized as an additional high-risk group that is prone to severe RSV infection (Bosco et al. 2021; Thompson et al. 2003). Reinfections with RSV continue throughout life, and the general clinical scenario has been worsened by the lack of a licensed vaccine to immunize young children, and other high-risk groups and the limited availability of cost-effective antiviral drugs. An improved understanding of both the human immunology associated with RSV infection and the biology of the RSV will aid in the development of new antiviral strategies to prevent and treat RSV infection. In this context, understanding the molecular events that lead to RSV particle assembly will aid in vaccine development and facilitate the development of new drugs that prevent RSV infection by, for example, blocking virus transmission. In this article, we review the current understanding of the RSV assembly process and the role that host cell factors play in this process.

## The Genetic Structure of RSV

The RSV is grouped with the *Pneumoviridae* family of viruses, and this family is divided into two subgenera, the *Orthopneumovirus* (e.g., human RSV) and the *Metapneomovirus* (e.g., human metapneumovirus). The RSV genome (vRNA) consists of a single-stranded RNA molecule of negative sense that is approximately 15 kDa in size. The RSV vRNA contains 10 virus genes that are arranged contiguously along the length of the vRNA, with the gene order from the 3′ start being 3′-NS1-NS2-N-P-M-SH-F-G-M2-L-5′ (Fig. 9.1a). Each gene in the vRNA is separated from the adjacent genes by genetic regulatory elements referred to as intergenic regions. The individual virus proteins that are known to be expressed from these genes can be grouped based on their location within mature virus particles and the role that they play during virus replication.

The SH, F, and G genes encode for the three virus integral membrane proteins, the small hydrophobic (SH) protein, the fusion (F) protein, and the attachment (G) protein, respectively. The G protein mediates cell attachment, while the F protein mediates membrane fusion during cell entry. The SH protein is not essential for virus

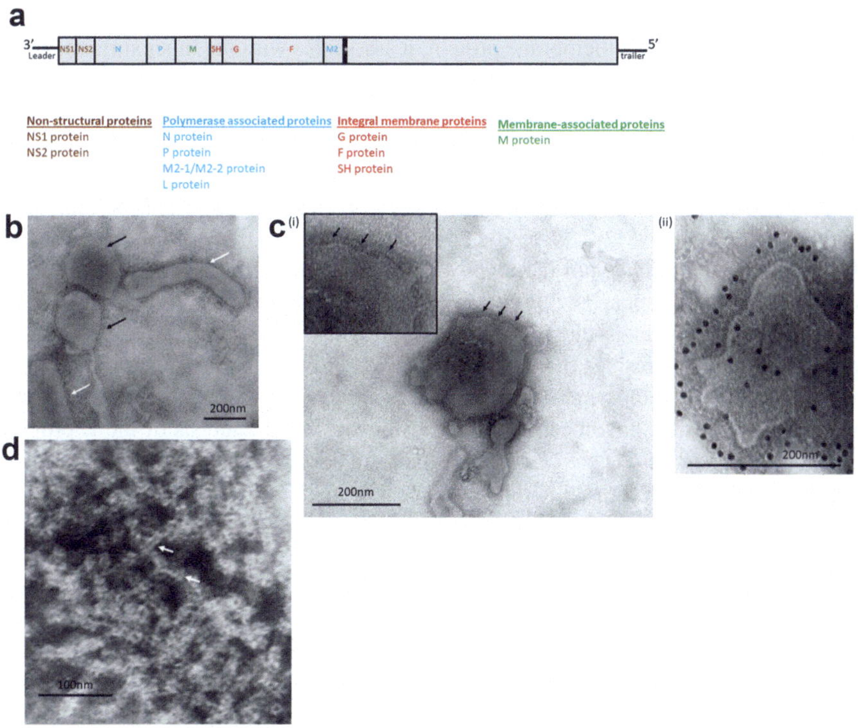


**Fig. 9.1** Organization of the respiratory syncytial virus (RSV) genome. (**a**) The virus genome (vRNA) from the beginning at the 3′ leader region to the 5′ trailer region is shown. The location of the genes that express the nonstructural proteins (brown), the proteins that form the ribonucleoprotein (RNP) complex (light blue), the membrane-associated protein (green), and the virus integral membrane proteins (red) are highlighted in the vRNA. The overlap between the M2 and L genes is also highlighted (*). The individual genes are drawn approximately to scale. Examination of purified RSV particles by transmission electron microscopy (**b**) reveals the appearance of pleomorphic-shaped RSV particles. The round (black arrows) and the more filamentous shaped (white arrows) virus particles are highlighted. (**c**) (**i**) At higher magnification, the location of the virus spike proteins (highlighted by black arrows) that protrude from the virus envelope. Inset is an enlarged image showing these structures. (**ii**) after immunostaining with anti-G. The black spots are the presence of 10 nm colloidal gold which indicates the location of the bound anti-G on the virus particles. (**d**) The appearance of the RNP (highlighted by white arrows) that are released from disrupted RSV particles that are present in a virus preparation

infection, and it is believed to play a role in immune evasion during the early stages of infection. Examination of cell-free RSV particles using electron microscopy to image negative-stained virus particles shows them to have a typical pleomorphic morphology, with a diameter ranging from 150 nm to upwards of 300 nm (Fig. 9.1b). The RSV particles are surrounded by a lipid envelope that is derived from the host cell in which the virus integral membrane proteins are inserted. Closer inspection of these particles by electron microscopy reveals that the G and F glycoproteins protrude from the virus envelope, and collectively they are referred to as the virus spike proteins (Fig. 9.1c). The M gene encodes the matrix (M) protein, and although itis not an integral virus membrane protein, it associates with the inner surface of the virus envelope (Money et al. 2009). It is therefore not detected when electron microscopy is used to image negative-stained intact virus particles.

The N, L, P, and M2 genes encode the nuclear (N) protein, the large (L) protein, the phospho (P) protein, and the M2-1 protein, respectively. The M2 gene contains two open reading frames (ORF), one ORF that encodes the M2-1 protein, and another ORF called M2-2 is predicted to encode for the M2-2 protein. The N, L, P, and M2-1 proteins and the vRNA form into a larger distinct helical structure called the virus nucleocapsid (NC) (Cao et al. 2020; Gilman et al. 2019; Decool et al. 2021). The N protein is the RNA binding protein that coats the vRNA, and the polymerase activity resides in the L protein. The P protein interacts directly with the L protein and functions as a co-factor to facilitate polymerase activity, and the precise binding sites have been identified using structural analysis of the L protein (Gilman et al. 2019). The M2-1 protein also functions as part of the polymerase complex, and its role in facilitating virus gene transcription is established (Fearns and Collins 1999). The fully functional RSV NC requires the presence of all these different factors to achieve a full activity in virus-infected cells. The virus NC can also be detected in virus preparations using electron microscopy once the virus particles have been disrupted to reveal the internal structures, and under these conditions, the NC appear as smaller but distinct helical arrays (Fig. 9.1d).

The RSV vRNA also contains the NS1 and NS2 genes that encode two nonstructural proteins called the NS1 and NS2 proteins, respectively, and these proteins are thought to play a role in immune evasion (Sedeyn et al. 2019). While there is some similarity in the gene order and encoded virus proteins in the RSV and human metapneumovirus (HMPV) genomes, there are also some distinct differences in the genome structure between the two viruses. In the HMPV genome the NS1 and NS2 genes are absent, and there is a genetic rearrangement in the position of the M2 gene to give 3′-N-P-M-F-M2-SH-G-L-5′(van den Hoogen et al. 2002). However, the HMPV expresses a set of structural proteins with similar properties to those displayed by the RSV structural proteins, and these HMPV proteins play analogous roles in the HMPV replication cycle.

## RSV Particle Morphogenesis at the Assembleome

The virus membrane-associated proteins are trafficked to the sites of RSV particle assembly at the plasma membrane, and their presence is an important determinant of the architecture of the mature RSV particles. In addition, these sites consist of an array of specific cellular factors that are required for RSV assembly and they also contribute to the formation of the virus envelope. The available evidence (discussed in this review) suggests that virus infection modifies the preexisting cell membrane microdomains that are ultimately used during RSV assembly, and it is proposed that these changes create unique membrane domains that are used for virus assembly. In order to distinguish these modified cell surface membrane sites in virus-infected cells from the corresponding sites in non-infected cells, in this review, the sites of RSV assembly that form during RSV infection are referred to as the RSV assembleome. The RSV assembleome can be considered to be analogous to the budozone that has been described in influenza virus-infected cells during influenza virus assembly (Schmitt and Lamb 2005).

## *Interactions between the Virus Membrane Structural Proteins at the Assembleome*

The G protein is a type II integral membrane protein, and its primary role is mediating the attachment of the virus to the host cell. The G protein is initially expressed as a polypeptide chain that is subsequently extensively modified by N-linked and O-linked glycosylation in the C-terminus ectodomain (Collins 1990; Collins and Mottet 1992). While the cytoplasmic N-terminus and the transmembrane domain are highly conserved between circulating viruses, the C-terminal ectodomain contains a region of greater sequence variability. The ectodomain contains two mucin-like domains that are separated by a substructure called the cysteine noose, which is a short sequence stabilized by two intra-chain disulfide bonds (Langedijk et al. 1996; 1998). The cell receptor binding sites for heparin sulfate (HS) (Feldman et al. 1999; Hallak et al. 2000a, b) and the CX3CR1 chemokine fractalkine receptor (Jeong et al. 2015; Johnson et al. 2015; Tripp et al. 2001) are located in the ectodomain, and these binding sites are conserved in non-tissue culture adapted circulating RSV stains (Kumaria et al. 2011).

The F protein is a type III integral membrane fusion protein that mediates the fusion of the virus envelope and cell membrane during virus entry into the host cell. The mature F protein exists as a homotrimer, and every single chain within the trimer is initially expressed as an inactive precursor (F0). The F0 is subsequently cleaved by furin into the F1 and F2 subunits (Anderson et al. 1992b) as it transits through the Golgi complex, and the available evidence suggests that the F0 protein is cleaved at two furin cleavage sites (Sugrue et al. 2001; Zimmer et al. 2001; González-Reyes et al. 2001). The F1 subunit contains the fusion peptide and heptad repeat (HR) regions, which play pivotal roles in mediating the membrane fusion process during cell entry of the virus (McLellan et al. 2013). The F1 and F2 subunits are covalently attached by inter-chain disulfide bridges (Scheid and Choppin 1977), and both subunits undergo N-linked glycosylation (Leemans et al. 2019; Rixon et al. 2002; McDonald and Sugrue 2007; Olmsted et al. 1986).

In RSV-infected cells, the SH protein is expressed as a small pentameric trans-membrane protein (Collins and Mottet 1993; Rixon et al. 2005), and the presence of nonglycosylated, N-linked glycosylated and polylactosaminoglycan-modified SH protein species can be detected (Anderson et al. 1992a; Olmsted and Collins 1989). Although the functional significance of the different SH protein species is still currently unclear, the SH protein does not appear to be required for establishing a virus infection in permissive cell lines (Bukreyev et al. 1997). A role for the SH protein viroporin activity in mediating immune evasion during the early stages of infection has been proposed (Triantafilou et al. 2013; Russell et al. 2015; Gan et al. 2008; Bukreyev et al. 1997).

The virus membrane-associated proteins can be trafficked to the plasma membrane independently of each other, but the existence of larger protein complexes involving these proteins within the virus envelope is now becoming established. Evidence suggests that the F protein interacts with the G protein to form a larger protein complex on the surface of infected cells (Low et al. 2008). This protein complex is also detected on purified virus particles (Ravi, Iyer, and Sugrue, unpublished observations) and in VLPs that are formed by recombinant expression of the F and G proteins (Ravi et al. 2021). It has recently been suggested that the interaction between the F and G proteins may stabilize the F protein in its prefusion form, which may have implications for RSV vaccine design (Cullen et al. 2022). The molecular cues that lead to the conversion of the RSV F protein from its prefusion conformation into the fusogenic conformation during cell entry are currently unclear. In several other paramyxoviruses, the conversion of the corresponding fusion protein into its fusogenic form is initiated by the engagement of the attachment protein with its cell receptor (reviewed in (Azarm and Lee 2020; Navaratnarajah et al. 2020)). The identification of a larger protein complex involving the RSV F and G proteins on infectious virus particles suggests that a similar mechanism may also be employed during RSV cell entry, although this possibility will require further investigation.

An early study provided evidence for a functional association between F, G, and SH proteins, suggesting the existence of a single protein complex involving all three virus proteins that were relevant to membrane fusion (Heminway et al. 1994). There have been no subsequent studies describing this protein complex, suggesting that if such a protein complex exists in RSV-infected cells, it may only exist transiently. Although the SH protein localizes at lipid raft microdomains in the Golgi complex (Olmsted and Collins 1989; Rixon et al. 2004), a stable protein complex involving the SH and G proteins on the surface of RSV-infected cells has been described (Rixon et al. 2005; Low et al. 2008). The functional significance of this interaction is currently unclear, but the presence of this stable complex on the surface of infected cells and the presence of low levels of the SH protein in the tissue culture supernatant of RSV-infected cells (Olmsted and Collins 1989; Low et al. 2008) had suggested that the SH protein may be incorporated into the virus envelope. Recent direct evidence has demonstrated that the SH protein is transported into the lipid envelope of mature virus particles that form on virus-infected cells (Huong et al. 2023).

The M protein undergoes dimerization during RSV infection (Bajorek et al. 2014; Förster et al. 2015), and although it is not an integral membrane protein, this peripheral membrane protein interacts with the inner surface of the virus envelope (Money et al. 2009). The M protein is thought to play an important role in virus particle assembly, but during the early stages of the virus replication cycle, the M protein is also transported to the nucleus of infected cells. The functional significance of the nuclear localization of the M protein is less well defined, but it has been postulated to play an additional role in controlling host gene expression (Li et al. 2021). The M protein interacts with the virus envelope at its inner surface (Money et al. 2009), and an interaction between the M protein and the N-terminal cytoplasmic of the G protein domain has also been described (Ghildyal et al. 2005). The M protein also interacts with the NC via the M2-1 protein, suggesting that the M protein may inhibit the activities associated with the virus polymerase complex (Ghildyal et al. 2002; Li et al. 2008). In the context of RSV particle assembly, since the NC is located beneath the virus envelope in virus particles, in the mature virus particle the M protein may act as a stable link between the virus envelope (via the G protein) and the underlying virus NC (via the M2-1 protein).

## *RSV Particles Form as Filamentous Structures at the Assembleome on Infected Cells*

The infectious RSV particles form as distinct filamentous projections on the surface of RSV-infected cells (Roberts et al. 1995; Jeffree et al. 2003), and these filamentous projections are referred to as virus filaments (Fig. 9.2a). In RSV-infected cells, the nucleocapsids accumulate in the cytoplasm, and they form part of a second distinct structure called the cytoplasmic inclusion body (Fig. 9.2a). Although several virus proteins are located in these structures, the co-expression of the N and P proteins appears to be the minimum requirement for inclusion body formation (García et al. 1993). The inclusion bodies can be detected in close proximity to the location where the virus filaments form (Fig. 9.2b and c), and current evidence suggests that nucleocapsids are trafficked from the inclusion bodies to the site of RSV assembly (Santangelo et al. 2006; Santangelo and Bao 2007). This paradigm in nucleocapsid packaging is supported by the spatial proximity of the virus filaments and cytoplasmic inclusion bodies in virus-infected cells (Radhakrishnan et al. 2010). It has been proposed that the virus utilizes molecular motors to achieve the short-range cellular movement from the inclusion bodies to the site of virus assembly, where the virus filaments form (Santangelo and Bao 2007).

The larger virus filaments that form on the surface of infected cells can be clearly distinguished from the shorter microvilli that are present on the surface of non-infected cells. This can be seen by using high-resolution immuno-scanning electron microscopy (I-SEM) to examine the surface topology of non-infected and RSV-infected cells that allow detection of the immunolabelled virus filaments (Fig. 9.3a). Although these virus filaments do not conform to the pleomorphic morphology exhibited by cell-free RSV particles, there is now a general acceptance that these filamentous structures are mature infectious virus particles. The virus filaments are observed on almost all cell types that are permissive to RSV infection, including primary cell types and tissues that are representative of the upper airway (Jumat et al. 2015). Analysis of virus-infected cells using I-SEM has revealed a high density of G protein incorporation into the virus filament, and the distribution of the

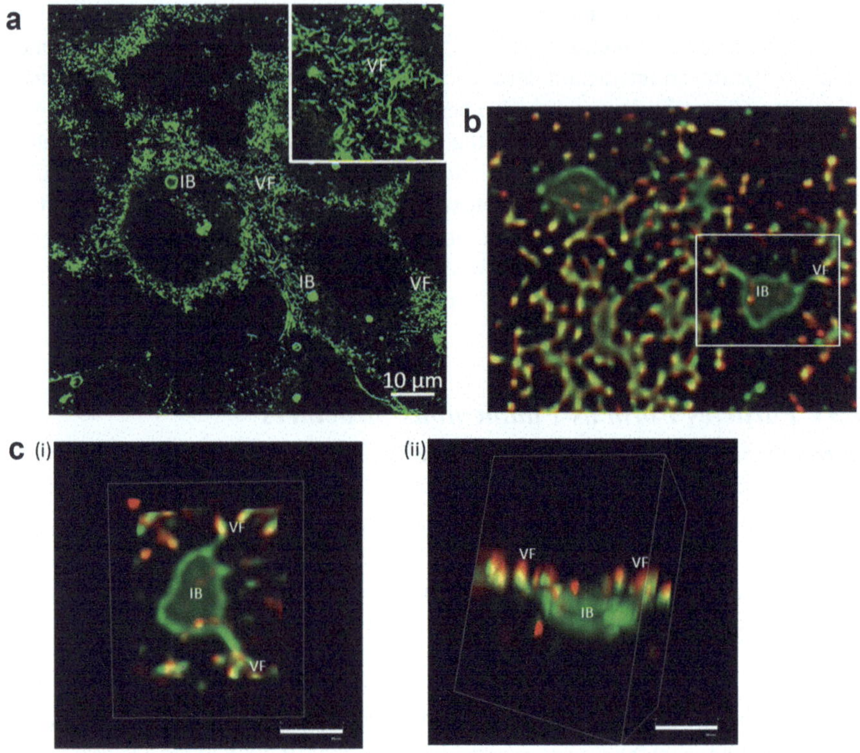


**Fig. 9.2** The presence of inclusion bodies and virus filaments in respiratory syncytial virus (RSV)-infected cells. (**a**) At 20 hrs post-infection RSV-infected cells were labeled using an anti-N antibody. These cells were examined in a Zeiss LSM 510 scanning confocal microscope to visualize the virus filament (VF) and cytoplasmic inclusion body (IB). Inset is an enlarged segment of the image showing the presence of the virus filaments. (**b** and **c**) RSV-infected cells were labeled using antibodies to the P protein (green) and F protein (red) and an image series was obtained from the same cell at different focal planes in the Z-axis using confocal microscopy, and the data were processed into three-dimensional representations. (**b**) It is a low-magnification image showing a region near the surface of the cell viewed from above the cell. The IB highlighted by the open white box is viewed at higher magnification from (**c** (**i**)) above and (**c** (**ii**)) in cross-section. While only the IB is stained with anti-P, the VF is labeled by both anti-P and anti-F, and co-staining is shown by the yellow staining

G protein staining has suggested that the G protein is closely packed along the length of the virus filament (Fig. 9.3b). This is supported by more recent evidence employing high-resolution imaging to examine the spatial distribution of the virus glycoproteins in the envelope of the virus filaments (Conley et al. 2022). In cells infected with the closely related HMPV, the infectious virus particles also form on the surface of infected cells with a similar filamentous morphology (Jumat et al. 2014). This suggests a similar mechanism of virus particle assembly for both RSV and HMPV, and further suggests that virus filament formation may be a common feature of the virus assembly process in the *Pneumoviridae* family of viruses.

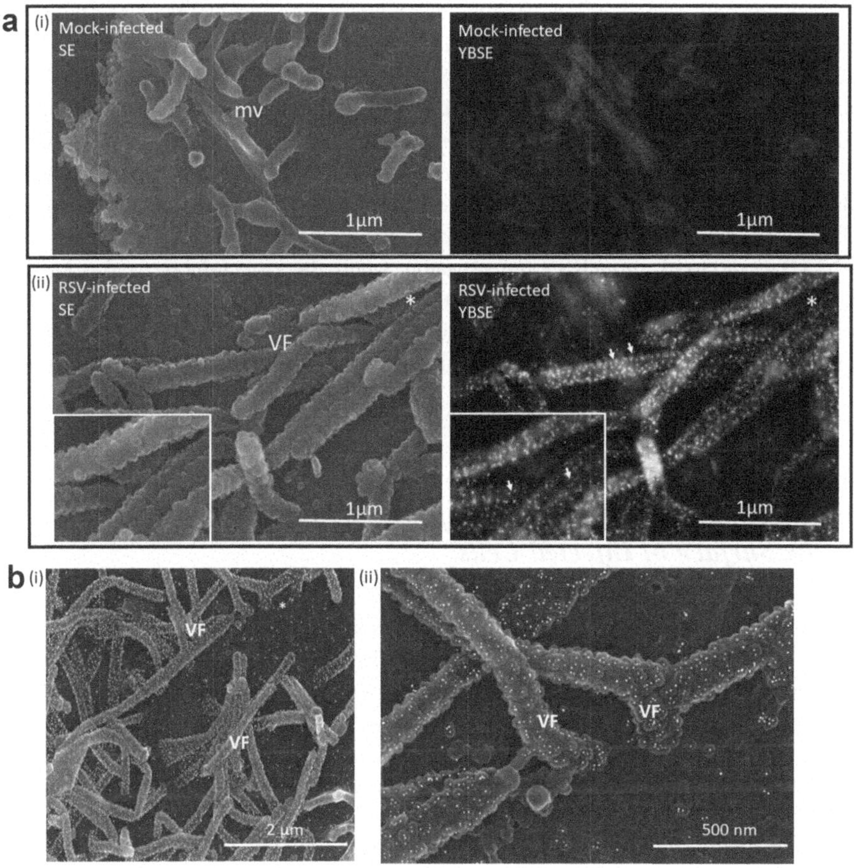


**Fig. 9.3** The Surface topology of respiratory syncytial virus (RSV)-infected cells revealed by immune-scanning electron microscopy. (**a**) (**i**) Mock-infected cells (non-infected cells) and (**ii**) RSV-infected cells were labeled with an antibody against the G protein and a second antibody conjugated to 10 nm colloidal gold. The cells were then examined using a Hitachi F4700 scanning electron microscope. The surface topology of the cell surface was imaged using a secondary emitter (SE) detector, and the presence of the bound antibody (and gold particles) using the YAG back scatter (YBSE) detector. The colloidal gold is visualized as small white spots in the images (highlighted by short white arrows) obtained using the YBSE detector. Inset in (**a**) (**ii**) image is an enlarged segment of the image taken obtained from the region indicated by *. (**b**) (**i**) Image was taken from RSV-infected cells where the images using the SE and YBSE detectors in the same field of view are superimposed. (**ii**) is a higher magnification image taken from the area (*) highlighted in (**i**). The microvilli (mv) and virus filaments (VF) are indicated

A low multiplicity of infection model was previously used to examine RSV multiple cycle infection in several permissive cell lines, and using this infection model, two distinct phases during multiple cycle RSV infection were identified (Huong et al. 2016). An earlier phase involved direct and localized cell-to-cell virus transmission in the cell monolayer, during which the virus infectivity remained exclusively cell-associated. A second phase was identified at the later stages of infection, during which the localized transmission continued, but it was characterized by virus-induced changes in cell physiology that correlated with the appearance of low levels of cell-free virus infectivity. Experimental conditions that inhibited virus filament formation led to an inhibition of both this localized virus transmission and the levels of recoverable virus infectivity, providing evidence that the virus filaments played a direct role in mediating this localized cell-to-cell transmission. This phenomenon was also observed when using the human nasal epithelial cell (hNEC) model to examine RSV transmission, where the virus infectivity was largely cell-associated. Low levels of cell-free infectivity that were shed into the overlying mucus in the hNEC model were associated with changes in the physiology of the infected ciliated cells at the later stages of infection (Huong et al. 2018).

## *Warning: Processing-instruction not allowed here!!!The Assembleome Forms at Lipid Raft Microdomains on the Surface of Infected Cells*

Specialized lipid domains that are referred to as lipid raft microdomains are present in the plasma membrane of many different cell types. Lipid raft microdomains are characterized by the increased concentration of specific lipid classes, which include cholesterol, sphingolipids such as sphingomyelin, and glycosphingolipid GM1. These lipid classes concentrate in the plasma membrane, and they provide localized stabilized membrane microdomains within the more mobile and fluid bulk lipid membrane (Sezgin et al. 2017). Specific proteins are also known to be trafficked into the lipid raft domains, and several of these raft-associated proteins play a role in mediating cell signaling events in the cell, for example, Rho GTPases. Since viruses are large multicomponent biological structures, these lipid raft microdomains are ideal platforms where the different cellular activities that are associated with the formation of viruses can be orchestrated. They thus provide ideal cellular platforms to drive the process of virus particle assembly for an array of different viruses, including RSV.

Several early studies employing high-resolution I-SEM revealed that the virus glycoproteins were trafficked into the virus filaments as they form at the cell surface (e.g., Jeffree et al. 2003). These data provided direct evidence that the virus structural proteins that ultimately were associated with the virus envelope were trafficked to specific microdomains on the plasma membrane. Early studies using imaging demonstrated that RSV particle assembly occurred at sites on the cell that were enriched in the established lipid raft marker protein caveolin-1 (cav-1), and that the cav-1 protein was incorporated into the virus filaments on the surface of virus-infected cells (Brown et al. 2002a). This was followed by the demonstration that the sites of RSV particle assembly could be identified using fluorescence probes that allow the detection of raft-associated lipid classes (Brown et al. 2002b; McCurdy and Graham 2003; Jeffree et al. 2003). Lipid raft microdomains can be experientially distinguished from the bulk lipid membrane based on their insolubility in nonionic detergents such as Triton X100 at 4 °C, giving rise to an operational definition of lipid rafts as detergent-resistant membranes (DRM). The imaging analayses were further supported by using this biochemical approach that demonstrated that in RSV-infected cells the virus structural proteins were found to partition into the DRM fraction (Brown et al. 2004; McCurdy and Graham 2003; McDonald et al. 2004; Marty et al. 2004). RSV particle assembly is now established to occur at lipid raft microdomains on the surface of virus-infected cells. In this context, drugs that solubilize membrane-associated cholesterol and destabilize lipid raft microdomains also lead to the inhibition of RSV particle assembly (Yeo et al. 2009). However, a detailed comparison of the lipid composition of lipid rafts isolated from non-infected and RSV-infected cells indicated that virus infection-induced changes in their lipid composition (Yeo et al. 2009). These changes in the lipid composition would be expected to alter the biological properties of these lipid raft microdomains in virus-infected cells.

The presence of raft-associated cellular proteins such as the cav-1 protein in the envelope of RSV particles suggested that the lipid envelope of RSV can be considered to have raft-like properties. This suggestion was supported by the detailed analysis of the lipid profile of isolated purified RSV particles which has revealed a high level of lipid classes that are associated with the formation of lipid raft membrane domains were also present in the virus envelope (Chan, Wenk and Sugrue, unpublished observations). An additional consequence of the use of lipid raft microdomains during RSV assembly is that specific raft-associated cell factors that have biological activities of clinical relevance may also be incorporated into the virus envelope, and which may potentially impart important functional properties to the virus envelope. In this context, the raft-associated CD55 and CD59 complement regulatory proteins were both detected in the envelope of mature virus particles (Brown et al. 2004). It was originally proposed that their presence may have a clinical context, since the presence of these complement regulatory proteins in the virus envelope could potentially impact on how the viruses interact with the complement system. This suggestion is supported by more recent research that has confirmed the association of CD55 protein with mature RSV particles, and that has assessed the consequence of virus-associated CD55 protein in relation to opsonization by the complement system (Kuppan et al. 2021).

Although differential detergent solubility at 4 °C is a robust operational definition to identify proteins that are associated with lipid raft microdomains, this can refer to a broad range of different types of lipid raft microdomain that are only defined by their detergent solubility properties. Consequently, in many cases the specific type of lipid raft microdomain under investigation that is isolated using detergent treatment remains uncharacterized. Caveolae are established as a specific class of lipid raft microdomain that plays a variety of different roles in cell homeostasis. Although several cellular proteins are involved in the formation of the caveolae coat complex (Ludwig et al. 2013), the cav-1 protein is a major structural determinant of caveolae. Therefore, in the earlier observations that described the presence of the cav-1 protein in the envelope of virus filaments it was unclear if the cav-1 protein was specifically recruited to the site of RSV assembly, or if the presence of cav-1 protein at these sites indicated the involvement of caveolae in the virus assembly process. More recent work has identified other caveolae-associated proteins (e.g., the cavins) at the site of RSV assembly, which confirmed the involvement of caveolae in the RSV assembly process (Ludwig et al. 2017). The distribution of the cav-1 protein in the virus particle was determined using advanced imaging techniques, and its presence in the virus envelope involved an interaction with the RSV G protein. At present the effect of cav-1 protein and the various cavin proteins in the virus envelope is unclear, but since these proteins are major structural determinants of the caveolae, it would be expected that their presence could impart specific biological properties to the RSV envelope (e.g., an increase in membrane rigidity). Although evidence for the involvement of lipid raft microdomains in the assembly of infectious HMPV particles has been presented (Jumat et al. 2014), it is currently unclear if caveolae are also involved in the HMPV assembly process.

Recombinant expression of combinations of the virus membrane-associated proteins is sufficient to lead to the formation of virus-like particles (VLPs) that display similar morphological features to the virus particles that form on RSV-infected cells. VLPs that resemble virus filaments can be detected on cells expressing only the recombinant G protein (Ravi et al. 2021), suggesting that the G protein may play an important role in virus particle assembly. The recombinant G protein is trafficked to lipid raft membrane domains that contain both the actin and filamin-1 proteins, host cell factors that are shown to be present in virus filaments isolated from infected cells (Radhakrishnan et al. 2010). This provides evidence that the virus glycoproteins contain inherent trafficking signals that allow their transport to the site of virus assembly. Although these trafficking signals have not been clearly defined, the F and G proteins are both modified by palmitoylation (Arumugham et al. 1989; Collins and Mottet 1992), and the addition of palmitoyl moieties to these proteins may facilitate their interaction with lipid raft microdomains. Evidence also suggests that an interaction between the M protein and the RSV glycoproteins may traffic the M protein into the DRM fraction (Henderson et al. 2002). This further suggests that during virus infection, the trafficking of the M protein to the assembleome may involve an interaction between the M protein and the virus glycoproteins (e.g. the G protein). In this context, stable interactions between the G protein and the other virus membrane-associated proteins and cell proteins at the assembleome may stabilize their interaction within the lipid envelope of the virus filaments.

## *Filamentous Actin (F-Actin) Remodeling at Lipid Raft Microdomains at the Assembleome Drives Virus Filament Formation*

Microvilli and similar cell surface projections are stabilized by the filamentous form of actin (F-actin) that forms into F-actin bundles, and on the surface of non-infected cells, the microvilli are the dominant topographical feature. However, on the surface of RSV-infected cells, there is a reduction in the levels of the microvilli, and the virus filaments become the dominant surface feature detected (Jeffree et al. 2003). Indeed, on RSV-infected cells the microvilli appear to be largely replaced by the virus filaments, and since the microvilli are stabilized by F-actin bundles, this change in surface topology is consistent with virus infection leading to remodeling of the cell surface of infected cells. It is currently unclear what effect the trafficking of the virus membrane proteins to the sites of virus assembly has on the biological properties of the microvilli and other surface projections that are initially present on the cell surface. However, virus-induced changes in cell surface topology suggest that the biological properties of the sites on the cell membrane that are used for virus particle assembly are modified via structural changes to the underlying the F-actin network. Early studies on RSV-infected cells have implicated the cortical cytoskeleton as playing a role in the morphogenesis of the mature RSV particles (Ulloa et al. 1998; Burke et al. 1998). The role of F-actin in RSV morphogenesis was supported by high-resolution imaging that has demonstrated a role for the F-actin network in the process of virus assembly (Jeffree et al. 2007), and in this context, the M protein has also been shown to directly interact with actin in RSV-infected cells (Shahriari et al. 2018). In addition, actin and several actin-binding proteins have been detected in purified virus preparations and virus filaments, and among these was the actin-binding protein filamin-1(Radhakrishnan et al. 2010; Ravi et al. 2021). The filamin-1 protein is an established actin-binding protein that interacts with the F-actin, and the filamin-1 protein is also able to cross-link F-actin bundles and facilitate F-actin remodeling (Popowicz et al. 2006). Collectively these observations by several groups have provided robust evidence for the role of the F-actin network in the morphogenesis of RSV particles.

The role of the F-actin network in regulating and orchestrating the clustering of lipid raft microdomains has been established (Chichili and Rodgers 2007). In the context of RSV particle assembly, the analysis of the lipid composition of lipid raft preparations isolated from RSV-infected cells has demonstrated that RSV infection led to the presence of increased levels of the signaling lipid phosphatidylinositol 4, 5-bisphosphate (PIP2) in the lipid raft membranes (Yeo et al. 2009). This event was consistent with earlier observations that RSV infection was associated with increased levels of phosphatidylinositol 3 kinase (PI3K) expression and PI3K activation during RSV infection (Thomas et al. 2002; Jeffree et al. 2007). The enzyme PI3K phosphorylates the substrate PIP2 to generate the product phosphatidylinositol-3, 4, 5-triphosphate (PIP3), and the presence of both PIP2 and PIP3 in inclusion bodies in RSV-infected cells suggests that these signalling lipids may have a role RSV particle assembly (Yeo et al. 2009). The cortical F-actin network can be remodeled by Rho GTPases, and the first evidence for the involvement of an activated rho GTPase in RSV particle assembly was the activation of the rhoA protein during RSV infection (Gower et al. 2005; Gower et al. 2001; McCurdy and Graham 2003). More recently, the rac1 protein has been shown to also play a major role in RSV particle assembly in virus-infected cells (Ravi et al. 2021; Malhi et al. 2021). Since the PI3K protein signaling pathway activates cellular proteins that in turn regulate the activity of the rac1 protein (Campa et al. 2015), this has suggested the contribution of a signaling pathway during RSV particle assembly that involves the PI3K and rac1 proteins. While the interaction between the F protein and the rhoA protein has been described (Gower et al. 2001), it is currently unclear if the rac1 protein similarly interacts with one or more virus proteins at the assembleome, and this will require further investigation.

Although the rac1 and rhoA proteins are distinct proteins, there is likely to be a degree of crosstalk between their respective signaling networks in the cell. Since inhibiting the activation of either protein leads to impaired RSV particle assembly (Gower et al. 2005; Jeffree et al. 2007), it is possible that the rhoA and rac1 proteins may be part of a larger signaling network that ultimately leads to RSV particle assembly, via the activation of common or related downstream effectors. Currently, the downstream effectors of these signaling networks that are activated during RSV assembly are still uncertain. Changes in F-actin staining and the altered membrane distribution of the filamin-1 protein during RSV infection support the role of F-actin remodeling in virus particle assembly (Ludwig et al. 2017; Ravi et al. 2021). The filamin-1 protein plays a role in the formation of the RSV particle architecture (Radhakrishnan et al. 2010), and a role for rac-1 protein in inducing F-actin remodeling by the filamin-1 protein has been described using other experimental systems (Bellanger et al. 2000; Ramírez-Ramírez et al. 2020). It is, therefore, possible that the rac-1 protein signaling pathway may also play a similar role by inducing structural changes in the F-actin network via filamin-1 protein. The actin-related protein 2/3 (Arp2/3) complex also plays an important role in mediating actin polymerization (Papalazarou and Machesky 2021), and the Arp2/3 complex facilitates RSV spread during the localized cell-to-cell transmission (Mehedi et al. 2016). The Arp2/3 complex is activated by upstream effectors such as the WAVE Regulatory Complex (WRC) (Papalazarou and Machesky 2021), and the WRC is in turn activated by the rac1 protein (Chen et al. 2017). This is consistent with the observations that inhibition of rac1 protein activation also inhibits localized RSV transmission in virus-infected cell monolayers (Huong et al. 2016). However, it remains to be ascertained by experimentation if during RSV infection the filamin-1 protein and the Arp2/3 complex function as downstream effectors of the activated rac1 protein during virus filament formation and virus transmission.

## *Trafficking of the rac1 and rhoA Proteins to the Assembleome Is Required for Virus Filament Formation*

Although the mevalonate pathway is established as a key metabolic pathway in the regulation of cholesterol biosynthesis, this metabolic pathway is also involved in the cellular process that provides the building blocks to make the isoprenoids that are used in protein prenylation. During protein prenylation, specific cytosolic proteins undergo a post-translation modification, whereby a small lipid molecule (e.g., geranylgeranyl isoprenoid) is attached to a c-terminal cysteine of the target protein via established consensus sequence motifs (reviewed in (Wang and Casey 2016)). The attachment of the lipid molecule to these normally cytosolic proteins provides a lipid anchor that allows their interaction with cellular membranes. The enzyme 3-hydroxy-3-methyl-glutaryl-CoA reductase (HMGCR) plays a role in regulating the activity of the mevalonate pathway, and the HMGCR protein can be inhibited using statin-based drugs such as lovastatin. As a consequence, these statin-based drugs inhibit both cholesterol biosynthesis and the formation of the lipid precursors that are involved in protein prenylation. Analysis of the changes in the host cell transcriptome that occur during the early stage of RSV infection indicated that prior to virus filament formation, there was an increased expression of several enzymes in the mevalonate pathway, including the HMGCR (Yeo et al. 2009). This indicated that prior to virus particle formation, virus infection induced the early increased expression of the mevalonate pathway, and since the HMGCR activity is transcriptionally regulated, this also suggested increases in the HMCGR activity. In this context, RSV filament formation is inhibited by lovastatin in a cholesterol-independent manner, suggesting that the involvement of the HMGCR in particle assembly was not related to cholesterol biosynthesis (Ravi et al. 2013; Gower and Graham 2001). Since lipid raft microdomains are composed of cholesterol, this also suggested that the inhibitory effect of lovastatin was not related to the destabilization of these membrane microdomains. This inhibition, however, could be reversed by providing exogenously added geranylgeranyl pyrophosphate (GGPP) to the cells, which is one of the isoprenoids that is generated via the mevalonate pathway (Malhi et al. 2021). This suggested that the antiviral effects of statins were related to the inhibition of the isoprenoid formation, and in this context, the membrane association of activated rhoA and rac1 proteins is mediated by protein prenylation (Wang and Casey 2016). Recent evidence has demonstrated that statins inhibit rac1 protein prenylation in RSV-infected cells (Malhi et al. 2021), suggesting that lovastatin, and other similar statin drugs, may inhibit virus filament formation by blocking the trafficking of the activated Rho GTPases to the cell membranes at the site of RSV particle assembly. Furthermore, lovastatin treatment inhibited localized RSV transmission (Gower and Graham 2001; Ravi et al. 2013), suggesting that F-actin remodeling via the membrane trafficking of Rho GTPases also plays a role in facilitating localized RSV transmission. Collectively, these data provide a functional link between RSV-induced changes in the expression of at least one cellular metabolic pathway and the trafficking of Rho GTPases to the site of RSV particle assembly during RSV infection.

In cells that produce RSV VLPs, the rac1 protein was also been shown to be present at the lipid raft microdomains where the VLPs formed (Ravi et al. 2021). It is suggested that the presence of the rac1 protein at the lipid raft microdomains that are used during virus particle assembly may be required to modulate F-actin remodeling at these sites during the process of virus filament formation. Although the RSV VLPs resemble virus filaments, these VLPs are not the same as the virus filaments that form on infected cells. It is proposed that other aspects of RSV infection would be required to initiate cell signaling events that lead to the activation of the Rho GTPases at these sites. These signaling events would ultimately lead to the modification of these cellular locations on the plasma membrane and lead to the formation of the RSV assembleome prior to virus filament formation.

## Future Perspectives

The virus filaments are a robust feature of RSV infection in permissive cells, and their presence in primary airway cells (e.g. hNEC) has reinforced their clinical relevance. Research efforts are being undertaken to understand both the structure of the virus filaments and the cellular processes that lead to their formation in RSV-infected cells. This information should facilitate the development of novel antivirus drugs that can block RSV particle assembly and prevent infection, and this information should also aid in the development of new RSV vaccine candidates, for example, VLPs. The work of several laboratories described in this review has highlighted the important role that the host cell factors play during RSV particle assembly. Several host cell factors identified at the assembleome also contribute to the formation of infectious and mature RSV particles. There have been several high-resolution structural analyses performed on cell-free RSV particles using cryo-electron microscopy (e.g., Liljeroos et al. 2013). However, in these analyses, it has been assumed that all the data that is revealed in these images of the purified virus particles is of virus origin. The research undertaken by several groups clearly indicates that host cell factors are also incorporated into the virus particles, but these non-virus factors are not featured in these high-resolution structures. It would be expected that these host cell factors will have an influence on the spatial organization of the virus proteins in the virus envelope, and consequently, may influence the architecture of the infectious mature virus particle. The challenge to understanding the complete architecture of RSV particles will be to identify and locate these host cell factors in the virus particles. This could be achieved, for example, by using a combination of high-resolution imaging and specialized labeling procedures, similar to the procedures that have been described for locating the presence of the cav-1 protein in the RSV filaments (Ludwig et al. 2017).

The collective body of evidence has established that virus filament formation involves the activation of cell signaling networks that are regulated by the rac1 and rhoA proteins. However, these proteins are upstream members of a larger signaling network, and future work will need to focus on identifying the important downstream proteins that are acted on by these signaling pathways during RSV infection. It is possible that during virus particle morphogenesis, these downstream effectors may act on one or more virus or host cell factors, and as a consequence, the biological properties of these factors may be modified. The identification of downstream effectors that are activated by the rac1 and rhoA proteins during RSV infection will be required to obtain a complete understanding of the molecular events that lead to RSV particle assembly. It may also be possible that the identification of these down steam effectors could also be used to better understand the process of virus particle assembly in other related paramyxoviruses.

Lastly, the body of work on RSV assembly has utilized cell systems that are permissive to RSV infection. This work has provided a proof-of-concept that cell factors play important roles in the RSV assembly process, and that several of the host cell factors identified are potentially druggable. This opens the possibility of using drug repurposing to target these cell factors using established drugs with existing safety profiles. This could shorten the time to administer effective and cost-effective treatments for RSV infection, particularly in low-income counties where many RSV-related childhood deaths occur (Nair et al. 2010). The future challenges will be to determine if these host cell factors that have been identified using permissive cells can form the basis for cost-effective antivirus strategies for use in patients. For example, statin-based drugs have known safety profiles and are used by many people to regulate blood cholesterol levels, such as for example, lovastatin (Mevacor®). These drugs are also effective at inhibiting RSV infection in both permissive cells and in animal models of infection, and the mechanism of its antiviral action is now being elucidated. Such drugs could be used to treat RSV infection, and it is presumed that some of the adverse side effects associated with the long-term use of statins will not be encountered during their short time use to treat RSV infection. It will therefore be interesting to determine if these and other cost-effective drugs that block RSV assembly can be translated from the bench to the bedside to control RSV infection in human patients.

**Acknowledgments** We would also wish to acknowledge previous group members at the MRC Virology Unit in Glasgow and at the Nanyang Technological University in Singapore, including previous colleagues and collaborators Gaie Brown and Jim Aitkin, (Institute of Virology, Glasgow UK) and Chris Jeffree (EMCB, Biological Sciences EM Facility, University of Edinburgh UK) for light and electron microscopy imaging of RSV. In addition, we acknowledge previous funding support to RJS from the Medical Research Council (UK), National Medical Research Council (Singapore), Singapore-MIT Alliance for Research and Technology (Singapore), and the Ministry of Education (Singapore).

# References

Anderson K, King AM, Lerch RA, Wertz GW (1992a) Polylactosaminoglycan modification of the respiratory syncytial virus small hydrophobic (SH) protein: a conserved feature among human and bovine respiratory syncytial viruses. Virology 191:417–430

Anderson K, Stott EJ, Wertz GW (1992b) Intracellular processing of the human respiratory syncytial virus fusion glycoprotein: amino acid substitutions affecting folding, transport and cleavage. J Gen Virol 73(Pt 5):1177–1188

Arumugham RG, Seid RC Jr, Doyle S, Hildreth SW, Paradiso PR (1989) Fatty acid acylation of the fusion glycoprotein of human respiratory syncytial virus. J Biol Chem 264:10339–10342

Azarm KD, Lee B (2020) Differential features of fusion activation within the Paramyxoviridae. Viruses 12(2):161

Bajorek M, Caly L, Tran KC, Maertens GN, Tripp RA, Bacharach E, Teng MN, Ghildyal R, Jans DA (2014) The Thr205 phosphorylation site within respiratory syncytial virus matrix (M) protein modulates M oligomerization and virus production. J Virol 88:6380–6393

Bellanger JM, Astier C, Sardet C, Ohta Y, Stossel TP, Debant A (2000) The Rac1- and RhoG-specific GEF domain of trio targets filamin to remodel cytoskeletal actin. Nat Cell Biol 2:888–892

Bosco E, van Aalst R, Mcconeghy KW, Silva J, Moyo P, Eliot MN, Chit A, Gravenstein S, Zullo AR (2021) Estimated cardiorespiratory hospitalizations attributable to influenza and respiratory syncytial virus among long-term care facility residents. JAMA Netw Open 4:e2111806

Brown G, Aitken J, Rixon HWM, Sugrue RJ (2002a) Caveolin-1 is incorporated into mature respiratory syncytial virus particles during virus assembly on the surface of virus-infected cells. J Gen Virol 83:611–621

Brown G, Rixon HWM, Sugrue RJ (2002b) Respiratory syncytial virus assembly occurs in GM1-rich regions of the host-cell membrane and alters the cellular distribution of tyrosine phosphorylated caveolin-1. J Gen Virol 83:1841–1850

Brown G, Jeffree CE, McDonald T, Rixon HW, Aitken JD, Sugrue RJ (2004) Analysis of the interaction between respiratory syncytial virus and lipid-rafts in Hep2 cells during infection. Virology 327:175–185

Bukreyev A, Whitehead SS, Murphy BR, Collins PL (1997) Recombinant respiratory syncytial virus from which the entire SH gene has been deleted grows efficiently in cell culture and exhibits site-specific attenuation in the respiratory tract of the mouse. J Virol 71:8973–8982

Burke E, Dupuy L, Wall C, Barik S (1998) Role of cellular actin in the gene expression and morphogenesis of human respiratory syncytial virus. Virology 252:137–148

Campa CC, Ciraolo E, Ghigo A, Germena G, Hirsch E (2015) Crossroads of PI3K and Rac pathways. Small GTPases 6:71–80

Cao D, Gao Y, Roesler C, Rice S, D'cunha P, Zhuang L, Slack J, Domke M, Antonova A, Romanelli S, Keating S, Forero G, Juneja P, Liang B (2020) Cryo-EM structure of the respiratory syncytial virus RNA polymerase. Nat Commun 11:368

Chen B, Chou HT, Brautigam CA, Xing W, Yang S, Henry L, Doolittle LK, Walz T, Rosen MK (2017) Rac1 GTPase activates the WAVE regulatory complex through two distinct binding sites. elife 6:e29795

Chichili GR, Rodgers W (2007) Clustering of membrane raft proteins by the actin cytoskeleton. J Biol Chem 282:36682–36691

Collins PL (1990) O glycosylation of glycoprotein G of human respiratory syncytial virus is specified within the divergent ectodomain. J Virol 64:4007–4012

Collins PL, Mottet G (1992) Oligomerization and post-translational processing of glycoprotein G of human respiratory syncytial virus: altered O-glycosylation in the presence of brefeldin a. J Gen Virol 73(Pt 4):849–863

Collins PL, Mottet G (1993) Membrane orientation and oligomerization of the small hydrophobic protein of human respiratory syncytial virus. J Gen Virol 74(Pt 7):1445–1450

Conley MJ, Short JM, Burns AM, Streetley J, Hutchings J, Bakker SE, Power BJ, Jaffery H, Haney J, Zanetti G, Murcia PR, Stewart M, Fearns R, Vijayakrishnan S, Bhella D (2022) Helical ordering of envelope-associated proteins and glycoproteins in respiratory syncytial virus. EMBO J 41:e109728

Cullen LM, Luo B, Wen Z, Zang L, Durr E, Morrison TG (2022) The respiratory syncytial Virus G protein enhances the immune responses to the RSV F protein in an enveloped virus-like particle vaccine candidate. bioRxiv, 2022.09.12.507712

Decool H, Gonnin L, Gutsche I, Sizun C, Eléouët JF, Galloux M (2021) Interactions between the nucleoprotein and the phosphoprotein of Pneumoviruses: structural insight for rational Design of Antivirals. Viruses 13(12):2449

Fearns R, Collins PL (1999) Role of the M2-1 transcription antitermination protein of respiratory syncytial virus in sequential transcription. J Virol 73:5852–5864

Feldman SA, Hendry RM, Beeler JA (1999) Identification of a linear heparin binding domain for human respiratory syncytial virus attachment glycoprotein G. J Virol 73:6610–6617

Förster A, Maertens GN, Farrell PJ, Bajorek M (2015) Dimerization of matrix protein is required for budding of respiratory syncytial virus. J Virol 89:4624–4635

Gan SW, Ng L, Lin X, Gong X, Torres J (2008) Structure and ion channel activity of the human respiratory syncytial virus (hRSV) small hydrophobic protein transmembrane domain. Protein Sci 17:813–820

García J, García-Barreno B, Vivo A, Melero JA (1993) Cytoplasmic inclusions of respiratory syncytial virus-infected cells: formation of inclusion bodies in transfected cells that coexpress the nucleoprotein, the phosphoprotein, and the 22K protein. Virology 195:243–247

Ghildyal R, Mills J, Murray M, Vardaxis N, Meanger J (2002) Respiratory syncytial virus matrix protein associates with nucleocapsids in infected cells. J Gen Virol 83:753–757

Ghildyal R, Li D, Peroulis I, Shields B, Bardin PG, Teng MN, Collins PL, Meanger J, Mills J (2005) Interaction between the respiratory syncytial virus G glycoprotein cytoplasmic domain and the matrix protein. J Gen Virol 86:1879–1884

Gilman MSA, Liu C, Fung A, Behera I, Jordan P, Rigaux P, Ysebaert N, Tcherniuk S, Sourimant J, Eléouët JF, Sutto-Ortiz P, Decroly E, Roymans D, Jin Z, Mclellan JS (2019) Structure of the respiratory syncytial virus polymerase complex. Cell 179:193–204.e14

González-Reyes L, Ruiz-Argüello MB, García-Barreno B, Calder L, López JA, Albar JP, Skehel JJ, Wiley DC, Melero JA (2001) Cleavage of the human respiratory syncytial virus fusion protein at two distinct sites is required for activation of membrane fusion. Proc Natl Acad Sci U S A 98: 9859–9864

Gower TL, Graham BS (2001) Antiviral activity of lovastatin against respiratory syncytial virus in vivo and in vitro. Antimicrob Agents Chemother 45:1231–1237

Gower TL, Peeples ME, Collins PL, Graham BS (2001) RhoA is activated during respiratory syncytial virus infection. Virology 283:188–196

Gower TL, Pastey MK, Peeples ME, Collins PL, McCurdy LH, Hart TK, Guth A, Johnson TR, Graham BS (2005) RhoA signaling is required for respiratory syncytial virus-induced syncytium formation and filamentous virion morphology. J Virol 79:5326–5336

Hallak LK, Collins PL, Knudson W, Peeples ME (2000a) Iduronic acid-containing glycosaminoglycans on target cells are required for efficient respiratory syncytial virus infection. Virology 271:264–275

Hallak LK, Spillmann D, Collins PL, Peeples ME (2000b) Glycosaminoglycan sulfation requirements for respiratory syncytial virus infection. J Virol 74:10508–10513

Heminway BR, Yu Y, Tanaka Y, Perrine KG, Gustafson E, Bernstein JM, Galinski MS (1994) Analysis of respiratory syncytial virus F, G, and SH proteins in cell fusion. Virology 200:801–805

Henderson G, Murray J, Yeo RP (2002) Sorting of the respiratory syncytial virus matrix protein into detergent-resistant structures is dependent on cell-surface expression of the glycoproteins. Virology 300:244–254

Huong TN, Iyer Ravi L, Tan BH, Sugrue RJ (2016) Evidence for a biphasic mode of respiratory syncytial virus transmission in permissive HEp2 cell monolayers. Virol J 13:12

Huong TN, Yan Y, Jumat MR, Lui J, Tan BH, Wang Y, Sugrue RJ (2018) A sustained antiviral host response in respiratory syncytial virus infected human nasal epithelium does not prevent progeny virus production. Virology 521:20–32

Huong TN, Iyer LR, Lui J, DY Wang, Tan BH, Sugrue RJ (2023) The respiratory syncytial virus SH protein is incorporated into infectious virus particles that form on virus-infected cells. Virology 580:28–40. https://doi.org/10.1016/j.virol.2023.01.013

Jeffree CE, Rixon HW, Brown G, Aitken J, Sugrue RJ (2003) Distribution of the attachment (G) glycoprotein and GM1 within the envelope of mature respiratory syncytial virus filaments revealed using field emission scanning electron microscopy. Virology 306:254–267

Jeffree CE, Brown G, Aitken J, Su-Yin DY, Tan BH, Sugrue RJ (2007) Ultrastructural analysis of the interaction between F-actin and respiratory syncytial virus during virus assembly. Virology 369:309–323

Jeong KI, Piepenhagen PA, Kishko M, Dinapoli JM, Groppo RP, Zhang L, Almond J, Kleanthous H, Delagrave S, Parrington M (2015) CX3CR1 is expressed in differentiated human ciliated airway cells and co-localizes with respiratory syncytial virus on cilia in a G protein-dependent manner. PLoS One 10:e0130517

Johnson SM, Mcnally BA, Ioannidis I, Flano E, Teng MN, Oomens AG, Walsh EE, Peeples ME (2015) Respiratory syncytial virus uses CX3CR1 as a receptor on primary human airway epithelial cultures. PLoS Pathog 11:e1005318

Jumat MR, Nguyen Huong T, Wong P, Loo LH, Tan BH, Fenwick F, Toms GL, Sugrue RJ (2014) Imaging analysis of human metapneumovirus-infected cells provides evidence for the involvement of F-actin and the raft-lipid microdomains in virus morphogenesis. Virol J 11:198

Jumat MR, Yan Y, Ravi LI, Wong P, Huong TN, Li C, Tan BH, Wang Y, Sugrue RJ (2015) Morphogenesis of respiratory syncytial virus in human primary nasal ciliated epithelial cells occurs at surface membrane microdomains that are distinct from cilia. Virology 484:395–411

Kumaria R, Iyer LR, Hibberd ML, Simões EA, Sugrue RJ (2011) Whole genome characterization of non-tissue culture adapted HRSV strains in severely infected children. Virol J 8:372

Kuppan JP, Mitrovich MD, Vahey MD (2021) A morphological transformation in respiratory syncytial virus leads to enhanced complement deposition. elife 10:e70575

Langedijk JP, Schaaper WM, Meloen RH, Van Oirschot JT (1996) Proposed three-dimensional model for the attachment protein G of respiratory syncytial virus. J Gen Virol 77 (Pt 6):1249–1257

Langedijk JP, de Groot BL, Berendsen HJ, van Oirschot JT (1998) Structural homology of the central conserved region of the attachment protein G of respiratory syncytial virus with the fourth subdomain of 55-kDa tumor necrosis factor receptor. Virology 243:293–302

Leemans A, Boeren M, van der Gucht W, Martinet W, Caljon G, Maes L, Cos P, Delputte P (2019) Characterization of the role of N-glycosylation sites in the respiratory syncytial virus fusion protein in virus replication, syncytium formation and antigenicity. Virus Res 266:58–68

Li D, Jans DA, Bardin PG, Meanger J, Mills J, Ghildyal R (2008) Association of respiratory syncytial virus M protein with viral nucleocapsids is mediated by the M2-1 protein. J Virol 82: 8863–8870

Li HM, Ghildyal R, Hu M, Tran KC, Starrs LM, Mills J, Teng MN, Jans DA (2021) Respiratory syncytial virus matrix protein-chromatin association is key to transcriptional inhibition in infected cells. Cells 10(10):2786

Li Y, Wang X, Blau DM, Caballero MT, Feikin DR, Gill CJ, Madhi SA, Omer SB, Simões EAF, Campbell H, Pariente AB, Bardach D, Bassat Q, Casalegno JS, Chakhunashvili G, Crawford N, Danilenko D, Do LAH, Echavarria M, Gentile A, Gordon A, Heikkinen T, Huang QS, Jullien S, Krishnan A, Lopez EL, Markić J, Mira-Iglesias A, Moore HC, Moyes J, Mwananyanda L, Nokes DJ, Noordeen F, Obodai E, Palani N, Romero C, Salimi V, Satav A, Seo E, Shchomak Z, Singleton R, Stolyarov K, Stoszek SK, von Gottberg A, Wurzel D, Yoshida LM, Yung CF, Zar HJ, Nair H (2022) Global, regional, and national disease burden estimates of acute lower respiratory infections due to respiratory syncytial virus in children younger than 5 years in 2019: a systematic analysis. Lancet 399:2047–2064

Liljeroos L, Krzyzaniak MA, Helenius A, Butcher SJ (2013) Architecture of respiratory syncytial virus revealed by electron cryotomography. Proc Natl Acad Sci U S A 110:11133–11138

Low KW, Tan T, Ng K, Tan BH, Sugrue RJ (2008) The RSV F and G glycoproteins interact to form a complex on the surface of infected cells. Biochem Biophys Res Commun 366:308–313

Ludwig A, Howard G, Mendoza-Topaz C, Deerinck T, Mackey M, Sandin S, Ellisman MH, Nichols BJ (2013) Molecular composition and ultrastructure of the caveolar coat complex. PLoS Biol 11:e1001640

Ludwig A, Nguyen TH, Leong D, Ravi LI, Tan BH, Sandin S, Sugrue RJ (2017) Caveolae provide a specialized membrane environment for respiratory syncytial virus assembly. J Cell Sci 130: 1037–1050

Malhi M, Norris MJ, Duan W, Moraes TJ, Maynes JT (2021) Statin-mediated disruption of rho GTPase prenylation and activity inhibits respiratory syncytial virus infection. Commun Biol 4: 1239

Marty A, Meanger J, Mills J, Shields B, Ghildyal R (2004) Association of matrix protein of respiratory syncytial virus with the host cell membrane of infected cells. Arch Virol 149:199–210

McCurdy LH, Graham BS (2003) Role of plasma membrane lipid microdomains in respiratory syncytial virus filament formation. J Virol 77:1747–1756

McDonald TP, Sugrue RJ (2007) The use of two-dimensional SDS-PAGE to analyze the glycan heterogeneity of the respiratory syncytial virus fusion protein. Methods Mol Biol 379:97–108

McDonald TP, Pitt AR, Brown G, Rixon HW, Sugrue RJ (2004) Evidence that the respiratory syncytial virus polymerase complex associates with lipid rafts in virus-infected cells: a proteomic analysis. Virology 330:147–157

Mclellan JS, Ray WC, Peeples ME (2013) Structure and function of respiratory syncytial virus surface glycoproteins. Curr Top Microbiol Immunol 372:83–104

Mehedi M, Mccarty T, Martin SE, Le Nouën C, Buehler E, Chen YC, Smelkinson M, Ganesan S, Fischer ER, Brock LG, Liang B, Munir S, Collins PL, Buchholz UJ (2016) Actin-related protein 2 (ARP2) and virus-induced Filopodia facilitate human respiratory syncytial virus spread. PLoS Pathog 12:e1006062

Money VA, Mcphee HK, Mosely JA, Sanderson JM, Yeo RP (2009) Surface features of a Mononegavirales matrix protein indicate sites of membrane interaction. Proc Natl Acad Sci U S A 106:4441–4446

Nair H, Nokes DJ, Gessner BD, Dherani M, Madhi SA, Singleton RJ, O'Brien KL, Roca A, Wright PF, Bruce N, Chandran A, Theodoratou E, Sutanto A, Sedyaningsih ER, Ngama M, Munywoki PK, Kartasasmita C, Simões EA, Rudan I, Weber MW, Campbell H (2010) Global burden of acute lower respiratory infections due to respiratory syncytial virus in young children: a systematic review and meta-analysis. Lancet 375:1545–1555

Navaratnarajah CK, Generous AR, Yousaf I, Cattaneo R (2020) Receptor-mediated cell entry of paramyxoviruses: mechanisms, and consequences for tropism and pathogenesis. J Biol Chem 295:2771–2786

Olmsted RA, Collins PL (1989) The 1A protein of respiratory syncytial virus is an integral membrane protein present as multiple, structurally distinct species. J Virol 63:2019–2029

Olmsted RA, Elango N, Prince GA, Murphy BR, Johnson PR, Moss B, Chanock RM, Collins PL (1986) Expression of the F glycoprotein of respiratory syncytial virus by a recombinant vaccinia virus: comparison of the individual contributions of the F and G glycoproteins to host immunity. Proc Natl Acad Sci U S A 83:7462–7466

Papalazarou V, Machesky LM (2021) The cell pushes back: the Arp2/3 complex is a key orchestrator of cellular responses to environmental forces. Curr Opin Cell Biol 68:37–44

Popowicz GM, Schleicher M, Noegel AA, Holak TA (2006) Filamins: promiscuous organizers of the cytoskeleton. Trends Biochem Sci 31:411–419

Radhakrishnan A, Yeo D, Brown G, Myaing MZ, Iyer LR, Fleck R, Tan BH, Aitken J, Sanmun D, Tang K, Yarwood A, Brink J, Sugrue RJ (2010) Protein analysis of purified respiratory syncytial virus particles reveals an important role for heat shock protein 90 in virus particle assembly. Mol Cell Proteomics 9:1829–1848

Ramírez-Ramírez D, Salgado-Lucio ML, Roa-Espitia AL, Fierro R, González-Márquez H, Cordero-Martínez J, Hernández-González EO (2020) Rac1 is necessary for capacitation and acrosome reaction in Guinea pig spermatozoa. J Cell Biochem 121:2864–2876

Ravi LI, Liang L, Wong PS, Brown G, Tan BH, Sugrue RJ (2013) Increased hydroxymethylglutaryl coenzyme a reductase activity during respiratory syncytial virus infection mediates actin dependent inter-cellular virus transmission. Antivir Res 100:259–268

Ravi LI, Tan TJ, Tan BH, Sugrue RJ (2021) Virus-induced activation of the rac1 protein at the site of respiratory syncytial virus assembly is a requirement for virus particle assembly on infected cells. Virology 557:86–99

Rixon HWM, Brown C, Brown G, Sugrue RJ (2002) Multiple glycosylated forms of the respiratory syncytial virus fusion protein are expressed in virus-infected cells. J Gen Virol 83:61–66

Rixon HWM, Brown G, Aitken J, McDonald T, Graham S, Sugrue RJ (2004) The small hydrophobic (SH) protein accumulates within lipid-raft structures of the Golgi complex during respiratory syncytial virus infection. J Gen Virol 85:1153–1165

Rixon HWM, Brown G, Murray JT, Sugrue RJ (2005) The respiratory syncytial virus small hydrophobic protein is phosphorylated via a mitogen-activated protein kinase p38-dependent tyrosine kinase activity during virus infection. J Gen Virol 86:375–384

Roberts SR, Compans RW, Wertz GW (1995) Respiratory syncytial virus matures at the apical surfaces of polarized epithelial cells. J Virol 69:2667–2673

Russell RF, McDonald JU, Ivanova M, Zhong Z, Bukreyev A, Tregoning JS (2015) Partial attenuation of respiratory syncytial virus with a deletion of a small hydrophobic gene is associated with elevated interleukin-1β responses. J Virol 89:8974–8981

Santangelo PJ, Bao G (2007) Dynamics of filamentous viral RNPs prior to egress. Nucleic Acids Res 35:3602–3611

Santangelo P, Nitin N, Laconte L, Woolums A, Bao G (2006) Live-cell characterization and analysis of a clinical isolate of bovine respiratory syncytial virus, using molecular beacons. J Virol 80:682–688

Scheid A, Choppin PW (1977) Two disulfide-linked polypeptide chains constitute the active F protein of paramyxoviruses. Virology 80:54–66

Schmitt AP, Lamb RA (2005) Influenza virus assembly and budding at the viral budozone. Adv Virus Res 64:383–416

Sedeyn K, Schepens B, Saelens X (2019) Respiratory syncytial virus nonstructural proteins 1 and 2: exceptional disrupters of innate immune responses. PLoS Pathog 15:e1007984

Sezgin E, Levental I, Mayor S, Eggeling C (2017) The mystery of membrane organization: composition, regulation and roles of lipid rafts. Nat Rev Mol Cell Biol 18:361–374

Shahriari S, Wei KJ, Ghildyal R (2018) Respiratory syncytial virus matrix (M) protein interacts with actin in vitro and in cell culture. Viruses 10(10):535

Shi T, Mcallister DA, O'Brien KL, Simoes EAF, Madhi SA, Gessner BD, Polack FP, Balsells E, Acacio S, Aguayo C, Alassani I, ALI A, Antonio M, Awasthi S, Awori JO, Azziz-Baumgartner-E, Baggett HC, Baillie VL, Balmaseda A, Barahona A, Basnet S, Bassat Q, Basualdo W, Bigogo G, Bont L, Breiman RF, Brooks WA, Broor S, Bruce N, Bruden D, Buchy P, Campbell S, Carosone-Link P, Chadha M, Chipeta J, Chou M, Clara W, Cohen C, de Cuellar E, Dang DA, Dash-Yandag B, Deloria-Knoll M, Dherani M, Eap T, Ebruke BE, Echavarria M, de Freitas Lázaro Emediato CC, Fasce RA, Feikin DR, Feng L, Gentile A, Gordon A, Goswami D, Goyet S, Groome M, Halasa N, Hirve S, Homaira N, Howie SRC, Jara J, Jroundi I, Kartasasmita CB, Khuri-Bulos N, Kotloff KL, Krishnan A, Libster R, Lopez O, Lucero MG, Lucion F, Lupisan SP, Marcone DN, Mccracken JP, Mejia M, Moisi JC, Montgomery JM, Moore DP, Moraleda C, Moyes J, Munywoki P, Mutyara K, Nicol MP, Nokes DJ, Nymadawa P, Da Costa Oliveira MT, Oshitani H, Pandey N, Paranhos-Baccalà G, Phillips LN,

Picot VS, Rahman M, Rakoto-Andrianarivelo M, Rasmussen ZA, Rath BA, Robinson A, Romero C, Russomando G, Salimi V, Sawatwong P, Scheltema N, Schweiger B et al (2017) Global, regional, and national disease burden estimates of acute lower respiratory infections due to respiratory syncytial virus in young children in 2015: a systematic review and modelling study. Lancet 390:946–958

Sugrue RJ, Brown C, Brown G, Aitken J, Mc LRHW (2001) Furin cleavage of the respiratory syncytial virus fusion protein is not a requirement for its transport to the surface of virus-infected cells. J Gen Virol 82:1375–1386

Thomas KW, Monick MM, Staber JM, Yarovinsky T, Carter AB, Hunninghake GW (2002) Respiratory syncytial virus inhibits apoptosis and induces NF-kappa B activity through a phosphatidylinositol 3-kinase-dependent pathway. J Biol Chem 277:492–501

Thompson WW, Shay DK, Weintraub E, Brammer L, Cox N, Anderson LJ, Fukuda K (2003) Mortality associated with influenza and respiratory syncytial virus in the United States. JAMA 289:179–186

Triantafilou K, Kar S, Vakakis E, Kotecha S, Triantafilou M (2013) Human respiratory syncytial virus viroporin SH: a viral recognition pathway used by the host to signal inflammasome activation. Thorax 68:66–75

Tripp RA, Jones LP, Haynes LM, Zheng H, Murphy PM, Anderson LJ (2001) CX3C chemokine mimicry by respiratory syncytial virus G glycoprotein. Nat Immunol 2:732–738

Ulloa L, Serra R, Asenjo A, Villanueva N (1998) Interactions between cellular actin and human respiratory syncytial virus (HRSV). Virus Res 53:13–25

van den Hoogen BG, Bestebroer TM, Osterhaus AD, Fouchier RA (2002) Analysis of the genomic sequence of a human metapneumovirus. Virology 295:119–132

Wang M, Casey PJ (2016) Protein prenylation: unique fats make their mark on biology. Nat Rev Mol Cell Biol 17:110–122

Yeo DS, Chan R, Brown G, Ying L, Sutejo R, Aitken J, Tan BH, Wenk MR, Sugrue RJ (2009) Evidence that selective changes in the lipid composition of raft-membranes occur during respiratory syncytial virus infection. Virology 386:168–182

Zimmer G, Budz L, Herrler G (2001) Proteolytic activation of respiratory syncytial virus fusion protein. Cleavage at two furin consensus sequences. J Biol Chem 276:31642–31650

# Chapter 10
# Japanese Encephalitis Virus-Infected Cells

**Kiran Bala Sharma, Simran Chhabra, and Manjula Kalia**

**Abstract** RNA virus infections have been a leading cause of pandemics. Aided by global warming and increased connectivity, their threat is likely to increase over time. The *flaviviruses* are one such RNA virus family, and its prototypes such as the Japanese encephalitis virus (JEV), Dengue virus, Zika virus, West Nile virus, etc., pose a significant health burden on several endemic countries. All viruses start off their life cycle with an infected cell, wherein a series of events are set in motion as the virus and host battle for autonomy. With their remarkable capacity to hijack cellular systems and, subvert/escape defence pathways, viruses are able to establish infection and disseminate in the body, causing disease. Using this strategy, JEV replicates and spreads through several cell types such as epithelial cells, fibroblasts, monocytes and macrophages, and ultimately breaches the blood-brain barrier to infect neurons and microglia. The neurotropic nature of JEV, its high burden on the paediatric population, and its lack of any specific antivirals/treatment strategies emphasise the need for biomedical research-driven solutions. Here, we highlight the latest research developments on Japanese encephalitis virus-infected cells and discuss how these can aid in the development of future therapies.

**Keywords** Autophagy · Blood-brain barrier · Cell death · ER stress · Flavivirus · Innate immunity · Japanese encephalitis virus · Neuroinflammation · Neurotropic · Unfolded protein response

## Introduction

Japanese encephalitis virus (JEV) remains one of the leading global causes of viral encephalitis. It poses a major threat to more than 2 billion people living in endemic regions like southeast Asian countries (van den Hurk et al. 2009; Pan et al. 2011) and is still evolving to new ecological niches of Europe, northern Australia and Africa

K. B. Sharma · S. Chhabra · M. Kalia (✉)
Regional Centre for Biotechnology, NCR Biotech Science Cluster, Faridabad, Haryana, India
e-mail: kiran.bala@rcb.res.in; simran.chhabra@rcb.res.in; manjula@rcb.res.in



S. Vijayakrishnan et al. (eds.), *Virus Infected Cells*, Subcellular Biochemistry 106,
https://doi.org/10.1007/978-3-031-40086-5_10

(Simon-Loriere et al. 2017; Gao et al. 2019). According to a 2019 WHO report, JEV causes almost 68,000 cases with 13,600–20,400 deaths annually. The virus is neurotropic, and its clinical manifestations range from febrile illness to central nervous system (CNS) disorders and death (Sips et al. 2012). A majority of JEV infections remain asymptomatic, and less than 1% of infections develop into the disease, which is either mild or neuroinvasive. Of the diseased cases, one-third recover completely, one-third develop severe lifelong neurological complications, and one-third ultimately succumb to the disease (Solomon 2004). Children aged 0–15 years are the most affected group and are likely to have more neurological complications than adults (Campbell et al. 2011). Currently, there is no antiviral therapy available and existing vaccines are struggling to control the JEV burden due to lack of long-term protection and cross-protection against newly emerging genotypes. Treatment is only supportive and limited clinical trials have been conducted for testing antiviral and drug therapies (Turtle and Solomon 2018).

JEV is an arthropod-borne flavivirus, which is transmitted in an enzootic life cycle between birds, pigs and other vertebrates by *Culex* mosquitoes. Humans are dead-end hosts due to low viremia. Following inoculation upon a blood-feeding mosquito bite, the virus first replicates in skin keratinocytes and then propagates to nearby lymph nodes and other tissues or organs like the liver and kidney, causing transient low viremia. If the virus is not restricted to the periphery, it can cross the blood-brain barrier (BBB) to gain entry into the CNS, which results in the neurological manifestations of the disease. The viral tropism and subsequent host responses govern disease pathogenesis and severity. JEV is known to infect diverse cell types such as epithelial cells, fibroblasts, monocytes, macrophages, dendritic cells (DCs), endothelial cells, brain resident microglial and neuronal cells, and activates an array of cellular responses. A detailed understanding of virus-host crosstalk is important for delineating crucial host responses and identifying cellular factors involved in disease pathogenesis and antiviral development. Herein we review the interactions of JEV with the mammalian host at the cellular and system level, and their role in disease biology.

## Epidemiology

JE was first reported in Japan, with more than 6000 cases during the 1924 epidemic. The prototype Nakayama strain was isolated from the brain of a fatal case in 1935, and since then, the disease has been recognised across Asia (Miyake 1964; Solomon 2003). Genetic studies have proposed that JEV evolved from an African ancestral virus that spread to the Indonesia-Malaysia region many centuries ago, from where it further spread throughout Asia (Solomon et al. 2003). In the first half of the twentieth century, JEV was recognised in the temperate regions of Asia such as Japan, Korea, Taiwan and mainland China; and continued to spread to southeast Asia, India, Bangladesh, Sri Lanka and Nepal over the next decades. By the 1990s, JEV showed its presence even in the non-Asian regions, Saipan and Australia (Filgueira and Lannes 2019; Mulvey et al. 2021).

JEV has a total of five genotypes (I–V), arising from the ancestor virus of the Indonesia-Malaysia region. Genotype III was reported to be responsible for most human cases in Asia up to the 1990s, whereas genotype I is now likely to become the dominant strain and the major cause of JE disease in the region (Pan et al. 2011). In a 2022 outbreak in Australia, a JEV genotype IV was identified (Sikazwe et al. 2022). Epidemiological and genetic studies have reported geographical expansion of different genotypes and JEV emergence in non-epidemic regions (van den Hurk et al. 2009; Gao et al. 2019; Mulvey et al. 2021). The wide distribution of the virus over the years has been due to changes in climate, ecology, agricultural and animal practices. Migratory bird patterns and population shifts can potentially further contribute to virus expansion to non-endemic regions. Increased surveillance and reporting of JEV infections need to be undertaken to assess the true burden of JE.

## Transmission Cycle

The *Culex tritaeniorrhynchus* mosquito which breeds in stagnant water (such as rice paddy fields), is the most important vector for human infections (Solomon et al. 2003). Other domestic (cows, dogs, chickens, goats and horses) and wild animals (flying foxes, frogs, snakes and ducks) can also be infected with the virus, but ones with high viral loads (birds and pigs) maintain the virus (Mansfield et al. 2017). Due to brief and low viremia, humans do not transmit the disease further (Turtle and Solomon 2018). However, a recent report suggested possible JEV transmission via blood transfusion in humans (Cheng et al. 2018). Birds maintain and amplify the virus in the environment, and migratory/seasonal birds are responsible for JEV spread/expansion to new geographical areas (Johnsen et al. 1974; Rodrigues et al. 1981; Yoshikawa et al. 2016; Bae et al. 2018; Preziuso et al. 2018; Turtle and Solomon 2018; Mulvey et al. 2021). Pigs are the natural hosts with prolonged and high viremia, and the virus can also transmit directly via the intra-nasal route in pigs (Ricklin et al. 2016; Garcia-Nicolas et al. 2018). The virus replicates and remains in the porcine tonsils for up to 25 days enabling its persistence in seasons when mosquitoes are inactive (Garcia-Nicolas et al. 2018). JEV can also persist in vaginal mucosa for several days and is shed in vaginal secretions in pigs, suggesting a potential for sexual transmission (Chapagain et al. 2022).

## Clinical Features

The majority of the JEV infections in humans are either asymptomatic or cause febrile illness with mild flu-like symptoms such as fever, sore throat, headache, muscle pain, diarrhoea and vomiting, that lasts for 5–15 days. Neurologic manifestations depend upon the site of infection in the CNS. Patients who develop symptoms of encephalitis suffer significant morbidity and mortality. Encephalitis is characterised by neck stiffness, disorientation, seizures, paralysis, coma and in severe cases, leads to death (Misra and Kalita 2010; Salimi et al. 2016).

## JEV Molecular Biology

JEV belongs to the *Flaviviridae* family that also contains several other pathogenic arboviruses such as West Nile virus (WNV), Zika virus (ZIKV), Dengue virus (DENV), Yellow fever virus (YFV), Murray valley Encephalitis (MVE), St. Louis encephalitis virus (SLEV) and Tick-borne encephalitis virus (TBEV). The enveloped virus contains a single-stranded positive-sense RNA genome of ~11 kb (Vashist et al. 2011), which is a single open reading frame (ORF), flanked by 5′ and 3′ non-coding regions (NCR). The viral polyprotein (~3400 aa) is cleaved into three structural proteins – Nucleocapsid (C), Membrane (M) and Envelope protein (E), and seven non-structural proteins (NS1, NS2A, NS2B, NS3, NS4A, NS4B and NS5), by the action of viral and host proteases (Chambers et al. 1990). The viral nucleocapsid is enclosed in a membrane containing the envelope (E) glycoprotein. A near-atomic structure of JEV has revealed various structural determinants associated with virus stability and neurovirulence (Wang et al. 2017b, c). The non-structural viral proteins are an integral part of the replication complex and also interact with diverse host factors involved in multiple cellular pathways to create an infection-supportive environment.

## Infection Route: A Cellular Overview

The JEV virus life cycle is an orchestration of six major steps: receptor binding, entry, polyprotein translation, genome replication, assembly and egress. Infection begins with the non-specific binding of the viral glycoprotein E to one or more cellular attachment factors, that enhance the avidity and facilitates specific interaction with the receptor. Studies in different cell types have identified attachment factors (Heparan sulphate proteoglycans, glycosaminoglycans) and several potential receptors: Heat shock protein 70, vimentin, laminin receptor, CD4, $\alpha 5\beta 3$ integrin, Dendritic Cell-Specific Intercellular adhesion molecule-3-Grabbing Non-integrin (DC-SIGN), Glucose-regulated protein 78 (GRP78), T cell immunoglobulin and mucin domain 1 (TIM-1), C type lectin member 5A (CLEC5A), plasmalemma vesicle-associated protein and gastrokine 3 (Chiou et al. 2005; Chien et al. 2008; Chen et al. 2012b; Nain et al. 2016; Nain et al. 2017; Mukherjee et al. 2018; Niu et al. 2018).

Virus-receptor interaction results in receptor-mediated endocytosis. JEV is seen to exploit different endocytic routes in a cell-type-dependent manner. Studies have now established that the virus utilises clathrin-mediated endocytosis (CME) to infect fibroblasts and epithelial cells, and under conditions of CME inhibition, it can employ clathrin-independent endocytosis (CIE) to infect neuronal cells (Zhu et al. 2012; Kalia et al. 2013; Yang et al. 2013; Xu et al. 2016; Liu et al. 2017; Khasa et al. 2019; Khasa et al. 2020). RNA interference-based screens have identified several membrane trafficking proteins such as the ARP2/3 complex, RhoA, Cdc42, Pak1, Rab5, Rab11, ezrin and valosin-containing protein (VCP) to be involved in JEV entry (Xu et al. 2016; Khasa et al. 2019; Khasa et al. 2020; Liu et al. 2020; Sehrawat et al. 2021; Zhou et al. 2021). Characterisation of JEV entry in neuronal cells, and its interaction with host factors, is an important research domain and can generate potential therapeutic targets to combat the virus at an early time of infection.

A low pH achieved in the endosome induces conformational changes in the viral E glycoprotein, which triggers the fusion of viral and host endosomal membranes and releases the viral RNA genome into the cytoplasm. The positive sense RNA is directly translated via the host translational machinery into two precursor polyproteins (with or without a ribosomal frameshifting at the beginning of NS2A-coding region), that are cleaved into three structural (C, prM and E) and seven non-structural (NS1 to NS5) proteins, along with NS1'.

The virus non-structural proteins: NS4A, NS4B, NS1, NS2A, NS2B, NS3 and NS5 are known to interact/associate with several host factors, majorly ER-associated proteins and lipids to form the ER-derived replication complexes (Arakawa and Morita 2019). Distinct structures referred to as convoluted membranes (CMs) and vesicle packets (VPs) can be seen in flavivirus-infected cells and are the sites for polyprotein translation/processing and viral RNA replication, which proceeds through the formation of the viral dsRNA intermediate. The signal peptidase complex subunit 1 (SPCS1) is a crucial host factor that interacts with NS2B and impacts the post-translational processing of JEV proteins and virus assembly (Ma et al. 2018). The ER-associated degradation pathway (ERAD) proteins such as microtubule-associated protein 1 light chain 3-I (MAP1LC3-I, hereafter LC3-I), EDEM1 and Sel1L are also found enriched in the JEV VPs and CMs (Sharma, Bhattacharyya et al. 2014; Sarkar et al. 2020). The replication complex has a pore opening into the cytosol for entry of nucleotides and exit of the positive-strand RNA for packaging. Interestingly, the production of the plus strands is 10–100 fold higher than minus strand RNA, showing an asymmetric and semi-conservative process of replication (Uchil and Satchidanandam 2003). The viral capsid protein, along with LC3-I (autophagy-independent form), is observed to be concentrated on lipid droplets (LDs), and this is likely to be the site of RNA packaging (Sarkar et al. 2020). Interestingly, the number of LDs decreases in JEV-infected cells suggesting a negative regulation of lipid metabolism, an observation which is also supported by proteome studies describing the down-modulation of lipid metabolic proteins in JEV-infected cells (Sarkar et al. 2020; Sharma et al. 2021a).

Virion assembly in the ER lumen follows as the viral genome and proteins assemble abundantly. The nucleocapsid is further enclosed by E and prM proteins decorated on ER membranes to form immature virus particles. These undergo maturation before budding out of the cell membrane, via furin protease activity that cleaves prM to M during exocytosis via the trans-golgi network (Stadler et al. 1997; Li et al. 2008; Yu et al. 2008). Mapping the virus-host interactome and developing inhibitors to block the virus-receptor/host dependency factor interaction are attractive targets for antiviral drug development.

## Cellular Response to Infection

Viruses have evolved strategies to interfere with cellular signalling pathways and exploit organellar compartments, which perturbs cellular homeostasis and triggers the activation of stress responses in infected cells. Flaviviruses are shown to induce stress responses such as the formation of stress granules, oxidative stress, and ER stress leading to the activation of the unfolded protein response (UPR), autophagy and activation of innate immunity. Cross-communication between these pathways regulates the antiviral and cell survival response, in addition to other cellular functions such as translation, metabolism, cytoskeletal organisation and inflammation, and therefore influences the viral pathogenesis and disease outcome.

### *Alteration of Signalling Pathways*

Activation of PI3-kinase/Akt signalling has been observed early during JEV infection and is thus likely to be a result of virus-receptor interaction (Das et al. 2010). JEV attachment can also specifically activate EGFR-PI3K signalling (Xu et al. 2016), resulting in phosphorylation of EGFR, and infection can be blocked by using EGFR inhibitors (Zhang et al. 2022). MAPK signalling including ERK, p38, MAPK and JNK plays an important role in JEV-induced caspase activation (Gupta et al. 2011) and neuroinflammation (Ye et al. 2016; He et al. 2017). JEV has also been shown to modulate several tyrosine phosphorylation-mediated signalling events in infected cells (Raung et al. 2005, 2007; Yang et al. 2012).

### *PKR Activation and Formation of Stress Granules*

Recognition of viral dsRNA activates the interferon (IFN)-induced protein kinase R (PKR), which phosphorylates the eukaryotic translation initiation factor 2$\alpha$ subunit (eIF2$\alpha$), and results in a block of protein translation. PKR activation also results in the sequestration of actively transcribing mRNA into cytoplasmic foci called stress granules (SG), and this process is circumvented by nearly all viruses to enable their propagation. The JEV NS2A protein can counteract PKR activation and eIF2-$\alpha$-phosphorylation (Tu et al. 2012). JEV capsid protein has been shown to inhibit the SG formation by binding to the RNA-binding protein, Caprin-1 which is an essential component of the stress granules (Katoh et al. 2013). JEV NS4B can also recruit the VCP-NPL4 complex and block stress granule formation (Arakawa et al. 2022).

## *Induction of Oxidative Stress*

Oxidative stress is generated when there is an imbalance between the production and neutralisation of reactive oxygen species (ROS). In general, ROS is produced as a by-product of normal aerobic metabolism, by a variety of enzymes present in mitochondria, ER, and peroxisomes, which are simultaneously taken care of by the antioxidant system (Go and Jones 2008; Roy et al. 2017; Zhang et al. 2019). ROS acts as a double-edged sword in cell health, as its maintenance to a certain level is necessary for signalling activation and survival. However, stress conditions such as pathogen infection, accumulate intracellular ROS, resulting in cell injury or death.

JEV has been demonstrated to trigger oxidative stress in infected cells together with the production of toxic oxygen species in neutrophils (Srivastava et al. 1999), human astrocytoma and astroglioma cell lines (Mishra et al. 2008), the superoxide anion and nitric oxide species in rat cortical glial cells (Liao et al. 2002), and ROS intermediates in murine neuroblastoma cells (Raung et al. 2001). The elevated ROS level during JEV infection is implicated in virus-induced cell death (Ghoshal et al. 2007; Ghosh and Basu 2009; Kumar et al. 2009b; Yang et al. 2010). Alterations in mitochondrial health (Lin et al. 2004), and high levels of proinflammatory mediators secreted by activated microglia (Ghoshal et al. 2007), are major drivers of JEV-induced oxidative stress. Even UV-inactivated JEV (replication incompetent) damages actively growing neuronal cells through a ROS-mediated pathway (Lin et al. 2004). ROS is thus a major contributor of JEV pathogenesis and therapeutic modulation of JEV-induced oxidative stress could be beneficial for the host (Zhang et al. 2014). Oxidative stress has also been linked to UPR activation, autophagy, immunity, inflammation and cell death pathways (Olagnier et al. 2014; Chen et al. 2018a; Sharma et al. 2018; de Almeida et al. 2020).

## *Activation of the Unfolded Protein Response*

The ER is a ubiquitous and versatile organelle involved in multiple cellular functions including protein production, folding, trafficking and turnover; lipid synthesis and distribution, calcium homeostasis, cell signalling and innate immunity. The ER is central to the JEV life cycle as it provides both a scaffold for viral protein translation and ER resident proteins and lipids for virus replication complex biogenesis and virion assembly. This poses a significant burden on the organelle, resulting in the induction of ER stress and activation of UPR (Yu et al. 2006; Blazquez et al. 2014). The three main pathways that modulate UPR are protein kinase-like ER resident kinase (PERK), the activating transcription factor 6 (ATF6), and the inositol-requiring enzyme 1 (IRE-1) (Liu and Kaufman 2003). The activation of the PERK pathway follows eIF2α-phosphorylation and causes global translational arrest. The insufficiency of translational arrest in reducing ER stress, leads to the activation of the IRE1 and ATF6 pathways, which upregulate the expression of ER chaperons and ERAD machinery components, to boost protein folding capacity and degrade terminally misfolded proteins. The UPR signalling attempts to restore ER homeostasis, however, prolonged ER stress and high-level signalling through PERK and eIF2α, results in ATF4 activation and expression of transcription factor GADD153 (CHOP), which altogether arrests the cell-cycle and induces apoptosis (Rozpedek et al. 2016). Activation of ER stress has also been linked to other crucial cellular processes such as lipid metabolism, autophagy, innate immunity and differentiation (McLean et al. 2011; Blazquez et al. 2014; Datan et al. 2016; Chan and Ou 2017; Sharma et al. 2017; Carletti et al. 2019), expanding its role in generating the integrative stress response against viral infections.

Studies have demonstrated that JEV infection generates ER stress and triggers the activation of all three sensors (PERK, ATF6 and IRE-1) of the UPR. PERK activation and CHOP expression have been linked to virus-induced apoptosis and disease pathogenesis (Su et al. 2002; Wang et al. 2019). The regulated IRE1-dependent decay (RIDD) pathway has been shown to benefit viral replication and enhance cell death (Bhattacharyya et al. 2014). In contrast, the XBP1 and ATF6-mediated UPR pathways exert a protective role against JEV-induced cell death via upregulating autophagy in neuronal cells (Yu et al. 2006; Sharma et al. 2017). The cellular ERAD pathway degrades the extra membrane-anchored JEV NS proteins in convoluted membranes and this process is essential for optimal virus replication (Tabata et al. 2021).

## *Upregulation of Autophagy*

Autophagy is a highly conserved, multi-step degradative process which is greatly involved in maintaining cellular homeostasis by degrading misfolded/faulty proteins and damaged organelles through lysosomal compartments. Autophagy has basal housekeeping functions and is induced by stress conditions such as hypoxia, pathogen infection, ER stress, oxidative stress and accumulation of aggregated proteins and damaged organelles. In the context of viruses, autophagy can either restrict or enhance infection, depending on the virus and cell type (Ahmad et al. 2018). To restrict virus proliferation, autophagy either directly targets the viral components for degradation or indirectly modulates other host antiviral and survival pathways (Lennemann and Coyne 2015; Abdoli et al. 2018; Choi et al. 2018; Sharma et al. 2019).

During JEV infection, autophagy has been shown to be induced in both in vitro and in vivo model systems. JEV-induced autophagy has been shown to support virus replication by suppressing the cellular immune environment (Li et al. 2012a; Jin et al. 2013), however, some studies have suggested otherwise (Sharma et al. 2014; Xu et al. 2017). Significant enhancement of JEV replication and titres has been observed in autophagy-deficient (ATG5/ATG7 depleted) mouse fibroblasts and neuronal cells (Sharma et al. 2014). The E3 ubiquitin ligase Nedd4 protein also restricts JEV-induced autophagy and facilitates JEV replication in human neuroblastoma cells (Xu et al. 2017). Autophagy thus appears to play an antiviral role for JEV by restricting virus replication and virus-induced cell death (Sharma et al. 2014). JEV-infected cells also show enhanced mitophagy flux and a decrease in mitochondria number through the interaction of the viral NS4A protein with PTEN-induced kinase 1 (Agarwal et al. 2022). Interestingly, based on the anti-inflammatory and neuroprotective properties, autophagy activators have the potential to be repurposed as antivirals against JEV infection.

## *Innate Immune Activation*

The cell recognises a virus infection through its pathogen recognition receptors (PRRs) that bind specific pathogen-associated molecular patterns (PAMPs). The PRR-PAMP association triggers downstream effectors and the production of type-I interferons (IFN-α and IFN-β) and inflammatory cyto/chemokines, which further initiates the JAK-STAT signalling in an autocrine and paracrine fashion. The IFN-driven JAK-STAT pathway ultimately induces the expression of a wide array of genes collectively referred to as interferon-stimulated genes (ISGs) that function to inhibit virus infection by directly acting on the virus itself or by enhancing the cellular antiviral state.

Various cell line and animal model studies have established the crucial role of numerous PRRs such as RIG-I, MDA-5, MyD88 (Kato et al. 2006), TLR3 (Han et al. 2014) and TLR7 (Nazmi et al. 2014; Awais et al. 2017), in sensing JEV components for innate immune activation. The quality and magnitude of the antiviral response of brain resident cells govern JE pathogenesis. Activation of RIG-I, MDA-5 and TLR-3 in JEV-infected neuronal and microglial cells is critical for virus inhibition, as their deletion compromised antiviral immunity and increased viral load (Nazmi et al. 2011; Jiang et al. 2014). Various reports have demonstrated the activation of a wide variety of ISGs including PKR, OAS, TRIM21, ISG15, IFITs, IFITMs, GBPs, MX1, etc., upon JEV infection (Clarke et al. 2014; Sharma et al. 2021a, b) (Fig. 10.1). Some of these such as IFNα, ISG15, MX2 and OAS-L have been shown to have an antiviral role against JEV (Hsiao et al. 2010; Liu et al. 2013; Zheng et al. 2016). Type-I IFN production by astrocytes restricts viral spread in the CNS and virus-induced cytopathic effects (Lindqvist et al. 2016).

The virus also engages to counteract the IFN response and establish a replication niche. JEV NS5 is a potent antagonist of IFN-induced Jak-STAT signalling through abrogation of nuclear translocation and tyrosine phosphorylation of Tyk2 and STAT1 (Lin et al. 2006). JEV NS4A also functions as an IFN antagonist through inhibition of STAT phosphorylation (Lin et al. 2008). JEV NS1' has been shown to

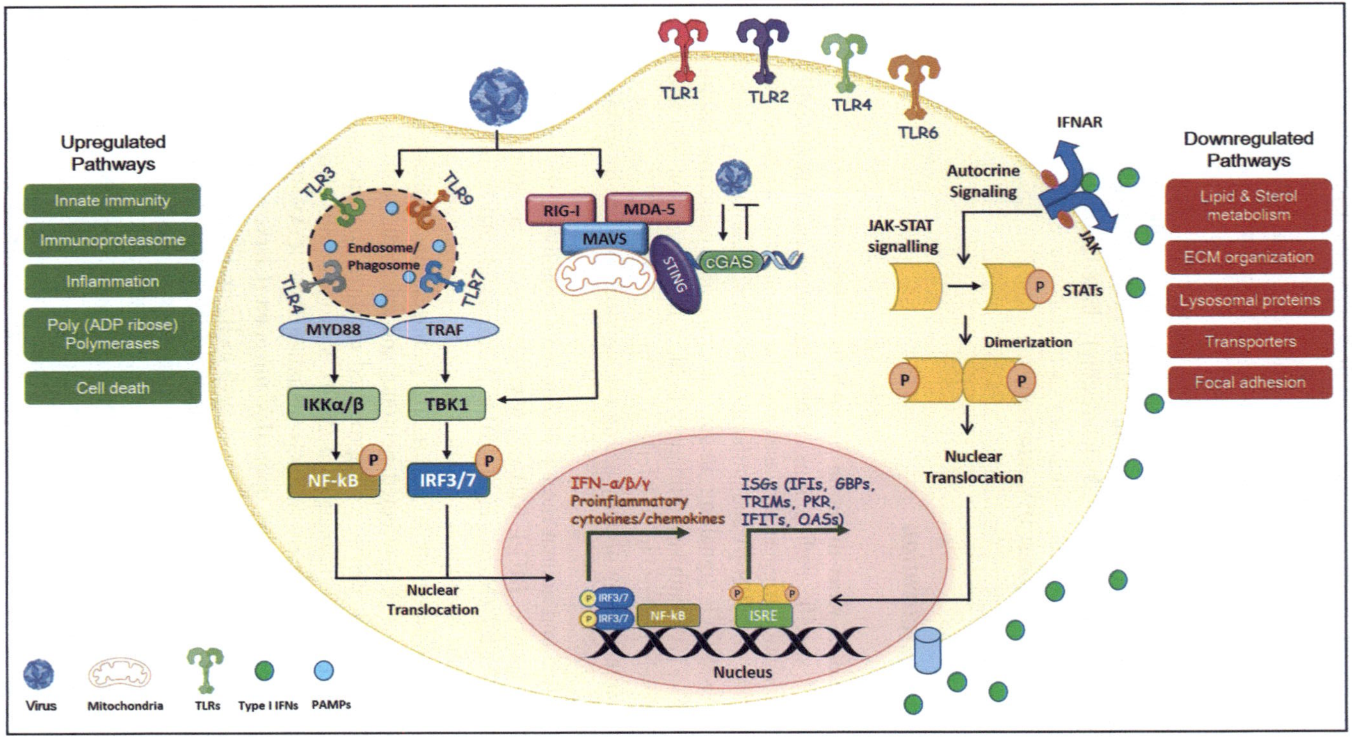


**Fig. 10.1** JEV infection-induced changes in fibroblasts. Upon binding, the virus is recognised by distinct pathogen recognition receptors (PRRs), initiating the downstream signalling by phosphorylation and nuclear translocation of IRF-3 and IRF-7 thereby resulting in the production of type 1-IFN. The newly synthesised IFN binds to IFNAR in an autocrine fashion and activates the downstream JAK-STAT signalling pathway. Nuclear translocation of STAT dimmers induces several ISGs that encode for different antiviral effector molecules thereby building a virus-resistant cellular state. Virus infection in the fibroblast results in remodelling of cellular proteome, causing activation of innate immune response (including the enhanced levels of PRR, IRFs, production of type-I interferon & ISGs and MHC presentation), cell death pathways and enhanced immunoproteasome and PARPs. Virus infection also induces the activation of DNA sensor

**Fig. 10.1** (continued) cGAS which plays a crucial role in restricting virus replication. Different cellular pathways modulating lipid and sterol metabolism, focal adhesion, transporters, ECM organisation and lysosomal proteins get downregulated in the infected cell, thereby impacting the cellular metabolism

inhibit mitochondrial antiviral signalling protein (MAVS) mediated IFNβ induction by blocking dephosphorylation of CDK1 (Li et al. 2021b).

The transcription factor IRF8 modulates microglial activation during infection and enhances IFN-γ production resulting in reduced viral loads in the brain (Tripathi et al. 2021). Activated microglial cells and astrocytes also produce a strong wave of pro/anti-inflammatory cytokines including RANTES, TNF-α, IL1-α, IL-6, IL-12, IL-18, IL1-β, CCL2, CXCL9, CXCL10, CXCL11 and pro-inflammatory enzymes like cyclooxygenase-2 and iNOS during JEV infection (Chen et al. 2004; Bhowmick et al. 2007; Ghoshal et al. 2007; Das et al. 2008; Gupta et al. 2010a, b; Chen et al. 2011a, b; Kaushik et al. 2011; Fadnis et al. 2013; Lannes et al. 2017a; Yu et al. 2019), which is thought to be the major driver of JEV-induced neuroinflammation and bystander neuronal cell death (Ye et al. 2016; He et al. 2017; Singh et al. 2020; Ashraf et al. 2021). The suppression of anti-inflammatory cytokine IL-10 has also been observed during JEV infection (Swarup et al. 2007). A quantitative phosphoproteomic analysis identified JNK1 cascade activation upon JEV infection as a major contributor to virus-induced encephalitis and lethality (Ye et al. 2016).

Viral infections including JEV, also activate an epithelial-mesenchymal transition (EMT)-like process as an antiviral strategy through activation of the Snail transcription factor, via coregulation with type-I IFN (Vedagiri et al. 2021).

All flaviviruses, including JEV, produce a unique subgenomic flavivirus RNA (sfRNA) as a result of incomplete degradation of genomic RNA by the cellular exoribonuclease XRN1. This consists of the highly structured 3′UTR and plays an important role in viral pathogenesis through the regulation of cellular mRNA decay and IFN responses (Clarke et al. 2015). Activation of the RNA decay pathways are an important aspect of the innate immune response to virus infection. The monocyte chemoattractant protein 1-induced protein 1 (MCPIP1) ribonuclease binds JEV RNA and targets it for degradation (Lin et al. 2013). The zinc finger protein ZFP36L1 recognises the AUUUA motif in the 3′-UTR of the JEV genome and destabilises it by degrading the viral RNA through the 5′-3′ XRN1 and 3′-5′RNA-exosome RNA decay pathways (Chiu et al. 2022).

These studies highlight the importance of active innate immunity hubs in virus restriction and necessitate their detailed exploration for understanding disease severity. An optimal innate immune response is also crucial for limiting JEV at the periphery and blocking viral entry to CNS, a decisive checkpoint in JE pathogenesis.

## *Cell Death Pathways*

An infected cell exposed to unrecoverable intracellular perturbations activates cell death pathways. Depending on the cell type and host response, viruses lead to three major regulated cell death modalities: apoptosis and inflammation programmed – necroptosis and pyroptosis (Dhuriya and Sharma 2018; Imre 2020). Cell death may repress virus replication and alter the local and systemic immune responses by releasing a variety of death-associated molecular patterns (DAMPS). Viruses have evolved strategies to inhibit or activate cell death pathways to escape the defence mechanism or to kill certain cell populations to dysregulate host immune responses.

Direct neuronal infection and prolonged microglial activation (gliosis) driven inflammatory environment trigger diverse markers of cell death pathways in JEV-infected neuronal cells (Chen et al. 2012a, b; Mukherjee et al. 2019). Several studies have supported the role of prolonged ER stress and virus-induced UPR pathways in neuronal cell death (Su et al. 2002; Mukherjee et al. 2017; Wang et al. 2019). JEV also activates mitochondrial stress leading to ROS production and cytochrome-c release in the cytoplasm and activation of Caspase-8, -9 and -3 (Tsao et al. 2008; Yang et al. 2009, 2010; Wongchitrat et al. 2019). Recent evidence points to the role of inflammation-driven necroptosis and pyroptosis in the pathogenesis of JEV infection. The levels of the MLKL protein, a marker of necroptosis, increase in neuronal cells upon JEV infection, and its deletion reduces JE progression and inflammatory cytokines in the mouse model (Bian et al. 2017). JEV-infected fibroblasts and mouse brain showed upregulated levels of Gasdermin D and MLKL, along with increased transcript levels of genes of the necroptosis pathways (Sharma et al. 2021a). JEV-infected-pyroptotic macrophages release IL1-α, which is shown to be responsible for viral neuroinvasion (Wang et al. 2020a, b). Transcriptomic analysis of macrophages has also shown the launch of diverse programmed cell death pathways upon JEV infection (Wang et al. 2020a, b). A crucial role of necroptosis has been shown in neuroinflammation and cell death in other neurological disorders, signifying that targeting these cell death pathways can potentially reduce disease severity.

## *Downregulation of Cell Adhesion Molecules*

JEV infection downregulates several collagens, laminin and other cell adhesion proteins involved in proteoglycan binding and ECM organisation (Sharma et al. 2021a). This could be a host-driven immune activation strategy for the generation of potential ligands for T and NK cell activation (Fig. 10.1).

## *Metabolic Reprogramming*

All virus infections reprogramme the cellular metabolome to meet the high energy and resource demands of virus replication. In parallel, the cell also regulates its metabolism as an innate immune defence programme. A proteome wide-study has shown that JEV infection of fibroblasts down-modulates several metabolic enzymes of sterol and lipid biosynthetic pathways and transporters (solute carrier and ATP-binding cassette transporter) involved in shuttling metabolites and ions across membranes (Sharma et al. 2021a). Downmodulation of cholesterol/lipid biosynthetic activities is likely to be intimately linked with the IFN and inflammatory response.

Increased glycolysis and pentose phosphate pathway flux, impaired oxidative phosphorylation and catabolic patterns of lipid metabolism are hallmarks of JEV replication in neurons (Li et al. 2021a). The unique metabolic signature of JEV infection is still an underexplored area, but we are likely to see this field expanding rapidly in the coming years with antiviral treatment strategies based on targeted metabolic modulation.

## Infection Route: A Host Overview

### *Infection Route in the Periphery*

JEV enters through an infected-mosquito bite, where the dermis layer acts as a primary site of infection. The local dermal cells (fibroblasts, endothelial, tissue-resident dendritic cells and pericytes) surrounding the mosquito bite area become infected and spread the virus to local lymph nodes, resulting in primary asymptomatic viremia (Filgueira and Lannes 2019; Ashraf et al. 2021). The virus escapes through either a hematogenous route or efferent lymphatic system and might infect multiple organs (liver, spleen, heart, muscle, kidney), generating secondary symptomatic viremia. Very little is known about how JEV affects the cardiovascular, respiratory, digestive, reproductive and urinary systems in humans (Qi et al. 2020; Chapagain et al. 2022); however, mouse model studies have clearly indicated that JEV infects several visceral organs in addition to the brain (Li et al. 2017). In the periphery, JEV mainly replicates in the monocytes/macrophages and DCs (Aleyas et al. 2009; Terry et al. 2012; Wang et al. 2016; Garcia-Nicolas et al. 2019). The virus migrates through the body either as free virions or via migratory infected DCs. In most cases, virus infection is cleared by an effective peripheral immune response. However, the virus can utilise various immune evasion strategies to escape peripheral immune surveillance to cross the BBB (Aleyas et al. 2010; Aleyas et al. 2012; Adhya et al. 2013; Manocha et al. 2014; Sood et al. 2017; Wang et al. 2017a; Banerjee and Tripathi 2019).

### *Peripheral Immune Response to Infection*

Antigen-presenting cells (APCs) such as DCs and macrophages are the first cells to generate a robust immune response with the production of various anti-/pro-inflammatory cytokines and lowering of blood viremia (Solomon 2004). The IFN-β response in macrophages and their migration to the CNS is regulated by the RNA-binding protein quaking (QKI), which functions as an immune suppressor (Liao et al. 2021; Deng et al. 2022). DCs initiate the adaptive immune response by stimulating T-cell activation (Aleyas et al. 2009; Li et al. 2011; Sooryanarain et al.

2012; Sharma et al. 2021b). The virus infection can also result in inflammatory demyelination in the peripheral nervous system (Wang et al. 2022).

JEV also replicates in monocytes and upregulates various antiviral and immune factors, resulting in their activation and differentiation into monocyte-derived dendritic cells (MoDCs) and monocyte-derived macrophages (MoDMs) (Cao et al. 2011; Sooryanarain et al. 2012; Gupta et al. 2014; Garcia-Nicolas et al. 2019). JEV induces functional impairment of DCs through MyD88-dependent and -independent pathways, which leads to poor CD4(+) and CD8(+) T cell responses, and boosts viral survival and dissemination in the body (Aleyas et al. 2009). The virus also reduces the expression of co-stimulatory cytokines in human and mouse DCs, leading to suppressed T cell activation and enhanced Treg (T regulatory cell) differentiation (Cao et al. 2011; Gupta et al. 2014) (Fig. 10.2). Transcriptional profiling of JEV-infected human MoDCs has demonstrated the activation of antiviral and inflammatory pathways, and expansion of Tregs in an allogenic response (Chauhan et al. 2021). Overall, the enhanced Tregs response can exert a neuroprotective effect by reducing excessive inflammatory response as seen in another virus-induced encephalitis (Lund et al. 2008; Anghelina et al. 2009; Lanteri et al. 2009; James et al. 2016). Numerous mouse studies have exhibited the contribution of DCs in protection against JE via T cell-dependent and -independent mechanisms. $CD11c^{hi}$ DCs regulate the innate $CD11^{b+}Ly\text{-}6C^{hi}$ monocyte differentiation to protect immune-privileged CNS during JEV infection (Kim et al. 2015). The ablation of $CD11c^{hi}$ DC is seen to lead to a higher ratio of $CD4^{+}$ Th17/Treg cells and $CD11b^{+}Ly\text{-}6C^{hi}/Ly\text{-}6C^{lo}$ monocytes in the lymphoid tissue and CNS leading to enhanced permeability of the BBB (Choi et al. 2017).

The exact function of T cells in JEV pathogenesis is still unclear as diverse mouse studies suggested different outcomes with partial or complete protection against JE. T cell activation leads to the production of various antiviral cytokines like IFN-γ, generation of T cell memory response, and humoral response (Sharma et al. 2021b) (Fig. 10.2). The involvement of human memory T cells in protection against human JE has been reported (Turtle et al. 2016). Interestingly, the responses of T cell subsets including CD4 (+) and CD8 (+) T cells are found to be associated with different clinical outcomes of JEV infection (Aleyas et al. 2009, 2010, 2012; Adhya et al. 2013; Turtle et al. 2016). The JEV-mediated cytolytic CD8 + T cell activation and associated IFN-γ response provide complete protection against JEV-induced morbidity (Larena et al. 2013; Jain et al. 2017). The CD4 (+) effector T cells are essential for B cell activation and in generating humoral response (Li et al. 2012b; Tarlinton 2019). In patients, IgM antibody response specific to JEV infection reaches a maximum within 7 days of infection and can be detected both in the serum and cerebrospinal fluid (Burke et al. 1985). JEV is also seen to modulate humoral response in mice by increasing myeloid-derived suppressor cell (MDSC) populations, which suppresses CD4+ T cell function and thus diminishes the splenic B cells (CD19+) and blood plasma cells (CD19 + CD138+) (Wang et al. 2017b, c). The generation of neutralising antibody response specific to JEV infection is shown to provide long-term immunity and protection (Lee et al. 1995; Lin et al. 1998; Konishi et al. 1999; Gupta et al. 2003; Plotkin 2010; Lee et al. 2016; Qiu et al. 2018).

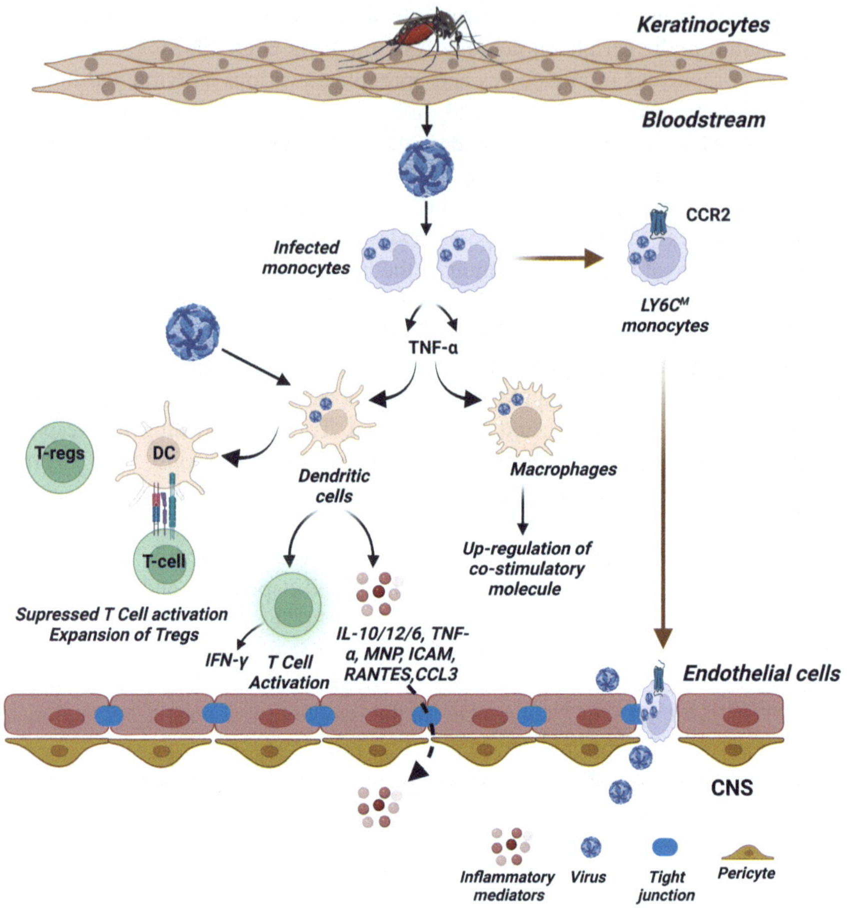


**Fig. 10.2** Peripheral immune response to JEV infection. JEV enters the human body through the bite of an infected *Culex* mosquito. The virus first replicates in the skin and local lymph nodes and then infects the peripheral immune cells, primarily monocytes resulting in the production of TNF-α, which causes activation and differentiation of monocytes into dendritic cells and macrophages. These cells then result in T cell activation, secretion of inflammatory cytokines (IL-6 and TNF-α) and upregulation of co-stimulatory molecules. JEV also directly impairs DCs function leading to suppressed T cell activation and enhanced Tregs. After systemic infection, the virus crosses the blood-brain barrier and enters the brain either directly or through trans-endothelial migration of virus-infected monocytes

Animal studies have demonstrated that passive transfer of JEV-specific monoclonal antibodies could provide protection against JE (Kimura-Kuroda and Yasui 1988; Zhang et al. 1989; Beasley et al. 2004; Goncalvez et al. 2008; Van Gessel et al. 2011; Fernandez et al. 2018), and this has potential as an early treatment strategy.

## *Virus Entry into the CNS*

The CNS is protected from peripheral contaminants by the BBB, which tightly regulates the selective transport of soluble factors and immune cells from blood to the brain. The BBB is comprised of tightly packed brain microvascular endothelial cells (BMECs) supported through interactions with microglial cells, astrocytes, pericytes and mast cells in the neurovascular unit that maintain the CNS microenvironment and neuronal function (Keaney and Campbell 2015; Villabona-Rueda et al. 2019).

In individuals who develop disease the virus crosses the BBB and replicates efficiently in the CNS. Mouse model studies suggest four possible mechanisms; (1) transcellular transport: virus infection in endothelial cells followed by passive transport of viral particles without affecting the cell viability; (2) diapedesis of virus-infected peripheral immune cells through endothelial cell junctions ("Trojan Horse" mechanism) or entry of infected immune cells through known physiological ways such as via the choroid plexus into the ventricular space; (3) virus transport via the peripheral nervous system (retrograde neuronal transport); and (4) virus entry upon BBB disruption due to virus-induced inflammatory mediators produced from cells present in both apical (blood) and basolateral (brain) sides of the BBB (Hsieh and St John 2020; Patabendige et al. 2018; Filgueira and Lannes 2019; Sharma et al. 2021b). A recent study using a BBB model of human brain endothelial cells and astrocytes suggested that JEV infection triggers the production of diverse host mediators, which regulate JEV production but disrupt BBB integrity, thus allowing virus to breach into the brain (Patabendige et al. 2018) (Fig. 10.3). Several inflammatory cytokines and metalloproteases produced from JEV-infected astrocytes and microglial cells trigger the proteasomal degradation of tight junction proteins (claudin-5 and ZO-1), leading to subsequent dysfunction of the endothelial barrier which promotes BBB leakage (Chen et al. 2014; Chang et al. 2015; Lannes et al. 2017b). Several studies also show that the virus gains CNS entry before BBB disruption (Li et al. 2015; Wang et al. 2018), and the breach is a fallout of a massive neuroinflammatory response in the brain.

In the JEV-infected brain, the basal ganglia, thalamus and nuclei of the brainstem are the most affected regions (Kumar et al. 2009a). Virus-induced damage to the midbrain, brain stem, motor neurons in the spinal cord, periventricular tissue damage, etc., may result in different clinical pathologies (Misra and Kalita 2010; Suman et al. 2016).

## *CNS Response to Infection*

Encephalitis is the hallmark of JEV pathogenesis. JEV infection of microglia (Thongtan et al. 2012; Gupta et al. 2017), astrocytes (Chen et al. 2011a, b) and neurons (Chen et al. 2018b; Yu et al. 2019), and subsequent upregulation of cell

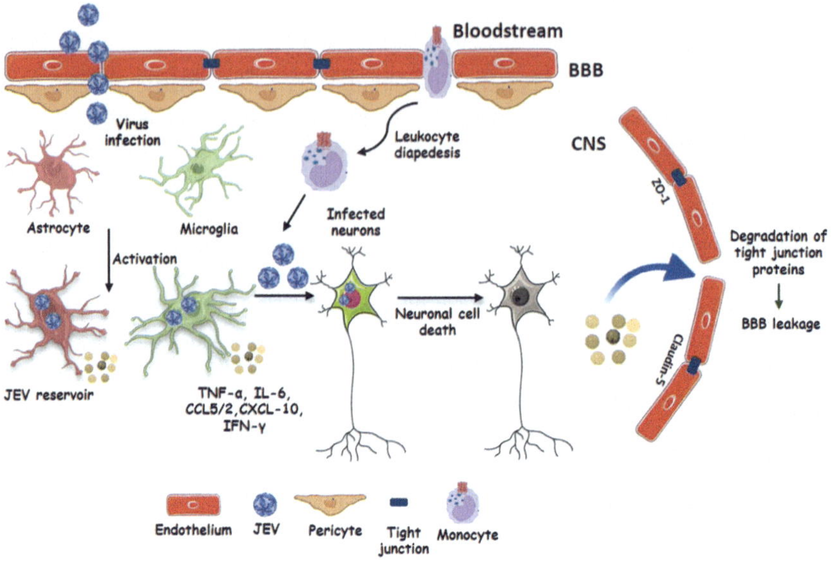


**Fig. 10.3** CNS response to JEV infection. After systemic infection, JEV crosses the BBB and enters into CNS either directly or by Trojan horse mechanism (transmigration of monocytes containing virus). Virus infection then leads to the activation of pericytes, astrocytes and microglial cells (acting as virus reservoir), subsequently leading to the release of certain inflammatory cytokines and new virion particles. Virus infection of neurons causes neuronal cell death either directly or by causing excessive neuroinflammation resulting in neuronal damage. The increased levels of cytokines result in enhanced level of metalloproteases (MMP2/9) causing degradation of tight junction proteins including ZO-1 and Claudin-5 and disruption of BBB. This increases the permeability of BBB facilitating enhanced migration of JEV-infected leukocytes and JEV particles into the CNS, thereby increasing the neuroinflammation and causing excessive neuronal tissue damage

death and inflammatory responses contributes to virus-induced neuroinflammation. Genes associated with glutamate signalling are downregulated in JEV-infected mouse brains suggesting a potential negative impact on neurotransmission as well (Clarke et al. 2014). The virus can also modulate dopamine levels and can use dopamine-mediated neuronal communication to enhance infection of D2R neurons (Simanjuntak et al. 2017). Microglial activation in response to JEV PAMPS or DAMPs triggers various inflammatory factors and cytokines such as TNF-α, IL-1β, IL6, RANTES, MCP-1, etc., which upon overproduction leads to neuronal damage (Ghoshal et al. 2007; Chen et al. 2011a, b; Yang et al. 2012; Lannes et al. 2017a; Chen et al. 2018b) (Fig. 10.3). During infection, the initial microglial activation and subsequent production of cyto-/chemokines is necessary to eliminate the pathogen which can be executed either by directly targeting the virus or by recruiting immune cells (Chen et al. 2011a, b). In response to cytokines, the immune cells including inflammatory monocytes (Terry et al. 2012) and JEV-specific T cells, may also be recruited to the infected brain. However, prolonged microglial activation is detrimental as it leads to a magnified proinflammatory response and enhanced immune cell infiltration which causes bystander neuronal cell death (Ghoshal et al. 2007; Wang et al. 2019; Singh et al. 2020). In several cases, the virus is cleared from the brain with minimal collateral damage, but in rare cases, heightened inflammation and direct infection to neurons may lead to neuronal cell death and damage to key centres in the brain with long-term deficiencies or a fatal outcome (Sarkari et al. 2012; Shirai et al. 2015). Chronic JEV infection of microglial cells (Thongtan et al. 2010; Lannes et al. 2017a) and lymphocytes (Sharma et al. 1991) has been reported, which increases the possibility of virus reactivation.

## Perspectives

Significant progress has been made in understanding the complex interplay of the JEV-host interaction at the cellular level. This has been augmented by high throughput omics studies that enable a holistic view of virus-driven changes in diverse cellular systems. The identification of crucial host dependency factors and pathways has also fuelled antiviral drug discovery. In addition, the cellular and animal model studies have given significant insights into the host immune response and disease pathogenesis. More epidemiological and molecular studies are required in the amplifying animal reservoirs (pigs and birds) and the transmitting insect vectors to better understand the virus propagation and spread. In the near future, we can expect to see advances driven by lipidomic, metabolomics, genomic and epigenomic studies that should enable biomarker development and enhance our understanding of their association with disease severity. These research-driven efforts will supplement disease management strategies and foster therapeutic development.

## References

Abdoli A, Alirezaei M, Mehrbod P, Forouzanfar F (2018) Autophagy: the multi-purpose bridge in viral infections and host cells. Rev Med Virol 28(4):e1973

Adhya D, Dutta K, Kundu K, Basu A (2013) Histone deacetylase inhibition by Japanese encephalitis virus in monocyte/macrophages: a novel viral immune evasion strategy. Immunobiology 218(10):1235–1247

Agarwal A, Alam MF, Basu B, Pattanayak S, Asthana S, Syed GH, Kalia M, Vrati S (2022) Japanese encephalitis virus NS4A protein interacts with PTEN-induced kinase 1 (PINK1) and promotes Mitophagy in infected cells. Microbiol Spectr 10(3):e0083022

Ahmad L, Mostowy S, Sancho-Shimizu V (2018) Autophagy-virus interplay: from cell biology to human disease. Front Cell Dev Biol 6:155

Aleyas AG, George JA, Han YW, Rahman MM, Kim SJ, Han SB, Kim BS, Kim K, Eo SK (2009) Functional modulation of dendritic cells and macrophages by Japanese encephalitis virus through MyD88 adaptor molecule-dependent and -independent pathways. J Immunol 183(4): 2462–2474

Aleyas AG, Han YW, George JA, Kim B, Kim K, Lee CK, Eo SK (2010) Multifront assault on antigen presentation by Japanese encephalitis virus subverts CD8+ T cell responses. J Immunol 185(3):1429–1441

Aleyas AG, Han YW, Patil AM, Kim SB, Kim K, Eo SK (2012) Impaired cross-presentation of CD8alpha+ CD11c+ dendritic cells by Japanese encephalitis virus in a TLR2/MyD88 signal pathway-dependent manner. Eur J Immunol 42(10):2655–2666

Anghelina D, Zhao J, Trandem K, Perlman S (2009) Role of regulatory T cells in coronavirus-induced acute encephalitis. Virology 385(2):358–367

Arakawa M, Morita E (2019) Flavivirus replication organelle biogenesis in the endoplasmic reticulum: comparison with other single-stranded positive-sense RNA viruses. Int J Mol Sci 20(9):2336

Arakawa M, Tabata K, Ishida K, Kobayashi M, Arai A, Ishikawa T, Suzuki R, Takeuchi H, Tripathi LP, Mizuguchi K, Morita E (2022) Flavivirus recruits the valosin-containing protein-NPL4 complex to induce stress granule disassembly for efficient viral genome replication. J Biol Chem 298(3):101597

Ashraf U, Ding Z, Deng S, Ye J, Cao S, Chen Z (2021) Pathogenicity and virulence of Japanese encephalitis virus: Neuroinflammation and neuronal cell damage. Virulence 12(1):968–980

Awais M, Wang K, Lin X, Qian W, Zhang N, Wang C, Wang K, Zhao L, Fu ZF, Cui M (2017) TLR7 deficiency leads to TLR8 compensative regulation of immune response against JEV in mice. Front Immunol 8:160

Bae W, Kim JH, Kim J, Lee J, Hwang ES (2018) Changes of epidemiological characteristics of Japanese encephalitis viral infection and birds as a potential viral transmitter in Korea. J Korean Med Sci 33(9):e70

Banerjee A, Tripathi A (2019). Recent advances in understanding Japanese encephalitis. F1000Res 8: F1000 Faculty Rev-1915.

Beasley DW, Li L, Suderman MT, Guirakhoo F, Trent DW, Monath TP, Shope RE, Barrett AD (2004) Protection against Japanese encephalitis virus strains representing four genotypes by passive transfer of sera raised against ChimeriVax-JE experimental vaccine. Vaccine 22(27–28):3722–3726

Bhattacharyya S, Sen U, Vrati S (2014) Regulated IRE1-dependent decay pathway is activated during Japanese encephalitis virus-induced unfolded protein response and benefits viral replication. J Gen Virol 95(Pt 1):71–79

Bhowmick S, Duseja R, Das S, Appaiahgiri MB, Vrati S, Basu A (2007) Induction of IP-10 (CXCL10) in astrocytes following Japanese encephalitis. Neurosci Lett 414(1):45–50

Bian P, Zheng X, Wei L, Ye C, Fan H, Cai Y, Zhang Y, Zhang F, Jia Z, Lei Y (2017) MLKL mediated necroptosis accelerates JEV-induced Neuroinflammation in mice. Front Microbiol 8: 303

Blazquez AB, Escribano-Romero E, Merino-Ramos T, Saiz JC, Martin-Acebes MA (2014) Stress responses in flavivirus-infected cells: activation of unfolded protein response and autophagy. Front Microbiol 5:266

Burke DS, Nisalak A, Ussery MA, Laorakpongse T, Chantavibul S (1985) Kinetics of IgM and IgG responses to Japanese encephalitis virus in human serum and cerebrospinal fluid. J Infect Dis 151(6):1093–1099

Campbell GL, Hills SL, Fischer M, Jacobson JA, Hoke CH, Hombach JM, Marfin AA, Solomon T, Tsai TF, Tsu VD, Ginsburg AS (2011) Estimated global incidence of Japanese encephalitis: a systematic review. Bull World Health Organ 89(10):766–774. 774A-774E

Cao S, Li Y, Ye J, Yang X, Chen L, Liu X, Chen H (2011) Japanese encephalitis virus wild strain infection suppresses dendritic cells maturation and function, and causes the expansion of regulatory T cells. Virol J 8:39

Carletti T, Zakaria MK, Faoro V, Reale L, Kazungu Y, Licastro D, Marcello A (2019) Viral priming of cell intrinsic innate antiviral signaling by the unfolded protein response. Nat Commun 10(1): 3889

Chambers TJ, Hahn CS, Galler R, Rice CM (1990) Flavivirus genome organization, expression, and replication. Annu Rev Microbiol 44:649–688

Chan ST, Ou JJ (2017) Hepatitis C virus-induced autophagy and host innate immune response. Viruses 9(8):224

Chang CY, Li JR, Chen WY, Ou YC, Lai CY, Hu YH, Wu CC, Chang CJ, Chen CJ (2015) Disruption of in vitro endothelial barrier integrity by Japanese encephalitis virus-infected astrocytes. Glia 63(11):1915–1932

Chapagain S, Pal Singh P, Le K, Safronetz D, Wood H, Karniychuk U (2022) Japanese encephalitis virus persists in the human reproductive epithelium and porcine reproductive tissues. PLoS Negl Trop Dis 16(7):e0010656

Chauhan S, Rathore DK, Sachan S, Lacroix-Desmazes S, Gupta N, Awasthi A, Vrati S, Kalia M (2021) Japanese encephalitis virus infected human monocyte-derived dendritic cells activate a transcriptional network leading to an antiviral inflammatory response. Front Immunol 12: 638694

Chen CJ, Chen JH, Chen SY, Liao SL, Raung SL (2004) Upregulation of RANTES gene expression in neuroglia by Japanese encephalitis virus infection. J Virol 78(22):12107–12119

Chen CJ, Ou YC, Chang CY, Pan HC, Liao SL, Raung SL, Chen SY (2011a) TNF-alpha and IL-1beta mediate Japanese encephalitis virus-induced RANTES gene expression in astrocytes. Neurochem Int 58(2):234–242

Chen CJ, Ou YC, Chang CY, Pan HC, Lin SY, Liao SL, Raung SL, Chen SY, Chang CJ (2011b) Src signaling involvement in Japanese encephalitis virus-induced cytokine production in microglia. Neurochem Int 58(8):924–933

Chen CJ, Ou YC, Chang CY, Pan HC, Liao SL, Chen SY, Raung SL, Lai CY (2012a) Glutamate released by Japanese encephalitis virus-infected microglia involves TNF-alpha signaling and contributes to neuronal death. Glia 60(3):487–501

Chen ST, Liu RS, Wu MF, Lin YL, Chen SY, Tan DT, Chou TY, Tsai IS, Li L, Hsieh SL (2012b) CLEC5A regulates Japanese encephalitis virus-induced neuroinflammation and lethality. PLoS Pathog 8(4):e1002655

Chen CJ, Ou YC, Li JR, Chang CY, Pan HC, Lai CY, Liao SL, Raung SL, Chang CJ (2014) Infection of pericytes in vitro by Japanese encephalitis virus disrupts the integrity of the endothelial barrier. J Virol 88(2):1150–1161

Chen Y, Zhou Z, Min W (2018a) Mitochondria, oxidative stress and innate immunity. Front Physiol 9:1487

Chen Z, Wang X, Ashraf U, Zheng B, Ye J, Zhou D, Zhang H, Song Y, Chen H, Zhao S, Cao S (2018b) Activation of neuronal N-methyl-D-aspartate receptor plays a pivotal role in Japanese encephalitis virus-induced neuronal cell damage. J Neuroinflammation 15(1):238

Cheng VCC, Sridhar S, Wong SC, Wong SCY, Chan JFW, Yip CCY, Chau CH, Au TWK, Hwang YY, Yau CSW, Lo JYC, Lee CK, Yuen KY (2018) Japanese encephalitis virus transmitted via blood transfusion, Hong Kong, China. Emerg Infect Dis 24(1):49

Chien YJ, Chen WJ, Hsu WL, Chiou SS (2008) Bovine lactoferrin inhibits Japanese encephalitis virus by binding to heparan sulfate and receptor for low density lipoprotein. Virology 379(1): 143–151

Chiou SS, Liu H, Chuang CK, Lin CC, Chen WJ (2005) Fitness of Japanese encephalitis virus to neuro-2a cells is determined by interactions of the viral envelope protein with highly sulfated glycosaminoglycans on the cell surface. J Med Virol 76(4):583–592

Chiu H, Chiu HP, Yu HP, Lin LH, Chen ZP, Lin YL, Lin RJ (2022) Zinc finger protein ZFP36L1 inhibits Flavivirus infection by both 5′-3' XRN1 and 3′-5' RNA-exosome RNA decay pathways. J Virol 96(1):e0166521

Choi JY, Kim JH, Patil AM, Kim SB, Uyangaa E, Hossain FMA, Eo SK (2017) Exacerbation of Japanese encephalitis by CD11c(hi) dendritic cell ablation is associated with an imbalance in regulatory Foxp3(+) and IL-17(+)CD4(+) Th17 cells and in Ly-6C(hi) and Ly-6C(lo) monocytes. Immune Netw 17(3):192–200

Choi Y, Bowman JW, Jung JU (2018) Autophagy during viral infection - a double-edged sword. Nat Rev Microbiol 16(6):341–354

Clarke P, Leser JS, Bowen RA, Tyler KL (2014) Virus-induced transcriptional changes in the brain include the differential expression of genes associated with interferon, apoptosis, interleukin 17 receptor a, and glutamate signaling as well as flavivirus-specific upregulation of tRNA synthetases. MBio 5(2):e00902–e00914

Clarke BD, Roby JA, Slonchak A, Khromykh AA (2015) Functional non-coding RNAs derived from the flavivirus 3′ untranslated region. Virus Res 206:53–61

Das S, Mishra MK, Ghosh J, Basu A (2008) Japanese encephalitis virus infection induces IL-18 and IL-1beta in microglia and astrocytes: correlation with in vitro cytokine responsiveness of glial cells and subsequent neuronal death. J Neuroimmunol 195(1–2):60–72

Das S, Chakraborty S, Basu A (2010) Critical role of lipid rafts in virus entry and activation of phosphoinositide 3′ kinase/Akt signaling during early stages of Japanese encephalitis virus infection in neural stem/progenitor cells. J Neurochem 115(2):537–549

Datan E, Roy SG, Germain G, Zali N, McLean JE, Golshan G, Harbajan S, Lockshin RA, Zakeri Z (2016) Dengue-induced autophagy, virus replication and protection from cell death require ER stress (PERK) pathway activation. Cell Death Dis 7:e2127

de Almeida A, de Almeida Rezende MS, Dantas SH, de Lima Silva S, de Oliveira J, de Lourdes Assuncao Araujo de Azevedo F, Alves R, de Menezes GMS, Santos PFD, Goncalves TAF, Schini-Kerth VB, de Medeiros IA (2020) Unveiling the role of inflammation and oxidative stress on age-related cardiovascular diseases. Oxidative Med Cell Longev 2020:1954398

Deng L, Wang W, Bian P, Wu M, Wang L, Lei Y, Lu Z, Zhai D (2022) QKI deficiency in macrophages protects mice against JEV infection by regulating cell migration and antiviral response. Mol Immunol 148:34–44

Dhuriya YK, Sharma D (2018) Necroptosis: a regulated inflammatory mode of cell death. J Neuroinflammation 15(1):199

Fadnis PR, Ravi V, Desai A, Turtle L, Solomon T (2013) Innate immune mechanisms in Japanese encephalitis virus infection: effect on transcription of pattern recognition receptors in mouse neuronal cells and brain tissue. Viral Immunol 26(6):366–377

Fernandez E, Kose N, Edeling MA, Adhikari J, Sapparapu G, Lazarte SM, Nelson CA, Govero J, Gross ML, Fremont DH, Crowe JE Jr, Diamond MS (2018) Mouse and human monoclonal antibodies protect against infection by multiple genotypes of Japanese encephalitis virus. MBio 9(1):e00008-18

Filgueira L, Lannes N (2019) Review of emerging Japanese encephalitis virus: new aspects and concepts about entry into the brain and inter-cellular spreading. Pathogens 8(3):111

Gao X, Liu H, Li X, Fu S, Cao L, Shao N, Zhang W, Wang Q, Lu Z, Lei W, He Y, Cao Y, Wang H, Liang G (2019) Changing geographic distribution of Japanese encephalitis virus genotypes, 1935-2017. Vector Borne Zoonotic Dis 19(1):35–44

Garcia-Nicolas O, Braun RO, Milona P, Lewandowska M, Dijkman R, Alves MP, Summerfield A (2018) Targeting of the nasal mucosa by Japanese encephalitis virus for non-vector-borne transmission. J Virol 92(24):e01091-18

Garcia-Nicolas O, Lewandowska M, Ricklin ME, Summerfield A (2019) Monocyte-derived dendritic cells as model to evaluate species tropism of mosquito-borne Flaviviruses. Front Cell Infect Microbiol 9:5

Ghosh D, Basu A (2009) Japanese encephalitis-a pathological and clinical perspective. PLoS Negl Trop Dis 3(9):e437

Ghoshal A, Das S, Ghosh S, Mishra MK, Sharma V, Koli P, Sen E, Basu A (2007) Proinflammatory mediators released by activated microglia induces neuronal death in Japanese encephalitis. Glia 55(5):483–496

Go YM, Jones DP (2008) Redox compartmentalization in eukaryotic cells. Biochim Biophys Acta 1780(11):1273–1290

Goncalvez AP, Chien CH, Tubthong K, Gorshkova I, Roll C, Donau O, Schuck P, Yoksan S, Wang SD, Purcell RH, Lai CJ (2008) Humanized monoclonal antibodies derived from chimpanzee Fabs protect against Japanese encephalitis virus in vitro and in vivo. J Virol 82(14):7009–7021

Gupta AK, Lad VJ, Koshy AA (2003) Protection of mice against experimental Japanese encephalitis virus infections by neutralizing anti-glycoprotein E monoclonal antibodies. Acta Virol 47(3):141–145

Gupta N, Lomash V, Rao PV (2010a) Expression profile of Japanese encephalitis virus induced neuroinflammation and its implication in disease severity. J Clin Virol 49(1):4–10

Gupta N, Santhosh SR, Babu JP, Parida MM, Rao PV (2010b) Chemokine profiling of Japanese encephalitis virus-infected mouse neuroblastoma cells by microarray and real-time RT-PCR: implication in neuropathogenesis. Virus Res 147(1):107–112

Gupta N, Bhaskar AS, Lakshmana Rao PV (2011) Transcriptional regulation and activation of the mitogen-activated protein kinase pathway after Japanese encephalitis virus infection in neuroblastoma cells. FEMS Immunol Med Microbiol 62(1):110–121

Gupta N, Hegde P, Lecerf M, Nain M, Kaur M, Kalia M, Vrati S, Bayry J, Lacroix-Desmazes S, Kaveri SV (2014) Japanese encephalitis virus expands regulatory T cells by increasing the expression of PD-L1 on dendritic cells. Eur J Immunol 44(5):1363–1374

Gupta MK, Behera SK, Dehury B, Mahapatra N (2017) Identification and characterization of differentially expressed genes from human microglial cell samples infected with Japanese encephalitis virus. J Vector Borne Dis 54(2):131–138

Han YW, Choi JY, Uyangaa E, Kim SB, Kim JH, Kim BS, Kim K, Eo SK (2014) Distinct dictation of Japanese encephalitis virus-induced neuroinflammation and lethality via triggering TLR3 and TLR4 signal pathways. PLoS Pathog 10(9):e1004319

He W, Zhao Z, Anees A, Li Y, Ashraf U, Chen Z, Song Y, Chen H, Cao S, Ye J (2017) p21-activated kinase 4 signaling promotes Japanese encephalitis virus-mediated inflammation in astrocytes. Front Cell Infect Microbiol 7:271

Hsiao NW, Chen JW, Yang TC, Orloff GM, Wu YY, Lai CH, Lan YC, Lin CW (2010) ISG15 overexpression inhibits replication of the Japanese encephalitis virus in human medulloblastoma cells. Antivir Res 85(3):504–511

Hsieh JT, St John AL (2020) Japanese encephalitis virus and its mechanisms of neuroinvasion. PLoS Pathog 16(4):e1008260

Imre G (2020) Cell death signalling in virus infection. Cell Signal 76:109772

Jain N, Oswal N, Chawla AS, Agrawal T, Biswas M, Vrati S, Rath S, George A, Bal V, Medigeshi GR (2017) CD8 T cells protect adult naive mice from JEV-induced morbidity via lytic function. PLoS Negl Trop Dis 11(2):e0005329

James EA, Gates TJ, LaFond RE, Yamamoto S, Ni C, Mai D, Gersuk VH, O'Brien K, Nguyen QA, Zeitner B, Lanteri MC, Norris PJ, Chaussabel D, Malhotra U, Kwok WW (2016) Neuroinvasive West Nile infection elicits elevated and atypically polarized T cell responses that promote a pathogenic outcome. PLoS Pathog 12(1):e1005375

Jiang R, Ye J, Zhu B, Song Y, Chen H, Cao S (2014) Roles of TLR3 and RIG-I in mediating the inflammatory response in mouse microglia following Japanese encephalitis virus infection. J Immunol Res 2014:787023

Jin R, Zhu W, Cao S, Chen R, Jin H, Liu Y, Wang S, Wang W, Xiao G (2013) Japanese encephalitis virus activates autophagy as a viral immune evasion strategy. PLoS One 8(1):e52909

Johnsen DO, Edelman R, Grossman RA, Muangman D, Pomsdhit J, Gould DJ (1974) Study of Japanese encephalitis virus in Chiangmia Valley, Thailand. V. Animal infections. Am J Epidemiol 100(1):57–68

Kalia M, Khasa R, Sharma M, Nain M, Vrati S (2013) Japanese encephalitis virus infects neuronal cells through a clathrin-independent endocytic mechanism. J Virol 87(1):148–162

Kato H, Takeuchi O, Sato S, Yoneyama M, Yamamoto M, Matsui K, Uematsu S, Jung A, Kawai T, Ishii KJ, Yamaguchi O, Otsu K, Tsujimura T, Koh CS, Reis C, Sousa YM, Fujita T, Akira S (2006) Differential roles of MDA5 and RIG-I helicases in the recognition of RNA viruses. Nature 441(7089):101–105

Katoh H, Okamoto T, Fukuhara T, Kambara H, Morita E, Mori Y, Kamitani W, Matsuura Y (2013) Japanese encephalitis virus core protein inhibits stress granule formation through an interaction with Caprin-1 and facilitates viral propagation. J Virol 87(1):489–502
Kaushik DK, Gupta M, Basu A (2011) Microglial response to viral challenges: every silver lining comes with a cloud. Front Biosci (Landmark Ed) 16:2187–2205
Keaney J, Campbell M (2015) The dynamic blood-brain barrier. FEBS J 282(21):4067–4079
Khasa R, Vaidya A, Vrati S, Kalia M (2019) Membrane trafficking RNA interference screen identifies a crucial role of the clathrin endocytic pathway and ARP2/3 complex for Japanese encephalitis virus infection in HeLa cells. J Gen Virol 100(2):176–186
Khasa R, Sharma P, Vaidya A, Vrati S, Kalia M (2020) Proteins involved in actin filament organization are key host factors for Japanese encephalitis virus life-cycle in human neuronal cells. Microb Pathog 149:104565
Kim JH, Choi JY, Kim SB, Uyangaa E, Patil AM, Han YW, Park SY, Lee JH, Kim K, Eo SK (2015) CD11c(hi) dendritic cells regulate Ly-6C(hi) monocyte differentiation to preserve immune-privileged CNS in lethal Neuroinflammation. Sci Rep 5:17548
Kimura-Kuroda J, Yasui K (1988) Protection of mice against Japanese encephalitis virus by passive administration with monoclonal antibodies. J Immunol 141(10):3606–3610
Konishi E, Yamaoka M, Khin Sane W, Kurane I, Takada K, Mason PW (1999) The anamnestic neutralizing antibody response is critical for protection of mice from challenge following vaccination with a plasmid encoding the Japanese encephalitis virus premembrane and envelope genes. J Virol 73(7):5527–5534
Kumar S, Kalita J, Saxena V, Khan MY, Khanna VK, Sharma S, Dhole TN, Misra UK (2009a) Some observations on the tropism of Japanese encephalitis virus in rat brain. Brain Res 1268: 135–141
Kumar S, Misra UK, Kalita J, Khanna VK, Khan MY (2009b) Imbalance in oxidant/antioxidant system in different brain regions of rat after the infection of Japanese encephalitis virus. Neurochem Int 55(7):648–654
Lannes N, Neuhaus V, Scolari B, Kharoubi-Hess S, Walch M, Summerfield A, Filgueira L (2017a) Interactions of human microglia cells with Japanese encephalitis virus. Virol J 14(1):8
Lannes N, Summerfield A, Filgueira L (2017b) Regulation of inflammation in Japanese encephalitis. J Neuroinflammation 14(1):158
Lanteri MC, O'Brien KM, Purtha WE, Cameron MJ, Lund JM, Owen RE, Heitman JW, Custer B, Hirschkorn DF, Tobler LH, Kiely N, Prince HE, Ndhlovu LC, Nixon DF, Kamel HT, Kelvin DJ, Busch MP, Rudensky AY, Diamond MS, Norris PJ (2009) Tregs control the development of symptomatic West Nile virus infection in humans and mice. J Clin Invest 119(11):3266–3277
Larena M, Regner M, Lobigs M (2013) Cytolytic effector pathways and IFN-gamma help protect against Japanese encephalitis. Eur J Immunol 43(7):1789–1798
Lee T, Komiya T, Watanabe K, Aizawa C, Hashimoto H (1995) Immune response in mice infected with the attenuated Japanese encephalitis vaccine strain SA14-14-2. Acta Virol 39(3):161–164
Lee EJ, Cha GW, Ju YR, Han MG, Lee WJ, Jeong YE (2016) Prevalence of neutralizing antibodies to Japanese encephalitis virus among high-risk age groups in South Korea, 2010. PLoS One 11(1):e0147841
Lennemann NJ, Coyne CB (2015) Catch me if you can: the link between autophagy and viruses. PLoS Pathog 11(3):e1004685
Li L, Lok SM, Yu IM, Zhang Y, Kuhn RJ, Chen J, Rossmann MG (2008) The flavivirus precursor membrane-envelope protein complex: structure and maturation. Science 319(5871):1830–1834
Li Y, Ye J, Yang X, Xu M, Chen L, Mei L, Zhu J, Liu X, Chen H, Cao S (2011) Infection of mouse bone marrow-derived dendritic cells by live attenuated Japanese encephalitis virus induces cells maturation and triggers T cells activation. Vaccine 29(4):855–862
Li JK, Liang JJ, Liao CL, Lin YL (2012a) Autophagy is involved in the early step of Japanese encephalitis virus infection. Microbes Infect 14(2):159–168
Li K, Li NL, Wei D, Pfeffer SR, Fan M, Pfeffer LM (2012b) Activation of chemokine and inflammatory cytokine response in hepatitis C virus-infected hepatocytes depends on toll-like

receptor 3 sensing of hepatitis C virus double-stranded RNA intermediates. Hepatology 55(3): 666–675

Li F, Wang Y, Yu L, Cao S, Wang K, Yuan J, Wang C, Wang K, Cui M, Fu ZF (2015) Viral infection of the central nervous system and Neuroinflammation precede blood-brain barrier disruption during Japanese encephalitis virus infection. J Virol 89(10):5602–5614

Li XF, Li XD, Deng CL, Dong HL, Zhang QY, Ye Q, Ye HQ, Huang XY, Deng YQ, Zhang B, Qin CF (2017) Visualization of a neurotropic flavivirus infection in mouse reveals unique viscerotropism controlled by host type I interferon signaling. Theranostics 7(4):912–925

Li M, Yang J, Ye C, Bian P, Yang X, Zhang H, Luo C, Xue Z, Lei Y, Lian J (2021a) Integrated metabolomics and transcriptomics analyses reveal metabolic landscape in neuronal cells during JEV infection. Virol Sin 36(6):1554–1565

Li Q, Zhou D, Jia F, Zhang L, Ashraf U, Li Y, Duan H, Song Y, Chen H, Cao S, Ye J (2021b) Japanese encephalitis virus NS1' protein interacts with host CDK1 protein to regulate antiviral response. Microbiol Spectr 9(3):e0166121

Liao SL, Raung SL, Chen CJ (2002) Japanese encephalitis virus stimulates superoxide dismutase activity in rat glial cultures. Neurosci Lett 324(2):133–136

Liao KC, Chuo V, Fagg WS, Modahl CM, Widen S, Garcia-Blanco MA (2021) The RNA binding protein quaking represses splicing of the fibronectin EDA exon and downregulates the interferon response. Nucleic Acids Res 49(17):10034–10045

Lin YL, Chen LK, Liao CL, Yeh CT, Ma SH, Chen JL, Huang YL, Chen SS, Chiang HY (1998) DNA immunization with Japanese encephalitis virus nonstructural protein NS1 elicits protective immunity in mice. J Virol 72(1):191–200

Lin RJ, Liao CL, Lin YL (2004) Replication-incompetent virions of Japanese encephalitis virus trigger neuronal cell death by oxidative stress in a culture system. J Gen Virol 85(Pt 2):521–533

Lin RJ, Chang BL, Yu HP, Liao CL, Lin YL (2006) Blocking of interferon-induced Jak-Stat signaling by Japanese encephalitis virus NS5 through a protein tyrosine phosphatase-mediated mechanism. J Virol 80(12):5908–5918

Lin CW, Cheng CW, Yang TC, Li SW, Cheng MH, Wan L, Lin YJ, Lai CH, Lin WY, Kao MC (2008) Interferon antagonist function of Japanese encephalitis virus NS4A and its interaction with DEAD-box RNA helicase DDX42. Virus Res 137(1):49–55

Lin RJ, Chien HL, Lin SY, Chang BL, Yu HP, Tang WC, Lin YL (2013) MCPIP1 ribonuclease exhibits broad-spectrum antiviral effects through viral RNA binding and degradation. Nucleic Acids Res 41(5):3314–3326

Lindqvist R, Mundt F, Gilthorpe JD, Wolfel S, Gekara NO, Kroger A, Overby AK (2016) Fast type I interferon response protects astrocytes from flavivirus infection and virus-induced cytopathic effects. J Neuroinflammation 13(1):277

Liu CY, Kaufman RJ (2003) The unfolded protein response. J Cell Sci 116(Pt 10):1861–1862

Liu K, Liao X, Zhou B, Yao H, Fan S, Chen P, Miao D (2013) Porcine alpha interferon inhibit Japanese encephalitis virus replication by different ISGs in vitro. Res Vet Sci 95(3):950–956

Liu CC, Zhang YN, Li ZY, Hou JX, Zhou J, Kan L, Zhou B, Chen PY (2017) Rab5 and Rab11 are required for Clathrin-dependent endocytosis of Japanese encephalitis virus in BHK-21 cells. J Virol 91(19):e01113-17

Liu K, Xiao C, Xi S, Hameed M, Wahaab A, Shao D, Li Z, Li B, Wei J, Qiu Y, Miao D, Zhu H, Ma Z (2020) Mosquito defensins enhance Japanese encephalitis virus infection by facilitating virus adsorption and entry within the mosquito. J Virol 94(21):e01164-20

Lund JM, Hsing L, Pham TT, Rudensky AY (2008) Coordination of early protective immunity to viral infection by regulatory T cells. Science 320(5880):1220–1224

Ma L, Li F, Zhang JW, Li W, Zhao DM, Wang H, Hua RH, Bu ZG (2018) Host factor SPCS1 regulates the replication of Japanese encephalitis virus through interactions with transmembrane domains of NS2B. J Virol 92(12):e00197-18

Manocha GD, Mishra R, Sharma N, Kumawat KL, Basu A, Singh SK (2014) Regulatory role of TRIM21 in the type-I interferon pathway in Japanese encephalitis virus-infected human microglial cells. J Neuroinflammation 11:24

Mansfield KL, Hernandez-Triana LM, Banyard AC, Fooks AR, Johnson N (2017) Japanese encephalitis virus infection, diagnosis and control in domestic animals. Vet Microbiol 201: 85–92

McLean JE, Wudzinska A, Datan E, Quaglino D, Zakeri Z (2011) Flavivirus NS4A-induced autophagy protects cells against death and enhances virus replication. J Biol Chem 286(25): 22147–22159

Mishra MK, Kumawat KL, Basu A (2008) Japanese encephalitis virus differentially modulates the induction of multiple pro-inflammatory mediators in human astrocytoma and astroglioma cell-lines. Cell Biol Int 32(12):1506–1513

Misra UK, Kalita J (2010) Overview: Japanese encephalitis. Prog Neurobiol 91(2):108–120

Miyake M (1964) The pathology of Japanese encephalitis. A review. Bull World Health Organ 30: 153–160

Mukherjee S, Singh N, Sengupta N, Fatima M, Seth P, Mahadevan A, Shankar SK, Bhattacharyya A, Basu A (2017) Japanese encephalitis virus induces human neural stem/ progenitor cell death by elevating GRP78, PHB and hnRNPC through ER stress. Cell Death Dis 8(1):e2556

Mukherjee S, Sengupta N, Chaudhuri A, Akbar I, Singh N, Chakraborty S, Suryawanshi AR, Bhattacharyya A, Basu A (2018) PLVAP and GKN3 are two critical host cell receptors which facilitate Japanese encephalitis virus entry into neurons. Sci Rep 8(1):11784

Mukherjee S, Akbar I, Kumari B, Vrati S, Basu A, Banerjee A (2019) Japanese encephalitis virus-induced let-7a/b interacted with the NOTCH-TLR7 pathway in microglia and facilitated neuronal death via caspase activation. J Neurochem 149(4):518–534

Mulvey P, Duong V, Boyer S, Burgess G, Williams DT, Dussart P, Horwood PF (2021) The ecology and evolution of Japanese encephalitis virus. Pathogens 10(12):1534

Nain M, Abdin MZ, Kalia M, Vrati S (2016) Japanese encephalitis virus invasion of cell: allies and alleys. Rev Med Virol 26(2):129–141

Nain M, Mukherjee S, Karmakar SP, Paton AW, Paton JC, Abdin MZ, Basu A, Kalia M, Vrati S (2017) GRP78 is an important host factor for Japanese encephalitis virus entry and replication in mammalian cells. J Virol 91(6):e02274-16

Nazmi A, Dutta K, Basu A (2011) RIG-I mediates innate immune response in mouse neurons following Japanese encephalitis virus infection. PLoS One 6(6):e21761

Nazmi A, Mukherjee S, Kundu K, Dutta K, Mahadevan A, Shankar SK, Basu A (2014) TLR7 is a key regulator of innate immunity against Japanese encephalitis virus infection. Neurobiol Dis 69:235–247

Niu J, Jiang Y, Xu H, Zhao C, Zhou G, Chen P, Cao R (2018) TIM-1 promotes Japanese encephalitis virus entry and infection. Viruses 10(11):630

Olagnier D, Peri S, Steel C, van Montfoort N, Chiang C, Beljanski V, Slifker M, He Z, Nichols CN, Lin R, Balachandran S, Hiscott J (2014) Cellular oxidative stress response controls the antiviral and apoptotic programs in dengue virus-infected dendritic cells. PLoS Pathog 10(12):e1004566

Pan XL, Liu H, Wang HY, Fu SH, Liu HZ, Zhang HL, Li MH, Gao XY, Wang JL, Sun XH, Lu XJ, Zhai YG, Meng WS, He Y, Wang HQ, Han N, Wei B, Wu YG, Feng Y, Yang DJ, Wang LH, Tang Q, Xia G, Kurane I, Rayner S, Liang GD (2011) Emergence of genotype I of Japanese encephalitis virus as the dominant genotype in Asia. J Virol 85(19):9847–9853

Patabendige A, Michael BD, Craig AG, Solomon T (2018) Brain microvascular endothelial-astrocyte cell responses following Japanese encephalitis virus infection in an in vitro human blood-brain barrier model. Mol Cell Neurosci 89:60–70

Plotkin SA (2010) Correlates of protection induced by vaccination. Clin Vaccine Immunol 17(7): 1055–1065

Preziuso S, Mari S, Mariotti F, Rossi G (2018) Detection of Japanese encephalitis virus in bone marrow of healthy young wild birds collected in 1997-2000 in Central Italy. Zoonoses Public Health 65(7):798–804

Qi ZL, Sun LY, Bai J, Zhuang HZ, Duan ML (2020) Japanese encephalitis following liver transplantation: a rare case report. World J Clin Cases 8(2):337–342

Qiu X, Lei Y, Yang P, Gao Q, Wang N, Cao L, Yuan S, Huang X, Deng Y, Ma W, Ding T, Zhang F, Wu X, Hu J, Liu SL, Qin C, Wang X, Xu Z, Rao Z (2018) Structural basis for neutralization of Japanese encephalitis virus by two potent therapeutic antibodies. Nat Microbiol 3(3):287–294

Raung SL, Kuo MD, Wang YM, Chen CJ (2001) Role of reactive oxygen intermediates in Japanese encephalitis virus infection in murine neuroblastoma cells. Neurosci Lett 315(1–2):9–12

Raung SL, Chen SY, Liao SL, Chen JH, Chen CJ (2005) Tyrosine kinase inhibitors attenuate Japanese encephalitis virus-induced neurotoxicity. Biochem Biophys Res Commun 327(2): 399–406

Raung SL, Chen SY, Liao SL, Chen JH, Chen CJ (2007) Japanese encephalitis virus infection stimulates Src tyrosine kinase in neuron/glia. Neurosci Lett 419(3):263–268

Ricklin ME, Garcia-Nicolas O, Brechbuhl D, Python S, Zumkehr B, Nougairede A, Charrel RN, Posthaus H, Oevermann A, Summerfield A (2016) Vector-free transmission and persistence of Japanese encephalitis virus in pigs. Nat Commun 7:10832

Rodrigues FM, Guttikar SN, Pinto BD (1981) Prevalence of antibodies to Japanese encephalitis and West Nile viruses among wild birds in the Krishna-Godavari Delta, Andhra Pradesh, India. Trans R Soc Trop Med Hyg 75(2):258–262

Roy J, Galano JM, Durand T, Le Guennec JY, Lee JC (2017) Physiological role of reactive oxygen species as promoters of natural defenses. FASEB J 31(9):3729–3745

Rozpedek W, Pytel D, Mucha B, Leszczynska H, Diehl JA, Majsterek I (2016) The role of the PERK/eIF2alpha/ATF4/CHOP signaling pathway in tumor progression during endoplasmic reticulum stress. Curr Mol Med 16(6):533–544

Salimi H, Cain MD, Klein RS (2016) Encephalitic arboviruses: emergence, clinical presentation, and Neuropathogenesis. Neurotherapeutics 13(3):514–534

Sarkar R, Sharma KB, Kumari A, Asthana S, Kalia M (2020) Japanese encephalitis virus capsid protein interacts with non-lipidated MAP1LC3 on replication membranes and lipid droplets. J Gen Virol 102(1):1–14

Sarkari NB, Thacker AK, Barthwal SP, Mishra VK, Prapann S, Srivastava D, Sarkari M (2012) Japanese encephalitis (JE) part II: 14 years' follow-up of survivors. J Neurol 259(1):58–69

Sehrawat S, Khasa R, Deb A, Prajapat SK, Mallick S, Basu A, Surjit M, Kalia M, Vrati S (2021) Valosin-containing protein/p97 plays critical roles in the Japanese encephalitis virus life cycle. J Virol 95(11):e02336-20

Sharma S, Mathur A, Prakash V, Kulshreshtha R, Kumar R, Chaturvedi UC (1991) Japanese encephalitis virus latency in peripheral blood lymphocytes and recurrence of infection in children. Clin Exp Immunol 85(1):85–89

Sharma M, Bhattacharyya S, Nain M, Kaur M, Sood V, Gupta V, Khasa R, Abdin MZ, Vrati S, Kalia M (2014) Japanese encephalitis virus replication is negatively regulated by autophagy and occurs on LC3-I- and EDEM1-containing membranes. Autophagy 10(9):1637–1651

Sharma M, Bhattacharyya S, Sharma KB, Chauhan S, Asthana S, Abdin MZ, Vrati S, Kalia M (2017) Japanese encephalitis virus activates autophagy through XBP1 and ATF6 ER stress sensors in neuronal cells. J Gen Virol 98(5):1027–1039

Sharma M, Sharma KB, Chauhan S, Bhattacharyya S, Vrati S, Kalia M (2018) Diphenyleneiodonium enhances oxidative stress and inhibits Japanese encephalitis virus induced autophagy and ER stress pathways. Biochem Biophys Res Commun 502(2):232–237

Sharma KB, Sharma M, Aggarwal S, Yadav AK, Bhatnagar S, Vrati S, Kalia M (2019) Quantitative proteome analysis of Atg5-deficient mouse embryonic fibroblasts reveals the range of the autophagy-modulated basal cellular proteome. mSystems 4(6):e00481-19

Sharma KB, Chhabra S, Aggarwal S, Tripathi A, Banerjee A, Yadav AK, Vrati S, Kalia M (2021a) Proteomic landscape of Japanese encephalitis virus-infected fibroblasts. J Gen Virol 102(9): 001657

Sharma KB, Vrati S, Kalia M (2021b) Pathobiology of Japanese encephalitis virus infection. Mol Asp Med 81:100994

Shirai K, Hayasaka D, Kitaura K, Takasaki T, Morita K, Suzuki R, Kurane I (2015) Qualitative differences in brain-infiltrating T cells are associated with a fatal outcome in mice infected with Japanese encephalitis virus. Arch Virol 160(3):765–775
Sikazwe C, Neave MJ, Michie A, Mileto P, Wang J, Cooper N, Levy A, Imrie A, Baird RW, Currie BJ, Speers D, Mackenzie JS, Smith DW, Williams DT (2022) Molecular detection and characterisation of the first Japanese encephalitis virus belonging to genotype IV acquired in Australia. PLoS Negl Trop Dis 16(11):e0010754
Simanjuntak Y, Liang JJ, Lee YL, Lin YL (2017) Japanese encephalitis virus exploits dopamine D2 receptor-phospholipase C to target dopaminergic human neuronal cells. Front Microbiol 8:651
Simon-Loriere E, Faye O, Prot M, Casademont I, Fall G, Fernandez-Garcia MD, Diagne MM, Kipela JM, Fall IS, Holmes EC, Sakuntabhai A, Sall AA (2017) Autochthonous Japanese encephalitis with yellow fever coinfection in Africa. N Engl J Med 376(15):1483–1485
Singh S, Singh G, Tiwari S, Kumar A (2020) CCR2 inhibition reduces neurotoxic microglia activation phenotype after Japanese encephalitis viral infection. Front Cell Neurosci 14:230
Sips GJ, Wilschut J, Smit JM (2012) Neuroinvasive flavivirus infections. Rev Med Virol 22(2): 69–87
Solomon T (2003) Recent advances in Japanese encephalitis. J Neurovirol 9(2):274–283
Solomon T (2004) Flavivirus encephalitis. N Engl J Med 351(4):370–378
Solomon T, Ni H, Beasley DW, Ekkelenkamp M, Cardosa MJ, Barrett AD (2003) Origin and evolution of Japanese encephalitis virus in Southeast Asia. J Virol 77(5):3091–3098
Sood V, Sharma KB, Gupta V, Saha D, Dhapola P, Sharma M, Sen U, Kitajima S, Chowdhury S, Kalia M, Vrati S (2017) ATF3 negatively regulates cellular antiviral signaling and autophagy in the absence of type I interferons. Sci Rep 7(1):8789
Sooryanarain H, Ayachit V, Gore M (2012) Activated CD56(+) lymphocytes (NK+NKT) mediate immunomodulatory and anti-viral effects during Japanese encephalitis virus infection of dendritic cells in-vitro. Virology 432(2):250–260
Srivastava S, Khanna N, Saxena SK, Singh A, Mathur A, Dhole TN (1999) Degradation of Japanese encephalitis virus by neutrophils. Int J Exp Pathol 80(1):17–24
Stadler K, Allison SL, Schalich J, Heinz FX (1997) Proteolytic activation of tick-borne encephalitis virus by furin. J Virol 71(11):8475–8481
Su HL, Liao CL, Lin YL (2002) Japanese encephalitis virus infection initiates endoplasmic reticulum stress and an unfolded protein response. J Virol 76(9):4162–4171
Suman V, Roy U, Panwar A, Raizada A (2016) Japanese encephalitis complicated with obstructive hydrocephalus. J Clin Diagn Res 10(2):OD18-20
Swarup V, Ghosh J, Duseja R, Ghosh S, Basu A (2007) Japanese encephalitis virus infection decrease endogenous IL-10 production: correlation with microglial activation and neuronal death. Neurosci Lett 420(2):144–149
Tabata K, Arakawa M, Ishida K, Kobayashi M, Nara A, Sugimoto T, Okada T, Mori K, Morita E (2021) Endoplasmic reticulum-associated degradation controls virus protein homeostasis, which is required for Flavivirus propagation. J Virol 95(15):e0223420
Tarlinton D (2019) B cells still front and Centre in immunology. Nat Rev Immunol 19(2):85–86
Terry RL, Getts DR, Deffrasnes C, van Vreden C, Campbell IL, King NJ (2012) Inflammatory monocytes and the pathogenesis of viral encephalitis. J Neuroinflammation 9:270
Thongtan T, Cheepsunthorn P, Chaiworakul V, Rattanarungsan C, Wikan N, Smith DR (2010) Highly permissive infection of microglial cells by Japanese encephalitis virus: a possible role as a viral reservoir. Microbes Infect 12(1):37–45
Thongtan T, Thepparit C, Smith DR (2012) The involvement of microglial cells in Japanese encephalitis infections. Clin Dev Immunol 2012:890586
Tripathi A, Singh Rawat B, Addya S, Surjit M, Tailor P, Vrati S, Banerjee A (2021) Lack of interferon (IFN) regulatory factor 8 associated with restricted IFN-gamma response augmented Japanese encephalitis virus replication in the mouse brain. J Virol 95(21):e0040621

Tsao CH, Su HL, Lin YL, Yu HP, Kuo SM, Shen CI, Chen CW, Liao CL (2008) Japanese encephalitis virus infection activates caspase-8 and -9 in a FADD-independent and mitochondrion-dependent manner. J Gen Virol 89(Pt 8):1930–1941

Tu YC, Yu CY, Liang JJ, Lin E, Liao CL, Lin YL (2012) Blocking double-stranded RNA-activated protein kinase PKR by Japanese encephalitis virus nonstructural protein 2A. J Virol 86(19): 10347–10358

Turtle L, Solomon T (2018) Japanese encephalitis - the prospects for new treatments. Nat Rev Neurol 14(5):298–313

Turtle L, Bali T, Buxton G, Chib S, Chan S, Soni M, Hussain M, Isenman H, Fadnis P, Venkataswamy MM, Satishkumar V, Lewthwaite P, Kurioka A, Krishna S, Shankar MV, Ahmed R, Begum A, Ravi V, Desai A, Yoksan S, Fernandez S, Willberg CB, Kloverpris HN, Conlon C, Klenerman P, Satchidanandam V, Solomon T (2016) Human T cell responses to Japanese encephalitis virus in health and disease. J Exp Med 213(7):1331–1352

Uchil PD, Satchidanandam V (2003) Characterization of RNA synthesis, replication mechanism, and in vitro RNA-dependent RNA polymerase activity of Japanese encephalitis virus. Virology 307(2):358–371

van den Hurk AF, Ritchie SA, Mackenzie JS (2009) Ecology and geographical expansion of Japanese encephalitis virus. Annu Rev Entomol 54:17–35

Van Gessel Y, Klade CS, Putnak R, Formica A, Krasaesub S, Spruth M, Cena B, Tungtaeng A, Gettayacamin M, Dewasthaly S (2011) Correlation of protection against Japanese encephalitis virus and JE vaccine (IXIARO((R))) induced neutralizing antibody titers. Vaccine 29(35): 5925–5931

Vashist S, Bhullar D, Vrati S (2011) La protein can simultaneously bind to both 3′- and 5′--noncoding regions of Japanese encephalitis virus genome. DNA Cell Biol 30(6):339–346

Vedagiri D, Gupta D, Mishra A, Krishna G, Bhaskar M, Sah V, Basu A, Nayak D, Kalia M, Valiya Veettil M, Harshan KH (2021) Retinoic acid-inducible gene I-like receptors activate snail to limit RNA viral infections. J Virol 95(21):e0121621

Villabona-Rueda A, Erice C, Pardo CA, Stins MF (2019) The evolving concept of the blood brain barrier (BBB): from a single static barrier to a heterogeneous and dynamic relay center. Front Cell Neurosci 13:405

Wang P, Hu K, Luo S, Zhang M, Deng X, Li C, Jin W, Hu B, He S, Li M, Du T, Xiao G, Zhang B, Liu Y, Hu Q (2016) DC-SIGN as an attachment factor mediates Japanese encephalitis virus infection of human dendritic cells via interaction with a single high-mannose residue of viral E glycoprotein. Virology 488:108–119

Wang C, Zhang N, Qi L, Yuan J, Wang K, Wang K, Ma S, Wang H, Lou W, Hu P, Awais M, Cao S, Fu ZF, Cui M (2017a) Myeloid-derived suppressor cells inhibit T follicular helper cell immune response in Japanese encephalitis virus infection. J Immunol 199(9):3094–3105

Wang P, Li M, Lu W, Zhang D, Hu Q, Liu Y (2017b) DC-SIGN promotes Japanese encephalitis virus transmission from dendritic cells to T cells via virological synapses. Virol Sin 32(6): 495–502

Wang X, Li SH, Zhu L, Nian QG, Yuan S, Gao Q, Hu Z, Ye Q, Li XF, Xie DY, Shaw N, Wang J, Walter TS, Huiskonen JT, Fry EE, Qin CF, Stuart DI, Rao Z (2017c) Near-atomic structure of Japanese encephalitis virus reveals critical determinants of virulence and stability. Nat Commun 8(1):14

Wang K, Wang H, Lou W, Ma L, Li Y, Zhang N, Wang C, Li F, Awais M, Cao S, She R, Fu ZF, Cui M (2018) IP-10 promotes blood-brain barrier damage by inducing tumor necrosis factor alpha production in Japanese encephalitis. Front Immunol 9:1148

Wang Q, Xin X, Wang T, Wan J, Ou Y, Yang Z, Yu Q, Zhu L, Guo Y, Wu Y, Ding Z, Zhang Y, Pan Z, Tang Y, Li S, Kong L (2019) Japanese encephalitis virus induces apoptosis and encephalitis by activating the PERK pathway. J Virol 93(17):e00887-19

Wang ZY, Zhen ZD, Fan DY, Qin CF, Han DS, Zhou HN, Wang PG, An J (2020a) Axl deficiency promotes the Neuroinvasion of Japanese encephalitis virus by enhancing IL-1alpha production from Pyroptotic macrophages. J Virol 94(17):e00602–e0e620

Wang ZY, Zhen ZD, Fan DY, Wang PG, An J (2020b) Transcriptomic analysis suggests the M1 polarization and launch of diverse programmed cell death pathways in Japanese encephalitis virus-infected macrophages. Viruses 12(3):356

Wang X, Wang G, Yang H, Fu S, He Y, Li F, Wang H, Wang Z (2022) A mouse model of peripheral nerve injury induced by Japanese encephalitis virus. PLoS Negl Trop Dis 16(11): e0010961

Wongchitrat P, Samutpong A, Lerdsamran H, Prasertsopon J, Yasawong M, Govitrapong P, Puthavathana P, Kitidee K (2019) Elevation of cleaved p18 Bax levels associated with the kinetics of neuronal cell death during Japanese encephalitis virus infection. Int J Mol Sci 20(20): 5016

Xu Q, Cao M, Song H, Chen S, Qian X, Zhao P, Ren H, Tang H, Wang Y, Wei Y, Zhu Y, Qi Z (2016) Caveolin-1-mediated Japanese encephalitis virus entry requires a two-step regulation of actin reorganization. Future Microbiol 11:1227–1248

Xu Q, Zhu N, Chen S, Zhao P, Ren H, Zhu S, Tang H, Zhu Y, Qi Z (2017) E3 ubiquitin ligase Nedd4 promotes Japanese encephalitis virus replication by suppressing autophagy in human neuroblastoma cells. Sci Rep 7:45375

Yang TC, Shiu SL, Chuang PH, Lin YJ, Wan L, Lan YC, Lin CW (2009) Japanese encephalitis virus NS2B-NS3 protease induces caspase 3 activation and mitochondria-mediated apoptosis in human medulloblastoma cells. Virus Res 143(1):77–85

Yang TC, Lai CC, Shiu SL, Chuang PH, Tzou BC, Lin YY, Tsai FJ, Lin CW (2010) Japanese encephalitis virus down-regulates thioredoxin and induces ROS-mediated ASK1-ERK/p38 MAPK activation in human promonocyte cells. Microbes Infect 12(8–9):643–651

Yang CM, Lin CC, Lee IT, Lin YH, Yang CM, Chen WJ, Jou MJ, Hsiao LD (2012) Japanese encephalitis virus induces matrix metalloproteinase-9 expression via a ROS/c-Src/PDGFR/PI3K/Akt/MAPKs-dependent AP-1 pathway in rat brain astrocytes. J Neuroinflammation 9:12

Yang S, He M, Liu X, Li X, Fan B, Zhao S (2013) Japanese encephalitis virus infects porcine kidney epithelial PK15 cells via clathrin- and cholesterol-dependent endocytosis. Virol J 10:258

Ye J, Zhang H, He W, Zhu B, Zhou D, Chen Z, Ashraf U, Wei Y, Liu Z, Fu ZF, Chen H, Cao S (2016) Quantitative phosphoproteomic analysis identifies the critical role of JNK1 in neuroinflammation induced by Japanese encephalitis virus. Sci Signal 9(448):ra98

Yoshikawa A, Nabeshima T, Inoue S, Agoh M, Morita K (2016) Molecular and serological epidemiology of Japanese encephalitis virus (JEV) in a remote Island of western Japan: an implication of JEV migration over the East China Sea. Trop Med Health 44:8

Yu CY, Hsu YW, Liao CL, Lin YL (2006) Flavivirus infection activates the XBP1 pathway of the unfolded protein response to cope with endoplasmic reticulum stress. J Virol 80(23): 11868–11880

Yu IM, Zhang W, Holdaway HA, Li L, Kostyuchenko VA, Chipman PR, Kuhn RJ, Rossmann MG, Chen J (2008) Structure of the immature dengue virus at low pH primes proteolytic maturation. Science 319(5871):1834–1837

Yu SP, Ong KC, Perera D, Wong KT (2019) Neuronal transcriptomic responses to Japanese encephalitis virus infection with a special focus on chemokine CXCL11 and pattern recognition receptors RIG-1 and MDA5. Virology 527:107–115

Zhang MJ, Wang MJ, Jiang SZ, Ma WY (1989) Passive protection of mice, goats, and monkeys against Japanese encephalitis with monoclonal antibodies. J Med Virol 29(2):133–138

Zhang Y, Wang Z, Chen H, Chen Z, Tian Y (2014) Antioxidants: potential antiviral agents for Japanese encephalitis virus infection. Int J Infect Dis 24:30–36

Zhang Z, Rong L, Li YP (2019) Flaviviridae viruses and oxidative stress: implications for viral pathogenesis. Oxidative Med Cell Longev 2019:1409582

Zhang YG, Chen HW, Zhang HX, Wang K, Su J, Chen YR, Wang XR, Fu ZF, Cui M (2022) EGFR activation impairs antiviral activity of interferon signaling in brain microvascular endothelial cells during Japanese encephalitis virus infection. Front Microbiol 13:894356

Zheng S, Zhu D, Lian X, Liu W, Cao R, Chen P (2016) Porcine 2′, 5′-oligoadenylate synthetases inhibit Japanese encephalitis virus replication in vitro. J Med Virol 88(5):760–768

Zhou M, Wang S, Guo J, Liu Y, Cao J, Lan X, Jia X, Zhang B, Xiao G, Wang W (2021) RNA interference screening reveals requirement for platelet-derived growth factor receptor Beta in Japanese encephalitis virus infection. Antimicrob Agents Chemother 65(6):e00113-21

Zhu YZ, Xu QQ, Wu DG, Ren H, Zhao P, Lao WG, Wang Y, Tao QY, Qian XJ, Wei YH, Cao MM, Qi ZT (2012) Japanese encephalitis virus enters rat neuroblastoma cells via a pH-dependent, dynamin and caveola-mediated endocytosis pathway. J Virol 86(24):13407–13422

# Chapter 11
# African Swine Fever Virus Host–Pathogen Interactions

**Christopher L. Netherton, Gareth L. Shimmon, Joshua Y. K. Hui, Samuel Connell, and Ana Luisa Reis**

**Abstract** African swine fever virus is a complex double-stranded DNA virus that exhibits tropism for cells of the mononuclear phagocytic system. Virus replication is a multi-step process that involves the nucleus of the host cell as well the formation of large perinuclear sites where progeny virions are assembled prior to transport to, and budding through, the plasma membrane. Like many viruses, African swine fever virus reorganises the cellular architecture to facilitate its replication and has evolved multiple mechanisms to avoid the potential deleterious effects of host cell stress response pathways. However, how viral proteins and virus-induced structures trigger cellular stress pathways and manipulate the subsequent responses is still relatively poorly understood. African swine fever virus alters nuclear substructures, modulates autophagy, apoptosis and the endoplasmic reticulum stress response pathways. The viral genome encodes for at least 150 genes, of which approximately 70 are incorporated into the virion. Many of the non-structural genes have not been fully characterised and likely play a role in host range and modifying immune responses. As the field moves towards approaches that take a broader view of the effect of expression of individual African swine fever genes, we summarise how the different steps in virus replication interact with the host cell and the current state of knowledge on how it modulates the resulting stress responses.

**Keywords** Virus entry · Virus replication · Apoptosis · Autophagy · Cellular stress · Mitochondria · Endoplasmic reticulum · Interferon

## Introduction

African swine fever virus (ASFV) is a large double-stranded DNA virus that encodes for at least 150 open reading frames depending on the virus isolate. The virus is a monophyletic member of the nucleocytoplasmic large DNA viruses that includes the

C. L. Netherton (✉) · G. L. Shimmon · J. Y. K. Hui · S. Connell · A. L. Reis
The Pirbright Institute, Woking, UK
e-mail: christopher.netherton@pirbright.ac.uk; joshua.hui@pirbright.ac.uk; ana.reis@pirbright.ac.uk

 
S. Vijayakrishnan et al. (eds.), *Virus Infected Cells*, Subcellular Biochemistry 106,
https://doi.org/10.1007/978-3-031-40086-5_11

poxviruses, iridoviruses and the giant mimiviruses. Its closest relative is the uncategorised abalone African swine fever like virus with which it shares approximately 40% identity at the amino acid level for some orthologous genes. Most field isolates cause a lethal haemorrhagic fever in domestic pigs and wild boar (*Sus scrofa*), called African swine fever (ASF). The disease was first identified in Kenya and Zambia at the beginning of the twentieth century and was since discovered to be endemic in sub-Saharan Africa, although the greatest genetic diversity appears to be in Southern and Eastern Africa. ASFV is established in a sylvatic cycle between warthogs (*Phacochoerus africanus*) and different soft tick species of the *Ornithodoros* genus that infest warthog burrows. *Ornithodoros porcinus* and *Carios erraticus* (formerly *Ornithodoros erraticus*) are biological vectors for ASFV and as such the virus is currently the only known DNA arbovirus. ASFV is also able to infect bushpigs (*Potamochoerus larvatus*), and virus has also been isolated from red river hogs (*Potamochoerus porcus*) and a single giant forest hog (*Hylochoerus meinertzhageni*). Warthogs and bushpigs do not exhibit clinical signs after experimental infection with virus isolates that would be fatal in the domestic pig. Given the range of different hosts that ASFV can replicate in, the virus has likely evolved multiple mechanisms to antagonise host responses and exploit the cell to benefit replication. Prediction of gene function has been problematic due to the dissimilarity of ASFV and its genes to those of other viruses; however, since the reappearance and spread of the disease since 2007 a wealth of functional studies have gleaned more insights into the way this unusual virus interacts with its hosts. Here we summarise ASFV-host interactions with a particular focus on viral manipulation of intracellular response pathways.

## Virus Replication

### *Virion Structure*

Intracellular virions are icosahedral structures approximately 200 nm in diameter formed from a series of distinct layers (Wang et al. 2019; Liu et al. 2019; Andres et al. 2020). Starting at the centre of the particle, these are the nucleoid that contains the genomic DNA, surrounded by a matrix or core which are comprised of the mature products of two polypeptides, pp220 (*CP2475L* gene[1]) and pp62 (*CP530R*

[1]ASFV gene nomenclature. Current ASFV gene nomenclature has evolved from three separate naming systems. The majority of the genes are named following the system that was developed for the Ba71v strain, the first virus for which a full genome was sequenced (Yáñez et al. 1995). These are in the form, for example, *B646L*, or *EP402R*, with the first letters indicating the *Eco*RI restriction fragment the ATG is located in (with a P indicating '), followed by the number of codons in the gene and finally, L or R indicating the direction of the gene on the genome. As more genomes became available, the five multigene families were renamed according so that orthologous copies of each gene family member were named consistently (Chapman et al. 2008; Imbery and

gene). The core/matrix is in turn surrounded by an internal envelope upon which the protein capsid is built (Hawes et al. 2008). Proteomic analysis identified 68 viral and 21 cellular proteins in purified virions (Alejo et al. 2018). Cryo-EM reconstructions have revealed that the capsid is built from pseudo-hexameric capsomers comprised of the major capsid protein p72 (*B646L* gene) and pentameric capsomers comprised of the penton protein (*H240R* gene). These capsomers form 20 trisymmetrons and 12 pentasymmetrons which together constitute the capsid, and interactions within symmetrons are thought to be secured through the p17 protein (*D117L* gene), whereas tri- and pentasymmetrons are connected in zipper regions by interactions with the product of the *M1249L* gene (Liu et al. 2019; Wang et al. 2019) and either p17 (Wang et al. 2019) or another unidentified minor capsid protein (Liu et al. 2019). P17 is a transmembrane protein located in the inner envelope and plays a role in linking the capsid to the internal viral membrane (Suárez et al. 2010a). Extracellular virions gain an additional external envelope when virions bud through the plasma membrane, at present the CD2v protein (*EP402R* gene) is the only viral protein to be confirmed to localise to the external envelope, although cellular proteins such as tetraspannin CD9 and integrin beta are also present (Alejo et al. 2018). Both intracellular virions and extracellular virions are infectious (Andrés et al. 2001).

## *Virus Entry*

Extracellular virions can enter porcine macrophages via macropinocytosis or clathrin-mediated endocytosis (Hernáez et al. 2016). Virions haemadsorbed to red blood cells can also be detected in phagosomes of tick midgut digestive cells (Kleiboeker et al. 1998) and it is possible that virus may enter splenic macrophages in a similar manner (Fig. 11.1). Vaccinia virus can enter cells by fusing directly with the plasma membrane (Quemin et al. 2018) and ASFV can also be driven to enter cells in this manner at low pH (Valdeira and Geraldes 1985). Intracellular virions that lack the external envelope are also infectious; however, their mechanism of entry is unclear. The route of entry into cultured cells is dependent on the virus preparation. Highly purified extracellular virus enters through receptor-mediated endocytosis alone, but cruder preparations that contain cell debris can also enter via macropinocytosis (Sánchez et al. 2012; Hernáez et al. 2016; Hernaez and Alonso 2010). Therefore, it seems likely that the intracellular virus is taken up by

---

Upton 2017). Finally, a small number of genes that were not present in Ba71v and were not assigned to multigene families retained nomenclature based on the *Sal*I restriction map of the Malawi Lil 20/1 isolate (Vydelingum et al. 1993). These were all from the *Sal*I fragment of the Malawi genome, and unfortunately, some of these have been mistranscribed into I, complicating the nomenclature. For example, l8L is also referred to as I8L or sometimes L8L. Proteins are distinguished from genes by the prefix p, except for some structural proteins, which are named after their apparent molecular mass on reducing SDS-PAGE gels or the product of the *EP402R* gene, which is named after its cellular homolog followed by the suffix v, that is CD2v.

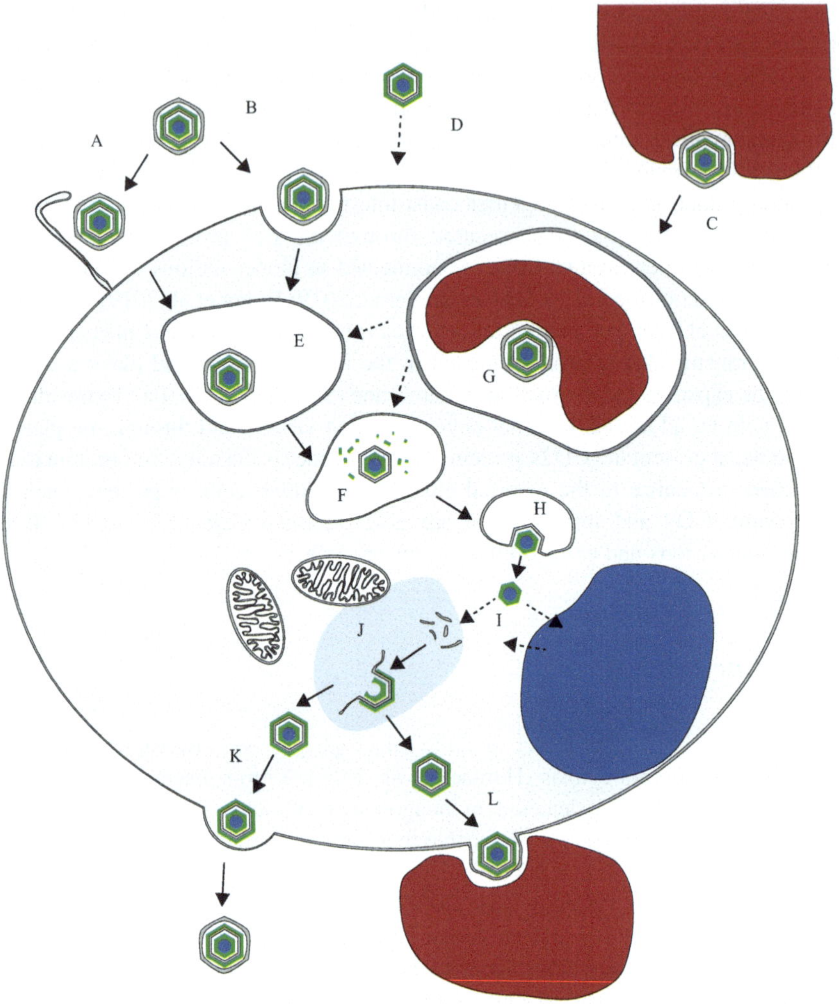


**Fig. 11.1** African swine fever virus replication cycle: Extracellular virions enter cells through macropinocytosis (**a**), receptor-mediated endocytosis (**b**) or by phagocytosis of haemadsorped red blood cells (**c** and **g**). Entry pathway(s) of infectious intracellular virus are unknown, but are likely through macropinocyotsis or endocytosis (**d**). Intact extracellular viruses can be seen in early endosomes (**e**); however, the extracellular envelope is lost as the virions transition through to late endosomes and lysosomes (**f**). The internal envelope fuses with the membranes of multivesicular endosomes releasing the viral core into the cytoplasm (**h**). Genome replication appears to require a poorly understood nuclear step (**i**) before virus factories are formed in perinuclear regions of the cytoplasm. Precursor viral membranes, viral proteins and assembling virions are concentrated in virus factories (**j**) and cellular components such as mitochondria are recruited to the periphery. Assembled intracellular viruses are trafficked to the plasma membrane by microtubules and gain the external envelope as they bud into the extracellular space (**k**) or onto the surface of haemadsorbed red blood cells (**l**)

macropinocytosis, a constitutive function of macrophages (Doodnauth et al. 2019), but it remains to be determined whether other mechanisms can also mediate entry of intracellular virus.

A number of ASFV genes have been shown to be important for virus entry either directly through biochemical means or indirectly via their gene products being targets of neutralising antibodies. Antibodies against p72 (*B646L* gene), p30 (*CP204L* gene) and p54 (*E183L* gene) can neutralise virus (Borca et al. 1994; Gomez-Puertas et al. 1998; Gómez-Puertas et al. 1996). Sera from pigs immunised with p72 and p54 is capable of inhibiting virus attachment, while sera from pigs immunised with p30 are capable of inhibiting internalisation. Interestingly antibodies against the p12 (*O61R* gene) do not affect virus entry despite strong biochemical evidence in support of this protein mediating virus attachment (Angulo et al. 1993). Recombinant p12 can saturate binding of purified virus to Vero cells at 4°C and vice versa. The situation is further complicated by the observation that none of these proteins would be predicted to be exposed on the surface of external virions. P72 is the major capsid protein and so should be hidden by the external envelope of the virus, and p54 and p12 are internal envelope proteins, therefore, would be predicted to reside under the capsid. It is plausible that the unidentified minor capsid protein suggested by cryo-EM reconstructions (Liu et al. 2019) may represent either p12 or p54. Alternatively, virion interactions with the surface of target cells induce confirmational changes in the surface of the virion that expose p72, p54, p30 and/or p12, which can account for the biochemical data described above. Lastly, experiments with p12 were performed using the highly purified virus in Vero cells, whereas those with p72, p30 and p54 were demonstrated using standard preparations in macrophages, and so the differences in results may reflect the different models employed.

Cellular receptors for the different infectious forms of ASFV have not been identified. Expression of CD163 correlated with the susceptibility of monocyte-derived cell populations and monoclonal antibodies against CD163 reduced infection of alveolar macrophages (Sánchez-Torres et al. 2003). However, infectivity did not correlate with CD163 expression in other model macrophage systems (Lithgow et al. 2014), and crucially genetically modified pigs that do not express CD163 are fully susceptible to ASF (Popescu et al. 2017). Intriguingly CD163 was detected by proteomic analysis in virions purified from porcine macrophages (Wang et al. 2019), which may explain the ability of CD163 monoclonals to restrict ASF replication.

After internalisation, virions transit through the endosomal-lysosomal complex before viral cores exit into the cytoplasm. Elevating lysosomal pH blocks virus core exit, and infectivity can be maintained for 3 days using reversible inhibitors such as chloroquine (Geraldes and Valdeira 1985). Virions arrested in early endosomes by drug treatments are essentially indistinguishable from extracellular particles and can be disassembled in vitro by progressive acidification (Hernáez et al. 2016). After entry, ASFV virions can be detected within early endosomes 10 minutes after infection. After 30 minutes, they can be observed within late endosomes and lysosomes, and from 45 minutes post-infection viral cores can be detected in the cytoplasm. Intact ASFV particles including the external envelope were observed in early endosomes; however, in late endosomes, the external envelope and capsid became disrupted (Hernáez et al. 2016). Viral cores are delivered to the cytoplasm after fusion between the internal envelope and the membrane of multivesicular late endosomes. The *E248R* and *E199L* genes encode for transmembrane proteins that are distant homologues of the vaccina virus proteins that form part of the entry fusion complex. pE199L and pE284R both localise to the internal envelope, and experiments with inducible mutants show that both pE199L and pE248R deficient particles are unable to exit the lysosomal endosomal pathway, strongly suggesting a role in fusion (Rodríguez et al. 2009; Hernáez et al. 2016; Matamoros et al. 2020). Correct function of both pE199L and pE248R is likely dependent on the formation of intramolecular disulphide bonds that are mediated by an ASFV redox pathway consisting of the *A151R* gene and the Erv1p/Alrp-like *B119L* gene (Rodríguez et al. 2006; Matamoros et al. 2020).

Although the ASFV receptor(s) have so far proven elusive, recent data has identified cellular components required for endosomal exit. Fusion is sensitive to both cholesterol depletion (Bernardes et al. 1998) and inhibition of cholesterol transport from endosomes (Cuesta-Geijo et al. 2016; Cuesta-Geijo et al. 2022). GFP-tagged pE248R and pE199L interact with the endosomal membrane proteins Niemann Pick C1 (NPC1) and lysosome-associated membrane protein-2 (Lamp2/CD107b) when overexpressed in HEK cells and pE199L also interacts with lysosome-associated membrane protein-1 (Lamp1/CD107a). Unlike Ebola or Lassa virus, ASFV growth is reduced, rather than being completely abolished in NPC1 Vero KO cells, and, therefore, it is likely that the ASFV fusion is not dependent on the interaction with a single cellular protein (Cuesta-Geijo et al. 2022). Entry/fusion is also a target for the anti-viral effect of interferon. Overexpression of human interferon-induced transmembrane protein 2 (IFITM2) in Vero cells inhibits virus replication and reduces the number of visible decapsidated virions in confocal assays (Munoz-Moreno et al. 2016). Removal of endogenous IFITMs from Veros enhances virus replication tenfold and overexpression of swine IFITMs in these cells inhibited ASFV replication, with IFITM3 overexpression having the greatest effect (Cai et al. 2022). Interestingly, IFITM3 is a cholesterol-binding protein and antiviral activity of this protein is mediated by interactions with sterols (Das et al. 2022).

## Early Steps in Replication

The events in the ASFV replication cycle after the deposition of the viral core into the cytoplasm until the appearance of perinuclear viral factories are poorly understood. Inner envelope protein p54 interacts with the light chain of cytoplasmic dynein (LC8), and this has been suggested to play a role in delivering virions to initial sites of replication consistent with the treatment of cells with microtubule depolymerising agents blocking early replication (Heath et al. 2001; Alonso et al. 2001; Hernaez et al. 2006; Hernáez et al. 2010; Netherton and Wileman 2013;

Quetglas et al. 2012). As the inner envelope is lost after exit from endosomes (Hernáez et al. 2016), this suggests that p54 remains associated with viral cores after fusion or that the observations relate to the dependence of endosomal maturation on microtubules (Driskell et al. 2007; Granger et al. 2014) and the interaction between p54 and LC8 relates to a hitherto unidentified step in replication.

Unlike poxviruses which replicate entirely in the cytoplasm, there is considerable evidence for a nuclear step in the ASFV replication cycle. Viral DNA has been consistently localised to the cell nucleus by in situ hybridisation (García-Beato et al. 1992; Rojo et al. 1999; Eulálio et al. 2007; Ballester et al. 2010). In macrophages where all cellular DNA synthesis was abrogated chemically, DNA can be detected in the nucleus of infected cells early in during replication and can be followed biochemically into the cytoplasm (García-Beato et al. 1992), furthermore viral DNA isolated from the nucleus is smaller than that seen in the cytoplasm (García-Beato et al. 1992). Experiments with the thymidine analogue 5′-bromo-2′-deoxyuridine show DNA synthesis in the nucleus of swine macrophages, as well as G2/M arrested Vero cells early during the replication cycle before appearing exclusively cytoplasmic from 8 hours onwards (Simões et al. 2015a). Taken together, this supports the current model that viral DNA synthesis is primed in the nucleus in the early stages of the replicative cycle and full-length genomic DNA is then synthesised in viral factories. ASFV infection induces profound changes in nuclear substructure. Within the first four hours of infection, the lamin network is broken down, and proteins normally localised to the nuclear envelope can be detected in the cytoplasm (Ballester et al. 2011). Markers for the nucleolus are dispersed, those of nuclear speckles, promyelocytic leukaemia protein nuclear bodies (PML-NBs) and Cajal bodies are enlarged, and RNA polymerase II is rapidly recruited to defined areas within the nucleus (Ballester et al. 2011; Simões et al. 2015a). Disruption of nuclear substructures is concomitant with the activation of the host DNA damage response probably by signalling through the Ataxia telangiectasia mutated (ATM) Rad 3-related (ATR) pathway (Simões et al. 2015a), but not the ATM pathway itself, nor the DNA protein kinase catalytic subunit (DNA-PKcs) pathway. Antibodies specific to different methylated forms of histone H3 and different heterochromatin protein 1 (HP1) isoforms suggest that ASFV infection may also affect chromatin remodelling (Simões et al. 2015b). Proteins related to the ATR pathway become juxtaposed with PML-NBs in ASFV-infected cells (Simões et al. 2015b), and ASFV replication is repressed in cells were PML is knocked down and increased in cells where the ATR pathway is activated. The mechanisms by which ASFV manipulates the nucleus appear different to other DNA viruses as polyomaviruses activate both ATM and ATR pathways, whereas adenoviruses and herpesviruses suppress the ATR pathway (Blackford et al. 2008; Verhalen et al. 2015; Wilkinson and Weller 2006). Contrary to what is observed with ASFV, herpesviruses replication is significantly enhanced in the absence of proteins associated with PML-NBs (Ma et al. 2022; Everett et al. 2006; Everett et al. 2008) and the interaction between PML-NBs and adenoviruses are more nuanced (Rosa-Calatrava et al. 2003; Hofmann et al. 2021; Müller and Dejean 1999). ASFV deregulates more than 1000 host genes in infected macrophages, with more than 8000 not affected (Cackett et al. 2022), and so the significant rearrangements observed in the architecture of nucleus do not lead to drastic changes in host transcription. It will be interesting to see how ASFV manipulates the host nucleus to support early stages of the replication cycle while maintaining some form of integrity to host transcription. There is limited information on the role of individual viral genes in the manipulation of the host genome. Both *A104R* and *K78R* encode for proteins that localise to both the nucleus and viral factories (Nunes-Correia et al. 2008; Netherton et al. 2009); however, these are late genes and are therefore expressed after the changes in the nucleus occur. Both are present in purified virions and, therefore, could play a role after entry, but immunofluorescence experiments do not detect pK78R (p10) early during the replication cycle (Nunes-Correia et al. 2008), and *A104R* is not an essential gene (Ramirez-Medina et al. 2022). In contrast, the viral core protein p37 that is derived from pp220 can be detected within the nucleus from 2 hours post-infection. This localisation was not dependent on viral gene expression and therefore is most likely related to a post-fusion event after delivery of the viral cores to the cytoplasm (Eulálio et al. 2007). P37 has multiple nuclear export signals, and a p37-GFP fusion is not imported into the nucleus, suggesting the p37 does not contain nuclear import signals (Eulálio et al. 2006) and consistent with this, there is no p37 nuclear signal during late stages of replication (Eulálio et al. 2007). This suggests that p37 is maintained within the nucleus during early stages by extrinsic factors rather than sequences present in the protein itself; however, as yet, there is no evidence that viral cores or other viral substructures locate to or are formed within the nucleus.

## *Virus Factories*

Viral DNA, proteins and membranes accumulate at perinuclear sites where nascent intracellular virions are assembled prior to transit to the plasma membrane and occur at a single structure in the cytoplasm termed viral replication sites, or virus factories (Aicher et al. 2021; Brookes et al. 1996; Moura Nunes et al. 1975; Rouiller et al. 1998). Virus factories are found close to the nucleus and appear to align with the microtubule organising centre (Heath et al. 2001). The early stages of factory formation are poorly understood but may involve the recruitment of the intermediate filament vimentin (Stefanovic et al. 2005). Immunofluorescence staining for cytoplasmic proteins such as p30 (*CP204L* gene) tends to show diminished signal in perinuclear areas before DNA replication begins (Stefanovic et al. 2005; Netherton et al. 2009) consistent with the electronlucent appearance of early virus factories (Aicher et al. 2021). Small, almost circular factories can be observed close to the nucleus (Aicher et al. 2021), but at later times viral proteins, viral membrane fragments, assembly intermediates as well as immature and mature intracellular virions can be seen in large irregularly shaped factories in the electron microscope, concurrent with the appearance of intense staining for individual viral proteins and viral DNA in the confocal microscope (Suarez et al. 2015; Aicher et al. 2021). Bioorthogonal labelling with a methionine analogue revealed that virus factories accumulate newly synthesised proteins coincident with pleomorphic ribbons that labelled with anti-p54 (*E183L* gene) and anti-p34 (*CP2475L* gene) upon which newly assembled virions were observed (Aicher et al. 2021). Immunogold labelling suggested these structures are consistent with viral membranes seen in the electron microscope. Proteins associated with the translation machinery, such as ribosomal and eukaryotic initiation factors, are recruited to replication sites and viral RNAs are recruited to the periphery of ASFV factories (Castelló et al. 2009). STED imaging of factories showed that topoisomerase encoded by *P1192R* (Coelho et al. 2015) has two distinct localisations, one coincident with the signal for cytoplasmic viral DNA and another that co-localised with assembling virions. It is tempting to speculate that the different localisations of pP1192R might correspond to separate functions in DNA replication and packaging. P1192R interacts with pA104R to effect DNA supercoiling (Frouco et al. 2017), and pA104R is a virion component, although pP1192R itself is not (Alejo et al. 2018). Taken together, these results demonstrate that specialised domains form within the ASFV replication sites, as seen in vaccinia virus replication sites (Katsafanas and Moss 2007). How the subdivision of the factory is maintained is, for the most part, unknown, but the overall location of the viral replication site is dependent on the cytoskeleton, as microtubule depolymerisation causes the formation of fragmented factories (Carvalho et al. 1988; Alves de Matos and Carvalho 1993; Alonso et al. 2001; Heath et al. 2001). Cells infected with a lethal conditional mutant of p54 form factories that are devoid of viral membranes, and the polyproteins derived from the *CP2475L* and *CP530R* genes fail to correctly localise to the factory, forming zipper-like structures that associate with the endoplasmic reticulum (Rodríguez et al. 2004). By analogy with the prototype poxvirus vaccinia virus, is it likely that multiple ASFV genes are involved in the correct localisation of viral membranes to replication sites, as conditional lethal mutants of the A6, A11, F10, H7 and L2 genes all lead to virus factories that lack viral membranes (Traktman et al. 1995; Resch et al. 2005; Satheshkumar et al. 2009; Maruri-Avidal et al. 2011; Meng et al. 2012) and A14 and A17 lead to accumulation of viral membranes at the periphery (Rodríguez et al. 1997, 1998). Repression of ASFV p17 (*D117L* gene) leads to a dislocation between viral membranes and the polyproteins, but viral membrane precursors are recruited to the factory (Suárez et al. 2010a) although not as consistently as observed in the absence of p54. As p54 interacts with dynein (Alonso et al. 2001; Hernáez et al. 2010) and microtubule depolymerisation disrupts factory formation it is conceivable that recruitment of membranes to virus factories is dependent on microtubule motors.

Virus factories morphologically resemble cellular structures called aggresomes which are an intracellular response to an accumulation of misfolded proteins. Proteins that exceed the cell's capacity for degradation are directed to the microtubule organising centre and chaperones, proteasomes and mitochondria are recruited to aid in refolding and degradation (Johnston et al. 1998). In addition, intermediate filaments such as vimentin are recruited to aggresomes, where they form a cage-like structure that sequesters the misfolded proteins away from the rest of the cytoplasm. With the exception of proteasomes, all of these are associated with ASFV viral factories (Castelló et al. 2009; Heath et al. 2001; Rojo et al. 1998), and redistribution of cytoskeletal elements occurs early in the ASFV replication cycle (Stefanovic et al. 2005). However, transport of ubiquitinated and non-ubiquitinated proteins to cellular aggresomes are dependent on BAG3 and HDAC6, respectively, and these proteins are not recruited to replication sites in the same way as they are in aggresomes, suggesting functional differences between the two structures (Muñoz-Moreno et al. 2015).

## *Membranes*

The origin of the viral membranes that go on to form the inner envelope of the virion has been an intriguing question ever since ASFV viral factories were first imaged in the electron microscope. As membranes in replication sites did not have obvious continuity with cellular structures, it was initially thought they were generated de novo. However, immunogold labelling showed that viral membranes were positive for both endoplasmic reticulum and viral proteins (Alejo et al. 1997; Rouiller et al. 1998; Andrés et al. 1998). Structural proteins co-sedimented with luminal endoplasmic reticulum on sucrose gradients (Cobbold et al. 1996) leading to a model of heavily modified ER cisternae upon which capsid and matrix proteins were recruited which drove the formation of the characteristic hexagonal-shaped virions (Rouiller et al. 1998; Andrés et al. 1998). Detailed electron microscopy work showed that the internal envelope consisted of a single lipid bilayer (Hawes et al. 2008) which was confirmed by the cryo-EM reconstructions of purified virions (Andres et al. 2020; Liu et al. 2019; Wang et al. 2019). Experiments with vaccinia suggest the internal envelope was derived from existing cellular membranes that are stabilised by viral proteins (Chlanda et al. 2009, 2011), and it appears that a similar mechanism occurs in ASFV-infected cells (Suarez et al. 2015). Viral membrane precursors can be seen as small discrete fragments or curls of membrane and can be detected in early factories that lack significant numbers of assembling virions (Suarez et al. 2015; Aicher et al. 2021). Later in infection, viral membranes form more complex interwoven structures that resemble skeleton viruses and may represent pre-virions. These more complex membrane structures are also connected to assembling virions. Although ER proteins were located in viral membranes (Rouiller et al. 1998; Andrés et al. 2001), interconnections between viral membranes and ER cisternae were not found, and intriguingly the ER appears to be leaky in ASFV-infected cells, with luminal proteins detected in the cytosol (Suarez et al. 2015). Taken together, this suggests that precursor membranes are generated from disrupted ER membranes (Suarez et al. 2015). The mechanism of ER disruption/collapse is unclear, but overexpression of p54 causes the collapse of ER cisterna, most likely through an intermolecular disulphide bond formed across the lumen of the ER by pairs of p54 (Windsor et al. 2012). Overexpression of p54 does cause the proliferation of smooth ER, possibly due to the induction of ER stress responses, which does not occur to a noticeable degree in ASFV-infected cells. Therefore, it is likely that other viral proteins, probably including p17 are responsible for the generation of viral precursors, and it is likely that their formation and localisation to viral factories are linked.

Zipper structures that appear as a proteinaceous layer that resembles the viral core sandwiched between two ER cisternae can also be detected in virus factories (Andrés et al. 1998). Visually similar structures can be seen at the plasma membrane in cells co-expressing pp220 and pp62, suggesting they are linked to the formation of viral membranes; however, zippers form in the absence of both p54 and p17 (Rodríguez et al. 2004; Suárez et al. 2010a) and viruses lacking either pp220 or pp62 (Andrés et al. 2002; Suárez et al. 2010b) form empty virions that contain the capsid and an internal envelope. Therefore, it seems unlikely that the zippers are direct descendants of the internal envelope and may represent another undetermined part of ASFV biology. Nonetheless, they demonstrate the myriad effects that ASFV has on the host cell during its replication and morphogenesis.

## *Cytoskeleton*

As well as a role in entry and formation/maintenance of virus factories, ASFV utilises the cytoskeleton to transport nascent viruses from the virus factory to the plasma membrane and then to project from the infected cells. ASFV particles bind to microtubules in vitro (Alves de Matos and Carvalho 1993), and microtubule depolymerisation causes the accumulation of virions within virus factories (Carvalho et al. 1988; Jouvenet et al. 2004). A similar phenotype is observed in cells infected with an ASFV mutant, where the expression of the *E120R* gene is repressed (Andrés et al. 2001). Interestingly, immunogold labelling localised pE120R to the outer capsid, consistent with the role of transporting intracellular virus to the plasma membrane; however, the structural solutions of the virus do not appear to show pE120R as part of the capsid (Wang et al. 2019; Liu et al. 2019; Andres et al. 2020). As viruses lacking pE120R are fully infectious (Andrés et al. 2001), it may suggest that the association between pE120R and the capsid is transitory and perhaps is lost as the virus buds. Anterograde transport along microtubules is driven by the kinesin family of motor proteins, and tetratricopeptide repeats in kinesin mediate interactions between the motor and cargo. Virion egress is inhibited by overexpression of the tetratricopeptide repeat domain of kinesin (Jouvenet et al. 2004; Jouvenet et al. 2006), demonstrating a link between anterograde transport and movement of nascent virions to the plasma membrane, as yet however there is no evidence of direct interaction between pE120R and kinesin.

Once at the plasma membrane, ASFV induces filopodia-like structures at the surface of both cultured cells and macrophages (Jouvenet et al. 2006), which are clearly visible as actin projections that also contain fascin. The projections associated with virions are dependent on the transport of virions to the plasma membrane as well as an unknown host or viral factor that is delivered by the secretory pathway. This likely provides a mechanism by which virions can infect neighbouring cells, although it may also be linked to the transfer of virions to the surface of haemabsorbed erythrocytes. Interestingly CD2v (EP402R) binds the actin adaptor protein SH3P7 (Kay-Jackson et al. 2004); however, a deletion mutant that lacks this gene was also shown to induce filopodia, and therefore other viral genes are likely involved in the generation of these structures (Jouvenet et al. 2006).

## Programmed Cell Death

ASFV replication subverts a number of host pathways including apoptosis, pyroptosis and necrosis which are programmed pathways that lead to rapid cell death. Succesful replication of progeny virus likely requires maintaining cell viability through the amelioration of these pathways. Conversely, once the virus has generated a sufficient number of progeny, activation of cell death and lysis may be beneficial as phagocytic cells, the principal targets of ASFV replication in vivo, are recruited to degrade apoptotic cell bodies.

ASFV induces apoptosis in cultured cells (Carrascosa et al. 2002; Hernáez et al. 2004) as well as in vivo (Ramiro-Ibáñez et al. 1996; Carrasco et al. 1996) and also appears to induce extensive apoptosis in non-target lymphocytes, particularly B-cells (Oura et al. 1998). In the absence of viral gene expression or replication, ASFV induces apoptosis after infection (Carrascosa et al. 2002), and therefore it is unsurprising that the virus encodes for a number of genes that modulate this including the *A179L*, *A224L* and *EP153R*. Overexpression of a number of ASFV structural proteins also appears to induce apoptosis (Hernáez et al. 2004; Li et al. 2021d), although as these proteins are normally compartmentalised within the virus factory, it is unclear if this occurs in infected cells.

The gene product of *A179L* (also *5-HL*) is a bcl-2 homologue with a highly conserved BH1 domain as well as BH2, BH3 and BH4 domains and can inhibit the induction of apoptosis by a range of stimuli in both mammalian and insect cells (Afonso et al. 1996; Revilla et al. 1997; Brun et al. 1998; Galindo et al. 2008; Shi et al. 2021). A179L shares a conserved glycine residue in the BH1 domain crucial for the anti-apoptotic activity of cellular bcl-2 proteins, and mutation of pA179L at this site also inhibits the activity of the ASFV homologue (Revilla et al. 1997; Hernaez et al. 2013; Banjara et al. 2019). pA179L is able to interact with a number of different cellular Bcl-2 homologues, with the highest affinity for a peptide derived from porcine Bim, followed by Bax, Bak, Puma and Bid (Banjara et al. 2017). The bax expression appears to be upregulated in ASFV-infected Vero cells, and Bim redistributes from the cytoplasm to the mitochondria suggesting activation of the mitochondrial pathway of apoptosis induction (Hernáez et al. 2004; Granja et al. 2004). Peptide interaction data broadly agreed with in vitro pull-down experiments (Galindo et al. 2008), as a peptide derived from Noxa showed tenfold lower affinity than other Bcl2 proteins, and no interaction was demonstrated in vitro. Nonetheless, the wide variety of binding partners is unusual (Rautureau et al. 2012; Banjara et al. 2017), which could be indicative of a requirement to affect a broad range of host responses or a need to be able to bind targets in both mammalian and arthropod hosts. Attempts to delete *A179L* from virulent viruses have so far been unsuccessful (Afonso et al. 1996); however, it is possible to disrupt the expression of *A179L* in a tissue culture-adapted virus (Barber 2015), suggesting it has a critical function in host range. Although A179L has a well-defined role in inhibiting apoptosis, overexpression enhances necroptosis, another major programmed cell death pathway, in a number of different model systems (Shi et al. 2021). Whether necroptosis is induced in cells infected with ASFV is still an open question.

The *A224L* gene (also *4CL* gene) encodes for a homologue of the cellular inhibitor of the apoptosis protein (IAP) family of genes that can bind and inhibit the action of caspases, one of the major classes of apoptosis effector proteins. Overexpression of A224L can inhibit the induction of apoptosis in Vero cells after co-treatment with tumour necrosis alpha and cycloheximide or treatment with staurosporine, and an A224L deletion mutant induces a higher level of caspase-3 activity in Vero cells leading to lower cell viability after infection. Like other IAP family members, pA224L can bind caspases as it co-precipitates with the active form of caspase-3. Deletion of A224L does not affect replication in Vero cells (Nogal et al. 2001) or macrophages (Neilan et al. 1997). In contrast to the work in Vero cells, macrophages infected with an A224L deletion mutant did not exhibit higher levels of apoptosis than those infected with wild-type virus 24 or 48 hours post-infection, suggesting redundancy in the ability of the virus to block apoptosis (Neilan et al. 1997). A224L is incorporated into virions (Chacón et al. 1995), suggesting a potential role immediately after infection, however as A224L is not essential for virulence in the pig (Neilan et al. 1997), it may be that the gene plays a more important role in modulating the host immune response in soft ticks or warthogs.

The *EP153R* gene (also referred to as *8CR*) encodes for a highly glycosylated protein with homology to C-type lectins that are non-essential for virus replication or virulence (Neilan et al. 1999; Galindo et al. 2000; Hurtado et al. 2004; Petrovan et al. 2021). *EP153R* may play a role in haemadsorption of red blood cells to the surface of infected cells (Galindo et al. 2000) but also appears to have a role in modulating apoptosis. An ASFV *EP153R* deletion mutant induces higher cell death and caspase-3 activity in both Vero cells and swine macrophages compared to wild-type virus, and similar results were obtained in cultured cells transiently expressing *EP153R* (Hurtado et al. 2004). The mechanism by which pEP153R modulates apoptosis is unknown, but the protein can also restrict the surface expression of MHC-I (Hurtado et al. 2011).

Activation of cellular inflammatory responses is in part controlled by inflammasomes, macromolecular complexes formed from different NOD-like receptor (NLR) proteins that interact with procaspase-1 via N-terminal caspase recruitment (CARD) domains, either directly or through the adaptor protein apoptosis-associated speck-like protein containing a CARD (ASC). Caspase-1 has a number of downstream effects including the processing of pro-IL-1β and pro-IL-18 into their pro-inflammatory active forms (Guo et al. 2015). Although it is unclear if ASFV induces inflammasomes in infected cells, several ASFV genes have been shown to modulate an inflammasome-dependent reporter system in vitro (Song et al.

2020), and both pMGF505-7R and pH240R interact with NLRP3 (Li et al. 2021c; Zhou et al. 2022; Huang et al. 2023) the scaffold for the NLRP3 inflammasome. Viral protease pS273R, which cleaves the polyproteins pp220 and pp62 during viral assembly (Alejo et al. 2003), also targets gasdermin D, a pyroptosis effector (Zhao et al. 2022). Pyroptosis is an alternative pathway of programmed cell death that is controlled by caspase-1 via inflammasomes. Caspase activity releases the n-terminal section of gasdermin D leading to its oligomerisation and insertion into the plasma membrane, where it forms a pore, ultimately leading to cell death via loss of osmotic pressure. The ASFV protease cleaves gasdermin D at different sites to cellular caspases rendering it inactive and unable to induce cytotoxicity. PS273R is a component of the viral particle and, therefore, may be able to act on gasdermin D, and potentially other targets, immediately after uncoating. Inflammasomes are in part regulated by autophagy, which is impaired during ASFV infection (see below); therefore, pS273R inhibition of pyroptosis may ameliorate any failure to degrade inflammasomes.

## Autophagy

Autophagy can be broadly described as three distinct but related processes. The incorporation of cytoplasmic contents directly into the lumen of lysosomes is termed microautophagy (Li et al. 2012). The delivery of proteins through the membrane of the lysosome by chaperones is called chaperone-mediated (Cuervo 2011). Macroautophagy refers to the bulk segregation of cytoplasm as originally described by Christian de Duve (De Duve 1963; Mortimore et al. 1983). Macroautophagy, more commonly referred to as simply autophagy, involves the direction of cytoplasmic cargo into specialised organelles autophagosomes, which ultimately fuse with lysosomes. In all cases, the cytoplasmic content, which can consist of proteins, parts of organelles and pathogens, is eventually degraded by the specialised acid hydrolases that lysosomes possess. Microautophagy and chaperone-mediated autophagy tends to degrade smaller objects or molecules compared to macroautophagy.

Macroautophagy (from here on referred to as autophagy) pathway is presented in Fig. 11.2 and comprises three main stages—initiation, vesicle elongation (formation of the complete autophagosome) and maturation (fusion with the lysosome). Autophagy can be induced by stress including amino acid starvation, ER stress, infection by pathogens and oxidative stress. Autophagy induction is highly regulated, and the mammalian target of rapamycin (mTOR) protein plays an important role (Sengupta et al. 2010). Under nutrient-rich conditions, mTOR forms an active complex that blocks autophagy; however, stimulation with rapamycin or amino acid starvation inhibits mTOR leading to an increase in autophagic flux (Noda and Ohsumi 1998). Uncoordinated 51-like kinases 1 and 2 (ULK1 and ULK2) mediate this function, and suppression of ULK1 expression inhibits autophagy below TOR in the pathway (Chan et al. 1999). FIP200, Atg13 and Atg101 are in complex with ULK1 and TOR controls Atg13 phosphorylation (Kamada et al. 2000). After TOR

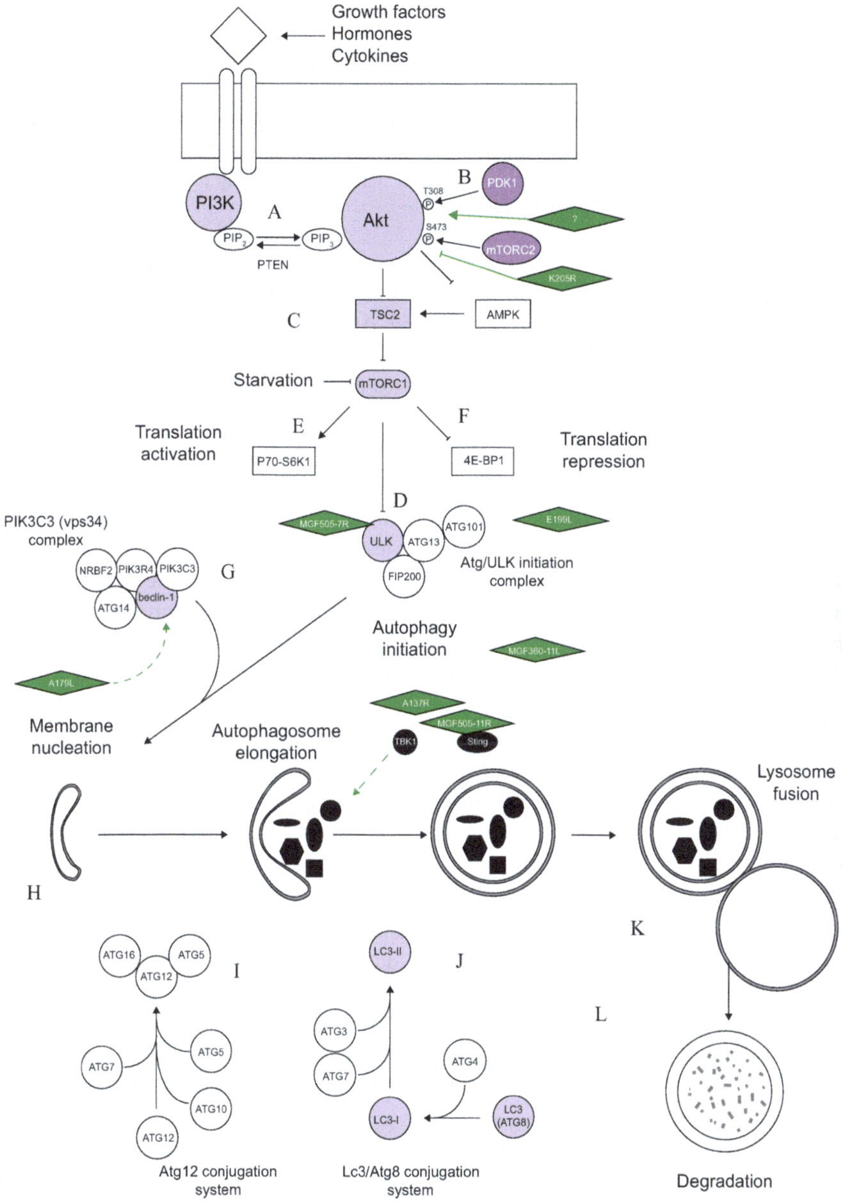


**Fig. 11.2** African swine fever virus and autophagy: Akt is activated on the cell surface by growth factors, hormones, or cytokines via the generation of $PIP_3$ (**a**) that recruits PDK1 to phosphorylate Akt residue T308 (**b**). mTORC2 phosphorylates Akt at residue S473 (**c**). mTORC1 is negatively regulated by TSC2, which in turn is inactivated by active Akt (**c**). mTORC1 negatively regulates the ULK complex, and therefore active Akt leads to active mTORC1 and autophagy suppression (**d**) as well as upregulation of translation by activating p70-S6K (**e**) and suppressing 4E-BP1 (**f**). Autophagy itself is initiated by the ULK1 complex action on the Vps34 complex, which includes the Bcl-2-like protein beclin-1 (**g**). Akt can act directly on the Vps34 complex by phosphorylating beclin-1. PtdIns3P generation by the Vps34 complex induces membrane nucleation, and the

**Fig. 11.2** (continued) recruitment of autophagy precursor proteins such as WIPI (**h**) and two ubiquitin-like conjugation processes are involved in the formation of autophagosomes. ATG5 and ATG12 are conjugated with each other and, once in complex with ATG16, specify the site of LC3 lipidation (**i**). LC3 is cleaved by ATG4 to form LC3-I, and ATG3 and ATG7 transfer PE to the c-terminus of LC3-I to form LC3-II (**j**). Completed autophagosomes fuse with lysosomes (**k**), where the contents are degraded (**l**). Steps within the autophagy pathway in which ASFV proteins interact are indicated and described in the text

inactivation, dephosphorylation of Atg13 increases the affinity for Atg17 and ULK1. Interactions between ULK1 and Atg13 and Atg17 induce the kinase activity of ULK1 kinase. In mammalian cells, mTOR exists as two complexes, mTORC1 and mTORC2, that are functionally distinct (Sengupta et al. 2010). In addition, mTORC1 is thought to be phosphorylated in direct response to nutrient signals (Long et al. 2005). mTORC1 is also activated by phosphatidylinositol 3 kinase (PI3K)/Akt signalling cascade after extracellular stimulation by growth factors, cytokines and hormones (Chan et al. 1999). The binding of receptor tyrosine kinases (RTKs) to these molecules at the cell surface leads to the recruitment of class I PI3K and the production of phosphatidylinositol (3,4,5)-trisphosphate (PIP3) from phosphatidylinositol (4,5)-bisphosphate (PIP2). Recruitment of protein kinase B (PKB)/Akt along with its activator phosphoinositide-dependent protein kinase 1 (PDK1) by the accumulation of PIP3 leads to the subsequent activation of Akt by phosphorylation at residue T308. However, complete activation of Akt requires phosphorylation at residue S473 by mTORC2 (Sarbassov et al. 2005). Once activated, Akt promotes inactivation of the tuberous sclerosis complex 2 (TSC2) protein via phosphorylation which is itself a negative regulator of mTORC1. Hence activated Akt leads to suppression of autophagy by mTORC1 activation through TSC2. Additionally, PTEN, a the 3′-phosphoinositide phosphatase, converts PIP3 back to PIP2 and so reduces the signalling from Akt and so upregulates autophagy (Arico et al. 2001). PI3K/Akt pathway also contributes to the regulation of protein translation as mTORC1 mediates the translation activator and repressors p70-S6K and 4E-BP1, respectively (Ma and Blenis 2009).

Translation is clearly crucial to the viral life cycle and like many viruses ASFV targets the Akt/mTORC pathway in order to ensure successful production of progeny viruses. ASFV leads to the rapid phosphorylation of both p70 and 4E-BP1 (Castelló et al. 2009; Shimmon et al. 2021), suggesting the activation of mTORC as early as 2 hours post-infection (Shimmon et al. 2021), concomitant with the exit of the viral core from the endosome. This occurs even after starvation, suggesting that ASFV infection is able to prevent the cell supressing mTORC1 activity in response to amino acid deprivation. ASFV infection also leads to an increase in the phosphorylation of Akt, particularly on S473 and is also able to inhibit drug-induced dephosphorylation of Akt (Shimmon et al. 2021). Overexpression of both *K205R* and *E199L* leads to the activation of autophagy. PK205R leads to dephosphorylation of AKT (S473) and mTOR and hence phosphorylation of ULK1 (Wang et al. 2022). E199L downregulates pyrroline-5-carboxylate reductase 2 (PYRC2) in Vero cells

(Chen et al. 2021) which may promote autophagy by a currently unknown mechanism.

Downstream of the ULK complex sits a class III phosphatidylinositol 3 kinase complex (PIK3C3, referred to as vps34 complex in yeast) which is also essential in the autophagy pathway (Funderburk et al. 2010). PIK3C3 is complexed with beclin-1 (vps30/Atg6 in yeast); PIK3R4 (Vps15), NRBF2 and Atg14. Beclin 1 function is regulated by Bcl-2 (B-cell lymphoma/leukaemia-2), which is an anti-apoptotic protein that prevents Beclin 1 from activating autophagy by binding to it under nutrient-replete conditions. Separation of Beclin 1 from Bcl-2 is necessary for the induction of autophagy (He and Klionsky 2009). Generation of phosphatidylinositol 3-phosphate (PtdIns3P) by the Vps34 complex is crucial for the biogenesis of the autophagosome (Petiot et al. 2000). The ASFV bcl-2 homologue has been shown to interact with beclin-1 in vitro (Banjara et al. 2017, 2019; Hernaez et al. 2013) and overexpression of *A179L* can inhibit stress-induced autophagy.

Ultimately the phagophore closes to form autophagosomes that fuse with lysosomes, where the cargo is degraded and recycled. Autophagy adaptor proteins can be used to target specific proteins or classes of proteins to the forming autophagosome in order to facilitate their degradation. For example, sequestosome 1 (SQSTM1/p62) binds both ubiquitin and LC3 (Atg8) a major structural protein of the autophagosome, thereby directing ubiquitinated proteins for degradation. Several ASFV proteins have either been shown to interact with components of the autophagy pathway or to direct other proteins for degradation by autophagy. pA137R and pMGF360-11L target TANK-binding kinase 1 (TBK1) for degradation through autophagy and pMGF360-11L and pMGF505-7R can also target interferon regulatory factor 7 (IRF7) (Yang et al. 2022a, b; Sun et al. 2022).

Stimulator of interferon genes (STING) and cyclic guanosine monophosphate/adenosine monophosphate synthase (cGAS) are important sensors of cytoplasmic DNA that lead to the induction of type I interferon production and other inflammatory cytokines (Zhang et al. 2021). After activation STING acts on TBK1 and IκKε to signal interferon induction (Sharma et al. 2003); however, activated STING also induces the formation of autophagosomes that appears to be a highly conserved function of the protein (Gui et al. 2019). ASFV pMGF505-11L[2] co-precipitates with STING and inhibits phosphorylation of TBK1 and interferon regulatory factor 3 (IRF3). Degradation of STING is blocked by both 3-methyladenine and $NH_4Cl$, suggesting a role for autophagy; however, the proteasome also plays a role as treatment with MG132 also blocks STING degradation in *MGF505-11L* expressing cells (Yang et al. 2021b). The N-terminal domain of pMGF505-7R also interacts with STING, and STING is degraded in *MGF505-7R* expressing cells (Li et al. 2021a). Conservation between the N-terminus of pMGF505-7R and pMGF505-11L is relatively low compared to other MGF505 members; therefore, it is likely that

[2]The authors report a study on *MGF505-11R*; however, there is no such annotated gene in the genomes of Georgia 2007/1 or SY18 isolates, and therefore we refer to this gene as *MGF505-11L*, which is present.

different regions of the protein interact with STING. Interestingly pMGF505-7R also interacts with ULK1, which may also play a role in STING degradation, and levels of STING were elevated in macrophages infected with an *MGF505-7R* deletion mutant (Li et al. 2021a). Degradation of STING by MGF505-7R and TBK1 by MGF360-11L was sensitive to treatment with 3-methyladenine, a drug that blocks both class I and class III phosphatidylinositol 3 kinases and can inhibit autophagy but not treatment with $NH_4Cl$ which elevates lysosomal pH and blocks degradation in lysosomes. As ASFV inhibits the formation of autophagosomes (Shimmon et al. 2021), it is likely that autophagy per se is not responsible for the degradation of STING or TBK1 but rather that components of the pathway are involved. For example, translocation of STING from the ER to cytoplasmic punctate where it assembles with TBK1 is dependent on Atg9a, but not Atg7 (Saitoh et al. 2009).

## Interferon Response

The innate response is found across all cellular life and is responsible for mounting the first response to infection. By deploying sentinels that recognise highly specific pathogen determinants, cells are able to initiate rapid changes within the cellular environment to combat the infection. These determinants are known as pathogen-associated molecular patterns (PAMP)s, and in the context of viral infection, they can be various forms of nucleic acids including single-stranded RNA (ssRNA), double-stranded RNA (dsRNA), viral DNA or viral proteins. The cellular machinery which recognises these PAMPs include several classes of germline-encoded receptors, known as pathogen recognition receptors (PRRs) (Akira et al. 2006; Goubau et al. 2013; Hertzog and Rehwinkel 2020). PRRs can be broadly divided into those that are transmembrane or cytosolic. The recognition of PAMPs by PRRs stimulates the innate immune system by activating a series of downstream pathways that ultimately culminate in the activation of transcription factors and induce the expression of type I interferon (IFN) and other inflammatory cytokines (Akira et al. 2006; Honda and Taniguchi 2006). This activation can be further divided into two primary pathways, the interferon regulator factor (IRF) and Nuclear Factor κB (NFκB) pathways. IFN acts as a signalling protein that functions in both an autocrine and paracrine manner. Its production leads to altered expression of over 300 different genes, known as IFN-stimulated genes (ISG)s, creating an antiviral environment in order to limit the replication of the invading viral pathogen (Schoggins et al. 2011). There is a strong selection of viruses to impede the innate immune response. The evolutionary arms race between virus and host has resulted in the vast majority of, if not all, viruses that infect metazoans deploying tactics to overcome and subvert these pathways.

## Type I Interferons

Response to viral infection by the innate immune system ultimately leads to the production of type I interferons. This is a class of cytokines that include 13 variants of IFNα in humans, a single IFNβ, and a number of poorly defined but related cytokines such as IFNκ and IFNδ (McNab et al. 2015). Interestingly swine, however, have 39 functional type I interferons, including 17 variants of IFNα (Sang et al. 2010; Zanotti et al. 2015).

## Induction of Type I Interferon Expression

Almost all cells are capable of inducing type I IFNβ following viral infection, whereas IFNα is primarily produced by dendritic cells. Induction of IFNβ expression occurs in a highly ordered process at the transcriptional level by activation of the IFNβ promoter. This promoter can be divided into at least 4 cis-regulatory element regions known as positive regulatory domains (PRD) I, II, III and IV (Kim and Maniatis 1997). When activated by upstream signalling pathways, transcription factors of the IRF family bind to PRDI/III (Fig. 11.3), whereas NFκB and AP1 (cJUN, ATF2 heterodimer) bind to regions PRDII or PRDIV respectively (Fig. 11.4). The binding of the transcription factors results in the formation of a higher order nucleoprotein complex termed the enhanceosome. Further recruitment of histone acetyl transferase cofactor proteins including cAMP-response element binding protein (CREB) binding protein (CBP), or paralog p300 is required (Merika et al. 1998). These cofactors aggregate at the promoter to target H3 and H4 histones of the nucleosome for acetylation. This relaxes the chromatin structure at the promoter, providing a platform for the cellular machinery to facilitate transcription (Gerritsen et al. 1997; Merika et al. 1998; Yoneyama et al. 1998).

## Interferon Regulatory Factors

The interferon regulatory factor (IRF) class of transcription factors contains up to 9 conserved members, IRF1-9. It appears that a robust IFN response to viral infection is conditional on the activation of the IRFs, specifically the constitutively expressed IRF3 (Honda et al. 2006). Inactive IRF3 remains dormant in the cytosol, until stimulation of signalling pathways following the detection of viral PAMPs leads to its activation. Phosphorylation of C-terminal regions of IRF3 by directly upstream kinases is key in the activation of the transcription factor. The closely related IκB kinase (IKK) homologs IκKε and TBK1 induce the phosphorylation of IRF3; however, TBK1 has been shown to act as the primary actor in the phosphorylation of serine residues 386 and 396 (Chen et al. 2008; Fitzgerald et al. 2003; Lin

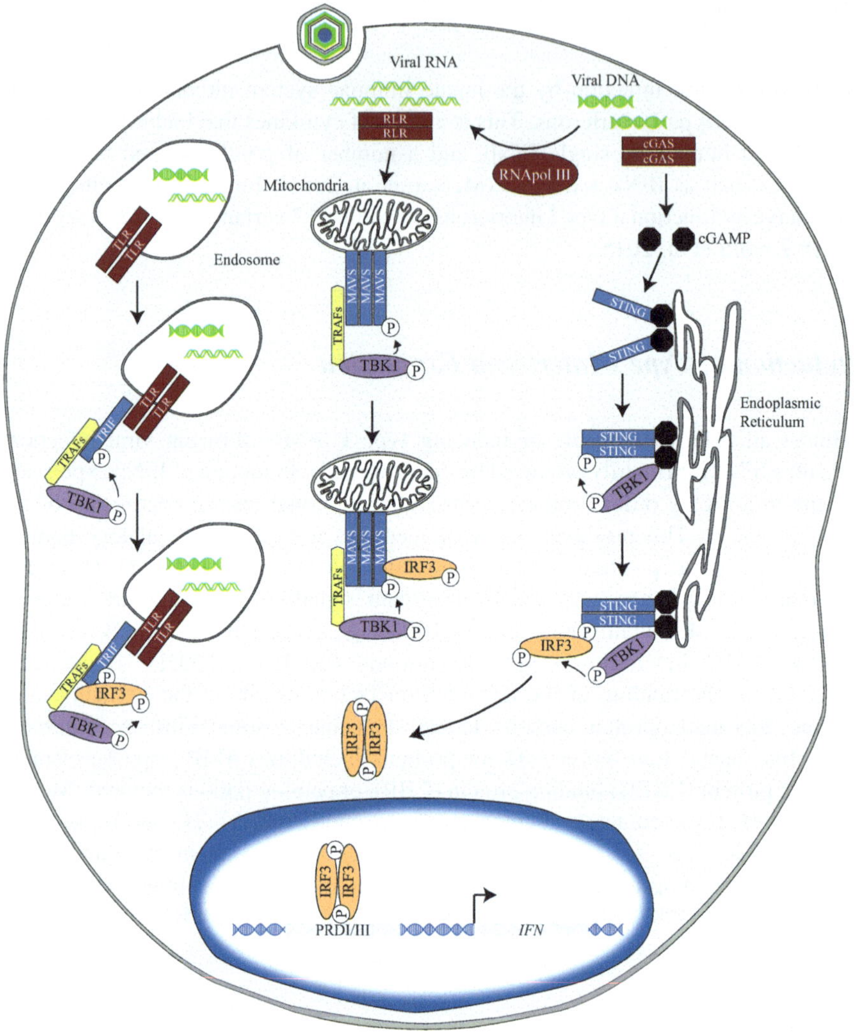


**Fig. 11.3** Interferon regulatory factor 3 activation. Virus infection leads to the presence of nucleic acids in endosomal compartments or the cytoplasm. Nucleic acids in the endosomal compartments are detected by toll-like receptors (TLR), leading to the recruitment of TRIF and TRAF3. This, in turn, recruits TBK1 leading to the phosphorylation of TRIF, allowing the binding of IRF3 to TRIF and phosphorylation of the transcription factor by TBK1. Viral RNA in the cytoplasm is recognised by RLR such as RIG-I or MDA-5 which leads to aggregation of MAVS at the mitochondria, which recruits TRAF3 followed by TBK1. TBK1 then phosphorylates MAVS allowing the recruitment of IRF3 for phosphorylation by TBK1. Recognition of cytoplasmic DNA by cGAS induces the production of cGAMP. cGAMP binds to STING at the ER inducing a conformational change which allows binding of TBK1. TBK1 phosphorylates STING enabling the recruitment and binding of IRF3 for phosphorylation by TBK1. Alternatively, RNApol III transcribes viral DNA into RNA for recognition by RLRs and signalling via the MAVS pathway. All three pathways ultimately lead to dimerisation of IRF3, nuclear translocation and binding to the PRDI/III domains in the interferon β promoter

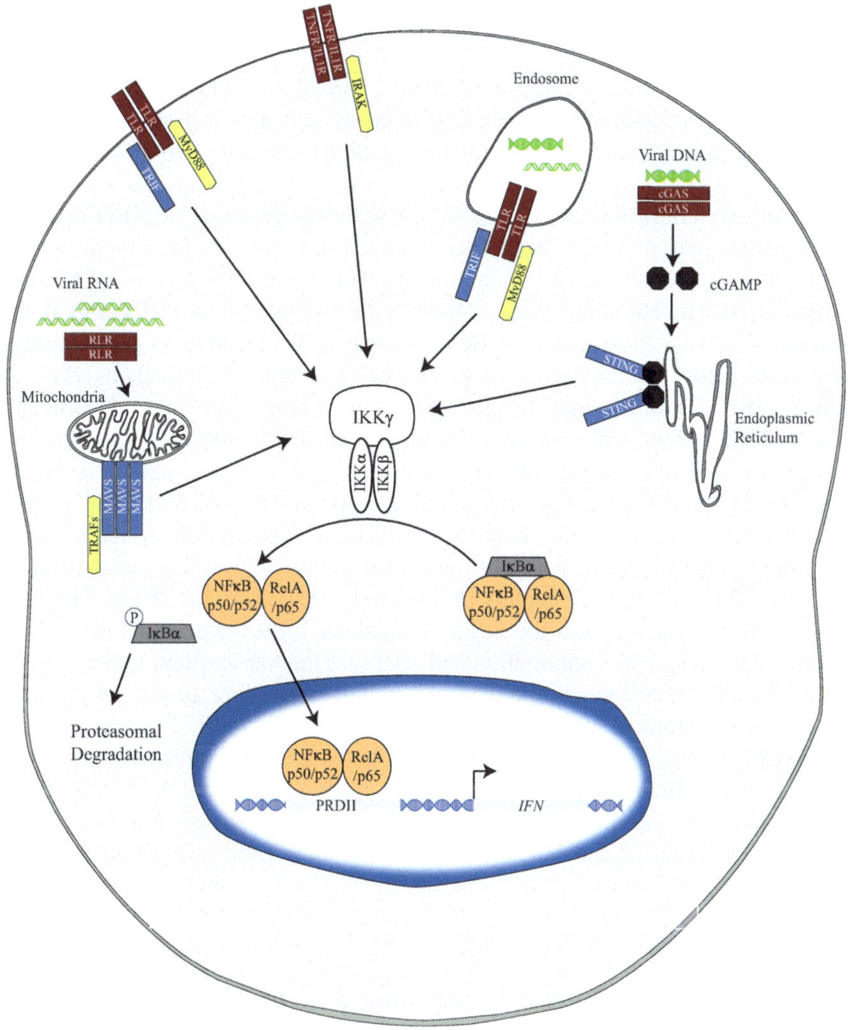


**Fig. 11.4** Nuclear factor kappa B activation: Stimulation of various PRRs or cytokine receptors leading to the activation of IKK complex and subsequent IκBα degradation. Active NFκB is released and translocates to the nucleus where it binds to the PRDII domains in the interferon β promoter

et al. 1998; Mori et al. 2004; Panne et al. 2007; Takahasi et al. 2003). The phosphorylation of the S386 residue leads to the formation of the IRF3 homodimer, a critical step in activation (Mori et al. 2004; Panne et al. 2007; Takahasi et al. 2003). This dimeric form of IRF3 subsequently translocates to the nucleus forming complexes with CBP/p300 cofactors and binds to the PRDI/III sites of the IFN promoter (Lin et al. 1998, 2000; Panne et al. 2007).

## *Nuclear Factor κB*

Although not the primary inducer of IFNβ expression, NFκB retains a crucial function in the development of an inflammatory and antiviral response. NFκB transcription factors are a family of proteins that bind to cis-acting κB DNA elements to regulate inflammatory gene transcription. NFκB is formed of either a homodimer or heterodimer consisting of proteins which share the presence of a Rel domain. These include p50, p52, p65 (RelA), RelB and cRel. In the canonical pathway, the NFκB p50-p65 heterodimer can be activated by Tumour necrosis factor-α (TNFα) or interleukin 1β (IL-1β) stimulation of their respective receptors TNFR and IL-1β receptor (IL1R), or PRR activation by relevant PAMPs. Inactive NFκB is sequestered in the cytosol by the negative regulator IκBα protein which binds to NFκB to induce a 180° conformational change, essentially locking it in a "closed" structure (Huxford et al. 1998). Stimulation of upstream signalling pathways results in the activation of the IKK kinase complex, which is formed of two catalytic subunits IKKα, IKKβ and the regulatory subunit IKKγ (also known as NEMO). The method of activation is not clear but may occur by either autophosphorylation, or by transforming growth factor β (TGFβ) activated kinase 1 (TAK1)-mediated phosphorylation of the catalytic subunits (Wang et al. 2001; Xu et al. 2011). The IKK complex then phosphorylates the negative regulator IκBα bound to NFκB, subsequently "unlocking" the confirmation and releasing the transcription factor, revealing a ubiquitination site on the negative regulator leading to its proteasomal destruction (Huxford et al. 1998; Karin and Ben-Neriah 2000; Traenckner et al. 1995). Despite the similarities between constituent catalytic proteins of the IKK complex, it has been revealed that IKKβ is the principal actor in NFκB signal transduction. Activation of NFκB eventually leads to the nuclear translocation of the transcription factor and binding to κB elements in promoters to induce gene expression.

## *Transmembrane Pathogen Recognition Receptors*

Transmembrane PRRs include Toll-like receptors (TLR)s. These are a class of 13 described PRRs (TLR1-13) with a broad range of specificity (Kawasaki and Kawai 2014). TLR3/7/9, for example, are located in endosomal membranes and recognise dsRNA, ssRNA and viral DNA, respectively, generated by the uncoating of viral particles. Activation of TLRs leads to the stimulation of downstream signalling pathways (Figs. 11.3 and 11.4), with the initial mediator being the intracellular Toll IL1R (TIR) domain that facilitates further signal transduction by binding adaptor proteins TIR domain-containing adapter-inducing interferon-β (TRIF) (in the case of TLR3) or MyD88 (TLR7 and TLR9). TLR3-associated TRIF interacts with the RING-domain E3 ubiquitin (Ub) ligase TNFα receptor-associated factor 6 (TRAF6) but also TRAF3 (Häcker et al. 2011). TRAF6-mediated signalling leads to IKK complex activation and, therefore, phosphoactivation of NFκB (Cusson-Hermance et al. 2005; Meylan et al. 2004; Wang et al. 2001). TRIF association with TRAF3 stimulates the IRF3 response (Yamamoto et al. 2003). Furthermore, there is a degree of TRAF-mediated crosstalk between IRF3 and NFκB pathways, with TRAF3 and TRAF6 being able to stimulate both (Fang et al. 2017).

TLR7 and TLR9 are expressed in plasmacytoid dendritic cells and are the primary sensors for viral ssRNA and unmethylated CpG-DNA, respectively. In contrast to TLR3, which mediates downstream signalling via TRIF, TLR7 and TLR9 facilitate signal transduction by interacting with the MyD88 adaptor protein at the TIR domain. Activation of these TLRs induces MyD88 interaction with a complex of IL1R-associated kinase 4 (IRAK)4, IRAK1 and TRAF6 that ultimately leads to the phosphoactivation of IKKβ and then NFκB (Kawasaki and Kawai 2014; Lin et al. 2010). Interestingly, although it is not thought to induce IRF3 activation, TLR7/9 MyD88 piloted signalling transduction also leads to TRAF6-mediated IRF7 activation via IRAK1 phosphorylation (Randall and Goodbourn 2008).

## *Cytosolic Pathogen Recognition Receptors*

Cytosolic PRRs include the RNA helicases retinoic acid-inducible gene-I (RIG-I)-like receptor (RLR) family including RIG-I, Melanoma differentiation-associated protein 5 (MDA5) and laboratory of genetics and physiology 2 (LGP2). These detect intracellular ssRNA and dsRNA and activate both IRF3 (Fig. 11.3) and NFκB (Fig. 11.4) pathways, thus inducing expression of type I IFN and inflammatory cytokines (Kato et al. 2006; Rehwinkel et al. 2010; Yoneyama et al. 2004). Both RIG-I and MDA5 share a high level of structural homology, containing two CARD domains and an RNA-binding DEAD-box helicase domain. Although both induce IFN response to viral dsRNA, it has been proposed that they are functionally distinct as there appears to be specificity in recognition of RNA dependant on the virus (Kato et al. 2006). RIG-I is stimulated by 5′ triphosphorylated RNA (Hornung et al. 2006); in contrast, MDA5 stimulation depends on the length of the RNA strands, with high molecular weight RNA inducing a response (Pichlmair et al. 2009). When activated, the CARD domains of RIG-I and MDA5 undergo K63 polyubiquitin modification forming oligomers (Jiang et al. 2012). RIG-I and MDA5 then associate with mitochondrial antiviral-signalling protein (MAVS), another CARD domain-containing protein, at the mitochondrial membrane resulting in the formation of MAVS oligomers, a crucial step in IRF3 activation (Hou et al. 2011; Zamorano Cuervo et al. 2018). These oligomers act as a platform for the recruitment of IRF3, and TRAF3, which in turn acts to recruit TBK1. This brings IRF3 into close proximity to TBK1, thus facilitating phosphoactivation of the transcription factor (Fang et al. 2017). Furthermore, stimulation of the RLRs induces cross-pathway activation. RIG-I was reported to induce NFκB via MAVS interaction with TRAF6 that subsequently leads to the activation of the IKK complex (Kawai et al. 2005; Seth et al. 2005; Xu et al. 2011; Yoneyama et al. 2004; Yoshida et al. 2008).

The discovery of the cGAS-STING-TBK1 pathway has revealed a new sentinel in the host cell repertoire, capable of stimulating IFN in response to viral DNA (Hertzog and Rehwinkel 2020; Sun et al. 2013; Zhang et al. 2014). cGAS binding of cytosolic DNA is independent of sequence. Binding to DNA forms a 2:2 complex and induces a conformational change in the PRR, enabling the catalysis of cyclic GMP-AMP (2′3′cGAMP) from ATP and GTP (Sun et al. 2013; Zhang et al. 2014). 2′3′cGAMP functions as a secondary messenger, binding to the ligand binding domain of downstream transmembrane protein STING, localised at the endoplasmic reticulum membrane (Ishikawa and Barber 2008; Shang et al. 2019). This interaction induces a conformational change in STING leading to oligomerisation, and recruitment of TBK1 (Shang et al. 2019; Zhang et al. 2019; Zhao et al. 2019). The exact mechanism of TBK1 activation by STING remains poorly defined. However, in a similar manner to MAVS-mediated signalling, phosphorylation of STING enables the recruitment of IRF3 to the adaptor protein. Like other cytosolic PRRs, cGAS-STING has also been implicated in the activation of NFκB mediated by TRAF6 (Abe and Barber 2014).

Notably, there is evidence that in certain cells induction of IFNβ by dsDNA is dependent on RIG-I and MAVS. This led to the identification of RNA polymerase III (Pol III) as an important sensor of cytosolic DNA. This enzyme acts by transcribing the DNA template into 5′-triphosphate dsRNA, which is then recognised by RIG-I, resulting in the activation of the RLR signalling cascade. Inhibition of Pol III activity was shown to markedly reduce IFNβ in cells infected with both herpes simplex virus type 1 (HSV-1) and adenovirus (Chiu et al. 2009).

## *IFN-Mediated Signalling*

Type I IFN expression activates further signalling cascades and the expression of over 300 genes, known as ISGs (Schoggins et al. 2011). The binding of secreted IFN to the IFNAR1/2 receptor heterodimer facilitates this action. Each subunit of the receptor dimer is associated with different intracellular proteins, IFNAR1 with tyrosine kinase 2 (Tyk2), and IFNAR2 with Janus kinase 1 (JAK1) but also signal transducers and activators of transcription STAT1 and STAT2. Ligand-induced activation of the receptor leads to its dimerisation and Tyk2-mediated phosphorylation of IFNAR1. This brings STAT2 into proximity to the receptor for its phosphorylation at the tyrosine 690 (Y690) residue. JAK1 then phosphorylates the tyrosine 701 (Y701) residue of STAT1. The phosphorylated STAT1 and 2 can form a heterodimer via SH domain interaction with phosphorylated tyrosine residues, in the process revealing a novel nuclear localisation signal that allows its nuclear translocation. The STAT1/2 heterodimer binds to IRF9 and forms a heterotrimer known as ISG factor 3 (ISGF3) that binds to the IFN-stimulated response element (ISRE) of ISG promoters to enhance transcription (Randall and Goodbourn 2008).

Interestingly cells produce a low basal level of ISGs. A recent model has been proposed indicating that a STAT2-IRF9 heterodimer formed independently of IFN stimulation is retained in the nucleus inducing homeostasis of this low level of basal ISG expression, until IFN stimulation-driven formation of ISGF3 acts as a molecular switch to enhance ISG expression (Platanitis et al. 2019).

## *ASFV Modulation of the IFN Response*

The ability of ASFV to modulate the IFN response was firstly demonstrated using a deletion mutant virus, PR4Δ35, lacking six MGF360 (9L, 10L, 11L, 12L, 13L and 14L) and two MGF505 (1R and 2R) genes. The comparison of host gene transcription between wild-type and mutant viruses by microarray analysis, revealed an upregulation of several ISG mRNAs when the cells were infected with the latter. These results suggested that MGF360 and/or MGF505 genes are involved in the inhibition of IFN response, and this hypothesis was confirmed by the observation that in contrast with the wild-type virus infection, the deletion mutant virus-infected culture supernatant contained significant amounts of IFN-α (Afonso et al. 2004). These genes were also implicated in the sensitivity of ASFV to type I IFN, indicating that ASFV is able to overcome activation of the JAK/STAT signalling and/or the antiviral state (Golding et al. 2016). In recent years, the molecular mechanisms underlying the ability of these and other ASFV-encoded genes to modify the IFN responses are finally being described. Interestingly, the identity of the host PRR (s) responsible for the recognition of ASFV infection remains controversial, and most likely, different receptors are involved. Phosphorylation of STING has been shown to occur following infection of porcine alveolar macrophages with both attenuated NH/P68 and highly virulent Arm/07 ASFV isolates although at different levels. Furthermore, a specific cGAS inhibitor, Ru521, was shown to decrease levels of STING phosphorylation in infected cells, arguing for an important role of the cGAS/STING pathway in the induction of IFN in response to ASFV (García-Belmonte et al. 2019). However, levels of IFN expression in infected cells treated with Ru521 were not measured, so other sensors of ASFV infection could not be ruled out. Another study provided evidence that ASFV is sensed by both cGAS and RNA Pol III-RIG-I pathways. Knockdown of either cGAS or RIG-I by siRNA resulted in reduced IFNβ transcription in infected alveolar macrophages, and a cumulative effect was observed in cells where both receptors were targeted. Importantly, treatment of cells with ML-60218, a specific inhibitor of RNA Pol III, decreased transcription of IFNβ, ISG15, ISG54 and MX1 following ASFV infection. Further experiments in transfected cells indicated that AT-rich islands in the ASFV genome are the main PAMP recognised by RNA Pol III, leading to the activation of the RLR pathway as described above (Ran et al. 2022). In addition to these pathways, the TLR/Myd88 axis was also involved in the recognition of ASFV infection. Knockdown of TLR1, TLR2, TLR3, TLR4, TLR9 (but not TLRs 5-8) and MyD88 significantly decreased pro-IL-1β transcription in cells infected with an

ASFV deletion mutant lacking the MGF505-7R gene. Since pro-IL-1β transcription is mainly controlled by NFκB-mediated signalling, it is reasonable to infer that TLRs play a role in ASFV sensing, at least in the induction of an inflammatory response (Li et al. 2021c).

## *Inhibition of IFN Induction at the Level of Transcription Factors or Their Activation*

A common theme has emerged regarding the inhibition of IFN induction by ASFV-encoded genes: most of them impact the level of shared components of the IRF and NFκB-mediated pathways. Although most of the recent publications have been focussing on inhibition of cGAS/STING-induced signalling, it is becoming apparent that ASFV evolved to manipulate key regulators of IFN induction, therefore having a broad impact including on those cascades initiated by RLR and TLR engagement.

ASFV pA137R directly interacts with TBK1, the central kinase involved in IRF3 phosphorylation, and leads to its degradation. As mentioned above, further evidence from experiments done in ATG5-KO cells suggests that this is mediated through an autophagy-dependent mechanism. As expected from a viral protein acting at this level, IRF3 translocation into the nucleus is inhibited following stimulation with both Sendai virus infection and poly dA:dT treatment (Sun et al. 2022). pDP96R, on the other hand, does not seem to interact with TBK1 but, through a yet unknown mechanism, inhibits induction of an IFNβ promoter in cells overexpressing TBK1. This was associated with a modest decrease in both TBK1 and IRF3 phosphorylation following cGAS/STING activation (Wang et al. 2018). pI215L, an ASFV encoded E2 ubiquitin conjugating enzyme, inhibits TBK1 activity by recruiting the E3 ubiquitin ligase RNF138 to degrade RNF128. This results in decreased K63-linked ubiquitination of the kinase, a crucial step for its full activation (Song et al. 2016). Importantly, pI215L is no longer able to impact on IFN induction in Rfn138$^{-/-}$ cells. In contrast, the cysteine 85 catalytic site, essential for pI215L E2 ubiquitin-conjugating activity, is not necessary for the described TBK1 inhibition (Huang et al. 2021). pM1249L, a structural protein, inhibits TBK1 phosphorylation in cells overexpressing cGAS/STING and impacts on IFNβ expression following induction with 2′3′cGAMP, poly I:C and Sendai virus infection (Cui et al. 2022). Both pMGF360-11L and pMGF505-7R interact with TBK1 and decrease its phosphorylation in cells overexpressing cGAS/STING. Transfection of cells with MGF360-11L or MGF505-7R also causes a marked decline in exogenously expressed TBK1. However, a decrease in expression of other transfected inducers such as cGAS, STING, IκKε, IRF3 and IRF7 is also observed, indicating that TBK1 might not be a specific target for degradation by these ASFV encoded proteins (Yang et al. 2022a, b).

The other main kinase of the IRF pathway, IκKε, is also a target for ASFV. Specifically, pS273R, a protein belonging to the SUMO-1-specific protease family, interacts with IκKε and inhibits its interaction with STING. Whether the interaction impacts on other IκKε dependent pathways remains to be tested. Remarkably, the pS273R protease activity seems to be required for IFN induction inhibition, although no changes in IκKε expression were observed in cells expressing the viral protein. An interesting possibility is that pS273R reduces IκKε sumoylation, by cleaving small ubiquitin-like modifier (SUMO) peptides from the kinase (Luo et al. 2022).

ASFV-encoded proteins can also directly impact on IRF3, the principal transcription factor involved in the "first wave" of type I IFN expression. pE120R interacts with IRF3, decreasing its interaction with TBK1 and subsequent phosphorylation. Notably, amino acids at positions 72 and 73 were shown to be essential for IFN inhibition. A mutant ASFV where these two residues were deleted induces higher levels of IFNβ than the wild type (Liu et al. 2021a). pE301R, on the other hand, inhibits IRF3 downstream of its phosphorylation, impeding the translocation into the nucleus (Liu et al. 2022). pM1249L, in addition to its effect on TBK1 described above, interacts with IRF3 targeting it for degradation (Cui et al. 2022). pMGF360-14L also causes IRF3 degradation by recruiting the tripartite motif protein 21 (TRIM21) to increase its K63-linked polyubiquitination (Wang et al. 2021). Interestingly, TRIM 21 has been shown to act as both a positive and negative regulator of IFNβ induction (Rajsbaum et al. 2014). In addition, IRF3 K63-linked polyubiquitination has been linked to its full activation rather than its degradation (Zeng et al. 2009). These findings highlight the complexity of IFN expression regulation and indicate that pMGF360-14L may use other mechanism(s) to inhibit IRF3 function.

Both pMGF360-11L and pMGF505-7R interact with IRF7 in transfected (Yang et al. 2022a, b), suggesting that this transcription factor is also targeted by ASFV-encoded proteins. IRF7 is itself an ISG with an important role in the positive feedback regulation of type I IFN expression (Sato et al. 1998), so it is not surprising that numerous viruses have evolved to modify its function.

The NFκB arm of the IFN induction pathway is frequently controlled by virus-encoded proteins, and ASFV is no exception. pI215L inhibits NFκB activation downstream of TRAF6 and TRAF2 but upstream of the IKK complex (Barrado-Gil et al. 2021). pH240R interacts with NEMO, the regulatory subunit of the IKK complex (Zhou et al. 2022; Huang et al. 2023) and disrupts its interaction with IKKβ (Huang et al. 2023). Furthermore, cells infected with ASFVΔH240R show significantly higher levels of NEMO when compared to cells infected with the wild-type virus (Zhou et al. 2022). pI226R, although not directly interacting with NEMO, increases its ubiquitination in transfected cells and leads to its degradation by the proteasome (Hong et al. 2022).

pD345L can interact with both catalytic subunits of the IKK complex, IKKα and IKKβ. This leads to the inhibition of IκBα phosphorylation and p65 translocation into the nucleus (Chen et al. 2022). pMGF505-7R interacts with IKKα in both transfected and infected cells. Accordingly, cells infected with ASFVΔMGF505-7R show higher levels of IκBα phosphorylation and degradation then cells infected with wild-type virus at 12 and 24 hours post-infection, respectively (Li et al. 2021c). pF317L co-precipitates with IKKβ in transfected cells and inhibits its phosphorylation. This in turn results in decreased levels of IκBα K48-linked ubiquitination following stimulation of cells with TNFα (Yang et al. 2021a).

One of the most studied ASFV-encoded proteins, pA238L, was first identified as an NFκB signalling inhibitor due to the presence of ankyrin motifs, a characteristic it shares with IκBα. pA238L binds to p65 (Revilla et al. 1998) and inhibits its binding to κB elements present in host promoters, including many regulating pro-inflammatory cytokines (Powell et al. 1996; Revilla et al. 1998). In addition, pA238L impairs p65-p300 interaction, preventing p65 acetylation and inhibiting its transactivation activity at the inducible nitric oxide synthase promoter (Granja et al. 2006b). Similarly, pA238L inhibits induction of TNFα by impacting on the recruitment of p300/CBP co-activators (Granja et al. 2006a). pMGF360-12L inhibits p50/p65 translocation into the nucleus in a process independent of IκBα phosphorylation and degradation. Instead, pMGF360-12L interacts with members of the importin-α family of nuclear transport proteins (karyopherin-α2, -α3 and -α4), competing with p65 for transport into the nucleus (Zhuo et al. 2021). pMGF505-7R binds to the p65 Rel homology domain and impairs the induction of an NFκB reporter in cells overexpressing p65. Interestingly, pMGF505-7R inhibits both the antiviral and the antibacterial effects induced by activation of the TLR8-NFκB axis (Liu et al. 2021b).

## *Inhibition of Specific PRRs Pathways*

As described above, the cGAS/STING pathway has been the subject of extensive research in the ASFV field. pE346R and pC129R both bind to and degrade the cell second messenger 2′3′cGAMP. Interestingly, this is mediated by their phosphodiesterase activity since the universal phosphodiesterase inhibitor IBMX reverts the degradation. Furthermore, pE346R also contains a sequence that resembles STING's 2′3′cGAMP-interacting motif. Mutations in two key residues of this sequence (tyrosine-76 and asparagine-78) abolish IFN inhibition. Additional evidence also suggests that pEP364R competes with STING for 2′3′cGAMP binding (Dodantenna et al. 2022). pL83L interacts with and degrades both cGAS and STING. Degradation of STING seems to be mediated by recruitment of Toll-interacting protein (Tollip). Tollip is an LC3-interacting protein with ubiquitin-binding domains that play an important role in selective autophagy (Cheng et al. 2023). The ASFV structural protein p17, encoded by the *D117L* gene, also interacts with STING precluding its association with the TBK1 and IκKε kinases. The p17 transmembrane domain (aa 39-59) is essential for co-localisation with STING and IFN inhibition (Zheng et al. 2022). pE184L also interacts with STING and inhibits its dimerisation, reducing both binding to TBK1 and IRF3. Cells infected with ASFVΔE184L show increased levels of STING dimers and IFNβ expression (Zhu et al. 2023). Finally, pMGF505-7R, another STING interactor, leads to its degradation via autophagy. This may be related to the ability of pMGF505-7R to bind to ULK1 and increase its expression (Li et al. 2021a).

pI329L inhibits IFNβ expression stimulated by TLR3, TLR4, TLR5, TLR8 and TLR9 (Correia et al. 2023). This ASFV protein was first described as a putative TLR homologue, containing the characteristic extracellular leucine-rich repeats and an intracellular region with weak similarity to TIR domains (de Oliveira et al. 2011). It was later proposed that pI329L prevents TLR dimerisation through a potential dimerisation domain present in the extracellular domain. The intracellular domain interacts with TRIF and impairs signalling mediated by this TLR3 and TLR4 adaptor (Correia et al. 2023).

pI267L is the only ASFV protein known to specifically inhibit RLR signalling. It interacts with Riplet, an E3 ubiquitin ligase that catalyses RIG-I K63-linked polyubiquitination. This in turn reduces interaction between RIG-I and MAVS, inhibiting cellular responses to both poly dA:dT, and poly I:C (Ran et al. 2022).

## *Inhibition of IFN-Mediated Signalling*

Despite its extensive arsenal of proteins able to inhibit IFN induction, ASFV has also evolved to modify the JAK/STAT signalling pathway following IFN stimulation. pMGF505-7R can inhibit cellular responses to both type I and type II IFNs (Correia et al. 2013). This suggests that the protein may target a host protein common to both signalling cascades (Fig. 11.5). Indeed, pMGF505-7R interacts with JAK1 but also with JAK2 resulting in their degradation. Interestingly, pMGF505-7R binds to the E3 ubiquitin ligase RNF125 and this increases its expression in transfected cells, an event that seems to be associated with JAK1 degradation. In contrast, JAK2 degradation is mediated by pMGF505-7R interaction with Hes5 and its downregulation (Li et al. 2021b). pMGF360-9L interacts with both STAT1 and STAT2 leading to their degradation via apoptosis and the ubiquitin-proteasome pathway respectively. ASFV infection causes a decrease in both STAT1 and STAT2 at late times post-infection, and this is less evident in cells infected with a virus lacking MGF360-9L (Zhang et al. 2022). Similarly, pI215L interacts with STAT2 and leads to its ubiquitination and subsequent degradation. pI215L ubiquitin-conjugating activity is not necessary for the interaction but is required for the degradation (Riera et al. 2022). pI215L also interacts and targets IRF9 for degradation following stimulation of cells with IFNα. In this case, the degradation is independent of its ubiquitin-conjugating activity, relying instead on the autophagy-lysosome pathway since the effect is abolished in ATG5 KO cells (Li et al. 2022).

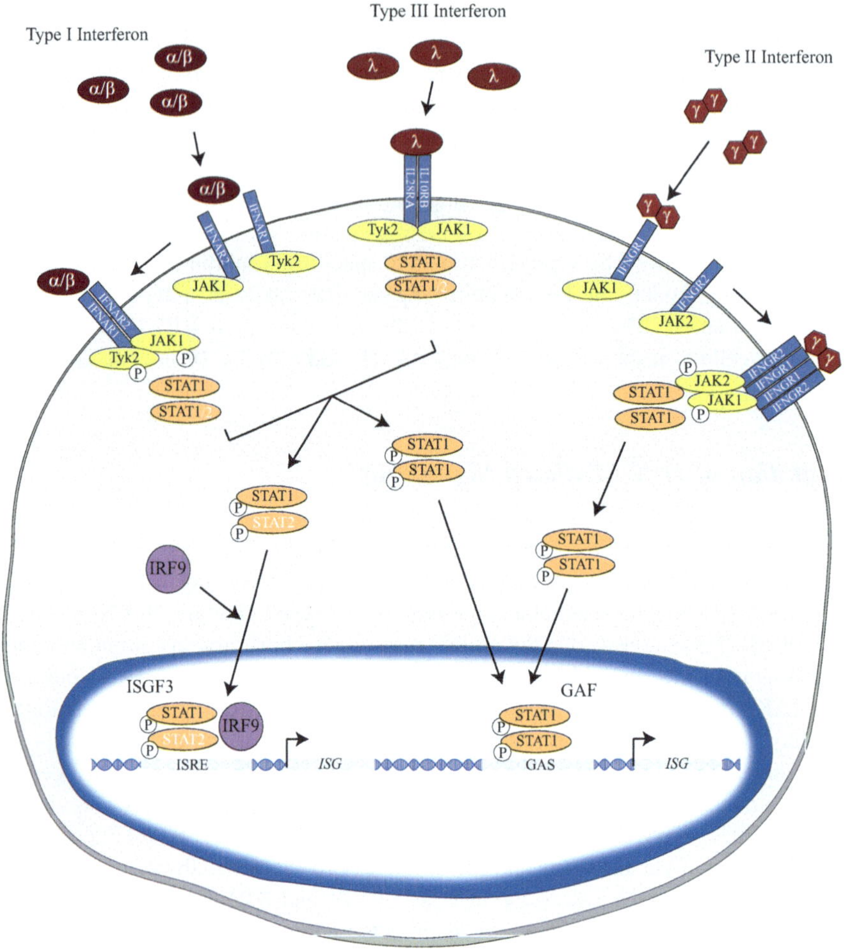


**Fig. 11.5** Interferon-stimulated genes. Type I interferons (interferon α/β) bind to the interferon alpha receptor subunits, which dimerise and lead to the activation and phosphorylation of the Janus kinases (JAK), JAK1 and Tyk2. JAK1 and Tyk 2, in turn, phosphorylate STAT1/STAT2 heterodimers which bind to IRF9 to form the ISGF3 complex, which translocates to the nucleus and binds to IFN-stimulated response elements (ISRE) in the promoters of interferon-stimulated genes and activate gene expression. Type III interferons (interferon λ) bind to the IL28 receptor alpha and IL10 receptor beta subunits and then induce the formation of the ISGF3 complex through the same JAK and STATs as interferon α/β. Type II interferon (interferon γ) binds to the high-affinity interferon-gamma receptor subunit 1, which dimerises and recruits the interferon-gamma receptor subunit 2. This leads to the phosphorylation and activation of JAK1 and JAK2, which in turn phosphorylate STAT1 homodimers to form the GAF complex. GAF translocates to the nucleus, where it activates the expression of genes containing a GAS sequence in their promoters. Both type I and type III interferon signalling can also lead to the formation of the GAF complex, as activated JAK1 and Tyk2 can also phosphorylate STAT1 homodimers

## Cellular Stress

### *Secretory Pathway*

ASFVs profound reorganisation of the cell to facilitate its replication would be predicted to induce various stress response pathways. The endoplasmic reticulum stress responses can detect overexpression of membrane proteins (so-called ER-overload responses), protein misfolding within the lumen of the ER (unfolded protein response) as well as changes in calcium levels within the ER. This can lead to an increase in the capacity of the ER to respond to stress via the generation of new membranes, upregulation of chaperones, as well reducing demand by restricting translation. ER stress can lead to the degradation of ER lumen through autophagy (termed reticulophagy) and, ultimately, the induction of apoptosis and cell death.

ASFV infection induces reorganisation of the later stages of the secretory pathway, particularly the trans-Golgi network (Andrés et al. 1998; McCrossan et al. 2001; Netherton et al. 2006) and the recruitment of endosomes to the periphery of virus factories (Netherton et al. 2006; Cuesta-Geijo et al. 2017). This reorganisation is concomitant with retardation in membrane trafficking of two model proteins in cultured cells (McCrossan et al. 2001; Netherton et al. 2006) and the down regulation of surface expression of both SLA-I (porcine equivalent of MHC class I) and CD16 on the surface of infected macrophages (Gonzalez Juarrero et al. 1992; Netherton et al. 2006; Lithgow et al. 2014; Franzoni et al. 2017; Franzoni et al. 2018). The effect of ASFV infection may be strain-specific as a virulent genotype I isolate did not downregulate the surface expression of SLA-I in two model systems (Franzoni et al. 2017; Franzoni et al. 2018); however, two studies reported a decrease in surface expression of SLA-I without a decrease in total protein levels suggesting a restriction within the secretory pathway (Gonzalez Juarrero et al. 1992; Netherton et al. 2006). The integrity of the Golgi and endosomal pathway are highly dependent on the cytoskeletal network (Allan et al. 2002; Brown et al. 2005; She et al. 2017; Soppina et al. 2009; Weller et al. 2010) and redistribution of endosomes to virus factories is blocked by microtubule depolymerising agents (Netherton et al. 2006). Disruption of the centrosomes, hyperacetylation of microtubules and the recruitment of kinesin motors to factories may therefore contribute to the redirection of the secretory pathway within ASFV-infected cells (Jouvenet et al. 2004; Jouvenet and Wileman 2005).

The cell responds to ER stress by upregulating the expression of chaperones which is controlled by the inositol response element 1 (IRE1) and activating transcription factor 6 (ATF6) pathways. In order to reduce the demand on the folding apparatus of the ER, the cell also restricts translation by phosphorylation of the eukaryotic initiation factor 2 alpha (eIF2α) by protein kinase RNA-activated (PKR) like ER kinase (PERK). Phosphorylation of eIF2α is a central control point in regulating translation and cellular stress as PKR and general control non-derepressible 2 (GCN2) restrict translation in response to sensing viral RNA and amino acid deprivation, respectively. Under conditions of translation repression, expression of activating transcription factor 4 (ATF4) is promoted, which can lead to the expression of a number of different proteins, including the pro-apoptotic transcription factor C/EBP homologous protein (CHOP/GADD153). IRE1 signalling is also linked to the activation of caspase-12 through the adaptor protein TRAF2 and c-Jun N-terminal inhibitory kinase (JIK) (Yoneda et al. 2001), as well as binding to the proapoptotic bcl-2 proteins Bax and Bak (Hetz et al. 2006) which interact with pA179L (see above).

Overexpression of the *K205R* and *D117L* genes (p17 protein) induces ER stress (Xia et al. 2020; Wang et al. 2022), and as mentioned above, overexpression of *E183L* (p54 protein) induces proliferation of ER membranes. Therefore, it is unsurprising that ASFV infection ultimately induces ER stress within cells by signalling through the ATF6 pathway (Galindo et al. 2012). ASFV does not induce processing of X-box binding protein 1 (XBP1), suggesting that infection does not induce phosphorylation and activation of IRE1, however conversely, at late times, caspase-12 activity can be detected in the absence of caspase-3 and caspase 8 activity suggesting signalling through IRE1(Galindo et al. 2012). Co-expression of *A224L*, the ASFV IAP homologue, and TRAF2 potentiates NFκB activity, and therefore, it is possible that the ER stress may be modulated by ASFV from the cytosol, it would be interesting to determine if IRE1 is hyperphosphorylated in infected cells. ASFV does not appear to induce phosphorylation of PERK (Netherton et al. 2006), and although ATF4 is translated at very late times (Galindo et al. 2012). CHOP does not appear to be expressed. ASFV is capable of inhibiting CHOP expression in cultured cells and primary macrophages in response to a number of different cellular insults (Netherton et al. 2006; Zhang et al. 2010). The *DP71L* gene (also *23-NL* and *NL-S*) is similar to the herpes simplex virus *ICP34.5* gene and cellular *GADD34* gene. GADD34 is part of the PP1 complex that dephosphorylates eIF2α and derepresses translation. pDP71L can interact with all three isoforms of PP1, and overexpression of the gene blocks induction of CHOP (Zhang et al. 2010). Mutations in the predicted PP1 and eIF2α binding domains blocked the ability of pDP71L to dephosphorylate eIF2α and inhibit CHOP induction (Barber et al. 2017). Two amino acids are crucial for interactions with PP1 isoforms, and wild-type pDP71L is capable of enhancing translation in vitro. There are likely other ASFV genes involved in the modulation of ER stress and translation as *DP71L* is a late gene, and CHOP signalling can be blocked in the presence of drugs that block late gene expression (Netherton et al. 2006), and ASFV deletion mutants lacking *DP71L* are still able to block CHOP induction (Zhang et al. 2010).

ASFV factories also recruit mitochondria to their periphery and appear to be actively respiring (Rojo et al. 1998), which may indicate they are generating ATP that the virus needs for replication (Cobbold et al. 2000). Expression of cpn60, mortalin (also known as mthsp70 or p74) and hsp70 are upregulated during ASFV infection which may indicate induction of the mitochondrial stress response (Rojo et al. 1998). CHOP was originally thought as a major regulator of the mitochondrial unfolded protein response (mtUPR); however, more recent data suggests that the mtUPR is controlled by the complex interplay between CHOP, C/EBPβ, ATF4 and ATF3 (Kaspar et al. 2021). Therefore, it would be of interest to see the effect of

ASFV on ATF3 and C/EBPβ as both CHOP and ATF4 expression are actively repressed in ASFV-infected cells.

## Concluding Remarks

African swine fever virus is a complex pathogen that encodes for numerous proteins that manipulate host responses, and the recent identification of multiple novel open reading frames further complicates the system (Cackett et al. 2020, 2022). The application of more modern techniques such as affinity-purification mass spectrometry are just beginning to describe the wealth of potential interactions that play a role in the virus life cycle. Combining interaction networks for individual viral proteins may identify novel pathways or proteins important for immune evasion and virus replication. Care will need to be taken with some proteins as expression outside the context of viral infection may result in interactions that are not relevant to infected cells. This is particularly relevant for proteins that are normally localised to the virus factory, but a number of ASFV proteins appear to be only weakly expressed in virus infected cells (Keßler et al. 2018; Wöhnke et al. 2021). The use of recombinant viruses expressing proteins fused to highly specific tags may solve these complications and will likely reveal additional detail, particularly interactions that depend on multiple viral proteins. This knowledge will build a much stronger and more sophisticated understanding of the host–virus interactions in ASFV-infected cells and may ultimately lead to novel targets for generating gene-deleted vaccine candidates. ASFV is highly divergent from other viruses, and therefore detailed understanding of the interactions between viral proteins and between viral and host proteins is likely to lead to the development of highly specific antivirals that can mimic and block interactions that are essential for virus propagation. Comparative analysis of the interactions between ASFV proteins and its different suid and arthropod hosts is likely to shed more light on how viruses have evolved to replicate and persist.

## References

Abe T, Barber GN (2014) Cytosolic-DNA-mediated, STING-dependent proinflammatory gene induction necessitates canonical NF-κB activation through TBK1. J Virol 88(10):5328–5341. https://doi.org/10.1128/jvi.00037-14

Afonso CL, Neilan JG, Kutish GF, Rock DL (1996) An African swine fever virus bcl-2 homolog, 5-HL, suppresses apoptotic cell death. J Virol 70(7):4858–4863

Afonso CL, Piccone ME, Zaffuto KM, Neilan J, Kutish GF, Lu Z, Balinsky CA, Gibb TR, Bean TJ, Zsak L, Rock DL (2004) African swine fever virus multigene family 360 and 530 genes affect host interferon response. J Virol 78(4):1858–1864. https://doi.org/10.1128/jvi.78.4.1858-1864.2004

Aicher SM, Monaghan P, Netherton CL, Hawes PC (2021) Unpicking the secrets of African swine fever viral replication sites. Viruses 13(1):77. https://doi.org/10.3390/v13010077

Akira S, Uematsu S, Takeuchi O (2006) Pathogen recognition and innate immunity. Cell 124(4): 783–801. https://doi.org/10.1016/j.cell.2006.02.015

Alejo A, Yáñez RJ, Rodríguez JM, Viñuela E, Salas ML (1997) African swine fever virus *trans*-prenyltransferase. J Biol Chem 272(14):9417–9423

Alejo A, Andrés G, Salas ML (2003) African swine fever virus proteinase is essential for core maturation and infectivity. J Virol 77(10):5571–5577

Alejo A, Matamoros T, Guerra M, Andres G (2018) A proteomic atlas of the African swine fever virus particle. J Virol. https://doi.org/10.1128/jvi.01293-18

Allan VJ, Thompson HM, McNiven MA (2002) Motoring around the Golgi. Nat Cell Biol 4(10): E236–E242. https://doi.org/10.1038/ncb1002-e236

Alonso C, Miskin J, Hernáez B, Fernandez-Zapatero P, Soto L, Cantó C, Rodríguez-Crespo I, Dixon L, Escribano JM (2001) African swine fever virus protein p54 interacts with the microtubular motor complex through direct binding to light-chain dynein. J Virol 75(20): 9819–9827

Alves de Matos AP, Carvalho ZG (1993) African swine fever virus interaction with microtubules. Biology of the Cell 78:229–234

Andrés G, García-Escudero R, Simón-Mateo C, Viñuela E (1998) African swine fever virus is enveloped by a two-membraned collapsed cisterna derived from the endoplasmic reticulum. J Virol 72(11):8988–9001

Andrés G, García-Escudero R, Viñuela E, Salas ML, Rodríguez JM (2001) African swine fever virus structural protein pE120R is essential for virus transport from assembly sites to plasma membrane but not for infectivity. J Virol 75(15):6758–6768

Andrés G, García-Escudero R, Salas ML, Rodríguez JM (2002) Repression of African swine fever virus polyprotein pp220-encoding gene leads to the assembly of icosahedral core-less particles. J Virol 76(6):2654–2666

Andres G, Charro D, Matamoros T, Dillard RS, Abrescia NGA (2020) The cryo-EM structure of African swine fever virus unravels a unique architecture comprising two icosahedral protein capsids and two lipoprotein membranes. J Biol Chem 295(1):1–12. https://doi.org/10.1074/jbc.AC119.011196

Angulo A, Viñuela E, Alcamí A (1993) Inhibition of African swine fever virus binding and infectivity by purified recombinant virus attachment protein p12. J Virol 67(9):5463–5471

Arico S, Petiot A, Bauvy C, Dubbelhuis PF, Meijer AJ, Codogno P, Ogier-Denis E (2001) The tumor suppressor PTEN positively regulates macroautophagy by inhibiting the phosphatidylinositol 3-kinase/protein kinase B pathway. J Biol Chem 276(38):35243–35246. https://doi.org/10.1074/jbc.C100319200

Ballester M, Galindo-Cardiel I, Gallardo C, Argilaguet JM, Segalés J, Rodríguez JM, Rodríguez F (2010) Intranuclear detection of African swine fever virus DNA in several cell types from formalin-fixed and paraffin-embedded tissues using a new in situ hybridisation protocol. Journal of virological methods 168(1-2):38–43. https://doi.org/10.1016/j.jviromet.2010.04.013

Ballester M, Rodríguez-Cariño C, Pérez M, Gallardo C, Rodríguez JM, Salas ML, Rodriguez F (2011) Disruption of nuclear organization during the initial phase of African swine fever virus infection. J Virol 85(16):8263–8269. https://doi.org/10.1128/jvi.00704-11

Banjara S, Caria S, Dixon LK, Hinds MG, Kvansakul M (2017) Structural insight into African swine fever virus A179L-mediated inhibition of apoptosis. J Virol 91(6):e02228-16. https://doi.org/10.1128/jvi.02228-16

Banjara S, Shimmon GL, Dixon LK, Netherton CL, Hinds MG, Kvansakul M (2019) Crystal structure of African swine fever virus A179L with the autophagy regulator beclin. Viruses 11(9):789. https://doi.org/10.3390/v11090789

Barber C (2015) Stress modulators encoded by African swine fever virus. University of London, St George's

Barber C, Netherton C, Goatley L, Moon A, Goodbourn S, Dixon L (2017) Identification of residues within the African swine fever virus DP71L protein required for dephosphorylation of translation initiation factor eIF2alpha and inhibiting activation of pro-apoptotic CHOP. Virology 504:107–113. https://doi.org/10.1016/j.virol.2017.02.002

Barrado-Gil L, Del Puerto A, Galindo I, Cuesta-Geijo M, García-Dorival I, de Motes CM, Alonso C (2021) African swine fever virus ubiquitin-conjugating enzyme is an immunomodulator targeting NF-κB activation. Viruses 13(6):1160. https://doi.org/10.3390/v13061160

Bernardes C, Antonio A, Lima MCP, De Valdeira ML (1998) Cholesterol affects African swine fever virus infection. Biochim Biophys Acta 1393(1):19–25

Blackford AN, Bruton RK, Dirlik O, Stewart GS, Taylor AM, Dobner T, Grand RJ, Turnell AS (2008) A role for E1B-AP5 in ATR signaling pathways during adenovirus infection. J Virol 82(15):7640–7652. https://doi.org/10.1128/jvi.00170-08

Borca MV, Irusta P, Carrillo C, Afonso CL, Burrage T, Rock DL (1994) African swine fever virus structural protein p72 contains a conformational neutralizing epitope. Virology 201:413–418

Brookes SM, Dixon LK, Parkhouse RME (1996) Assembly of African swine fever virus: quantitative ultrastructural analysis *in vitro* and *in vivo*. Virology 224:84–92

Brown CL, Maier KC, Stauber T, Ginkel LM, Wordeman L, Vernos I, Schroer TA (2005) Kinesin-2 is a motor for late endosomes and lysosomes. Traffic 6(12):1114–1124. https://doi.org/10.1111/j.1600-0854.2005.00347.x

Brun A, Rodríguez F, Escribano JM, Alonso C (1998) Functionality and cell anchorage dependence of the African swine fever virus gene *A179L*, a viral *bcl-2* homolog, in insect cells. J Virol 72(12):10227–10233

Cackett G, Matelska D, Sýkora M, Portugal R, Malecki M, Bähler J, Dixon L, Werner F (2020) The African swine fever virus transcriptome. J Virol 94(9):e00119-20. https://doi.org/10.1128/jvi.00119-20

Cackett G, Portugal R, Matelska D, Dixon L, Werner F (2022) African swine fever virus and host response: transcriptome profiling of the georgia 2007/1 strain and porcine macrophages. J Virol 96(5):e0193921. https://doi.org/10.1128/jvi.01939-21

Cai S, Zheng Z, Cheng J, Zhong L, Shao R, Zheng F, Lai Z, Ou J, Xu L, Zhou P, Lu G, Zhang G (2022) Swine interferon-inducible transmembrane proteins potently inhibit African swine fever virus replication. Front Immunol 13:827709. https://doi.org/10.3389/fimmu.2022.827709

Carrasco L, Chacón-M de Lara F, Martín de las Mulas J, Gómez-Villamandos JC, Pérez J, Wilkinson PJ, Sierra MA (1996) Apoptosis in lymph nodes in acute African swine fever. J Comp Path 115:415–428

Carrascosa AL, Bustos MJ, Nogal ML, González de Buitrago G, Revilla Y (2002) Apoptosis in an early step of African swine fever virus entry into Vero cells does not require virus replication. Virology 294:372–382

Carvalho ZG, Alves De Matos AP, Rodrigues-Pousada C (1988) Association of African swine fever virus with the cytoskeleton. Virus Res 11:175–192

Castelló A, Quintas A, Sánchez EG, Sabina P, Nogal M, Carrasco L, Revilla Y (2009) Regulation of host translational machinery by African swine fever virus. PLoS pathogens 5(8):e1000562. https://doi.org/10.1371/journal.ppat.1000562

Chacón MR, Almazán F, Nogal ML, Viñuela E, Rodríguez JF (1995) The African swine fever virus IAP homolg is a late structural polypeptide. Virology 214:670–674

Chan TO, Rittenhouse SE, Tsichlis PN (1999) AKT/PKB and other D3 phosphoinositide-regulated kinases: kinase activation by phosphoinositide-dependent phosphorylation. Annu Rev Biochem 68:965–1014. https://doi.org/10.1146/annurev.biochem.68.1.965

Chapman DA, Tcherepanov V, Upton C, Dixon LK (2008) Comparison of the genome sequences of non-pathogenic and pathogenic African swine fever virus isolates. J Gen Virol 89(Pt 2):397–408. https://doi.org/10.1099/vir.0.83343-0

Chen W, Srinath H, Lam SS, Schiffer CA, Royer WE Jr, Lin K (2008) Contribution of Ser386 and Ser396 to activation of interferon regulatory factor 3. J Mol Biol 379(2):251–260. https://doi.org/10.1016/j.jmb.2008.03.050

Chen S, Zhang X, Nie Y, Li H, Chen W, Lin W, Chen F, Xie Q (2021) African swine fever virus protein E199L promotes cell autophagy through the interaction of PYCR2. Virol Sin 36(2): 196–206. https://doi.org/10.1007/s12250-021-00375-x
Chen H, Wang Z, Gao X, Lv J, Hu Y, Jung YS, Zhu S, Wu X, Qian Y, Dai J (2022) ASFV pD345L protein negatively regulates NF-κB signalling by inhibiting IKK kinase activity. Vet Res 53(1): 32. https://doi.org/10.1186/s13567-022-01050-z
Cheng M, Kanyema MM, Sun Y, Zhao W, Lu Y, Wang J, Li X, Shi C, Wang J, Wang N, Yang W, Jiang Y, Huang H, Yang G, Zeng Y, Wang C, Cao X (2023) African swine fever virus L83L negatively regulates the cGAS-STING-mediated IFN-I pathway by recruiting tollip to promote STING autophagic degradation. J Virol 97(5):e0192322. https://doi.org/10.1128/jvi.01923-22
Chiu YH, Macmillan JB, Chen ZJ (2009) RNA polymerase III detects cytosolic DNA and induces type I interferons through the RIG-I pathway. Cell 138(3):576–591. https://doi.org/10.1016/j.cell.2009.06.015
Chlanda P, Carbajal MA, Cyrklaff M, Griffiths G, Krijnse-Locker J (2009) Membrane rupture generates single open membrane sheets during vaccinia virus assembly. Cell Host Microbe 6(1): 81–90. https://doi.org/10.1016/j.chom.2009.05.021
Chlanda P, Carbajal MA, Kolovou A, Hamasaki M, Cyrklaff M, Griffiths G, Krijnse-Locker J (2011) Vaccinia virus lacking A17 induces complex membrane structures composed of open membrane sheets. Arch Virol 156(9):1647–1653. https://doi.org/10.1007/s00705-011-1012-1
Cobbold C, Whittle JT, Wileman T (1996) Involvement of the endoplasmic reticulum in the assembly and envelopment of African swine fever virus. J Virol 70(12):8382–8390
Cobbold C, Brookes SM, Wileman T (2000) Biochemical requirements of virus wrapping by the endoplasmic reticulum calcium store during envelopment of African swine fever virus. J Virol 74(5):2151–2160
Coelho J, Martins C, Ferreira F, Leitão A (2015) African swine fever virus ORF P1192R codes for a functional type II DNA topoisomerase. Virology 474:82–93. https://doi.org/10.1016/j.virol.2014.10.034
Correia S, Ventura S, Parkhouse RM (2013) Identification and utility of innate immune system evasion mechanisms of ASFV. Virus Res 173(1):87–100. https://doi.org/10.1016/j.virusres.2012.10.013
Correia S, Moura PL, Ventura S, Leitão A, Parkhouse RME (2023) I329L: a dual action viral antagonist of TLR activation encoded by the African swine fever virus (ASFV). Viruses 15(2): 445
Cuervo AM (2011) Chaperone-mediated autophagy: dice's 'wild' idea about lysosomal selectivity. Nat Rev Mol Cell Biol 12(8):535–541. https://doi.org/10.1038/nrm3150
Cuesta-Geijo M, Chiappi M, Galindo I, Barrado-Gil L, Muñoz-Moreno R, Carrascosa JL, Alonso C (2016) Cholesterol flux is required for endosomal progression of African swine fever virions during the initial establishment of infection. J Virol 90(3):1534–1543. https://doi.org/10.1128/jvi.02694-15
Cuesta-Geijo M, Barrado-Gil L, Galindo I, Muñoz-Moreno R, Alonso C (2017) Redistribution of endosomal membranes to the African swine fever virus replication site. Viruses 9(6):133. https://doi.org/10.3390/v9060133
Cuesta-Geijo M, García-Dorival I, Del Puerto A, Urquiza J, Galindo I, Barrado-Gil L, Lasala F, Cayuela A, Sorzano COS, Gil C, Delgado R, Alonso C (2022) New insights into the role of endosomal proteins for African swine fever virus infection. PLoS Pathogens 18(1):e1009784. https://doi.org/10.1371/journal.ppat.1009784
Cui S, Wang Y, Gao X, Xin T, Wang X, Yu H, Chen S, Jiang Y, Chen Q, Jiang F, Wang D, Guo X, Jia H, Zhu H (2022) African swine fever virus M1249L protein antagonizes type I interferon production via suppressing phosphorylation of TBK1 and degrading IRF3. Virus Res 319: 198872. https://doi.org/10.1016/j.virusres.2022.198872
Cusson-Hermance N, Khurana S, Lee TH, Fitzgerald KA, Kelliher MA (2005) Rip1 mediates the Trif-dependent toll-like receptor 3- and 4-induced NF-{kappa}B activation but does not

contribute to interferon regulatory factor 3 activation. J Biol Chem 280(44):36560–36566. https://doi.org/10.1074/jbc.M506831200

Das T, Yang X, Lee H, Garst EH, Valencia E, Chandran K, Im W, Hang HC (2022) S-palmitoylation and sterol interactions mediate antiviral specificity of IFITMs. ACS Chem Biol. https://doi.org/10.1021/acschembio.2c00176

De Duve C (1963) The lysosome. Sci Am 208:64–72. https://doi.org/10.1038/scientificamerican0563-64

de Oliveira VL, Almeida SC, Soares HR, Crespo A, Marshall-Clarke S, Parkhouse RM (2011) A novel TLR3 inhibitor encoded by African swine fever virus (ASFV). Arch Virol 156(4):597–609. https://doi.org/10.1007/s00705-010-0894-7

Dodantenna N, Ranathunga L, Chathuranga WAG, Weerawardhana A, Cha JW, Subasinghe A, Gamage N, Haluwana DK, Kim Y, Jheong W, Poo H, Lee JS (2022) African swine fever virus EP364R and C129R target cyclic GMP-AMP to inhibit the cGAS-STING signaling pathway. J Virol 96(15):e0102222. https://doi.org/10.1128/jvi.01022-22

Doodnauth SA, Grinstein S, Maxson ME (2019) Constitutive and stimulated macropinocytosis in macrophages: roles in immunity and in the pathogenesis of atherosclerosis. Philos Trans R Soc Lond B Biol Sci 374(1765):20180147. https://doi.org/10.1098/rstb.2018.0147

Driskell OJ, Mironov A, Allan VJ, Woodman PG (2007) Dynein is required for receptor sorting and the morphogenesis of early endosomes. Nat Cell Biol 9(1):113–120. https://doi.org/10.1038/ncb1525

Eulálio A, Nunes-Correia I, Carvalho AL, Faro C, Citovsky V, Salas J, Salas ML, Simões S, de Lima MC (2006) Nuclear export of African swine fever virus p37 protein occurs through two distinct pathways and is mediated by three independent signals. J Virol 80(3):1393–1404. https://doi.org/10.1128/jvi.80.3.1393-1404.2006

Eulálio A, Nunes-Correia I, Salas J, Salas ML, Simões S, Pedroso de Lima MC (2007) African swine fever virus p37 structural protein is localized in nuclear foci containing the viral DNA at early post-infection times. Virus Res 130(1-2):18–27. https://doi.org/10.1016/j.virusres.2007.05.009

Everett RD, Rechter S, Papior P, Tavalai N, Stamminger T, Orr A (2006) PML contributes to a cellular mechanism of repression of herpes simplex virus type 1 infection that is inactivated by ICP0. J Virol 80(16):7995–8005. https://doi.org/10.1128/jvi.00734-06

Everett RD, Parada C, Gripon P, Sirma H, Orr A (2008) Replication of ICP0-null mutant herpes simplex virus type 1 is restricted by both PML and Sp100. J Virol 82(6):2661–2672. https://doi.org/10.1128/jvi.02308-07

Fang R, Jiang Q, Zhou X, Wang C, Guan Y, Tao J, Xi J, Feng JM, Jiang Z (2017) MAVS activates TBK1 and IKKε through TRAFs in NEMO dependent and independent manner. PLoS pathogens 13(11):e1006720. https://doi.org/10.1371/journal.ppat.1006720

Fitzgerald KA, McWhirter SM, Faia KL, Rowe DC, Latz E, Golenbock DT, Coyle AJ, Liao SM, Maniatis T (2003) IKKepsilon and TBK1 are essential components of the IRF3 signaling pathway. Nat Immunol 4(5):491–496. https://doi.org/10.1038/ni921

Franzoni G, Graham SP, Giudici SD, Bonelli P, Pilo G, Anfossi AG, Pittau M, Nicolussi PS, Laddomada A, Oggiano A (2017) Characterization of the interaction of African swine fever virus with monocytes and derived macrophage subsets. Vet Microbiol 198:88–98. https://doi.org/10.1016/j.vetmic.2016.12.010

Franzoni G, Dei Giudici S, Oggiano A (2018) Infection, modulation and responses of antigen-presenting cells to African swine fever viruses. Virus Res 258:73–80. https://doi.org/10.1016/j.virusres.2018.10.007

Frouco G, Freitas FB, Coelho J, Leitao A, Martins C, Ferreira F (2017) DNA-binding properties of African swine fever virus pA104R, a histone-like protein involved in viral replication and transcription. J Virol 91(12):e02498-16. https://doi.org/10.1128/jvi.02498-16

Funderburk SF, Wang QJ, Yue Z (2010) The Beclin 1-VPS34 complex--at the crossroads of autophagy and beyond. Trends Cell Biol 20(6):355–362. https://doi.org/10.1016/j.tcb.2010.03.002

Galindo I, Almazán F, Bustos MJ, Viñuela E, Carrascosa AL (2000) African swine fever virus EP153R open reading frame encodes a glycoprotein involved in the hemadsorption of infected cells. Virology 266(2):340–351. https://doi.org/10.1006/viro.1999.0080

Galindo I, Hernaez B, Díaz-Gil G, Escribano JM, Alonso C (2008) A179L, a viral Bcl-2 homologue, targets the core Bcl-2 apoptotic machinery and its upstream BH3 activators with selective binding restrictions for Bid and Noxa. Virology 375(2):561–572. https://doi.org/10.1016/j.virol.2008.01.050

Galindo I, Hernáez B, Muñoz-Moreno R, Cuesta-Geijo MA, Dalmau-Mena I, Alonso C (2012) The ATF6 branch of unfolded protein response and apoptosis are activated to promote African swine fever virus infection. Cell Death Dis 3(7):e341. https://doi.org/10.1038/cddis.2012.81

García-Beato R, Salas ML, Viñuela E, Salas J (1992) Role of the host cell nucleus in the replication of African swine fever virus DNA. Virology 188:637–649

García-Belmonte R, Pérez-Núñez D, Pittau M, Richt JA, Revilla Y (2019) African swine fever virus Armenia/07 virulent strain controls interferon beta production through the cGAS-STING pathway. J Virol 93(12):e02298-18. https://doi.org/10.1128/jvi.02298-18

Geraldes A, Valdeira M (1985) Effect of chloroquine on African swine fever virus infection. J Gen Virol 66:1145–1148

Gerritsen ME, Williams AJ, Neish AS, Moore S, Shi Y, Collins T (1997) CREB-binding protein/p300 are transcriptional coactivators of p65. Proc Natl Acad Sci U S A 94(7):2927–2932. https://doi.org/10.1073/pnas.94.7.2927

Golding JP, Goatley L, Goodbourn S, Dixon LK, Taylor G, Netherton CL (2016) Sensitivity of African swine fever virus to type I interferon is linked to genes within multigene families 360 and 505. Virology 493:154–161. https://doi.org/10.1016/j.virol.2016.03.019

Gómez-Puertas P, Rodríguez F, Oviedo JM, Ramiro-Ibáñez F, Ruiz-Gonzalvo F, Alonso C, Escribano JM (1996) Neutralizing antibodies to different proteins of African swine fever virus inhibit both virus attachment and internalization. J Virol 70(8):5689–5694. https://doi.org/10.1128/JVI.70.8.5689-5694.1996

Gomez-Puertas P, Rodriguez F, Oviedo JM, Brun A, Alonso C, Escribano JM (1998) The African swine fever virus proteins p54 and p30 are involved in two distinct steps of virus attachment and both contribute to the antibody-mediated protective immune response. Virology 243(2):461–471. https://doi.org/10.1006/viro.1998.9068

Gonzalez Juarrero M, Mebus CA, Pan R, Revilla Y, Alonso JM, Lunney JK (1992) Swine leukocyte antigen and macrophage marker expression on both African swine fever virus-infected and non-infected primary porcine macrophage cultures. Vet Immunol Immunopathol 32(3-4):243–259. https://doi.org/10.1016/0165-2427(92)90049-v

Goubau D, Deddouche S, Reis e Sousa C (2013) Cytosolic sensing of viruses. Immunity 38(5):855–869. https://doi.org/10.1016/j.immuni.2013.05.007

Granger E, McNee G, Allan V, Woodman P (2014) The role of the cytoskeleton and molecular motors in endosomal dynamics. Semin Cell Dev Biol 31(100):20–29. https://doi.org/10.1016/j.semcdb.2014.04.011

Granja AG, Nogal ML, Hurtado C, Salas J, Salas ML, Carrascosa AL, Revilla Y (2004) Modulation of p53 cellular function and cell death by African swine fever virus. J Virol 78(13):7165–7174. https://doi.org/10.1128/jvi.78.13.7165-7174.2004

Granja AG, Nogal ML, Hurtado C, Del Aguila C, Carrascosa AL, Salas ML, Fresno M, Revilla Y (2006a) The viral protein A238L inhibits TNF-alpha expression through a CBP/p300 transcriptional coactivators pathway. J Immunol 176(1):451–462

Granja AG, Sabina P, Salas ML, Fresno M, Revilla Y (2006b) Regulation of inducible nitric oxide synthase expression by viral A238L-mediated inhibition of p65/RelA acetylation and p300 transactivation. J Virol 80(21):10487–10496. https://doi.org/10.1128/jvi.00862-06

Gui X, Yang H, Li T, Tan X, Shi P, Li M, Du F, Chen ZJ (2019) Autophagy induction via STING trafficking is a primordial function of the cGAS pathway. Nature 567(7747):262–266. https://doi.org/10.1038/s41586-019-1006-9

Guo H, Callaway JB, Ting JP (2015) Inflammasomes: mechanism of action, role in disease, and therapeutics. Nat Med 21(7):677–687. https://doi.org/10.1038/nm.3893

Häcker H, Tseng PH, Karin M (2011) Expanding TRAF function: TRAF3 as a tri-faced immune regulator. Nat Rev Immunol 11(7):457–468. https://doi.org/10.1038/nri2998

Hawes PC, Netherton CL, Wileman TE, Monaghan P (2008) The envelope of intracellular African swine fever virus is composed of a single lipid bilayer. J Virol 82(16):7905–7912. https://doi.org/10.1128/jvi.00194-08

He C, Klionsky DJ (2009) Regulation mechanisms and signaling pathways of autophagy. Annu Rev Genet 43:67–93. https://doi.org/10.1146/annurev-genet-102808-114910

Heath CM, Windsor M, Wileman T (2001) Aggresomes resemble sites specialized for virus assembly. J Cell Biol 153(3):449–456

Hernaez B, Alonso C (2010) Dynamin- and clathrin-dependent endocytosis in African swine fever virus entry. J Virol 84(4):2100–2109. https://doi.org/10.1128/jvi.01557-09

Hernáez B, Díaz-Gil G, García-Gallo M, Ignacio Quetglas J, Rodríguez-Crespo I, Dixon L, Escribano JM, Alonso C (2004) The African swine fever virus dynein-binding protein p54 induces infected cell apoptosis. FEBS Lett 569(1-3):224–228. https://doi.org/10.1016/j.febslet.2004.06.001

Hernaez B, Escribano JM, Alonso C (2006) Visualization of the African swine fever virus infection in living cells by incorporation into the virus particle of green fluorescent protein-p54 membrane protein chimera. Virology 350(1):1–14. https://doi.org/10.1016/j.virol.2006.01.021

Hernáez B, Tarragó T, Giralt E, Escribano JM, Alonso C (2010) Small peptide inhibitors disrupt a high-affinity interaction between cytoplasmic dynein and a viral cargo protein. J Virol 84(20):10792–10801. https://doi.org/10.1128/jvi.01168-10

Hernaez B, Cabezas M, Muñoz-Moreno R, Galindo I, Cuesta-Geijo MA, Alonso C (2013) A179L, a new viral Bcl2 homolog targeting Beclin 1 autophagy related protein. Curr Mol Med 13(2):305–316

Hernáez B, Guerra M, Salas ML, Andrés G (2016) African swine fever virus undergoes outer envelope disruption, capsid disassembly and inner envelope fusion before core release from multivesicular endosomes. PLoS pathogens 12(4):e1005595. https://doi.org/10.1371/journal.ppat.1005595

Hertzog J, Rehwinkel J (2020) Regulation and inhibition of the DNA sensor cGAS. EMBO Rep 21(12):e51345. https://doi.org/10.15252/embr.202051345

Hetz C, Bernasconi P, Fisher J, Lee AH, Bassik MC, Antonsson B, Brandt GS, Iwakoshi NN, Schinzel A, Glimcher LH, Korsmeyer SJ (2006) Proapoptotic BAX and BAK modulate the unfolded protein response by a direct interaction with IRE1alpha. Science 312(5773):572–576. https://doi.org/10.1126/science.1123480

Hofmann S, Stubbe M, Mai J, Schreiner S (2021) Double-edged role of PML nuclear bodies during human adenovirus infection. Virus Res 295:198280. https://doi.org/10.1016/j.virusres.2020.198280

Honda K, Taniguchi T (2006) IRFs: master regulators of signalling by Toll-like receptors and cytosolic pattern-recognition receptors. Nat Rev Immunol 6(9):644–658. https://doi.org/10.1038/nri1900

Honda K, Takaoka A, Taniguchi T (2006) Type I interferon [corrected] gene induction by the interferon regulatory factor family of transcription factors. Immunity 25(3):349–360. https://doi.org/10.1016/j.immuni.2006.08.009

Hong J, Chi X, Yuan X, Wen F, Rai KR, Wu L, Song Z, Wang S, Guo G, Chen JL (2022) I226R protein of African swine fever virus is a suppressor of innate antiviral responses. Viruses 14(3):575. https://doi.org/10.3390/v14030575

Hornung V, Ellegast J, Kim S, Brzózka K, Jung A, Kato H, Poeck H, Akira S, Conzelmann KK, Schlee M, Endres S, Hartmann G (2006) 5′-Triphosphate RNA is the ligand for RIG-I. Science 314(5801):994–997. https://doi.org/10.1126/science.1132505

Hou F, Sun L, Zheng H, Skaug B, Jiang QX, Chen ZJ (2011) MAVS forms functional prion-like aggregates to activate and propagate antiviral innate immune response. Cell 146(3):448–461. https://doi.org/10.1016/j.cell.2011.06.041

Huang L, Xu W, Liu H, Xue M, Liu X, Zhang K, Hu L, Li J, Liu X, Xiang Z, Zheng J, Li C, Chen W, Bu Z, Xiong T, Weng C (2021) African swine fever virus pI215L negatively regulates cGAS-STING signaling pathway through recruiting RNF138 to inhibit K63-linked ubiquitination of TBK1. J Immunol 207(11):2754–2769. https://doi.org/10.4049/jimmunol.2100320

Huang L, Liu H, Ye G, Liu X, Chen W, Wang Z, Zhao D, Zhang Z, Feng C, Hu L, Yu H, Zhou S, Zhang X, He X, Zheng J, Bu Z, Li J, Weng C (2023) Deletion of African Swine Fever Virus (ASFV) H240R gene attenuates the virulence of ASFV by enhancing NLRP3-mediated inflammatory responses. J Virol 97(2):e0122722. https://doi.org/10.1128/jvi.01227-22

Hurtado C, Granja AG, Bustos MJ, Nogal ML, Gonzalez de Buitrago G, de Yebenes VG, Salas ML, Revilla Y, Carrascosa AL (2004) The C-type lectin homologue gene (EP153R) of African swine fever virus inhibits apoptosis both in virus infection and in heterologous expression. Virology 326(1):160–170. https://doi.org/10.1016/j.virol.2004.05.019

Hurtado C, Bustos MJ, Granja AG, de León P, Sabina P, López-Viñas E, Gómez-Puertas P, Revilla Y, Carrascosa AL (2011) The African swine fever virus lectin EP153R modulates the surface membrane expression of MHC class I antigens. Arch Virol 156(2):219–234. https://doi.org/10.1007/s00705-010-0846-2

Huxford T, Huang DB, Malek S, Ghosh G (1998) The crystal structure of the IkappaBalpha/NF-kappaB complex reveals mechanisms of NF-kappaB inactivation. Cell 95(6):759–770. https://doi.org/10.1016/s0092-8674(00)81699-2

Imbery J, Upton C (2017) Organization of the multigene families of African swine fever virus. Fine Focus 3(2):155–171

Ishikawa H, Barber GN (2008) STING is an endoplasmic reticulum adaptor that facilitates innate immune signalling. Nature 455(7213):674–678. https://doi.org/10.1038/nature07317

Jiang X, Kinch LN, Brautigam CA, Chen X, Du F, Grishin NV, Chen ZJ (2012) Ubiquitin-induced oligomerization of the RNA sensors RIG-I and MDA5 activates antiviral innate immune response. Immunity 36(6):959–973. https://doi.org/10.1016/j.immuni.2012.03.022

Johnston JA, Ward CL, Kopito RR (1998) Aggresomes: a cellular response to misfolded proteins. J Cell Biol 143(7):1883–1898

Jouvenet N, Wileman T (2005) African swine fever virus infection disrupts centrosome assembly and function. J Gen Virol 86(3):589–594

Jouvenet N, Monaghan P, Way M, Wileman T (2004) Transport of African swine fever virus from assembly sites to the plasma membrane is dependent on microtubules and conventional kinesin. J Virol 78(15):7990–8001

Jouvenet N, Windsor M, Rietdorf J, Hawes P, Monaghan P, Way M, Wileman T (2006) African swine fever virus induces filopodia-like projections at the plasma membrane. Cell Microbiol 8(11):1803–1811. https://doi.org/10.1111/j.1462-5822.2006.00750.x

Kamada Y, Funakoshi T, Shintani T, Nagano K, Ohsumi M, Ohsumi Y (2000) Tor-mediated induction of autophagy via an Apg1 protein kinase complex. J Cell Biol 150(6):1507–1513. https://doi.org/10.1083/jcb.150.6.1507

Karin M, Ben-Neriah Y (2000) Phosphorylation meets ubiquitination: the control of NF-[kappa]B activity. Annu Rev Immunol 18:621–663. https://doi.org/10.1146/annurev.immunol.18.1.621

Kaspar S, Oertlin C, Szczepanowska K, Kukat A, Senft K, Lucas C, Brodesser S, Hatzoglou M, Larsson O, Topisirovic I, Trifunovic A (2021) Adaptation to mitochondrial stress requires CHOP-directed tuning of ISR. Sci Adv 7(22):eabf0971. https://doi.org/10.1126/sciadv.abf0971

Kato H, Takeuchi O, Sato S, Yoneyama M, Yamamoto M, Matsui K, Uematsu S, Jung A, Kawai T, Ishii KJ, Yamaguchi O, Otsu K, Tsujimura T, Koh CS, Reis e Sousa C, Matsuura Y, Fujita T, Akira S (2006) Differential roles of MDA5 and RIG-I helicases in the recognition of RNA viruses. Nature 441(7089):101–105. https://doi.org/10.1038/nature04734

Katsafanas GC, Moss B (2007) Colocalization of transcription and translation within cytoplasmic poxvirus factories coordinates viral expression and subjugates host functions. Cell Host Microbe 2(4):221–228. https://doi.org/10.1016/j.chom.2007.08.005

Kawai T, Takahashi K, Sato S, Coban C, Kumar H, Kato H, Ishii KJ, Takeuchi O, Akira S (2005) IPS-1, an adaptor triggering RIG-I- and Mda5-mediated type I interferon induction. Nat Immunol 6(10):981–988. https://doi.org/10.1038/ni1243

Kawasaki T, Kawai T (2014) Toll-like receptor signaling pathways. Front Immunol 5:461. https://doi.org/10.3389/fimmu.2014.00461

Kay-Jackson PC, Goatley LC, Cox L, Miskin JE, Parkhouse RME, Wienands J, Dixon LK (2004) The CD2v protein of African swine fever virus interacts with the actin-binding adaptor protein SH3P7. J Gen Virol 85(Pt 1):119–130. https://doi.org/10.1099/vir.0.19435-0

Keßler C, Forth JH, Keil GM, Mettenleiter TC, Blome S, Karger A (2018) The intracellular proteome of African swine fever virus. Sci Rep 8(1):14714. https://doi.org/10.1038/s41598-018-32985-z

Kim TK, Maniatis T (1997) The mechanism of transcriptional synergy of an in vitro assembled interferon-beta enhanceosome. Mol Cell 1(1):119–129. https://doi.org/10.1016/s1097-2765(00)80013-1

Kleiboeker SB, Burrage TG, Scoles GA, Fish D, Rock DL (1998) African swine fever virus infection in the argasid host, ornithodoros porcinus porcinus. J Virol 72(3):1711–1724

Li WW, Li J, Bao JK (2012) Microautophagy: lesser-known self-eating. Cell Mol Life Sci 69(7):1125–1136. https://doi.org/10.1007/s00018-011-0865-5

Li D, Yang W, Li L, Li P, Ma Z, Zhang J, Qi X, Ren J, Ru Y, Niu Q, Liu Z, Liu X, Zheng H (2021a) African swine fever virus MGF-505-7R negatively regulates cGAS-STING-mediated signaling pathway. J Immunol 206(8):1844–1857. https://doi.org/10.4049/jimmunol.2001110

Li D, Zhang J, Yang W, Li P, Ru Y, Kang W, Li L, Ran Y, Zheng H (2021b) African swine fever virus protein MGF-505-7R promotes virulence and pathogenesis by inhibiting JAK1- and JAK2-mediated signaling. J Biol Chem 297(5):101190. https://doi.org/10.1016/j.jbc.2021.101190

Li J, Song J, Kang L, Huang L, Zhou S, Hu L, Zheng J, Li C, Zhang X, He X, Zhao D, Bu Z, Weng C (2021c) pMGF505-7R determines pathogenicity of African swine fever virus infection by inhibiting IL-1β and type I IFN production. PLoS Pathogens 17(7):e1009733. https://doi.org/10.1371/journal.ppat.1009733

Li T, Zhao G, Zhang T, Zhang Z, Chen X, Song J, Wang X, Li J, Huang L, Wen L, Li C, Zhao D, He X, Bu Z, Zheng J, Weng C (2021d) African swine fever virus pE199L induces mitochondrial-dependent apoptosis. Viruses 13(11):2240. https://doi.org/10.3390/v13112240

Li L, Fu J, Li J, Guo S, Chen Q, Zhang Y, Liu Z, Tan C, Chen H, Wang X (2022) African swine fever virus pI215L inhibits type i interferon signaling by targeting interferon regulatory factor 9 for autophagic degradation. J Virol 96(17):e0094422. https://doi.org/10.1128/jvi.00944-22

Lin R, Heylbroeck C, Pitha PM, Hiscott J (1998) Virus-dependent phosphorylation of the IRF-3 transcription factor regulates nuclear translocation, transactivation potential, and proteasome-mediated degradation. Mol Cell Biol 18(5):2986–2996. https://doi.org/10.1128/mcb.18.5.2986

Lin R, Génin P, Mamane Y, Hiscott J (2000) Selective DNA binding and association with the CREB binding protein coactivator contribute to differential activation of alpha/beta interferon genes by interferon regulatory factors 3 and 7. Mol Cell Biol 20(17):6342–6353. https://doi.org/10.1128/mcb.20.17.6342-6353.2000

Lin SC, Lo YC, Wu H (2010) Helical assembly in the MyD88-IRAK4-IRAK2 complex in TLR/IL-1R signalling. Nature 465(7300):885–890. https://doi.org/10.1038/nature09121

Lithgow P, Takamatsu H, Werling D, Dixon L, Chapman D (2014) Correlation of cell surface marker expression with African swine fever virus infection. Vet Microbiol 168(2-4):413–419. https://doi.org/10.1016/j.vetmic.2013.12.001

Liu S, Luo Y, Wang Y, Li S, Zhao Z, Bi Y, Sun J, Peng R, Song H, Zhu D, Sun Y, Li S, Zhang L, Wang W, Sun Y, Qi J, Yan J, Shi Y, Zhang X, Wang P, Qiu HJ, Gao GF (2019) Cryo-EM structure of the african swine fever virus. Cell Host microbe 26(6):836–843.e833. https://doi.org/10.1016/j.chom.2019.11.004

Liu H, Zhu Z, Feng T, Ma Z, Xue Q, Wu P, Li P, Li S, Yang F, Cao W, Xue Z, Chen H, Liu X, Zheng H (2021a) African swine fever virus E120R protein inhibits interferon beta production by interacting with IRF3 To block its activation. J Virol 95(18):e0082421. https://doi.org/10.1128/jvi.00824-21

Liu X, Ao D, Jiang S, Xia N, Xu Y, Shao Q, Luo J, Wang H, Zheng W, Chen N, Meurens F, Zhu J (2021b) African swine fever virus A528R inhibits TLR8 mediated NF-κB activity by targeting p65 activation and nuclear translocation. Viruses 13(10):2046. https://doi.org/10.3390/v13102046

Liu X, Liu H, Ye G, Xue M, Yu H, Feng C, Zhou Q, Liu X, Zhang L, Jiao S, Weng C, Huang L (2022) African swine fever virus pE301R negatively regulates cGAS-STING signaling pathway by inhibiting the nuclear translocation of IRF3. Vet Microbiol 274:109556. https://doi.org/10.1016/j.vetmic.2022.109556

Long X, Ortiz-Vega S, Lin Y, Avruch J (2005) Rheb binding to mammalian target of rapamycin (mTOR) is regulated by amino acid sufficiency. J Biol Chem 280(25):23433–23436. https://doi.org/10.1074/jbc.C500169200

Luo J, Zhang J, Ni J, Jiang S, Xia N, Guo Y, Shao Q, Cao Q, Zheng W, Chen N, Zhang Q, Chen H, Chen Q, Zhu H, Meurens F, Zhu J (2022) The African swine fever virus protease pS273R inhibits DNA sensing cGAS-STING pathway by targeting IKKε. Virulence 13(1):740–756. https://doi.org/10.1080/21505594.2022.2065962

Ma XM, Blenis J (2009) Molecular mechanisms of mTOR-mediated translational control. Nat Rev Mol Cell Biol 10(5):307–318. https://doi.org/10.1038/nrm2672

Ma Y, Li J, Dong H, Yang Z, Zhou L, Xu P (2022) PML body component Sp100A restricts wild-type herpes simplex virus 1 infection. J Virol 96(8):e0027922. https://doi.org/10.1128/jvi.00279-22

Maruri-Avidal L, Domi A, Weisberg AS, Moss B (2011) Participation of vaccinia virus l2 protein in the formation of crescent membranes and immature virions. J Virol 85(6):2504–2511. https://doi.org/10.1128/jvi.02505-10

Matamoros T, Alejo A, Rodríguez JM, Hernáez B, Guerra M, Fraile-Ramos A, Andrés G (2020) African swine fever virus protein pE199L mediates virus entry by enabling membrane fusion and core penetration. mBio 11(4):10–128. https://doi.org/10.1128/mBio.00789-20

McCrossan M, Windsor M, Ponnambalam S, Armstrong J, Wileman T (2001) The *trans* Golgi network is lost from cells infected with African swine fever virus. J Virol 75(23):11755–11765

McNab F, Mayer-Barber K, Sher A, Wack A, O'Garra A (2015) Type I interferons in infectious disease. Nat Rev Immunol 15(2):87–103. https://doi.org/10.1038/nri3787

Meng X, Embry A, Rose L, Yan B, Xu C, Xiang Y (2012) Vaccinia virus A6 is essential for virion membrane biogenesis and localization of virion membrane proteins to sites of virion assembly. J Virol 86(10):5603–5613. https://doi.org/10.1128/jvi.00330-12

Merika M, Williams AJ, Chen G, Collins T, Thanos D (1998) Recruitment of CBP/p300 by the IFN beta enhanceosome is required for synergistic activation of transcription. Mol Cell 1(2): 277–287. https://doi.org/10.1016/s1097-2765(00)80028-3

Meylan E, Burns K, Hofmann K, Blancheteau V, Martinon F, Kelliher M, Tschopp J (2004) RIP1 is an essential mediator of Toll-like receptor 3-induced NF-kappa B activation. Nat Immunol 5(5): 503–507. https://doi.org/10.1038/ni1061

Mori M, Yoneyama M, Ito T, Takahashi K, Inagaki F, Fujita T (2004) Identification of Ser-386 of interferon regulatory factor 3 as critical target for inducible phosphorylation that determines activation. J Biol Chem 279(11):9698–9702. https://doi.org/10.1074/jbc.M310616200

Mortimore GE, Hutson NJ, Surmacz CA (1983) Quantitative correlation between proteolysis and macro- and microautophagy in mouse hepatocytes during starvation and refeeding. Proc Natl Acad Sci U S A 80(8):2179–2183. https://doi.org/10.1073/pnas.80.8.2179

Moura Nunes JF, Vigário JD, Terrinha AM (1975) Ultrastructural study of African swine fever virus replication in cultures of swine bone marrow cells. Arch Virol 49:59–66

Müller S, Dejean A (1999) Viral immediate-early proteins abrogate the modification by SUMO-1 of PML and Sp100 proteins, correlating with nuclear body disruption. J Virol 73(6):5137–5143. https://doi.org/10.1128/jvi.73.6.5137-5143.1999

Muñoz-Moreno R, Barrado-Gil L, Galindo I, Alonso C (2015) Analysis of HDAC6 and BAG3-aggresome pathways in African swine fever viral factory formation. Viruses 7(4):1823–1831. https://doi.org/10.3390/v7041823

Munoz-Moreno R, Cuesta-Geijo MA, Martinez-Romero C, Barrado-Gil L, Galindo I, Garcia-Sastre A, Alonso C (2016) Antiviral role of IFITM proteins in African swine fever virus infection. PloS one 11(4):e0154366. https://doi.org/10.1371/journal.pone.0154366

Neilan JG, Lu Z, Kutish GF, Zsak L, Burrage TG, Borca MV, Carrillo C, Rock DL (1997) A BIR motif containing gene of African swine fever virus, *4CL*, is nonessential for growth *in vitro* and viral virulence. Virology 230:252–264

Neilan JG, Borca MV, Lu Z, Kutish GF, Kleiboeker SB, Carrillo C, Zsak L, Rock DL (1999) An African swine fever virus ORF with similarity to C-type lectins is non-essential for growth in swine macrophages in vitro and for virus virulence in domestic swine. J Gen Virol 80(Pt 10): 2693–2697. https://doi.org/10.1099/0022-1317-80-10-2693

Netherton CL, Wileman TE (2013) African swine fever virus organelle rearrangements. Virus Res 173(1):76–86. https://doi.org/10.1016/j.virusres.2012.12.014

Netherton CL, McCrossan MC, Denyer M, Ponnambalam S, Armstrong J, Takamatsu HH, Wileman TE (2006) African swine fever virus causes microtubule-dependent dispersal of the trans-golgi network and slows delivery of membrane protein to the plasma membrane. J Virol 80(22):11385–11392. https://doi.org/10.1128/jvi.00439-06

Netherton CL, Simpson J, Haller O, Wileman TE, Takamatsu HH, Monaghan P, Taylor G (2009) Inhibition of a large double-stranded DNA virus by MxA protein. J Virol 83(5):2310–2320. https://doi.org/10.1128/jvi.00781-08

Noda T, Ohsumi Y (1998) Tor, a phosphatidylinositol kinase homologue, controls autophagy in yeast. J Biol Chem 273(7):3963–3966. https://doi.org/10.1074/jbc.273.7.3963

Nogal ML, González de Buitrago G, Rodríguez C, Cubelos B, Carrascosa AL, Salas ML, Revilla Y (2001) African swine fever virus IAP homologue inhibits caspase activation and promotes cell survival in mammalian cells. J Virol 75(6):2535–2543. https://doi.org/10.1128/jvi.75.6.2535-2543.2001

Nunes-Correia I, Rodríguez JM, Eulálio A, Carvalho AL, Citovsky V, Simões S, Faro C, Salas ML, Pedroso de Lima MC (2008) African swine fever virus p10 protein exhibits nuclear import capacity and accumulates in the nucleus during viral infection. Vet Microbiol 130(1-2):47–59. https://doi.org/10.1016/j.vetmic.2007.12.010

Oura CAL, Powell PP, Parkhouse RME (1998) African swine fever: a disease characterized by apoptosis. J General Virol 79:1427–1438

Panne D, McWhirter SM, Maniatis T, Harrison SC (2007) Interferon regulatory factor 3 is regulated by a dual phosphorylation-dependent switch. J Biol Chem 282(31):22816–22822. https://doi.org/10.1074/jbc.M703019200

Petiot A, Ogier-Denis E, Blommaart EF, Meijer AJ, Codogno P (2000) Distinct classes of phosphatidylinositol 3′-kinases are involved in signaling pathways that control macroautophagy in HT-29 cells. J Biol Chem 275(2):992–998. https://doi.org/10.1074/jbc.275.2.992

Petrovan V, Rathakrishnan A, Islam M, Goatley LC, Moffat K, Sanchez-Cordon PJ, Reis AL, Dixon LK (2021) Role of African swine fever virus (ASFV) proteins EP153R and EP402R in reducing viral persistence in blood and virulence in pigs infected with BeninΔDP148R. J Virol 96(1):e01340-21. https://doi.org/10.1128/jvi.01340-21

Pichlmair A, Schulz O, Tan CP, Rehwinkel J, Kato H, Takeuchi O, Akira S, Way M, Schiavo G, Reis e Sousa C (2009) Activation of MDA5 requires higher-order RNA structures generated during virus infection. J Virol 83(20):10761–10769. https://doi.org/10.1128/jvi.00770-09

Platanitis E, Demiroz D, Schneller A, Fischer K, Capelle C, Hartl M, Gossenreiter T, Müller M, Novatchkova M, Decker T (2019) A molecular switch from STAT2-IRF9 to ISGF3 underlies interferon-induced gene transcription. Nat Commun 10(1):2921. https://doi.org/10.1038/s41467-019-10970-y

Popescu L, Gaudreault NN, Whitworth KM, Murgia MV, Nietfeld JC, Mileham A, Samuel M, Wells KD, Prather RS, Rowland RRR (2017) Genetically edited pigs lacking CD163 show no resistance following infection with the African swine fever virus isolate, Georgia 2007/1. Virology 501:102–106. https://doi.org/10.1016/j.virol.2016.11.012

Powell PP, Dixon LK, Parkhouse RME (1996) An IkB homolog encoded by African swine fever virus provides a novel mechanism for downregulation of proinflammatory cytokine responses in host macrophages. J Virol 70(12):8527–8533

Quemin ER, Corroyer-Dulmont S, Krijnse-Locker J (2018) Entry and disassembly of large DNA viruses: electron microscopy leads the way. J Mol Biol 430(12):1714–1724. https://doi.org/10.1016/j.jmb.2018.04.019

Quetglas JI, Hernáez B, Galindo I, Muñoz-Moreno R, Cuesta-Geijo MA, Alonso C (2012) Small rho GTPases and cholesterol biosynthetic pathway intermediates in African swine fever virus infection. J Virol 86(3):1758–1767. https://doi.org/10.1128/jvi.05666-11

Rajsbaum R, García-Sastre A, Versteeg GA (2014) TRIMmunity: the roles of the TRIM E3-ubiquitin ligase family in innate antiviral immunity. J Mol Biol 426(6):1265–1284. https://doi.org/10.1016/j.jmb.2013.12.005

Ramirez-Medina E, Vuono EA, Pruitt S, Rai A, Espinoza N, Valladares A, Silva E, Velazquez-Salinas L, Borca MV, Gladue DP (2022) Deletion of African swine fever virus histone-like protein, A104R from the georgia isolate drastically reduces virus virulence in domestic pigs. Viruses 14(5):1112. https://doi.org/10.3390/v14051112

Ramiro-Ibáñez F, Ortega A, Brun A, Escribano JM, Alonso C (1996) Apoptosis: a mechanism of cell killing and lymphoid organ impairment during acute African swine fever virus infection. J General Virol 77:2209–2219

Ran Y, Li D, Xiong MG, Liu HN, Feng T, Shi ZW, Li YH, Wu HN, Wang SY, Zheng HX, Wang YY (2022) African swine fever virus I267L acts as an important virulence factor by inhibiting RNA polymerase III-RIG-I-mediated innate immunity. PLoS Pathogens 18(1):e1010270. https://doi.org/10.1371/journal.ppat.1010270

Randall RE, Goodbourn S (2008) Interferons and viruses: an interplay between induction, signalling, antiviral responses and virus countermeasures. J Gen Virol 89(Pt 1):1–47. https://doi.org/10.1099/vir.0.83391-0

Rautureau GJ, Yabal M, Yang H, Huang DC, Kvansakul M, Hinds MG (2012) The restricted binding repertoire of Bcl-B leaves Bim as the universal BH3-only prosurvival Bcl-2 protein antagonist. Cell Death Dis 3(12):e443. https://doi.org/10.1038/cddis.2012.178

Rehwinkel J, Tan CP, Goubau D, Schulz O, Pichlmair A, Bier K, Robb N, Vreede F, Barclay W, Fodor E, Reis e Sousa C (2010) RIG-I detects viral genomic RNA during negative-strand RNA virus infection. Cell 140(3):397–408. https://doi.org/10.1016/j.cell.2010.01.020

Resch W, Weisberg AS, Moss B (2005) Vaccinia virus nonstructural protein encoded by the A11R gene is required for formation of the virion membrane. J Virol 79(11):6598–6609. https://doi.org/10.1128/jvi.79.11.6598-6609.2005

Revilla Y, Cebrián A, Baixerás E, Martínez-A C, Viñuela E, Salas ML (1997) Inhibition of apoptosis by the African swine fever virus bcl-2 homologue: role of the BH1 domains. Virology 228:400–404. https://doi.org/10.1006/viro.1996.8395

Revilla Y, Callejo M, Rodríguez JM, Culebras E, Nogal ML, Salas ML, Viñuela E, Fresno M (1998) Inhibition of nuclear factor kappaB activation by a virus-encoded IkappaB-like protein. J Biol Chem 273(9):5405–5411. https://doi.org/10.1074/jbc.273.9.5405

Riera E, García-Belmonte R, Madrid R, Pérez-Núñez D, Revilla Y (2022) African swine fever virus ubiquitin-conjugating enzyme pI215L inhibits IFN-I signaling pathway through STAT2 degradation. Front Microbiol 13:1081035. https://doi.org/10.3389/fmicb.2022.1081035

Rodríguez JR, Risco C, Carrascosa JL, Esteban M, Rodríguez D (1997) Characterization of early stages in vaccinia virus membrane biogenesis: implications of the 21-kilodalton protein and a newly identified 15-kilodalton envelope protein. J Virol 71(3):1821–1833. https://doi.org/10.1128/jvi.71.3.1821-1833.1997

Rodríguez JR, Risco C, Carrascosa JL, Esteban M, Rodríguez D (1998) Vaccinia virus 15-kilodalton (A14L) protein is essential for assembly and attachment of viral crescents to virosomes. J Virol 72(2):1287–1296. https://doi.org/10.1128/jvi.72.2.1287-1296.1998

Rodríguez JM, García-Escudero R, Salas ML, Andrés G (2004) African swine fever virus structural protein p54 is essential for the recruitment of envelope precursors to assembly sites. J Virol 78(8):4299–1313

Rodríguez I, Redrejo-Rodríguez M, Rodríguez JM, Alejo A, Salas J, Salas ML (2006) African swine fever virus pB119L protein is a flavin adenine dinucleotide-linked sulfhydryl oxidase. J Virol 80(7):3157–3166. https://doi.org/10.1128/jvi.80.7.3157-3166.2006

Rodríguez I, Nogal ML, Redrejo-Rodríguez M, Bustos MJ, Salas ML (2009) The African swine fever virus virion membrane protein pE248R is required for virus infectivity and an early postentry event. J Virol 83(23):12290–12300. https://doi.org/10.1128/jvi.01333-09

Rojo G, Chamorro M, Salas ML, Viñuela E, Cuezva JM, Salas J (1998) Migration of mitochondria to viral assembly sites in African swine fever virus-infected cells. J Virol 72(9):7583–7588

Rojo G, García-Beato R, Viñuela E, Salas ML, Salas J (1999) Replication of African swine fever virus DNA in infected cells. Virology 257:524–536

Rosa-Calatrava M, Puvion-Dutilleul F, Lutz P, Dreyer D, de Thé H, Chatton B, Kedinger C (2003) Adenovirus protein IX sequesters host-cell promyelocytic leukaemia protein and contributes to efficient viral proliferation. EMBO Rep 4(10):969–975. https://doi.org/10.1038/sj.embor.embor943

Rouiller I, Brookes SM, Hyatt AD, Windsor M, Wileman T (1998) African swine fever virus is wrapped by the endoplasmic reticulum. J Virol 72(3):2373–2387

Saitoh T, Fujita N, Hayashi T, Takahara K, Satoh T, Lee H, Matsunaga K, Kageyama S, Omori H, Noda T, Yamamoto N, Kawai T, Ishii K, Takeuchi O, Yoshimori T, Akira S (2009) Atg9a controls dsDNA-driven dynamic translocation of STING and the innate immune response. Proc Natl Acad Sci U S A 106(49):20842–20846. https://doi.org/10.1073/pnas.0911267106

Sánchez EG, Quintas A, Pérez-Núñez D, Nogal M, Barroso S, Carrascosa ÁL, Revilla Y (2012) African swine fever virus uses macropinocytosis to enter host cells. PLoS pathogens 8(6):e1002754. https://doi.org/10.1371/journal.ppat.1002754

Sánchez-Torres C, Gómez-Puertas P, Gómez-del-Moral M, Alonso F, Escribano JM, Ezquerra A, Domínguez J (2003) Expression of porcine CD163 on monocytes/macrophages correlates with permissiveness to African swine fever infection. Arch Virol 148(12):2307–2323. https://doi.org/10.1007/s00705-003-0188-4

Sang Y, Rowland RR, Hesse RA, Blecha F (2010) Differential expression and activity of the porcine type I interferon family. Physiol Genomics 42(2):248–258. https://doi.org/10.1152/physiolgenomics.00198.2009

Sarbassov DD, Guertin DA, Ali SM, Sabatini DM (2005) Phosphorylation and regulation of Akt/PKB by the rictor-mTOR complex. Science 307(5712):1098–1101. https://doi.org/10.1126/science.1106148

Satheshkumar PS, Weisberg A, Moss B (2009) Vaccinia virus H7 protein contributes to the formation of crescent membrane precursors of immature virions. J Virol 83(17):8439–8450. https://doi.org/10.1128/jvi.00877-09

Sato M, Hata N, Asagiri M, Nakaya T, Taniguchi T, Tanaka N (1998) Positive feedback regulation of type I IFN genes by the IFN-inducible transcription factor IRF-7. FEBS Lett 441(1):106–110. https://doi.org/10.1016/s0014-5793(98)01514-2

Schoggins JW, Wilson SJ, Panis M, Murphy MY, Jones CT, Bieniasz P, Rice CM (2011) A diverse range of gene products are effectors of the type I interferon antiviral response. Nature 472(7344):481–485. https://doi.org/10.1038/nature09907

Sengupta S, Peterson TR, Sabatini DM (2010) Regulation of the mTOR complex 1 pathway by nutrients, growth factors, and stress. Mol Cell 40(2):310–322. https://doi.org/10.1016/j.molcel.2010.09.026

Seth RB, Sun L, Ea CK, Chen ZJ (2005) Identification and characterization of MAVS, a mitochondrial antiviral signaling protein that activates NF-kappaB and IRF 3. Cell 122(5):669–682. https://doi.org/10.1016/j.cell.2005.08.012

Shang G, Zhang C, Chen ZJ, Bai XC, Zhang X (2019) Cryo-EM structures of STING reveal its mechanism of activation by cyclic GMP-AMP. Nature 567(7748):389–393. https://doi.org/10.1038/s41586-019-0998-5

Sharma S, tenOever BR, Grandvaux N, Zhou GP, Lin R, Hiscott J (2003) Triggering the interferon antiviral response through an IKK-related pathway. Science 300(5622):1148–1151. https://doi.org/10.1126/science.1081315

She ZY, Pan MY, Tan FQ, Yang WX (2017) Minus end-directed kinesin-14 KIFC1 regulates the positioning and architecture of the Golgi apparatus. Oncotarget 8(22):36469–36483. https://doi.org/10.18632/oncotarget.16863

Shi J, Liu W, Zhang M, Sun J, Xu X (2021) The A179L gene of African swine fever virus suppresses virus-induced apoptosis but enhances necroptosis. Viruses 13(12):2490. https://doi.org/10.3390/v13122490

Shimmon GL, Hui JYK, Wileman TE, Netherton CL (2021) Autophagy impairment by African swine fever virus. J Gen Virol 102(8):001637. https://doi.org/10.1099/jgv.0.001637

Simões M, Martins C, Ferreira F (2015a) Early intranuclear replication of African swine fever virus genome modifies the landscape of the host cell nucleus. Virus Res 210:1–7. https://doi.org/10.1016/j.virusres.2015.07.006

Simões M, Rino J, Pinheiro I, Martins C, Ferreira F (2015b) Alterations of nuclear architecture and epigenetic signatures during African swine fever virus infection. Viruses 7(9):4978–4996. https://doi.org/10.3390/v7092858

Song G, Liu B, Li Z, Wu H, Wang P, Zhao K, Jiang G, Zhang L, Gao C (2016) E3 ubiquitin ligase RNF128 promotes innate antiviral immunity through K63-linked ubiquitination of TBK1. Nat Immunol 17(12):1342–1351. https://doi.org/10.1038/ni.3588

Song J, Li K, Li T, Zhao G, Zhou S, Li H, Li J, Weng C (2020) Screening of PRRSV- and ASFV-encoded proteins involved in the inflammatory response using a porcine iGLuc reporter. J Virol Methods 285:113958. https://doi.org/10.1016/j.jviromet.2020.113958

Soppina V, Rai AK, Ramaiya AJ, Barak P, Mallik R (2009) Tug-of-war between dissimilar teams of microtubule motors regulates transport and fission of endosomes. Proc Natl Acad Sci U S A 106(46):19381–19386. https://doi.org/10.1073/pnas.0906524106

Stefanovic S, Windsor M, Nagata KI, Inagaki M, Wileman T (2005) Vimentin rearrangement during African swine fever virus infection involves retrograde transport along microtubules and phosphorylation of vimentin by calcium calmodulin kinase II. J Virol 79(18):11766–11775. https://doi.org/10.1128/jvi.79.18.11766-11775.2005

Suárez C, Gutiérrez-Berzal J, Andrés G, Salas ML, Rodríguez JM (2010a) African swine fever virus protein p17 is essential for the progression of viral membrane precursors toward icosahedral intermediates. J Virol 84(15):7484–7499. https://doi.org/10.1128/jvi.00600-10

Suárez C, Salas ML, Rodríguez JM (2010b) African swine fever virus polyprotein pp62 is essential for viral core development. J Virol 84(1):176–187. https://doi.org/10.1128/jvi.01858-09

Suarez C, Andres G, Kolovou A, Hoppe S, Salas ML, Walther P, Krijnse Locker J (2015) African swine fever virus assembles a single membrane derived from rupture of the endoplasmic reticulum. Cell Microbiol 17(11):1683–1698. https://doi.org/10.1111/cmi.12468

Sun L, Wu J, Du F, Chen X, Chen ZJ (2013) Cyclic GMP-AMP synthase is a cytosolic DNA sensor that activates the type I interferon pathway. Science 339(6121):786–791. https://doi.org/10.1126/science.1232458

Sun M, Yu S, Ge H, Wang T, Li Y, Zhou P, Pan L, Han Y, Yang Y, Sun Y, Li S, Li LF, Qiu HJ (2022) The A137R protein of African swine fever virus inhibits type I interferon production via the autophagy-mediated lysosomal degradation of TBK1. J Virol 96(9):e0195721. https://doi.org/10.1128/jvi.01957-21

Takahasi K, Suzuki NN, Horiuchi M, Mori M, Suhara W, Okabe Y, Fukuhara Y, Terasawa H, Akira S, Fujita T, Inagaki F (2003) X-ray crystal structure of IRF-3 and its functional implications. Nat Struct Biol 10(11):922–927. https://doi.org/10.1038/nsb1001

Traenckner EB, Pahl HL, Henkel T, Schmidt KN, Wilk S, Baeuerle PA (1995) Phosphorylation of human I kappa B-alpha on serines 32 and 36 controls I kappa B-alpha proteolysis and NF-kappa B activation in response to diverse stimuli. Embo j 14(12):2876–2883. https://doi.org/10.1002/j.1460-2075.1995.tb07287.x

Traktman P, Caligiuri A, Jesty SA, Liu K, Sankar U (1995) Temperature-sensitive mutants with lesions in the vaccinia virus F10 kinase undergo arrest at the earliest stage of virion morphogenesis. J Virol 69(10):6581–6587. https://doi.org/10.1128/jvi.69.10.6581-6587.1995

Valdeira ML, Geraldes A (1985) Morphological study on the entry of ASFV into cells. Biol Cell 55: 35–40

Verhalen B, Justice JL, Imperiale MJ, Jiang M (2015) Viral DNA replication-dependent DNA damage response activation during BK polyomavirus infection. J Virol 89(9):5032–5039. https://doi.org/10.1128/jvi.03650-14

Vydelingum S, Baylis SA, Bristow C, Smith GL, Dixon LK (1993) Duplicated genes within the variable right end of the genome of a pathogenic isolate of African swine fever virus. J Gen Virol 74:2125–2130. https://doi.org/10.1099/0022-1317-74-10-2125

Wang C, Deng L, Hong M, Akkaraju GR, Inoue J, Chen ZJ (2001) TAK1 is a ubiquitin-dependent kinase of MKK and IKK. Nature 412(6844):346–351. https://doi.org/10.1038/35085597

Wang X, Wu J, Wu Y, Chen H, Zhang S, Li J, Xin T, Jia H, Hou S, Jiang Y, Zhu H, Guo X (2018) Inhibition of cGAS-STING-TBK1 signaling pathway by DP96R of ASFV China 2018/1. Biochem Biophys Res Commun 506(3):437–443. https://doi.org/10.1016/j.bbrc.2018.10.103

Wang N, Zhao D, Wang J, Zhang Y, Wang M, Gao Y, Li F, Wang J, Bu Z, Rao Z, Wang X (2019) Architecture of African swine fever virus and implications for viral assembly. Science 366(6465):640–644. https://doi.org/10.1126/science.aaz1439

Wang Y, Cui S, Xin T, Wang X, Yu H, Chen S, Jiang Y, Gao X, Jiang Y, Guo X, Jia H, Zhu H (2021) African swine fever virus MGF360-14L negatively regulates type I interferon signaling by targeting IRF3. Front Cell Infect Microbiol 11:818969. https://doi.org/10.3389/fcimb.2021.818969

Wang Q, Zhou L, Wang J, Su D, Li D, Du Y, Yang G, Zhang G, Chu B (2022) African swine fever virus K205R induces ER stress and consequently activates autophagy and the NF-κB signaling pathway. Viruses 14(2):394. https://doi.org/10.3390/v14020394

Weller SG, Capitani M, Cao H, Micaroni M, Luini A, Sallese M, McNiven MA (2010) Src kinase regulates the integrity and function of the Golgi apparatus via activation of dynamin 2. Proc Natl Acad Sci U S A 107(13):5863–5868. https://doi.org/10.1073/pnas.0915123107

Wilkinson DE, Weller SK (2006) Herpes simplex virus type I disrupts the ATR-dependent DNA-damage response during lytic infection. J Cell Sci 119(Pt 13):2695–2703. https://doi.org/10.1242/jcs.02981

Windsor M, Hawes P, Monaghan P, Snapp E, Salas ML, Rodríguez JM, Wileman T (2012) Mechanism of collapse of endoplasmic reticulum cisternae during African swine fever virus infection. Traffic 13(1):30–42. https://doi.org/10.1111/j.1600-0854.2011.01293.x

Wöhnke E, Fuchs W, Hartmann L, Blohm U, Blome S, Mettenleiter TC, Karger A (2021) Comparison of the proteomes of porcine macrophages and a stable porcine cell line after infection with african swine fever virus. Viruses 13(11):2198. https://doi.org/10.3390/v13112198

Xia N, Wang H, Liu X, Shao Q, Ao D, Xu Y, Jiang S, Luo J, Zhang J, Chen N, Meurens F, Zheng W, Zhu J (2020) African swine fever virus structural protein p17 inhibits cell proliferation through ER stress-ROS mediated cell cycle arrest. Viruses 13(1):21. https://doi.org/10.3390/v13010021

Xu G, Lo YC, Li Q, Napolitano G, Wu X, Jiang X, Dreano M, Karin M, Wu H (2011) Crystal structure of inhibitor of κB kinase β. Nature 472(7343):325–330. https://doi.org/10.1038/nature09853

Yamamoto M, Sato S, Hemmi H, Hoshino K, Kaisho T, Sanjo H, Takeuchi O, Sugiyama M, Okabe M, Takeda K, Akira S (2003) Role of adaptor TRIF in the MyD88-independent toll-like receptor signaling pathway. Science 301(5633):640–643. https://doi.org/10.1126/science.1087262

Yáñez RJ, Rodríguez JM, Nogal ML, Yuste L, Enríquez C, Rodríguez JF, Viñuela E (1995) Analysis of the complete nucleotide sequence of African swine fever virus. Virology 208: 259–278

Yang J, Li S, Feng T, Zhang X, Yang F, Cao W, Chen H, Liu H, Zhang K, Zhu Z, Zheng H (2021a) African swine fever virus F317L protein inhibits NF-κB activation to evade host immune response and promote viral replication. mSphere 6(5):e0065821. https://doi.org/10.1128/mSphere.00658-21

Yang K, Huang Q, Wang R, Zeng Y, Cheng M, Xue Y, Shi C, Ye L, Yang W, Jiang Y, Wang J, Huang H, Cao X, Yang G, Wang C (2021b) African swine fever virus MGF505-11R inhibits type I interferon production by negatively regulating the cGAS-STING-mediated signaling pathway. Vet Microbiol 263:109265. https://doi.org/10.1016/j.vetmic.2021.109265

Yang K, Xue Y, Niu H, Shi C, Cheng M, Wang J, Zou B, Wang J, Niu T, Bao M, Yang W, Zhao D, Jiang Y, Yang G, Zeng Y, Cao X, Wang C (2022a) African swine fever virus MGF360-11L negatively regulates cGAS-STING-mediated inhibition of type I interferon production. Vet Res 53(1):7. https://doi.org/10.1186/s13567-022-01025-0

Yang K, Xue Y, Niu T, Li X, Cheng M, Bao M, Zou B, Shi C, Wang J, Yang W, Wang N, Jiang Y, Yang G, Zeng Y, Cao X, Wang C (2022b) African swine fever virus MGF505-7R protein interacted with IRF7and TBK1 to inhibit type I interferon production. Virus Res 322:198931. https://doi.org/10.1016/j.virusres.2022.198931

Yoneda T, Imaizumi K, Oono K, Yui D, Gomi F, Katayama T, Tohyama M (2001) Activation of caspase-12, an endoplastic reticulum (ER) resident caspase, through tumor necrosis factor receptor-associated factor 2-dependent mechanism in response to the ER stress. J Biol Chem 276(17):13935–13940. https://doi.org/10.1074/jbc.M010677200

Yoneyama M, Suhara W, Fukuhara Y, Fukuda M, Nishida E, Fujita T (1998) Direct triggering of the type I interferon system by virus infection: activation of a transcription factor complex containing IRF-3 and CBP/p300. Embo J 17(4):1087–1095. https://doi.org/10.1093/emboj/17.4.1087

Yoneyama M, Kikuchi M, Natsukawa T, Shinobu N, Imaizumi T, Miyagishi M, Taira K, Akira S, Fujita T (2004) The RNA helicase RIG-I has an essential function in double-stranded RNA-induced innate antiviral responses. Nat Immunol 5(7):730–737. https://doi.org/10.1038/ni1087

Yoshida R, Takaesu G, Yoshida H, Okamoto F, Yoshioka T, Choi Y, Akira S, Kawai T, Yoshimura A, Kobayashi T (2008) TRAF6 and MEKK1 play a pivotal role in the RIG-I-like helicase antiviral pathway. J Biol Chem 283(52):36211–36220. https://doi.org/10.1074/jbc.M806576200

Zamorano Cuervo N, Osseman Q, Grandvaux N (2018) Virus infection triggers MAVS polymers of distinct molecular weight. Viruses 10(2):56. https://doi.org/10.3390/v10020056

Zanotti C, Razzuoli E, Crooke H, Soule O, Pezzoni G, Ferraris M, Ferrari A, Amadori M (2015) Differential biological activities of swine interferon-α subtypes. J Interferon Cytokine Res: the official journal of the International Society for Interferon and Cytokine Research 35(12): 990–1002. https://doi.org/10.1089/jir.2015.0076

Zeng W, Xu M, Liu S, Sun L, Chen ZJ (2009) Key role of Ubc5 and lysine-63 polyubiquitination in viral activation of IRF3. Mol Cell 36(2):315–325. https://doi.org/10.1016/j.molcel.2009.09.037

Zhang F, Moon A, Childs K, Goodbourn S, Dixon LK (2010) The African swine fever virus DP71L protein recruits the protein phosphatase 1 catalytic subunit to dephosphorylate eIF2alpha and inhibits CHOP induction but is dispensable for these activities during virus infection. J Virol 84(20):10681–10689. https://doi.org/10.1128/jvi.01027-10

Zhang X, Wu J, Du F, Xu H, Sun L, Chen Z, Brautigam CA, Zhang X, Chen ZJ (2014) The cytosolic DNA sensor cGAS forms an oligomeric complex with DNA and undergoes switch-like conformational changes in the activation loop. Cell Rep 6(3):421–430. https://doi.org/10.1016/j.celrep.2014.01.003

Zhang C, Shang G, Gui X, Zhang X, Bai XC, Chen ZJ (2019) Structural basis of STING binding with and phosphorylation by TBK1. Nature 567(7748):394–398. https://doi.org/10.1038/s41586-019-1000-2

Zhang K, Wang S, Gou H, Zhang J, Li C (2021) Crosstalk between autophagy and the cGAS-STING signaling pathway in Type I interferon production. Front Cell Dev Biol 9:748485. https://doi.org/10.3389/fcell.2021.748485

Zhang K, Yang B, Shen C, Zhang T, Hao Y, Zhang D, Liu H, Shi X, Li G, Yang J, Li D, Zhu Z, Tian H, Yang F, Ru Y, Cao WJ, Guo J, He J, Zheng H, Liu X (2022) MGF360-9L is a major virulence factor associated with the african swine fever virus by antagonizing the JAK/STAT signaling pathway. mBio 13(1):e0233021. https://doi.org/10.1128/mbio.02330-21

Zhao B, Du F, Xu P, Shu C, Sankaran B, Bell SL, Liu M, Lei Y, Gao X, Fu X, Zhu F, Liu Y, Laganowsky A, Zheng X, Ji JY, West AP, Watson RO, Li P (2019) A conserved PLPLRT/SD motif of STING mediates the recruitment and activation of TBK1. Nature 569(7758):718–722. https://doi.org/10.1038/s41586-019-1228-x

Zhao G, Li T, Liu X, Zhang T, Zhang Z, Kang L, Song J, Zhou S, Chen X, Wang X, Li J, Huang L, Li C, Bu Z, Zheng J, Weng C (2022) African swine fever virus cysteine protease pS273R inhibits pyroptosis by noncanonically cleaving gasdermin D. J Biol Chem 298(1):101480. https://doi.org/10.1016/j.jbc.2021.101480

Zheng W, Xia N, Zhang J, Cao Q, Jiang S, Luo J, Wang H, Chen N, Zhang Q, Meurens F, Zhu J (2022) African swine fever virus structural protein p17 inhibits cGAS-STING signaling pathway through interacting With STING. Front Immunol 13:941579. https://doi.org/10.3389/fimmu.2022.941579

Zhou P, Dai J, Zhang K, Wang T, Li LF, Luo Y, Sun Y, Qiu HJ, Li S (2022) The H240R protein of african swine fever virus inhibits interleukin 1β production by inhibiting NEMO expression and NLRP3 oligomerization. J Virol 96(22):e0095422. https://doi.org/10.1128/jvi.00954-22

Zhu Z, Li S, Ma C, Yang F, Cao W, Liu H, Chen X, Feng T, Shi Z, Tian H, Zhang K, Chen H, Liu X, Zheng H (2023) African swine fever virus E184L protein interacts with innate immune adaptor STING to block IFN production for viral replication and pathogenesis. J Immunol 210(4):442–458. https://doi.org/10.4049/jimmunol.2200357

Zhuo Y, Guo Z, Ba T, Zhang C, He L, Zeng C, Dai H (2021) African swine fever virus MGF360-12L inhibits type I interferon production by blocking the interaction of importin α and NF-κB signaling pathway. Virol Sin 36(2):176–186. https://doi.org/10.1007/s12250-020-00304-4

# Chapter 12
# Coronavirus and the Cytoskeleton of Virus-Infected Cells

**Yifan Xing, Qian Zhang, and Yaming Jiu**

**Abstract** The cytoskeleton, which includes actin filaments, microtubules, and intermediate filaments, is one of the most important networks in the cell and undertakes many fundamental life activities. Among them, actin filaments are mainly responsible for maintaining cell shape and mediating cell movement, microtubules are in charge of coordinating all cargo transport within the cell, and intermediate filaments are mainly thought to guard against external mechanical pressure. In addition to this, cytoskeleton networks are also found to play an essential role in multiple viral infections. Due to the COVID-19 epidemic, including SARS-CoV-2, SARS-CoV and MERS-CoV, so many variants have caused wide public concern, that any virus infection can potentially bring great harm to human beings and society. Therefore, it is of great importance to study coronavirus infection and develop antiviral drugs and vaccines. In this chapter, we summarize in detail how

---

Yifan Xing and Qian Zhang contributed equally with all other contributors.

---

Y. Xing
Shanghai Institute of Immunity and Infection (Formerly Institut Pasteur of Shanghai), Chinese Academy of Sciences, Shanghai, China

University of Chinese Academy of Sciences, Beijing, China
e-mail: yfxing@siii.cas.cn

Q. Zhang
Unit of Cell Biology and Imaging Study of Pathogen Host Interaction, The Center for Microbes, Development and Health, Key Laboratory of Molecular Virology and Immunology, Institut Pasteur of Shanghai, Chinese Academy of Sciences, Shanghai, China
e-mail: qzhang@siii.cas.cn

Y. Jiu (✉)
Shanghai Institute of Immunity and Infection (Formerly Institut Pasteur of Shanghai), Chinese Academy of Sciences, Shanghai, China

University of Chinese Academy of Sciences, Beijing, China

Unit of Cell Biology and Imaging Study of Pathogen Host Interaction, The Center for Microbes, Development and Health, Key Laboratory of Molecular Virology and Immunology, Institut Pasteur of Shanghai, Chinese Academy of Sciences, Shanghai, China
e-mail: ymjiu@siii.cas.cn



S. Vijayakrishnan et al. (eds.), *Virus Infected Cells*, Subcellular Biochemistry 106,
https://doi.org/10.1007/978-3-031-40086-5_12

the cytoskeleton responds and participates in coronavirus infection by analyzing the possibility of the cytoskeleton and its related proteins as antiviral targets, thereby providing ideas for finding more effective treatments.

**Keywords** Cytoskeleton · Coronavirus · Infection · SARS-CoV-2 · Actin · Microtubule · Intermediate filaments

# Introduction to the Cytoskeleton

## *Cytoskeleton*

The cytoskeleton constitutes a filamentous network in eukaryotic cells, which consists of actin filaments, microtubules, and intermediate filaments. Cytoskeleton not only functions in cell structure and behavior regulation including cell morphogenesis, epithelial–mesenchymal transition (EMT), and cell migration but also participates in important intracellular life activities including protein and virus replication and transport (Wang et al. 2021), vesicle trafficking (Shi et al. 2020), autophagy (Helfand et al. 2011), and so forth. Over the years, extensive studies have reported that many viruses hijack cytoskeleton networks to complete their own propaganda (Foo et al. 2015; Denes et al. 2018; Miranda-Saksena et al. 2018; Bedi et al. 2019; Zhang et al. 2019), showing the importance of this "cell scaffolding" to viral infection. In this chapter, we will introduce the relationship between host cytoskeleton and coronaviral infection, making a systematic summary of cytoskeleton-related pathogen–host interaction.

## *Actin Filaments*

Actin filaments, also known as microfilaments, are on average ~ 7 nm in diameter. They are mainly composed of multimers assembled from globular (G) actin and other components including actin-binding protein, myosin, tropomyosin, and α-actinin. On the one hand, the common biological functions of actin filaments are regulating cell attachment, spreading, motility, endocytosis, cell division, and the control of cytoplasmic circulation. Actin filaments can be aggregated into bundles to form stress fibers and are fixed to the extracellular matrix through focal adhesions, thereby providing shear resistance to cells. The dendritic structure formed by actin filaments supports the shape of villi on the surface of epithelial cells in the respiratory tract and digestive tract. However, because it does not contain myosin or tropomyosin, here it possesses no contractility. The actin filament backbone is highly dynamic. The formation of lamellipodia when aggregation occurs can regulate cell deformation and movement. Unassembled G-actin and fibrillar actin

(F-actin) are assembled and disassembled in a dynamic pattern to promote cell movement.

On the other hand, actin filaments are also involved in the pathological processes of cells, such as cell chemotaxis, adhesion, phagocytosis (May et al. 2001), T-lymphocyte activation (Billadeau et al. 2007), and so forth. For nerve cells, actin shows neuronal polarity and is essential for signal transduction and synaptic structure stability (Xu et al. 2013). In addition, increasing studies have reported that the actin filament system provides the driving force required for virus assembly (Wang et al. 2021), budding and releasing after virus infection, thereby participating in the regulation of the virus life cycle. In a study of the interaction of the rabies virus (RABV) multifunctional matrix protein (M) with the cytoskeleton, the authors revealed actin as a binding partner of the M protein, which may regulate the viral assembly and budding in the pathogenesis of rabies (Zandi et al. 2021). Another study on DENV infection found that viral infection significantly increased cell motility and disruption of the actin filaments (Jhan et al. 2017). There are also studies on IPEC-J2 cells showing that infected with porcine epidemic diarrhea virus (PEDV) and transmissible gastroenteritis virus (TGEV) affect the remodeling of actin filaments. Drugs that inhibit virus replication and release can act as anti-virus targets to regulate the remodeling of microfilaments (Zhao et al. 2014). The above findings demonstrate the significant involvement of the actin filaments system in virus infection-related diseases.

## *Microtubules*

Microtubules, as the backbone of the cytoskeleton, are composed of the ~55 kD α- and β-tubulin, with a diameter of ~25 nm. The usually found tubular forms are single microtubules, duplex microtubules, and triplet microtubules. The most common form exists as tubulin (αβ) dimers and multimers in a head-to-tail manner to form tubulin fibrils. The process of heterodimerization of α-tubulin and β-tubulin is dynamically reversible. Many cellular structures are known to be composed of microtubules, including cilia, flagella (which contribute to motility), and mitotic spindles (which control spindle production and motility during cell division). In addition, microtubules also possess kinetic properties of polymerization and depolymerization and can form the cytoskeleton together with other fibers to maintain cell morphology. Microtubules that have been reported so far play an important role in the processes of cell proliferation and division, signal and material transduction, and the localization and function of organelles depend on the stability of these structures.

Microtubule-associated proteins (MAPs) can promote the aggregation and stability of microtubules and are essential components of microtubule structure and function. Microtubule proteins undergo various posttranslational modifications, including detyrosinylation, acetylation, polyglutamylation, and polyglycinylation, which are also reversible. These modifications play a key role in proteolysis, signal transduction, gene expression regulation and protein interactions. Studies have shown that the knockdown of microtubule-associated protein 1S (MAP1S) also causes autophagy defects and promotes hepatocellular carcinoma. Overexpressed Tau (MAPT) impedes the transport of synaptic vesicles and organelles in vivo, and its reduction rescues the defect in axonal transport in a mouse model of Alzheimer's disease (Ukmar-Godec et al. 2020). Although the microtubule itself does not produce contractile force, it can induce the assembly and disassembly of actin stress fibers and focal adhesions by activating RhoA and then stimulating the downstream effector ROCK (Ezratty et al. 2005), to depolymerize and induce actin stress fiber formation and cell shrinkage.

Viruses often rely on an intact microtubule network during multiple stages of the replication cycle. Some viruses associate directly with microtubule-dependent motors for the transport of intact virions, capsids, individual viral proteins, or RNA, to sites of replication and plasma membranes (Wang et al. 2018). Viral particles move along microtubules within the cell, and the long-range transport speed can reach a very high level. It has been demonstrated that members of the Herpesviridae (Sodeik et al. 1997), Adenoviridae (Suomalainen et al. 1999), Parvoviridae (Suikkanen et al. 2003), Poxviridae and Baculoviridae utilize microtubules and the actin cytoskeleton for cytoplasmic transport or transport in the cytoplasm (Döhner and Sodeik 2005). During virion assembly and release, microtubules are also used to transport within extracellular vesicles or to transport capsids and nucleoprotein granules from the cytoplasm out of the budding compartment. In conclusion, the microtubule system plays a critical role in the transport of intracellular 'cargoes', cell shape, polarity, and motility. In the context of viral infection, the microtubule cytoskeleton can be hijacked by viruses for their directional transport to support their infection.

## *Intermediate Filaments*

Intermediate filaments, ~10 nm in diameter, vary widely in composition in different cells, including vimentin, keratin, desmin, neuronal fibers, and neuroglial filaments. Intermediate filaments display tissue-specific expression, such as 'acidic' keratins are specifically expressed in epithelial tissue, and desmin is expressed in smooth muscle, neurofilaments in the nervous system, etc. In terms of structure, they are arranged in opposite directions to each other during the assembly process and thus show no polarity. Nevertheless, they have stronger elasticity and can withstand higher mechanical pressure, which makes the intermediate filaments the most stable cytoskeletal system. Of note, vimentin is the most abundant component in the intermediate filament protein family, responsible for maintaining the integrity of the cell shape against external mechanical pressure.

Early research on intermediate filaments mainly focused on the aspect of mechanical support. But, with the progression of research, it was found that there was also a hub that regulated many signaling pathways. For instance, studies indicated that vimentin is involved in signaling pathways such as Raf-1 and RhoA and regulates

TNF-α mediated apoptosis. Moreover, vimentin also can bind to secretagogues to induce apoptosis in different types of cancer cells. These studies suggest that vimentin can be used as a potential target for antitumor therapy. The intermediate filaments are also involved in the regulation of 14–3-3 and mTOR signaling pathways. The deletion of keratin 17 can cause the 14–3-3 protein to fail to aggregate in the cytoplasm, thereby activating the mTOR signaling pathway and causing cell morphological changes. In addition, intermediate fibers also affect many signaling pathways, such as PKC, PKA, JNK, CaMK II, Akt, and phosphatase pathways.

In recent years, more studies have focused on the regulatory mechanism of the intermediate filament vimentin in viral infection. For example, vimentin affects pathogen invasion as a receptor and is involved in transcellular migration and immune responses (Döhner and Sodeik 2005). Another study found that surface vimentin in Neuro-2a cells interacts with the Chandipura virus (CHPV) and acts as a receptor to promote the binding of CHPV to cells, which in turn affects the replication process after virus entry (Döhner and Sodeik 2005). In addition to the regulation of virus entry and replication, vimentin also affects the release of the virus from the cell. Polly Roy's team (Bhattacharya et al. 2007) found that disruption of vimentin structure resulted in an increase in cell-associated bluetongue virus (BTV) and a decrease in the amount of virus released by infected cells by demonstrating that the association of BTV particles with intermediate filaments is driven by the interaction of VP2 with vimentin to promote viral efflux.

In summary, although early researchers identified the importance of cytoskeletal networks for viral infection, the mechanisms by which viruses invade host cytoskeleton, regulators, and dynein adapters have received particular attention due to the current worldwide mass COVID-19 infection. Future studies, especially the study of host signaling pathways and downstream effector mechanisms, will undoubtedly provide important new ideas for the underlying mechanisms by which the cytoskeletal system functions during viral infection.

## Coronavirus

### *The Structure and Life Cycle of Coronavirus*

The coronaviruses belong to the genus coronavirus in the family Coronaviridae of the order Nidovirales in the systematic taxonomy. The shape is spherical (~125 nm in diameter), and the surface is covered with stick-shaped spikes (S protein), showing the appearance of a corona, with a diameter of about 80–120 nm (Snijder et al. 2003; Fehr et al. 2015). In terms of structure, coronaviruses are enveloped viruses, and virus particles are composed of two parts, the outer envelope and the inner helical nucleocapsid. Its genome is about 27 ~ 32 kb in length, and it is a linear single-stranded positive RNA chain with a methylated cap-like structure at the 5′ end and a poly(A) tail at the 3′ end.

Coronaviruses are widespread and can infect humans, birds, and mammals including bats, felines, rodents, and pigs (Cui et al. 2019). Coronaviruses are divided into four genera (α, β, γ, and δ), and there are currently seven known coronaviruses that can infect human (Carod-Artal 2020). Human coronaviruses (HCoV) are divided into α-CoV and β-CoV (MERS-CoV, SARS-CoV, HCoVOC43, and so forth) (Malik 2020). Since various physiological functions of viruses are determined by proteins, people focus more on the proteins encoded by the coronavirus genome. The main structural proteins that regulate the assembly and are encoded in all coronavirus genomes are: the spike protein (S), membrane protein (M), envelope protein (E), and nucleocapsid protein (N) (Brian et al. 2005), and the genes encoded all appear in 5′ to 3′ order.

The coronavirus protruding surface densely glycosylated spike (S) protein binds to the host cell surface receptor angiotensin-converting enzyme 2 (ACE2) through plasma membrane fusion, phagocytosis, micropinocytosis, and clathrin-mediated or clathrin-independent endocytosis into infected cells. The protein sequence of the S1 domain of the virus is not conserved, and there are great differences between different coronavirus species, which limits the binding of the virus to the host cell surface receptor at the early stage of infection. Therefore, the type of coronavirus S protein receptor determines the species and tissue sensitivity of coronavirus. Upon entry into the host cell, the multiprotein translation process is activated, resulting in the production of smaller proteins that form a series of nonstructural proteins of the viral transcriptase-replicase complex. At the same time after entering the cell, the virion genome RNA is released into the cytoplasm, translated, replicated, and the viral replicase and other complexes are assembled. First, the positive-strand RNA is translated to generate the negative-strand RNA polymerase precursor protein, and then RNA polymerase is generated under the hydrolysis of the protein to generate the antisense negative-strand template. Next, subgenomic mRNAs are synthesized from the minus-strand subgenomic template, whose posttranslational products are the structural proteins of the virus. Following replication and subgenomic RNA synthesis, encapsulation occurs, leading to the formation of mature viruses, a process that takes place in the endoplasmic reticulum and Golgi apparatus. After assembly, virions are transported to the cell surface through vesicles and fused with the plasma membrane through exocytosis, and are then released into a new extracellular environment to infect other host cells (Cui et al. 2019; Malik 2020).

## *SARS-CoV-2 and COVID-19 Infection*

Severe acute respiratory syndrome coronavirus 2 (SARS-CoV-2) has been around since 2019, causing a disease called coronavirus disease 2019 (COVID-19). The virus spread rapidly, indeed spreading throughout the world in a short period, and maintained high contagiousness and continued outbreaks in countries around the world. As of June 2022, 529 million cases and 6.29 million deaths have been reported. SARS-CoV-2 is a large positive-stranded RNA genome belonging to the family Coronaviridae and is a β-CoV.

Studies have shown that the S protein plays an important role in SARS-CoV-2 evolution and spread, but so far, various genes that mainly encode S protein elements are frequently mutated (Hatmal et al. 2020). This has led directly to the creation of new, more aggressive mutants including Omicron, Delta, and alpha mutants since the discovery of the novel pathogenic coronavirus SARS-CoV-2 in 2019. The emergence and spread of these mutant strains have created new and ongoing challenges around the world. In another study, the authors demonstrated that the SARS-CoV-2 N protein promotes NLRP3 inflammasome activation to induce excessive inflammation. The N protein promotes the maturation of proinflammatory cytokines and is observed at the cellular level in vitro and in vivo in mice. More importantly, the N protein aggravated lung injury and accelerated death in mouse models of sepsis and acute inflammation (Pan et al. 2021). At present, research teams have emphasised the overproduction of inflammatory cytokines and chemokines in COVID-19 patients, including IL-6, IL-8, IL-1β, TNF-α, IFN-γ, MIP1α and 1β, CCL2, CCL5, CCL20, CXCL1, CXCL2, CXCL8, CXCL10, and CXCL17(X. Ren et al. 2021; Del Valle et al. 2020; Jamal et al. 2021; Xu et al. 2020). Moreover, SARS-CoV-2 has been reported to cause long-term respiratory and neurological sequelae in addition to inflammatory responses leading to lung damage (Wang et al. 2020a).

## *Hazards of Other Virus Infections*

We know that human-to-human transmission is primarily through respiratory droplet contact, but viruses are not always confined to the respiratory tract, liver and gut, and in some cases, they can invade the central nervous system and cause neurological systemic disease (Carod-Artal 2020). Examples include human coronaviruses (OC-43, 229E, MERS, and SARS) and some animal coronaviruses (porcine hemagglutinating encephalomyelitis coronavirus). Neurological symptoms such as headache, dizziness, myalgia, and loss of smell have been reported in patients affected by the coronavirus, as well as with encephalopathy, encephalitis, necrotizing hemorrhagic encephalopathy, stroke, seizures, rhabdomyolysis, Guillain–Barre syndrome, etc.

Here, this review will place more emphasis on the mechanism of action between SARS-CoV-2 and the host cytoskeleton, providing possibilities for mankind to conquer coronavirus-like viruses.

## Response of the Host Cell Cytoskeleton Network Following Coronavirus Infection

Viral infections alter cell morphology, as well as host gene expression and protein posttranslational modification. Like most positive-strand RNA viruses (Paul et al. 2013; den Boon et al. 2010; Zhang et al. 2019), coronaviruses infection causes visible remodeling of the morphology of cellular membranes, subcellular organelles, and cytoskeleton network (Snijder et al. 2006; Wen et al. 2020). Of note, the cytoskeleton, the most basic scaffold of cells, is involved in all aspects of cell morphological remodeling.

### *Coronavirus Infection Induces Expression Level or Posttranslational Modification Changes of the Cytoskeleton and its Related Proteins*

Numerous studies have indicated that coronavirus infection induces rapid changes to the host cytoskeleton network, by both omics analysis and bioinformatics predictive analysis. For instance, the phosphorylation level of several cytoskeleton proteins, such as vimentin (Ser39 and Ser56), stathmin (STMN1 Ser16 and Ser25), and α-catenin (CTNNA1 Ser641), was found to be regulated during SARS-CoV-2 infection (Bouhaddou et al. 2020). Furthermore, it has been predicted that the nonstructural protein 5 (NSP5) and 3C-like (3CL) protease of SARS-CoV-2 could cleave microtubule organization-associated proteins and cytoskeleton proteins, respectively (Scott et al. 2022; Chen et al. 2022). Moreover, the expression level of cytoskeleton-related proteins also showed marked changes after infectious bronchitis virus (IBV) and TGEV infection (Emmott et al. 2010; Cao et al. 2012; Cao et al. 2011; Zhang et al. 2013). Additionally, not only for mRNA, microarray analysis revealed the changes in noncoding RNA and found that miRNA target genes are associated with actin cytoskeleton regulation after IBV stimulation (Lin et al. 2019). Expression of individual viral proteins, such as N protein and nonstructural protein 7 (NS7) of Porcine deltacoronavirus (PDCoV) in cells also results in changes in the expression level of cytoskeleton-associated proteins including ezrin, α-actinin-4 and tubulin (Lee et al. 2015; Choi et al. 2019). In addition to infection-induced intracellular changes, cytoskeleton-associated proteins were also found to be incorporated into assembled virions. Using mass spectrometry analysis, actin, tubulin, vimentin, and many cytoskeleton binding proteins such as annexin A2 and destrin were all identified in association with the IBV virion (Dent et al. 2015; Kong et al. 2010). These findings provide strong supports that cytoskeletal proteins are closely associated with coronavirus infection.

As the pathogen causing the global COVID-19 pandemic, SARS-CoV-2 infection is highly cytopathic, which calls for more sophisticated insights into how this virus alters host cells. These changes are thought to create a more suitable environment conducive to the viral life cycle, despite inducing substantial cell perturbation and eventually causing cell death. One study provided a comprehensive repository of SARS-CoV-2-induced ultrastructural cell changes (Cortese et al. 2020). Specifically, they analyzed the alterations of three types of cytoskeleton network, actin filaments, microtubules, and intermediate filaments in SARS-CoV-2-infected A549-ACE2 cells. An accumulation of cortical actin at the plasma membrane was observed in infected cells. Moreover, actin in the cytoplasm was also found to wrap around the SARS-CoV-2 spike protein to form a ring-shaped signal, suggesting that actin may be involved in the viral life cycle. In addition, vimentin intermediate filaments were shown to form a cage-like structure in the perinuclear region that incorporated double-stranded RNA (dsRNA). Microtubules are located at the periphery of this 'cage' and have no spatial overlap with viral dsRNA. dsRNA is presumed to be a viral replication intermediate and located at the interface between the viral replication site and double-membrane vesicles (DMVs) (Klein et al. 2020; Snijder et al. 2020). Therefore, vimentin intermediate filaments might serve as a scaffold or boundary to compartmentalize viral replication organelles from others. In order to record the dynamics of vimentin cage formation, the authors performed live cell imaging and found that the majority of cage formation events were detected between 6 and 9 hours after infection, with time leading to massive cell death (Cortese et al. 2020). These findings provided compelling evidence that SARS-CoV-2 infection alters the cytoskeleton network, which potentially reflects an impact on the viral life cycle or virus-induced cell death (Fig. 12.1). During infection, all three types of cytoskeleton networks showed significant changes. Below, we summarized how coronavirus utilized and modified the cytoskeleton network during infection, in relation to actin filaments, microtubules, and intermediate filaments.

## *Coronavirus Infection Induces Actin Filament Rearrangement*

The most visible changes caused by coronavirus infection occur in the cell membrane region, which is mainly associated with actin filament remodeling. Using high-resolution scanning electron microscopy (SEM) and transmission electron microscopy (TEM), structural changes on the cell surface can be readily captured. Clearly, a ruffled host cell and thick-ended edge appeared after SARS-CoV-2 infection (Caldas et al. 2020). A similar phenomenon also occurs in SARS-CoV-infected Vero cells (Ng et al. 2004). These changes are considered to be more conducive to the early step of infection, especially adhesion of the virion particles to cells. Likewise, researchers also observed an increased cell surface projection in SARS-CoV-2 infected cells coexisting with SARS-CoV infection, which may facilitate the release of progeny virus as a result of the driving force offered by actin filaments (Caldas et al. 2020; Ng et al. 2004; Bouhaddou et al. 2020) (Fig. 12.2b). Moreover, a thin ($<0.7$ μm) strand of F-actin containing tunneling nanotubes (TNT) appeared between two cells in the case of SARS-CoV-2 infection, which may provide molecular information transfer

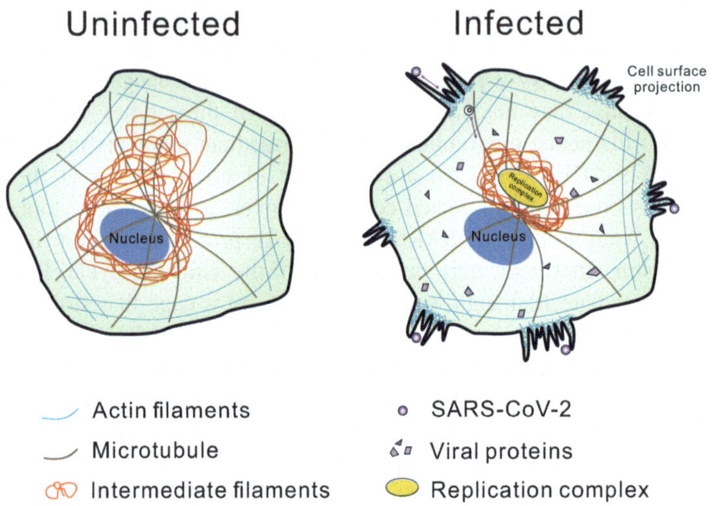


**Fig. 12.1** Changes of cytoskeleton network upon SARS-CoV-2 infection. In normal cells, stress fibers exist under the cell cortex to support cell shape. Cells with smooth edges and almost no protrusions (LHS). Microtubules are mainly responsible for the transport of cargos in cells to maintain normal life activities. Intermediated filaments vimentin is distributed in the cytoplasm, with aggregation near the nucleus. Upon SARS-CoV-2 infection (RHS), many cell surface projections stretch out, mediating virus entry or providing force for the release of progeny virus. Actin filaments are clearly rearranged in these projection areas. Meanwhile, microtubules are hijacked by viruses to transport viral proteins or virus-containing vesicles to complete their own life cycle. For intermediate filaments vimentin, the infection makes these filaments shrink in the perinuclear region, with a viral replication component (encircled), which may provide a location for SARS-CoV-2 to effectively replicate

as well as viral cell-to-cell transmission (Caldas et al. 2020). In addition, a study indicated that there is no need for direct viral infection, only a single viral protein, such as SARS-CoV nucleocapsid (N) protein, is capable of inducing p38 mitogen-activated protein kinase (MAPK) downstream actin reorganization (Surjit et al. 2004).

As for SARS-CoV-2 and SARS-CoV, the rearrangement of actin cytoskeleton is common in many other coronavirus infections. Porcine hemagglutinating encephalomyelitis virus (PHEV), transmissible gastroenteritis virus (TGEV) and porcine epidemic diarrhea virus (PEDV) infection cause acute enteritis in swine of all ages with high mortality in piglets (Chae et al. 2000). It has been reported that PHEV, TGEV and PEDV infection leads to active actin rearrangement (Li et al. 2017; Lv et al. 2019; Zhao et al. 2014; Sun et al. 2017). In order to test whether the internalization of virions is accompanied by actin remodeling, F-actin and DiD-PHEV were fixed and stained at different time points after infection in Neuro-2a cells. DiD-PHEVs were found to be associated with filopodia protrusions at the cell surface initially, then the bound viruses surf toward the foot of filopodia with actin retrograde flow. This process is accompanied by actin stress fiber depolymerization, resulting in a more rounded cell shape. Lastly, actin accumulated in

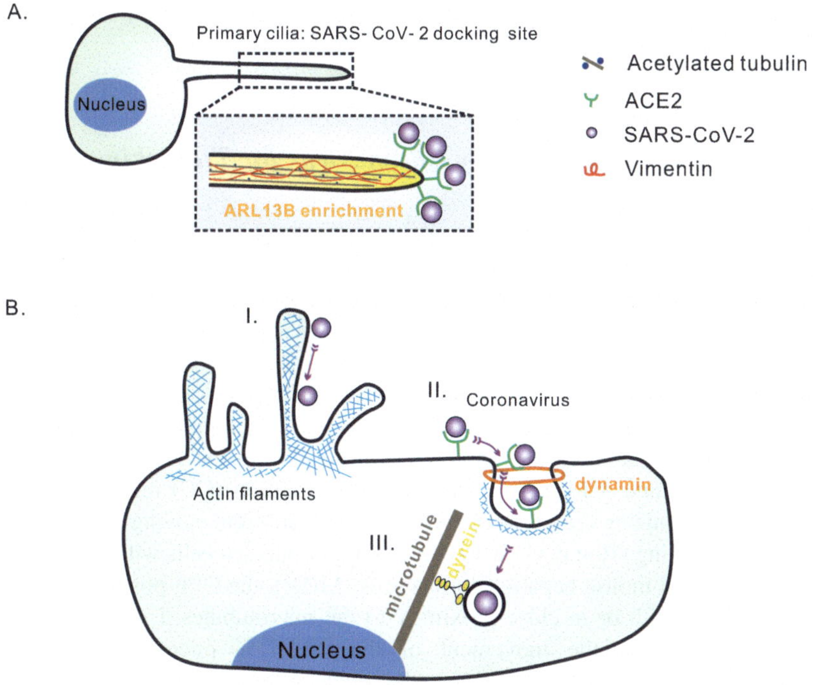


**Fig. 12.2** Cytoskeleton-related internalization mechanism of coronavirus. (**a**) Primary cilia serve as the SARS-CoV-2 docking site: vimentin, SARS-CoV-2 spike protein, and ACE2 are found to colocalized in ARL13B enrichment primary cilia structure. The colocalization of these three proteins is thought to cooperate together to mediate virus docking and leads to ciliary dysfunction, thus initiating infection (Lalioti et al. 2022). (**b**) Invasion model of coronavirus: Firstly, coronavirus usually binds to the cell surface projections and then surfs to the cell body. Actin filament rearrangement significantly changes below the cell membrane in these regions (Caldas et al. 2020; Ng et al. 2004; Bouhaddou et al. 2020). Secondly, the virus specifically binds with receptor ACE2, mediating the endocytosis process. The endocytosis of coronavirus usually is clathrin-dependent, and the segment of virus-containing vesicles with the plasma membrane is dependent on microtubule-associated protein dynamin (Van Hamme et al. 2008; Owczarek et al. 2018; Wang et al. 2020b; Milewska et al. 2018). Lastly, virus-containing vesicles move along microtubules to move into the cell, this process usually relies on the motor protein dynein (Hou et al. 2019; Hagemeijer et al. 2010; Pasick et al. 1994; Kalicharran et al. 1995)

flaky pseudopods after virions were transported to the cell body. This process gives us a whole picture of how the actin cytoskeleton flow movement occurs during PHEV uptake at a very early stage (Li et al. 2017). At the molecular level, PHEV infection stimulates the integrin α5β1-FAK (focal adhesion kinase)-Rac1 (Ras-related C3 botulinum toxin substrate 1)/Cdc42 (cell division cycle 42)-PAK (p21-activated protein kinases)-LIMK (LIM kinase) axis, resulting in the dynamic of cofilin activity and F-actin rearrangement (Lv et al. 2019). Similarly, TGEV actives phosphoinositide-3 kinase (PI3K) through viral spike protein binding to epidermal growth factor receptor (EGFR), leading to the activation of cofilin and F-actin reorganization by Rac1/Cdc42 GTPases (Hu et al. 2016).

## *Microtubules Are Hijacked by Coronavirus for Trafficking*

Microtubules are important for intracellular cargo transport, relying on the motor proteins dynein and kinesin (Hirokawa 1998; Welte 2004; Ross et al. 2008). In the context of viral infection, microtubules are usually employed by viruses to transport their own components or host factors which are beneficial to the viral life cycle (Döhner et al. 2005; Naghavi et al. 2017). To visualize the intracellular viral movement, the porcine epidemic diarrhea virus (PEDV) was successfully labeled by quantum dots without affecting its growth kinetics. The microtubule-dependent movements were classified into three types according to the localization in cells (near-cell membrane (CM) region, middle-cell cytoplasm (CC) region, and near-microtubule organizing center (MTOC) region) with different moving characteristics by live cell tracking (Hou et al. 2019). Similarly, in infected cells with recombinant nsp2-GFP tagged mouse hepatitis coronavirus (MHV), the GFP-positive foci were also associated with or in close proximity to the microtubules. Live cell imaging futher revealed that the movement of these foci is microtubule-dependent (Hagemeijer et al. 2010). Moreover, the nucleocapsid protein of JHM virus (JHMV), a neurotropic murine coronavirus, was found to interact with microtubule-associated protein tau, and the trafficking of JHMV protein was also dependent on microtubules (Pasick et al. 1994; Kalicharran et al. 1995) (Fig. 12.2b).

## *Detection of Intermediate Filament Changes upon Infection*

Despite much knowledge on the different types of intermediate filament proteins in cells, such as vimentin, nestin, lamin, etc., studies on intermediate filament proteins during coronavirus infection are limited. Only one previous report showed vimentin network rearrangement upon SARS-CoV-2 infection, which may contribute to its replication (Cortese et al. 2020). Besides, there are also some omics analyses indicating that the expression level of intermediate filaments proteins changes during coronavirus stimulation. Therefore, the effect of viral infection on intermediate filaments should receive more extensive attention and research.

In the following section, we summarize the effect of coronavirus infection on the host cytoskeleton network, showing a close interaction between the two. Briefly, actin filaments are the most significantly changed cytoskeleton element, with more protrusions formed on the cell surface, which are considered to be beneficial to viral entry and release. Microtubules are always hijacked by viruses to transport their own proteins and host factors for propaganda. For intermediate filaments, more attention should be paid during coronavirus infection. Next, we will consider how the cytoskeleton participates in the viral life cycle.

## Host Cytoskeleton Participates in Coronavirus Life Cycle

As we mentioned above, coronaviruses cause drastic cytoskeletal morphological changes, indicating that the cytoskeleton network plays an essential role in the process of infection. Here we summarize how the cytoskeleton and its related proteins in host cells regulate the coronaviral life cycle at different steps of infection.

### *Host Cytoskeleton Protein Facilitates Coronaviral Internalization*

Various compounds that alter the cytoskeleton network integrity and dynamics have been used to investigate the contribution of the cytoskeletal network to SARS-CoV-2 infection. In a recent study, Vero E6 cells were treated with Latrunculin A (pan actin filament disrupting drug) 2 hours postinfection. Surprisingly, although the actin network showed obvious changes upon coronavirus infection, prior disruption of the actin network did not show any apparent difference in viral dsRNA and extracellular infectivity (Cortese et al. 2020). However, another work showed an inhibition of the entry of SARS-CoV-2 pseudoviral particles into HEK293-ACE2-GFP cells, when cells were treated with the same Latrunculin A (Zhang et al. 2020). These two studies leading to contradictory results illustrate a problem in that the dosing time is critical, which may cause different consequences. To test the effect of drugs on virus invasion, adding drugs before or at the same time with the virus is much better than postinfection treatment. Consist with this idea, a study used four actin network-altering drugs, Latrunculin A, Jasplakinolide, Cytochalasin B, Cytochalasin D, and one microtubule-altering drug nocodazole, to identify the role of cytoskeleton during MHV infection. These workers found that only when they added all the drugs at an early time, rather than at a late stage of infection was MHV infection reduced, revealing the importance of the actin and tubulin network in the entry step of infection (Burkard et al. 2014).

Not only for SARS-CoV-2, but drugs interfering with actin networks also seem harmful to many other coronavirus infection establishments. For instance, Jasplakinolide (stabilization of actin cortex) treatment inhibits the entry of HCoV-OC43 (Owczarek et al. 2018). For the invasion of HCoV-NL63, the two actin inhibitors Cytochalasin D (inhibits actin polymerization) and Jasplakinolide (binds F-actin and stabilizes actin filaments) both block viral particles from penetrating the cell, while the microtubule interfering drug nocodazole shows no effect on viral entry (Milewska et al. 2018). Infection with infectious bronchitis virus (IBV) also causes actin network rearrangement as we mentioned above. However, a study indicated that the entry of IBV can be enhanced by Cytochalasin D or Jasplakinolide (Wang et al. 2019). Nevertheless, these results are self-contradictory and out of expectation, the author cannot give an explicit explanation, indicating the complex process of viral invasion.

After attachment, the internalization of many viruses is reported to rely on dynamin, a microtubule-associated protein and usually functions by pinching off the invaginated vesicles from the cell membrane. For example, both dynamin inhibitory peptide and dominant-negative dynamin significantly inhibited the internalization of feline infectious peritonitis virus (FIPV) (Van Hamme et al. 2008). During this process, others also indicated that myosin light chain kinase (MLCK) and myosin 1 play a vital role in FIPV internalization and subsequent transportation (Dewerchin et al. 2014). Inhibition of dynamin by its inhibitor MiTMAB (interacts with the lipid binding domain of dynamin) or dynasore (noncompetitively inhibits GTPase activity of dynamin) both significantly inhibited HCoV-OC43 entry efficiency to target cells (Owczarek et al. 2018). Similarly, MiTMAB also blocks the internalization of HCoV-NL63 (Milewska et al. 2018). Either by expression of the dynamin-1 dominant-negative mutation or knocking-down by siRNA-inhibited IBV infection, revealed the import role of dynamin-1 in IBV endocytosis (Wang et al. 2019). Furthermore, the internalization of TGEV was shown to be dynamin 2-dependent (Wang et al. 2020b). These studies reveal the general role of dynamin in the process of coronavirus infection, prompting us to take this protein family into consideration when studying other related viruses (Fig. 12.2b).

SARS-CoV and SARS-CoV-2 both utilize the host cell angiotensin-converting enzyme 2 (ACE2) for internalization. It is firmly believed that viral spike protein is important for the interaction between virions and host cell receptors. Recently, many works have indicated the role of extracellular vimentin in both SARS-CoV and SARS-CoV-2 infection, especially in the entry step (Suprewicz et al. 2021; Amraei et al. 2022; Yu et al. 2016). Early on, people revealed the direct interaction between the SARS-CoV spike protein and vimentin. By gene knocking-down and antibody neutralization assay, they further confirmed the vital role of cell surface vimentin function as a putative co-receptor in the uptake of SARS-CoV virus-like particles (Yu et al. 2016). Similarly, vimentin was also shown to bind to SARS-CoV-2 spike protein. Entry assay conducted with pseudotyped SARS-CoV-2 further confirmed the role of vimentin in facilitating virus invasion (Suprewicz et al. 2022; Amraei et al. 2022). Further, a study revealed that SARS-CoV-2 spike protein, ACE2 and vimentin appear to concur at a certain cellular structure, primary cilia, indicating that primary cilia may be the docking structure for SARS-CoV-2 invasion (Lalioti et al. 2022) (Fig. 12.2a). These results indicated that extracellular vimentin is involved in the spike protein-ACE2 complex, functioning as a critical component in mediating SARS-CoV and SARS-CoV-2 internalization, and vimentin-targeting agents may have significance for clinical treatment of infection.

In addition to ACE2, it has also been reported that the entry of SARS-CoV-2 is assisted by cell surface heparan sulfate (HS). Two drugs, BNTX and Sunitinib, targeting the HS ligand α-Syn fibril, not only block the entry of both SARS-CoV and

SARS-CoV-2 pseudoviral particles but also cause actin cytoskeleton rearrangement, revealing the essential role played by actin in the entry of virus (Zhang et al. 2020). Other proteins, such as ezrin, a membrane-actin linker, were also identified to interact with the SARS-CoV spike protein and function as a restricting host factor in the entry process (Millet et al. 2012).

## *Host Cytoskeleton Affects Coronaviral Replication*

To confirm the role of intermediate filaments in SARS-CoV-2 infection, a study applied Withaferin A (an intermediated filaments disrupting drug) to treat cells 2 hours postinfection. As expected, there was a robust reduction in viral dsRNA and supernatant infectious virions found after 6 hours of treatment with Withaferin A. Compared with the result that vimentin intermediate filaments form a cage-like structure that incorporates double-stranded (ds) RNA, these data reveal the importance of intermediate filaments in the SARS-CoV-2 replication step (Cortese et al. 2020). Vimentin also plays a role in TGEV replication. Using siRNA to knockdown vimentin in ST (swine testis) cells, it was found that there was a significant reduction of cell-associated virus by TCID50 assay, which reflected viral replication. Meanwhile, cellular vimentin is associated with the TGEV N protein, and this interaction may help virions to transport through a functional Golgi complex for viral maturation (Zhang et al. 2015).

As for MHV, a study indicated that two stains, RSA59 and RSMHV2, showed different responsiveness to microtubules. Disrupting microtubules with colchicine or vinblastine remarkably reduced the replication of RSA59, whereas it did not affect the RSMHV2 strain, indicating the vital role of microtubule-dependent axonal transport on RSA59 infection and replication. The only difference between these two strains is the spike gene, suggesting that the microtubule-dependent transportation might be a spike protein-mediated process (Biswas et al. 2014).

## *Host Cytoskeleton Participates in the Coronaviral Assembly Process*

A study by Wang et al. (2009) confirmed that actin interacts with infectious bronchitis virus (IBV) M protein at amino acids A159 and K160. Abolishing this interaction by A159-K160 mutation in full-length transcripts did not generate an infectious virus by electroporation using the IBV clone system, indicating that the interaction between actin and M protein is essential for progeny virus production in the late stage of infection. Moreover, Cytochalasin D treatment at the early but not late stage of replication resulted in no virion release to the supernatant, suggesting that actin may function in viral assembly and budding, but not for release.

### *Coronaviral Release Can Be Promoted by Cytoskeleton Proteins*

To detect the role of microtubules on SARS-CoV-2 infection, Cortese et al. (2020) used nocodazole and colcemid to induce microtubule depolymerization and found that these two drugs did not affect SARS-CoV-2 infection. On the contrary, inhibiting microtubule depolymerization or polymerization by paclitaxel or vinblastine, respectively, showed a strong effect on the production of infectious extracellular viruses. This finding emphasized the comprehensive role of the microtubule network in SARS-CoV-2 infection.

In order to detect the role of actin filaments in PEDV and TGEV infection, IPEC-J2 cells were treated with Cytochalasin D or Jasplakinolide. The results showed that neither Jasplakinolide nor Cytochalasin D affected the entry of these two viruses into IPEC-J2 cells. Instead, the replication and release steps of PEDV and TGEV were notably inhibited by either Jasplakinolide or Cytochalasin D (Zhao et al. 2014). This work indicated the role of the actin network in PEDV and TGEV infection, and microtubules also play a role in TGEV infection. To be more specific, the distribution of TGEV spike protein diffused throughout the cytoplasm after nocodazole treatment, compared to the DMSO-treated group near the nucleus. Moreover, the colocalization between the TGEV spike and membrane protein was also reduced in nocodazole-treated cells. Importantly, the microtubule targeting drug nocodazole results in a reduction of infectious virions release and with less spike protein incorporated into virions (Rüdiger et al. 2016).

In this section, we summarize how the cytoskeleton and its related proteins participate in the coronavirus life cycle. In some cases, the specific affected step of infection cannot be precisely defined. To sum up, actin is mainly involved in three steps of virus infection, internalization, assembly and release. Microtubule-related proteins, especially dynamin, primarily play a role in viral internalization. The microtubule network itself is involved in coronavirus replication and release. Intermediate filament vimentin participates in viral internalization and replication. All three cytoskeleton protein networks play an essential role in coronavirus infection. It is of great significance to summarize common rules, which may enable us to understand more about coronavirus infection.

## Cytoskeleton and Pathogenesis

After the SARS-CoV-2 virus enters the human body, it rapidly replicates and proliferates, blocking alveolar blood oxygen exchange and inducing a cytokine storm in the lungs. Concurrently, it interacts with neutrophils, monocytes, and immune response cells, resulting in diffuse alveolar damage (DAD), acute lung injury

(ALI), and acute respiratory distress syndrome (ARDS) (Huang et al. 2020; Jiang et al. 2020; Zhou et al. 2020), and consequently pulmonary fibrosis. These pathological manifestations are particularly severe at later stages (Li et al. 2012). It has been reported that the pathogenic mechanism of the coronavirus involves vascular permeability, airway epithelial ciliated cells, and nervous system regulation. Clarifying the pathogenic mechanism of coronavirus will help design inhibitory drugs for each key target, which is of great importance for human beings to conquer the virus. By summarizing the studies of many laboratories in recent years, we found that an increasing number of researchers have paid attention to the important role played by the cytoskeleton in the pathogenesis of this coronavirus.

## *The Ciliary Structure Is Involved in the Pathogenic Process*

Epithelial cells lining the airways of mammals, such as nasal and pulmonary cilia, play a key role in the defense against infection (Tilley et al. 2015). Loss of cilia is one of the most striking ultrastructural abnormalities in coronavirus-infected cells. The mucociliary clearance mechanism prevents the accumulation of coronavirus particles and mucus in the lungs. It has also been reported that in human nasal epithelial cells, coronavirus infects ciliated cells and thus causes cilia shedding, inducing anosmia (Nicholls et al. 2003). Analysis of autopsy samples from patients with COVID-19 found that SARS-CoV-2 replication was predominantly in airway epithelial cells and alveoli (Hou et al. 2020). Recent studies have found that SARS-CoV-2 has a preferential tropism for ciliated epithelial cells and occasionally infects transitional and secretory cells.

The best-characterized receptor we know of SARS-CoV-2 is ACE2, and the spike-ACE2 interaction has been extensively studied. In immunoassay studies, vimentin aggregates with acetylated tubulin and ARL13B proteins in cilia-like apical structures. The overall colocalization of vimentin, spike, and ACE2 proteins was observed. The spike protein exists in the outer layer of the virus and can mediate the binding of virus particles to cellular receptors and trigger membrane fusion. Ciliated cells have been identified as one of the selective targets of SARS-CoV-2 infection in human tissue studies. Therefore, the authors speculate that vimentin plays an active role in cooperation with ACE2 in the contact between virus and cilia, which may be a specific site of virus docking and leads to ciliary dysfunction, and then mediates virus infection (Lalioti et al. 2022). Therefore, researchers have speculated that many pathological changes of COVID-19 are related to ciliary dysfunction.

Some scholars have also provided evidence from the perspective of phosphoproteomics, proving that the cytoskeleton plays a regulatory role during coronavirus infection. It was also found that kinases downstream of Rho/Rac/Cdc42 GTPases and several cytoskeletal organization-related kinases and effector proteins such as phosphorylated vimentin at site Ser39 and phosphorylated vimentin at site

Ser56 were down-regulated during infection, and cytoskeletal proteins such as motor protein myosin IIa (MYH9 S1943) were down-regulated during infection (Bouhaddou et al. 2020). Primary cilia are microtubule-based organelles (Buqaileh et al. 2021) that reside on the cell surface and sense various environmental stimuli. Analysis of the interaction between the viral nonstructural protein Nsp13 and centrosome components from the perspective of molecular biology also provides a potential molecular mechanism (Li et al. 2020a).

Other coronaviruses have also shown pathological features that target airway-ciliated cells. After the virus infects the host, ciliated cells undergo physical and chemical damage, and at the same time, the levels of microtubule cytoskeleton system-related proteins change. Including SARS-CoV and common cold coronaviruses HCoV-NL63, -OC43, -HKU1, etc. (Zhu et al. 2020; Pizzorno et al. 2020; Hao et al. 2020). Previously, Chilvers et al. (2001) inoculated HCoV 229E into the human nasal cavity, and the respiratory tract epithelium was destroyed, resulting in ciliary motility disorders; epithelial integrity was significantly lost, and the number of ciliated cells was significantly reduced, together with peripheral and central region tubulin changes. These results suggest that the microtubule cytoskeleton system is involved in the regulation mechanism of cilia and virus infection (Hou et al. 2020). In another experiment using animals as hosts, it was found that in MERS-CoV-infected dromedary camels, MERS-COV colocalized with keratin18. The experiment detected the cilia in the airways of animals by anti-α-acetylated tubulin-specific antibodies. There was extensive loss of tubulin signaling in the turbinates and trachea of infected animals (Haverkamp et al. 2018). In addition, the disease caused by Canine respiratory coronavirus (CRCoV), loss, and damage of nasal and tracheal cilia was used to assess its potential impact on the mucociliary system (Priestnall et al. 2009; Mitchell et al. 2013).

Together, these studies fully demonstrate the importance of the cytoskeleton for the regulation of ciliated cells in the pathogenesis of coronavirus, unraveling these mechanistic studies provides the basis for virus-host interactions in protective immunity, host susceptibility, and viral pathogenesis.

## *Lung Injury Caused by SARS-CoV-2 Is Related to Endothelial Cell Permeability Regulation*

Defects in the integrity of the pulmonary vascular barrier are considered to be one of the hallmarks of pathological changes in SARS-CoV-2 infection. In general, vascular endothelial (VE) cells maintain vascular permeability at low levels (Rho et al. 2017).

However, when coronavirus infection induces inflammation, dynamic regulation of endothelial cells can limit vascular barrier function and induce immune cell extravasation, triggering the viral defense mechanism of the host. It was found that through the coiled-coil of Rho and its downstream protein kinase ROCK, nonmyosin cytoplasm-induced myosin contraction and the formation of radial stress fibers can induce dynamic reorganization of the actin cytoskeleton (Eisenhut et al. 2020), which in turn enables Rap1 signaling in vascular endothelial cells (Yamamoto et al. 2021). This process enhances VE-cadherin-mediated cell-cell adhesion function and vascular barrier function, which can help improve the clinicopathological status of patients.

In addition to the contractile regulation of actin, the polymerization regulator of actin is also required for cytoskeletal reorganization. Rap1 inhibits Rho activity as the actin backbone regulates the adhesion function and vascular permeability of VE-cadherin. Thus, activated Cdc42/Rac1 limits vascular permeability, and the dynamic regulation of cytoskeletal organization in vascular endothelial cells and enables cells to avoid hypertonic states (Yamamoto et al. 2021).

The permeability of the lung epithelial cell barrier also plays a crucial role in SARS-CoV-2-induced lung injury and lung homeostasis. It is found that MRCKα and its downstream activation of the myosin light chain are required for the NKA β1 subunit to regulate alveolar barrier function. The synergistic effect of myosin and the actin filament cytoskeleton system can promote cell movement so that endothelial cell permeability changes dynamically with different microenvironments. In contrast, MRCKα is expressed in both the human respiratory tract and alveoli, with reduced expression in SARS-CoV-2 infected patients (Bai et al. 2021).

## *Relationship with the Nervous System*

SARS-CoV-2 not only severely damages respiratory systems such as the trachea and lungs, at the same time neurological complications in patients with COVID-19 have also become one of the important causes of morbidity and mortality. The latest research progress points out that viral RNA appears in the brain and cerebrospinal fluid. Other evidence also illustrates the neurotropism of SARS-CoV-2 (Meinhardt et al. 2021). In the dynamic regulation of the nervous system and virus, the expression of cytoskeleton-related proteins such as fibronectin increases. This change was accompanied by elongation and contraction of the morphology of brain pericytes exposed to the S protein (DeOre et al. 2021).

The nervous system is divided into the central nervous system (CNS) (including the brain and spinal cord) and the peripheral nervous system (PNS) (including the cranial and spinal nerves). These two major parts are the dominant players in the regulation of physiological functions in the body. Clinical reports show that coronavirus can affect the central nervous system. Induction included cerebrovascular disease (Thakur et al. 2021), multifocal cerebral micro-occlusion, and stroke (Mao et al. 2020; Klok et al. 2020). Clinical symptoms include loss of smell and taste, headache, fatigue, nausea, etc. SARS-CoV-2 RNA was also shown to exist in the peripheral nervous system (Matschke et al. 2020; Schurink et al. 2020; Andalib et al. 2021). The infection causes, for example, nerve pain and skeletal muscle damage, Guillain-Barre syndrome, cranial polyneuritis, neuromuscular junction disorders, neuro-ophthalmic disorders, and autonomic dysfunction (Andalib et al. 2021). Thus, it is also important to focus on the role of viruses in the pathogenesis of these neurological complications.

Regarding viral entry, the degree of expression of ACE2 receptors in the nervous system may affect the neurovirulence of SARS-CoV-2. This receptor is expressed in endothelial smooth muscle cells. In addition, studies have found the RNA and S protein of SARS-CoV-2 in specific cells of the brain and nasopharynx. Other investigators have also detected intact virus particles in the nasopharynx (Bilinska et al. 2020). Thus, it is proposed that the virus can infect olfactory neurons, and its proteins can enter the nervous system from the PNS terminal and the olfactory epithelium, and even exist in the medulla oblongata of the brain. In addition, other studies have proposed three pathways by which SARS-CoV-2 enters the nervous system. One is a transsynaptic pathway from nasal epithelial cells to the brain via trigeminal branches (Ferreira et al. 2020). The second is through endocytosis or exocytosis, internalization at nerve terminals, and retrograde transport. Viruses can be transported back to neuronal cell bodies along microtubules by transsynaptic transfer as well as axonal transport mechanisms (vesicular trafficking) (Zubair et al. 2020). The last is a common route for consanguineous virus entry into the nervous system, via internalization and transport across the brain endothelium, thereby crossing the intact blood-brain barrier (BBB) (Iadecola et al. 2020; Iadecola et al. 2020; DeOre et al. 2021). Recent studies have confirmed that angiotensin-converting enzyme 2 (ACE2) at the SARS-CoV-2 binding site contributes to spike-induced barrier disruption through the activation of RhoA. The authors analyzed RhoA to identify it as a key molecule regulating endothelial cytoskeleton and tight junction complex dynamics.

Neuroimmune aspects are caused by viruses. Some researchers have found that SARS-CoV-2 deregulates the vascular and immune functions of brain pericytes through the spike protein. To cope with the viral infection and maintain the homeostasis of autoimmunity, a large number of cytokines are involved. It has been reported that SARS-CoV-2 induces ACE2 downregulation in the nervous system to activate the canonical RAS pathway. Cascade events lead to oxidative stress, neuroinflammation, vasodilation, coagulopathy, and thrombosis. In addition, SARS-CoV-2 binds to toll-like receptors and releases proinflammatory cytokines such as interleukin (IL)-1, (IL)-6, which in turn induce immune responses, leading to brain tissue damage and stroke. Most COVID-19 patients also show a trend toward increased IFN release, which leads to inflammation and immune system suppression (Andalib et al. 2021). These clinical manifestations are firstly innate immune hyperactivity and then immunosuppression. This can effectively protect the body's homeostasis. Our review of the reciprocal regulation between SARS-CoV-2 and the nervous system thus summarizes new mechanistic insights into the pathobiology of cerebrovascular disease associated with COVID-19.

## Possibilities of Cytoskeleton-Related Treatment of COVID-19

As of April 12th, 2022, the World Health Organization (WHO) has counted nearly 500 million confirmed cases of COVID-19 worldwide and more than six million deaths. Although global vaccination rates have reached an advanced level, the defense capacity of existing vaccines is inconclusive in the face of the continuous emergence of new virus variants (Basky et al. 2022). Numerous scholars are still exploring, aiming to determine the most effective treatment options.

As our understanding of the pandemic has grown, many different therapeutic avenues have been developed. For example, pharmacology, immunology, traditional Chinese medicine (TCM) (Ren et al. 2020), etc. At this stage, enzyme inhibitors and neutralizing antibody drugs are the main treatments for SARS-CoV-2 infection. Some clinicians have proposed anti-inflammatory agents: colchicine (Elshafei et al. 2020), as multiple sclerosis (MS) drugs: fingolimod and sipomod (Kloc et al. 2020) are currently new drugs discovered for COVID-19 treatment. Recently, many others have suggested that vimentin could serve as a new target for the treatment of COVID-19 (Li et al. 2020b; Ramos et al. 2020). It is speculated that drugs reducing vimentin expression can be used to treat patients with COVID-19.

In this review, we have paid special attention to the physiological and pathological processes of the cytoskeletal system closely related to coronaviruses such as SARS-CoV-2. Discovering the exact mechanism by which SARS-CoV-2 subverts host cells is critical for validating specific drug targets and effective treatments. In the case of actin filaments, TLRs, CLRs and other receptors (Ezrin and dipeptidyl peptidase 4) that enhance antiviral immunity and viral clearance may serve as therapeutic targets for COVID-19 (Gadanec et al. 2021). In lung disease associated with COVID-19, bradykinin and tumor necrosis factor-alpha (TNF$\alpha$) disrupt the actin cytoskeleton, which could be the leading cause of death of living organisms. Ezrin peptides can be used to inhibit SARS-CoV-2 entry, as well as other viral infections, including HIV-1, hepatitis C virus, human papillomavirus, etc. (Norris et al. 2021). TNF$\alpha$ disrupts the human lung epithelial cytoskeleton system and exerts its effects through Rho kinase. Dedicated to restoring the integrity of the pulmonary endothelial cytoskeleton in patients with COVID-19, thereby reducing the symptoms of patients, this method deserves further exploration by researchers.

In the case of microtubules, the use of microtubule-targeted drugs to treat coronavirus-infected individuals may be effective (Norris et al. 2021). Viral load can be reduced due to targeting microtubules by its inhibitors. The approved drug vinca alkaloid causes the breakdown of the microtubule network; in contrast, paclitaxel stabilizes the microtubule system. In addition, the anti-inflammatory agent mentioned above, colchicine inhibits microtubule polymerization and has performed preliminary studies on the safety of the treatment for patients with COVID-19 (Elshafei et al. 2020).

In the case of intermediate filaments, attenuating the role of vimentin in virus-induced infection could theoretically inhibit viral infection. It has been fully confirmed by many researchers: (1) Vimentin is a co-receptor receptor for SARS-CoV and SARS-CoV-2 entering new cells, except for ACE2; (2) Vimentin is involved in the replication of the virus life cycle (3) Vimentin exerts an anti-inflammatory effect in the cytokine storm caused by a viral infection, and causes the body's autoimmune response (Ramos et al. 2020); (4) The low expression of vimentin inhibits the epithelial–mesenchymal effect of the body transform. According to the above conclusions, it is found that the dual role of vimentin in viral infection may have synergistic advantages for patients. Therefore, some people speculate that drugs that reduce vimentin expression could be used to treat patients with COVID-19. The latest research suggests that ALD-R491 (an oral, noncytotoxic vimentin-targeting small molecule), by changing the physical properties of vimentin filaments, has preclinical efficacy against COVID-19. The authors confirmed through in vivo and in vitro experiments (Li et al. 2021) that ALD-R491 can hinder the entry and exit of viruses into and out of cells, and increase the microbicidal capacity of macrophages, thereby promoting pathogen clearance.

## Summary

In this chapter, we have summarized the connection between coronavirus infection and three major cytoskeleton networks (Table 12.1). This includes the dynamic changes in the cytoskeleton and its related proteins upon viral infection, how cytoskeleton proteins regulate the coronavirus life cycle, and cytoskeleton-related pathogenesis caused by coronavirus infection. It needs to be emphasized that all three cytoskeleton networks are heavily involved in every process of infection. Firstly, the invasion of coronavirus is mainly dependent on the actin filament rearrangement. Viruses are attached to the plasma membrane and interact with its specific receptors or co-receptors, inducing actin filament-dependent filopodia formation. Endocytosis starts after the virus moves along filopodia to the cell body. These processes involved actin filaments, vimentin filaments, and several actin-related proteins. The subcellular transport of a virus or its components relies on the microtubule network. The replication of coronavirus is reported to associate with tubulin and vimentin. We know little about viral assembly. Only actin filaments have been reported to be involved in this process. Finally, the release of progeny virus also requires actin filaments and microtubules to provide the 'driving force'. Most importantly, coronavirus infection-induced pathogenesis is highly correlated with cytoskeleton networks, indicating that drugs targeting cytoskeleton proteins may have an essential influence on treating diseases caused by viral infections. Although there is a substantial increase in the understanding of the regulation of cytoskeleton components and corresponding elaborate subcellular structures in the process of coronavirus infections, there are still many questions that remain future pursuing. In particular, conducting clinical research on drugs targeting cytoskeletons may help to inspire new strategies to control infection and infection-induced pathological damage.

**Table 12.1** Correlation between coronavirus infection and the cytoskeleton

| No. | Virus | Involved cytoskeletal system | Modes of action | Models | Reference |
|---|---|---|---|---|---|
| 1 | Human coronaviruses (HCoV) OC43 | Actin filaments | The virus employs caveolin-1 dependent endocytosis for the entry and the scission of virus-containing vesicles from the cell surface is dynamin-dependent. Furthermore, the vesicle internalization process requires actin cytoskeleton rearrangements. | Cell: HCT-8 | Owczarek et al. (2018) |
| 2 | Human coronavirus NL63 | Actin filaments | Subsequent vesicle scission by dynamin results in virus internalization, and the newly formed vesicle passes the actin cortex, which requires active cytoskeleton rearrangement. | Cell: LLC-Mk2 | Milewska et al. (2018) |
| 3 | Human coronavirus 229E | Microtubules | Inoculation of healthy volunteers with human coronavirus caused disruption of the ciliated epithelium and ciliary dyskinesia. This is likely to impair mucociliary clearance. | Clinical trials | Chilvers et al. (2001) |
| 4 | Avian coronavirus infectious bronchitis virus (IBV) | Actin filaments | The actin filaments might block virion assembly and budding, but not release of the virus particles. | Cell: H1299; Vero | Wang et al. (2009) |
| 5 | Avian coronavirus infectious bronchitis virus (IBV) | Intermediate filaments;<br>Actin filaments | Changes in intracellular mRNA levels after IBV infection by proteomic analysis. The proportions of many cytoskeletal proteins including MYO6, tubulin, vimentin, and acitn in the nucleus were changed. | Cell: Vero | Emmott et al. (2010) |
| 6 | Avian coronavirus infectious bronchitis virus (IBV) | Microtubules | Reproductive damage in hens after viral replication in the epithelium of the oviduct, resulting in loss of cilia and degeneration | Cell: Vero | Villarreal et al. (2007) |

(continued)

**Table 12.1** (continued)

| No. | Virus | Involved cytoskeletal system | Modes of action | Models | Reference |
|---|---|---|---|---|---|
| | | | and necrosis of the epithelial and glandular cells. | | |
| 7 | Transmissible gastroenteritis virus (TGEV) | Actin filaments | The EGFR-PI3K-Rac1/Cdc42-PAK-LIMK signaling pathway affects cofilin phosphorylation, affects actin rearrangement, and allows viruses to invade cells. | Cell: IPEC-J2; HEK293T | Hu et al. (2016) |
| 8 | Transmissible gastroenteritis virus (TGEV) | Actin filaments | TGEV can affect microfilament remodeling in IPEC-J2 cells, and drug-interfered microfilaments can inhibit virus replication and release. | Cell: IPEC-J2; ST; Vero | Zhao et al. (2014) |
| 9 | Transmissible gastroenteritis virus (TGEV) | Actin filaments; Microtubules; Intermediate filaments | Following TGEV infection, mass spectrometry analysis revealed changes in the composition of all three cytoskeletal proteins. It is speculated that changes in actin and tubulin are beneficial for virus replication. | Cell: ST | Zhang et al. (2013) |
| 10 | Porcine epidemic diarrhea virus (PEDV) | Microtubules | Microtubule, dynein, and kinesin-1 are involved in PEDV infection and can influence PEDV fusion and accumulation in the perinuclear region but cannot affect PEDV attachment or internalization. | Cell: Vero | Hou et al. (2021) |
| 11 | Porcine epidemic diarrhea virus (PEDV) | Actin filaments | Mutations in the S protein result in a higher titer. These phenomena such as CPE may be related to the rearrangement of actin. | Cell: Vero | Sun et al. (2017) |
| 12 | Porcine epidemic diarrhea virus (PEDV) | Cytoskeleton-associated protein | E-cadherin expression was decreased at the villus tip, whereas expression of N-cadherin, vimentin and snail was increased. Reduced actin expression in | Animal: Weaned pigs | Chen et al. (2021) |

| | | | | | |
|---|---|---|---|---|---|
| | | | microvilli and zonula occludens-1 (ZO-1). Increased protein concentration of transforming growth factor beta 1 (TGFβ1), which mediates EMT and cytoskeletal changes. | | |
| 13 | Porcine hemagglutinating encephalomyelitis virus (PHEV) | Actin filaments | After infection, actin cytoskeleton depolymerization occurred, with the number of actin stress fibers decreasing, and transient blebs filled with actin forming at the cell surface. | Cell: Neuro-2a | Li et al. (2017) |
| 14 | Porcine hemagglutinating encephalomyelitis virus (PHEV) | Actin filaments | α5β1-FAK-Cofilin pathway causes cytoskeletal rearrangement. | Cell: Neuro-2a | Lv et al. (2019) |
| 15 | Porcine deltacoronavirus (PDCoV) | Microtubules | Increased N protein upregulates ezrin expression, to organize the cortical cytoskeleton by mediating interactions between actin and the plasma membrane proteins. | Cell: HEK-293 T; PAM-pCD163-N | Lee et al. (2015) |
| 16 | Porcine deltacoronavirus (PDCoV) | Microtubules | A substantial downregulation of α-actinin-4 was confirmed in NS7-expressing and PDCoV-infected cells. | Cell: HEK-293 T; HeLa | Lee et al. (2015) |
| 17 | Feline infectious peritonitis virus (FIPV) | Microtubules | Microtubule-related protein dynamin entry Dynamin inhibitory peptide and DN dynamin effectively inhibited virus internalization. | Cell: CrFK | Dewerchin et al. (2014) |
| 18 | Feline infectious peritonitis virus (FIPV) | Microtubules | Intracellular trafficking over microtubules was mediated by MLCK, myosin 1 and a small actin tail. | Cell: Cat monocytes | Van Hamme et al. (2008) |
| 19 | Mouse hepatitis coronavirus (MHV) | Microtubules | The microtubule-associated motility proteins DYNC1H1 and DYNC2H1 are important for infection with MHV (siRNA). | Cell: LR7; FCWF; HEK293T; MDCK; Vero; HeLa; LR7; HeLa-mCC1a | Burkard et al. (2014) |

(continued)

**Table 12.1** (continued)

| No. | Virus | Involved cytoskeletal system | Modes of action | Models | Reference |
|---|---|---|---|---|---|
| 20 | Mouse hepatitis coronavirus (MHV) | Microtubules | Overexpression of nsp2-GFP fusion protein in MHV-infected cells and found to colocalize with tubulin. | Cell: LR7 | Hagemeijer et al. (2010) |
| 21 | Mouse hepatitis coronavirus (MHV) | Microtubules | Disrupting microtubules significantly blocks neuronal transport. | Cell: Neuro2a | Biswas et al. (2014) |
| 22 | Mouse hepatitis coronavirus (MHV) | Microtubules | MHV uses microtubules (MTs) as carriers to reach the cell surface, meantime restricted MT-mediated Cx43 delivery to the cell membrane. | Cell: Astrocyte; Oligodendrocyte; C57Bl/6 | Basu et al. (2017) |
| 23 | Neurotropic corona virus JHM (JHMV) | Microtubules | Specific interaction between antibodies to the microtubule-associated protein tau and the JHMV nucleocapsid protein (N). | Cell: OBL-21 | Pasick et al. (1994) |
| 24 | MERS-coronavirus (MERS-CoV) | Microtubule-associated epithelial ciliary system | *Animal culture tissue* | Animal: Dromedaries | Haverkamp et al. (2018) |
| 25 | Canine respiratory coronavirus (CRCoV) | Microtubule-associated epithelial ciliary system | Reduced mRNA levels of the proinflammatory cytokines TNF-α and IL-6 and the chemokine IL-8 within 72 hours of vaccination. | In vitro organ culture system | Mitchell et al. (2013) |

## References


Amraei R et al (2022) Extracellular vimentin is an attachment factor that facilitates SARS-CoV-2 entry into human endothelial cells. Proc Natl Acad Sci U S A 119(6):e2113874119

Andalib S et al (2021) Peripheral nervous system manifestations associated with COVID-19. Curr Neurol Neurosci Rep 21:9

Bai H et al (2021) The Na+, K+-ATPase β1 subunit regulates epithelial tight junctions via MRCKα. JCI Insight 6(4):e134881

Basky G et al (2022) XE, XD & XF: what to know about the omicron hybrid variants. CMAJ 194: E654–e655

Basu R et al (2017) Microtubule-assisted altered trafficking of astrocytic gap junction protein connexin 43 is associated with depletion of connexin 47 during mouse hepatitis virus infection. J Biol Chem 292:14747–14763

Bedi S et al (2019) Friend or foe: the role of the cytoskeleton in influenza a virus assembly. Viruses 11(1):46

Bhattacharya B et al (2007) Interaction between bluetongue virus outer capsid protein VP2 and vimentin is necessary for virus egress. Virol J 4:7

Bilinska K et al (2020) Expression of the SARS-CoV-2 entry proteins, ACE2 and TMPRSS2, in cells of the olfactory epithelium: identification of cell types and trends with age. ACS Chem Neurosci 11:1555–1562

Billadeau DD et al (2007) Regulation of T-cell activation by the cytoskeleton. Nat Rev Immunol 7: 131–143

Biswas K et al (2014) Effect of microtubule disruption on neuronal spread and replication of demyelinating and nondemyelinating strains of mouse hepatitis virus in vitro. J Virol 88:3043–3047

Bouhaddou M et al (2020) The global phosphorylation landscape of SARS-CoV-2 infection. Cell 182:685–712.e19

Brian DA et al (2005) Coronavirus genome structure and replication. Curr Top Microbiol Immunol 287:1–30

Buqaileh R et al (2021) Can cilia provide an entry gateway for SARS-CoV-2 to human ciliated cells? Physiol Genomics 53:249–258

Burkard C et al (2014) Coronavirus cell entry occurs through the endo−/lysosomal pathway in a proteolysis-dependent manner. PLoS Pathog 10:e1004502

Caldas LA et al (2020) Ultrastructural analysis of SARS-CoV-2 interactions with the host cell via high resolution scanning electron microscopy. Sci Rep 10:16099

Cao Z et al (2011) Proteomic analysis of chicken embryonic trachea and kidney tissues after infection in ovo by avian infectious bronchitis coronavirus. Proteome Sci 9:11

Cao Z et al (2012) Proteomics analysis of differentially expressed proteins in chicken trachea and kidney after infection with the highly virulent and attenuated coronavirus infectious bronchitis virus in vivo. Proteome Sci 10:24

Carod-Artal FJ (2020) Neurological complications of coronavirus and COVID-19. Rev Neurol 70: 311–322

Chae C et al (2000) Prevalence of porcine epidemic diarrhoea virus and transmissible gastroenteritis virus infection in Korean pigs. Vet Rec 147:606–608

Chen YM et al (2021) Epithelial-mesenchymal transition of absorptive enterocytes and depletion of Peyer's patch M cells after PEDV infection. Virology 552:43–51

Chen H et al (2022) Prediction of coronavirus 3C-like protease cleavage sites using machine-learning algorithms. Virol Sin 37:437–444

Chilvers MA et al (2001) The effects of coronavirus on human nasal ciliated respiratory epithelium. Eur Respir J 18:965–970

Choi S et al (2019) Functional characterization and proteomic analysis of porcine Deltacoronavirus accessory protein NS7. J Microbiol Biotechnol 29:1817–1829

Cortese M et al (2020) Integrative imaging reveals SARS-CoV-2-induced reshaping of subcellular morphologies. Cell Host Microbe 28:853–866.e5

Cui J et al (2019) Origin and evolution of pathogenic coronaviruses. Nat Rev Microbiol 17:181–192

Del Valle DM et al (2020) An inflammatory cytokine signature predicts COVID-19 severity and survival. Nat Med 26:1636–1643

den Boon JA et al (2010) Organelle-like membrane compartmentalization of positive-strand RNA virus replication factories. Annu Rev Microbiol 64:241–256

Denes CE et al (2018) Cytoskeletons in the closet-subversion in Alphaherpesvirus infections. Viruses 10(2):79

Dent SD et al (2015) The proteome of the infectious bronchitis virus beau-R virion. J Gen Virol 96: 3499–3506

DeOre BJ et al (2021) SARS-CoV-2 spike protein disrupts blood-brain barrier integrity via RhoA activation. J Neuroimmune Pharmacol 16:722–728

Dewerchin HL et al (2014) Myosins 1 and 6, myosin light chain kinase, actin and microtubules cooperate during antibody-mediated internalisation and trafficking of membrane-expressed viral antigens in feline infectious peritonitis virus infected monocytes. Vet Res 45:17

Döhner K, Sodeik B (2005) The role of the cytoskeleton during viral infection. Curr Top Microbiol Immunol 285:67–108

Döhner K et al (2005) Viral stop-and-go along microtubules: taking a ride with dynein and kinesins. Trends Microbiol 13:320–327

Eisenhut M et al (2020) Pathways in the pathophysiology of coronavirus 19 lung disease accessible to prevention and treatment. Front Physiol 11:872

Elshafei MN et al (2020) The efficacy of colchicine in the management of coronavirus disease 2019: a protocol for systematic review and meta-analysis. Medicine (Baltimore) 99:e21911

Emmott E et al (2010) Quantitative proteomics using stable isotope labeling with amino acids in cell culture reveals changes in the cytoplasmic, nuclear, and nucleolar proteomes in Vero cells infected with the coronavirus infectious bronchitis virus. Mol Cell Proteomics 9:1920–1936

Ezratty EJ et al (2005) Microtubule-induced focal adhesion disassembly is mediated by dynamin and focal adhesion kinase. Nat Cell Biol 7:581–590

Fehr AR et al (2015) Coronaviruses: an overview of their replication and pathogenesis. Methods Mol Biol 1282:1–23

Ferreira A et al (2020) COVID-19 and herpes zoster co-infection presenting with trigeminal neuropathy. Eur J Neurol 27:1748–1750

Foo KY et al (2015) Interaction between Flavivirus and cytoskeleton during virus replication. Biomed Res Int 2015:427814

Gadanec LK et al (2021) Can SARS-CoV-2 virus use multiple receptors to enter host cells? Int J Mol Sci 22(3):992

Hagemeijer MC et al (2010) Dynamics of coronavirus replication-transcription complexes. J Virol 84:2134–2149

Hao S et al (2020) Long-term modeling of SARS-CoV-2 infection of in vitro cultured polarized human airway epithelium. mBio 11(6):e02852-20

Hatmal MM et al (2020) Comprehensive structural and molecular comparison of spike proteins of SARS-CoV-2, SARS-CoV and MERS-CoV, and their interactions with ACE2. Cell 9(12):2638

Haverkamp AK et al (2018) Experimental infection of dromedaries with Middle East respiratory syndrome-coronavirus is accompanied by massive ciliary loss and depletion of the cell surface receptor dipeptidyl peptidase 4. Sci Rep 8:9778

Helfand BT et al (2011) Vimentin organization modulates the formation of lamellipodia. Mol Biol Cell 22:1274–1289

Hirokawa N (1998) Kinesin and dynein superfamily proteins and the mechanism of organelle transport. Science 279:519–526

Hou W et al (2019) Real-time analysis of quantum dot labeled single porcine epidemic diarrhea virus moving along the microtubules using single particle tracking. Sci Rep 9:1307

Hou YJ et al (2020) SARS-CoV-2 reverse genetics reveals a variable infection gradient in the respiratory tract. Cell 182:429–446.e14
Hou W et al (2021) Dynamic dissection of dynein and Kinesin-1 cooperatively mediated intercellular transport of porcine epidemic diarrhea coronavirus along microtubule using single virus tracking. Virulence 12:615–629
Hu W et al (2016) The epidermal growth factor receptor regulates cofilin activity and promotes transmissible gastroenteritis virus entry into intestinal epithelial cells. Oncotarget 7:12206–12221
Huang C et al (2020) Clinical features of patients infected with 2019 novel coronavirus in Wuhan, China. Lancet 395:497–506
Iadecola C et al (2020) Effects of COVID-19 on the nervous system. Cell 183:16–27.e1
Jamal M et al (2021) Immune dysregulation and system pathology in COVID-19. Virulence 12: 918–936
Jhan MK et al (2017) Dengue virus infection increases microglial cell migration. Sci Rep 7:91
Jiang F et al (2020) Review of the clinical characteristics of coronavirus disease 2019 (COVID-19). J Gen Intern Med 35:1545–1549
Kalicharran K et al (1995) Involvement of microtubules and the microtubule-associated protein tau in trafficking of JHM virus and components within neurons. Adv Exp Med Biol 380:57–61
Klein S et al (2020) SARS-CoV-2 structure and replication characterized by in situ cryo-electron tomography. Nat Commun 11:5885
Kloc M et al (2020) The multiple sclerosis (MS) drugs as a potential treatment of ARDS in COVID-19 patients. Mult Scler Relat Disord 45:102437
Klok FA et al (2020) Incidence of thrombotic complications in critically ill ICU patients with COVID-19. Thromb Res 191:145–147
Kong Q et al (2010) Proteomic analysis of purified coronavirus infectious bronchitis virus particles. Proteome Sci 8:29
Lalioti V et al (2022) Cell surface detection of vimentin, ACE2 and SARS-CoV-2 spike proteins reveals selective colocalization at primary cilia. Sci Rep 12:7063
Lee S et al (2015) Functional characterization and proteomic analysis of the nucleocapsid protein of porcine deltacoronavirus. Virus Res 208:136–145
Li SW et al (2012) Correlation between TGF-β1 expression and proteomic profiling induced by severe acute respiratory syndrome coronavirus papain-like protease. Proteomics 12:3193–3205
Li Z et al (2017) Porcine Hemagglutinating encephalomyelitis virus enters neuro-2a cells via Clathrin-mediated endocytosis in a Rab5-, cholesterol-, and pH-dependent manner. J Virol 91(23):e01083-17
Li W et al (2020a) COVID-19, cilia, and smell. FEBS J 287:3672–3676
Li Z et al (2020b) Vimentin as a target for the treatment of COVID-19. BMJ Open Respir Res 7(1): e000623
Li Z et al (2021) A vimentin-targeting Oral compound with host-directed antiviral and anti-inflammatory actions addresses multiple features of COVID-19 and related diseases. MBio 12:e0254221
Lin J et al (2019) Microarray analysis of infectious bronchitis virus infection of chicken primary dendritic cells. BMC Genomics 20:557
Lv X et al (2019) Porcine Hemagglutinating encephalomyelitis virus activation of the integrin α5β1-FAK-Cofilin pathway causes cytoskeletal rearrangement to promote its invasion of N2a cells. J Virol 93(5):e01736-18
Malik YA (2020) Properties of coronavirus and SARS-CoV-2. Malays J Pathol 42:3–11
Mao L et al (2020) Neurologic manifestations of hospitalized patients with coronavirus disease 2019 in Wuhan, China. JAMA Neurol 77:683–690
Matschke J et al (2020) Neuropathology of patients with COVID-19 in Germany: a post-mortem case series. Lancet Neurol 19:919–929
May RC et al (2001) Phagocytosis and the actin cytoskeleton. J Cell Sci 114:1061–1077

Meinhardt J et al (2021) Olfactory transmucosal SARS-CoV-2 invasion as a port of central nervous system entry in individuals with COVID-19. Nat Neurosci 24:168–175
Milewska A et al (2018) Entry of human coronavirus NL63 into the cell. J Virol 92(3):e01933-17
Millet JK et al (2012) Ezrin interacts with the SARS coronavirus spike protein and restrains infection at the entry stage. PLoS One 7:e49566
Miranda-Saksena M et al (2018) Infection and transport of herpes simplex virus type 1 in neurons: role of the cytoskeleton. Viruses 10(2):92
Mitchell JA et al (2013) Tropism and pathological findings associated with canine respiratory coronavirus (CRCoV). Vet Microbiol 162:582–594
Naghavi MH et al (2017) Microtubule regulation and function during virus infection. J Virol 91(16):e00538-17
Ng ML et al (2004) Topographic changes in SARS coronavirus-infected cells at late stages of infection. Emerg Infect Dis 10:1907–1914
Nicholls JM et al (2003) Lung pathology of fatal severe acute respiratory syndrome. Lancet 361: 1773–1778
Norris V et al (2021) Role of multifunctional cytoskeletal filaments in Coronaviridae infections: therapeutic opportunities for COVID-19 in a nutshell. Cell 10(17):1818
Owczarek K et al (2018) Early events during human coronavirus OC43 entry to the cell. Sci Rep 8: 7124
Pan P et al (2021) SARS-CoV-2 N protein promotes NLRP3 inflammasome activation to induce hyperinflammation. Nat Commun 12:4664
Pasick JM et al (1994) Distribution and trafficking of JHM coronavirus structural proteins and virions in primary neurons and the OBL-21 neuronal cell line. J Virol 68:2915–2928
Paul D et al (2013) Architecture and biogenesis of plus-strand RNA virus replication factories. World J Virol 2:32–48
Pizzorno A et al (2020) Characterization and treatment of SARS-CoV-2 in nasal and bronchial human airway epithelia. Cell Rep Med 1:100059
Priestnall SL et al (2009) Quantification of mRNA encoding cytokines and chemokines and assessment of ciliary function in canine tracheal epithelium during infection with canine respiratory coronavirus (CRCoV). Vet Immunol Immunopathol 127:38–46
Ramos I et al (2020) Vimentin as a multifaceted player and potential therapeutic target in viral infections. Int J Mol Sci 21(13):4675
Ren JL et al (2020) Traditional Chinese medicine for COVID-19 treatment. Pharmacol Res 155: 104743
Ren X et al (2021) COVID-19 immune features revealed by a large-scale single-cell transcriptome atlas. Cell 184:1895–1913.e19
Rho SS et al (2017) Dynamic regulation of vascular permeability by vascular endothelial cadherin-mediated endothelial cell-cell junctions. J Nippon Med Sch 84:148–159
Ross JL et al (2008) Cargo transport: molecular motors navigate a complex cytoskeleton. Curr Opin Cell Biol 20:41–47
Rüdiger AT et al (2016) Tubulins interact with porcine and human S proteins of the genus Alphacoronavirus and support successful assembly and release of infectious viral particles. Virology 497:185–197
Schurink B et al (2020) Viral presence and immunopathology in patients with lethal COVID-19: a prospective autopsy cohort study. Lancet Microbe 1:e290–e299
Scott BM et al (2022) Predicted coronavirus Nsp5 protease cleavage sites in the human proteome. BMC Genom Data 23:25
Shi X et al (2020) Unidirectional regulation of vimentin intermediate filaments to Caveolin-1. Int J Mol Sci 21(20):7436
Snijder EJ et al (2003) Unique and conserved features of genome and proteome of SARS-coronavirus, an early split-off from the coronavirus group 2 lineage. J Mol Biol 331:991–1004
Snijder EJ et al (2006) Ultrastructure and origin of membrane vesicles associated with the severe acute respiratory syndrome coronavirus replication complex. J Virol 80:5927–5940

Snijder EJ et al (2020) A unifying structural and functional model of the coronavirus replication organelle: tracking down RNA synthesis. PLoS Biol 18:e3000715
Sodeik B et al (1997) Microtubule-mediated transport of incoming herpes simplex virus 1 capsids to the nucleus. J Cell Biol 136:1007–1021
Suikkanen S et al (2003) Exploitation of microtubule cytoskeleton and dynein during parvoviral traffic toward the nucleus. J Virol 77:10270–10279
Sun M et al (2017) Identification of two mutation sites in spike and envelope proteins mediating optimal cellular infection of porcine epidemic diarrhea virus from different pathways. Vet Res 48:44
Suomalainen M et al (1999) Microtubule-dependent plus- and minus end-directed motilities are competing processes for nuclear targeting of adenovirus. J Cell Biol 144:657–672
Suprewicz Ł et al. (2021) Extracellular vimentin as a target against SARS-CoV-2 host cell invasion. bioRxiv
Suprewicz Ł et al (2022) Extracellular vimentin as a target against SARS-CoV-2 host cell invasion. Small 18:e2105640
Surjit M et al (2004) The SARS coronavirus nucleocapsid protein induces actin reorganization and apoptosis in COS-1 cells in the absence of growth factors. Biochem J 383:13–18
Thakur KT et al (2021) COVID-19 neuropathology at Columbia University Irving medical center/New York Presbyterian hospital. Brain 144:2696–2708
Tilley AE et al (2015) Cilia dysfunction in lung disease. Annu Rev Physiol 77:379–406
Ukmar-Godec T et al (2020) Biomolecular condensation of the microtubule-associated protein tau. Semin Cell Dev Biol 99:202–214
Van Hamme E et al (2008) Clathrin- and caveolae-independent entry of feline infectious peritonitis virus in monocytes depends on dynamin. J Gen Virol 89:2147–2156
Villarreal LY et al (2007) Orchitis in roosters with reduced fertility associated with avian infectious bronchitis virus and avian metapneumovirus infections. Avian Dis 51:900–904
Wang J et al (2009) Interaction of the coronavirus infectious bronchitis virus membrane protein with beta-actin and its implication in virion assembly and budding. PLoS One 4:e4908
Wang IH et al (2018) Imaging, tracking and computational analyses of virus entry and egress with the cytoskeleton. Viruses 10(4):166
Wang H et al (2019) Infectious bronchitis virus entry mainly depends on clathrin mediated endocytosis and requires classical endosomal/lysosomal system. Virology 528:118–136
Wang F et al (2020a) Long-term respiratory and neurological sequelae of COVID-19. Med Sci Monit 26:e928996
Wang J et al (2020b) Dynamics of transmissible gastroenteritis virus internalization unraveled by single-virus tracking in live cells. FASEB J 34:4653–4669
Wang JY et al (2021) An autoantigen profile of human A549 lung cells reveals viral and host etiologic molecular attributes of autoimmunity in COVID-19. J Autoimmun 120:102644
Welte MA (2004) Bidirectional transport along microtubules. Curr Biol 14:R525–R537
Wen Z et al (2020) Cytoskeleton-a crucial key in host cell for coronavirus infection. J Mol Cell Biol 12:968–979
Xu K et al (2013) Actin, spectrin, and associated proteins form a periodic cytoskeletal structure in axons. Science 339:452–456
Xu Z et al (2020) Pathological findings of COVID-19 associated with acute respiratory distress syndrome. Lancet Respir Med 8:420–422
Yamamoto K et al (2021) Rap1 small GTPase regulates vascular endothelial-cadherin-mediated endothelial cell-cell junctions and vascular permeability. Biol Pharm Bull 44:1371–1379
Yu YT et al (2016) Surface vimentin is critical for the cell entry of SARS-CoV. J Biomed Sci 23:14
Zandi F et al (2021) Rabies virus matrix protein targets host actin cytoskeleton: a protein-protein interaction analysis. Pathog Dis 79(1):ftaa075
Zhang X et al (2013) Identification of cellular proteome using two-dimensional difference gel electrophoresis in ST cells infected with transmissible gastroenteritis coronavirus. Proteome Sci 11:31

Zhang X et al (2015) Identification of the interaction between vimentin and nucleocapsid protein of transmissible gastroenteritis virus. Virus Res 200:56–63
Zhang Y et al (2019) The role of host cytoskeleton in Flavivirus infection. Virol Sin 34:30–41
Zhang Q et al (2020) Heparan sulfate assists SARS-CoV-2 in cell entry and can be targeted by approved drugs in vitro. Cell Discov 6(1):1–14
Zhao S et al (2014) Transmissible gastroenteritis virus and porcine epidemic diarrhoea virus infection induces dramatic changes in the tight junctions and microfilaments of polarized IPEC-J2 cells. Virus Res 192:34–45
Zhou F et al (2020) Clinical course and risk factors for mortality of adult inpatients with COVID-19 in Wuhan, China: a retrospective cohort study. Lancet 395:1054–1062
Zhu N et al (2020) Morphogenesis and cytopathic effect of SARS-CoV-2 infection in human airway epithelial cells. Nat Commun 11:3910
Zubair AS et al (2020) Neuropathogenesis and neurologic manifestations of the coronaviruses in the age of coronavirus disease 2019: a review. JAMA Neurol 77:1018–1027

# Chapter 13
# Viral RNA Is a Hub for Critical Host–Virus Interactions

**Alfredo Castello and Louisa Iselin**

**Abstract** RNA is a central molecule in the life cycle of viruses, acting not only as messenger (m)RNA but also as a genome. Given these critical roles, it is not surprising that viral RNA is a hub for host–virus interactions. However, the interactome of viral RNAs remains largely unknown. This chapter discusses the importance of cellular RNA-binding proteins in virus infection and the emergent approaches developed to uncover and characterise them.

**Keywords** RNA · RNA-Binding Proteins (RBPs) · RNA viruses · Interferon-Stimulated Genes (ISGs) · Riboregulation · RNA Interactome Capture (RIC)

## Introduction: RNA-Binding Proteins Are at the Core of Virus Infection

RNA is a central molecule in the life cycle of viruses. In RNA viruses, it does not only function as a blueprint for the synthesis of viral proteins (i.e. messenger [m] RNA) but also as storage of the genetic information as a genome (Garcia-Moreno et al. 2018). While DNA viruses employ DNA molecules to store their genetic information, they still need to produce mRNAs to synthesise the viral proteins. However, their larger coding capacity enables the expression of additional RNAs such as non-coding RNAs with regulatory roles (Tagawa et al. 2021). Despite the central importance of RNA for viruses, the interactions that viral RNA establishes with the host cell have remained largely unexplored. This chapter focuses on RNA

A. Castello (✉)
MRC University of Glasgow Centre for Virus Research, Glasgow, UK
e-mail: alfredo.castello@glasgow.ac.uk

L. Iselin
MRC University of Glasgow Centre for Virus Research, Glasgow, UK

Nuffield Department of Medicine, Peter Medawar Building for Pathogen Research, University of Oxford, Oxford, UK
e-mail: louisa.iselin@stcatz.ox.ac.uk

 
S. Vijayakrishnan et al. (eds.), *Virus Infected Cells*, Subcellular Biochemistry 106,
https://doi.org/10.1007/978-3-031-40086-5_13

viruses and the importance of their RNAs in establishing critical interactions with the host cell.

Viral RNA is known to engage with both viral and cellular RNA-binding proteins (RBPs) in infected cells, forming ribonucleoproteins (RNPs) with critical roles in the viral life cycle. These interactions can indeed mediate, facilitate or modulate virtually all the steps of the viral life cycle, including viral replication/transcription, protein synthesis, viral RNA stability and location, and viral particle assembly (Dicker et al. 2021; Garcia-Moreno et al. 2018). On the other hand, RBPs are also central to the cell's antiviral defences. Viral RNA often contains pathogen-associated molecular patterns (PAMPs) that can be recognised and bound by a specialised pool of cellular RBPs. These PAMPs include tri-phosphate 5′ ends, undermethylated cap structures, long double-stranded RNA tracks (dsRNA), sequence biases (CpG), or unusual codons (Jensen and Thomsen 2012). It is, therefore, not surprising that many interferon-stimulated genes (ISG) are endowed with RNA-binding activity, which is required to sense and function on viral RNAs in order to activate the cellular antiviral response. These antiviral RBPs include well-characterised ISGs such as RIGI, MDM5, IFITs, OASs and protein kinase R (PKR) amongst many others (Dias Junior et al. 2019; Jensen and Thomsen 2012; Rehwinkel and Gack 2020; Vladimer et al. 2014).

Most of the host–virus interactions involving RBPs have been discovered and characterised in case-by-case. However, recent technological advances have enabled the identification of viral RNA/cellular RBP interactions on a proteome-wide scale, revealing an unexplored interface between the virus and the host cell.

## RNA-Binding Proteins, Structure and Function

RBPs are highly heterogeneous in structure, function and modes of RNA binding. They can associate with RNA using specialised globular domains and/or intrinsically disordered regions, referred to here as RNA-binding domains (RBDs) (Castello et al. 2016; Jarvelin et al. 2016; Lunde et al. 2007), which mediate long-lived, transitory, specific or non-specific interactions. Affinity and specificity are driven by the type of amino acids at the interface (typically positively charged and aromatic side chains), the nucleotide sequence and the spatial configuration between these amino acids and nucleotides (Fig. 13.1a) (Lunde et al. 2007). Therefore, RBPs can bind to RNA specifically, and this can occur by recognition of a nucleotide sequence, a secondary structure or both (Lunde et al. 2007). Other RBPs, however, engage with RNA in a non-selective manner by interacting with the phosphate-backbone or/and through promiscuous binding with the nucleotide bases. Both specific and promiscuous RBPs can play critical roles in RNA metabolism. For example, the eukaryotic initiation factor (eIF)4E interacts with all capped mRNAs in the cytoplasm to aid protein synthesis (Jackson et al. 2010). This interaction, while global, is critical to mediate a central process such as translation initiation across the wide spectrum of cellular mRNAs.

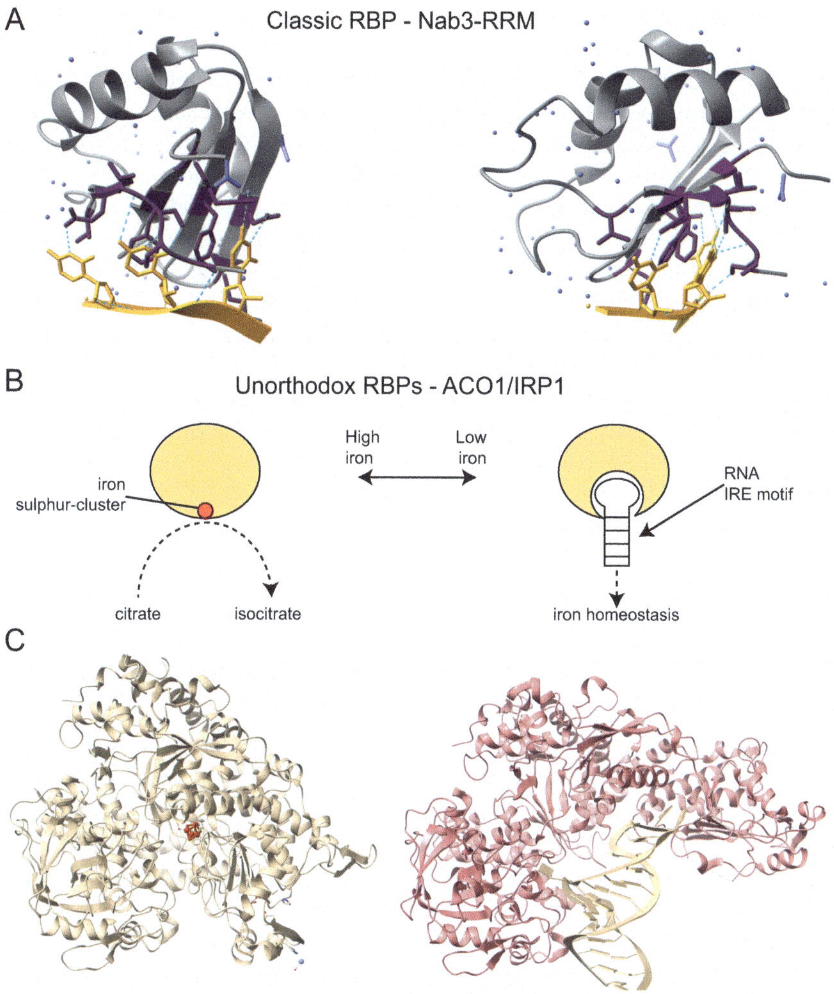


**Fig. 13.1** Classical vs unorthodox RBPs. (**a**) Crystal structure of the Nab3 RRM domain (grey and magenta) and the target RNA sequence (yellow) (PDB 2XNR). Amino acids in contanct with nucleotides are highlighted in magenta, revealing aromatic (Phe), polar (Ser) and positively charged (Arg and Lys) in contact with the nucleotide bases. (**b**) Schematic representation of the dual activity of aconitase 1 (ACO1)/iron regulatory protein 1 (IRP1). Its close (when associated to an iron-sulphur cluster, PDB 2B3X) or open (when bound to RNA PDB 3SNP) conformations are shown below. Structure display was done using the Chimera X software

Many RBPs have a modular design with multiple RBDs coordinated by flexible regions. Such modular architecture increases specificity and affinity by allowing cooperative binding to RNA of individual RBDs. Interestingly, it was believed that these 'modules' are composed of a limited pool of well-characterised RBDs. These domains, named 'classic', include the RNA recognition motif (RRM) (Fig. 13.1a), the K-homology domain (KH), double-stranded RNA-binding domain (dsRBD), cold shock domain (CSH) and the DEAD-box helicase domain amongst others (Castello et al. 2016; Lunde et al. 2007). However, an unexpected discovery in the early 1990s provided evidence that RBPs are more diverse and complex than previously anticipated. Aconitase 1 (ACO1) converts citrate into isocitrate in the tricarboxylic acid cycle, and this reaction is facilitated by an iron-sulphur cluster assembled in its active centre (Fig. 13.1b and c). When iron levels decrease in the cell, ACO1 is no longer able to assemble with an iron-sulphur cluster and, as apoprotein, becomes an RBP known as iron regulatory protein 1 (IRP1) (Hentze et al. 2010; Muckenthaler et al. 2017). As an RBP, IRP1/ACO1 interacts with mRNAs involved in iron homeostasis, adjusting gene expression to increase intracellular iron levels. This example laid the foundations for a new class of RBPs with unorthodox modes of RNA binding that play pivotal roles in connecting RNA metabolism and cellular homeostasis (Castello et al. 2015; Hentze et al. 2018; Hentze and Preiss 2010). In the last decade, a wide range of unorthodox RBPs have been identified and characterised (Castello et al. 2015; Hentze et al. 2018). Several of these have recently been associated with the regulation of virus infection and will be discussed in further detail below.

## The Expanding Universe of Cellular RNA-Binding Proteins

Cellular RBPs were initially discovered on a case-to-case basis using biochemical approaches. Later, in silico identification of classic RBPs was made possible by searching across the proteome for protein-encoding classic RBDs (Lunde et al. 2007). However, this approach fell short of identifying RBPs with unorthodox modes of RNA binding, such as ACO1/IRP1. In 2012, two independent studies reported a new approach to identifying RBPs systematically and comprehensively in living cells, called RNA interactome capture (RIC) (Fig. 13.2a) (Baltz et al. 2012; Castello et al. 2012). In brief, RIC employs ultraviolet light (UV) irradiation to promote crosslinks between RNA and proteins placed at 'zero' distance without inducing detectable protein–protein crosslinks (Pashev et al. 1991). Once protein–RNA complexes are frozen covalently, cells are lysed under denaturing conditions and polyadenylated RNA is isolated by hybridisation with oligo(dT) magnetic beads. Following very stringent washes, RBPs are released using RNase treatment and analysed by mass spectrometry (Fig. 13.2a). Applied to HeLa and HEK293 cells, RIC identified nearly a thousand cellular RBPs bound to cellular mRNAs, with more than 50% representing novel RNA binders with no known RBD. Consistent with classic RBPs, newly discovered RNA binders were enriched in intrinsically disordered regions and positively charged amino acids (Castello et al. 2012). The RNA-binding activity of many of these new RBPs was validated by orthogonal methods and further confirmed by the implementation of RIC and other complementary methods in different cell types, tissues and organisms (Bach-Pages et al.

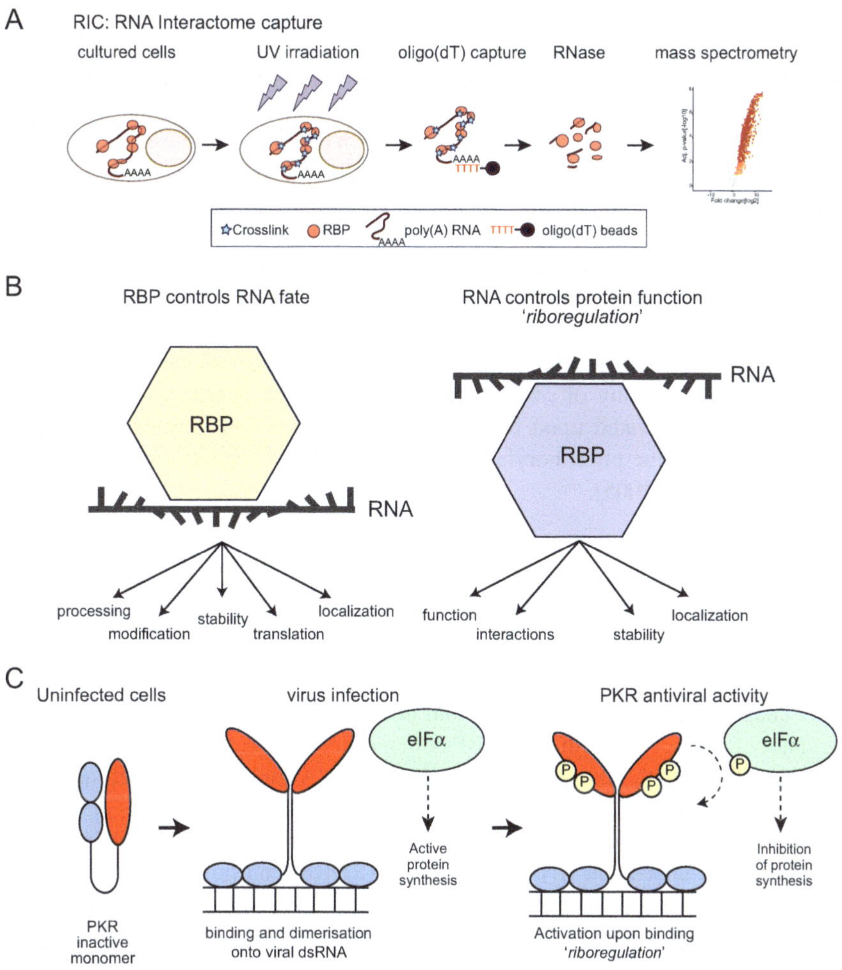


**Fig. 13.2** Discovery of unorthodox RBP and the concept of riboregulation. (**a**) Schematic representation of the RNA interactome capture (RIC) protocol. (**b**) Schematic representation of riboregulation. (**c**) Activation mechanism of the antiviral protein PKR by interaction with viral dsRNA

2020; Beckmann et al. 2015; Hentze et al. 2018; Matia-Gonzalez et al. 2015; Queiroz et al. 2019; Trendel et al. 2019).

An open question was how these newly discovered RBPs interact with RNA if they do not harbour any recognisable RBD. To identify the RNA-binding surfaces of these proteins, several approaches used UV crosslinking, controlled proteolysis and proteomics to determine the peptides within RBPs that crosslink with RNA on a proteome-wide scale (Castello et al. 2016; He et al. 2016; Hentze et al. 2018; Kramer et al. 2014). Interestingly, these studies identified more than a thousand binding sites, many of which mapped to enzymatic cores, protein–protein interaction surfaces and intrinsically disordered regions, representing prevalent platforms for RNA binding in unorthodox RBPs (Castello et al. 2016; He et al. 2016). The discovery of dozens of metabolic enzymes, particularly with NAD-binding activity, that interact with RNA highlighted that ACO1/IRP1 was only the tip of the iceberg (Castello et al. 2015; He et al. 2016). More recently, it was shown that one of these enzymes named enolase 1 (ENO1) is allosterically inhibited by RNA binding, reducing glycolysis (Castello et al. 2012; Huppertz et al. 2022). This discovery reinforced the idea that RBPs are not just limited to regulating RNA fate, but they can also be modulated by their interaction with RNA through a process known as 'riboregulation' (Fig. 13.2b) (Castello et al. 2015; Hentze et al. 2018; Hentze and Preiss 2010; Huppertz et al. 2022). Riboregulation thus makes possible the existence of 'molecular sensors' that bridge RNA metabolism with other apparently unrelated cellular processes. The prevalence and diversity of riboregulatory processes are beautifully illustrated by PKR, which is activated upon binding to viral dsRNA and inhibits viral protein synthesis through the phosphorylation of eIF2$\alpha$ (Fig. 13.2c) (Jensen and Thomsen 2012; Taylor et al. 2005).

## The Essential Roles of Viral RBPs in Infection

Viruses themselves encode a wide range of RBPs, including RNA polymerases, RNA chaperones, helicases, capping and decapping enzymes, nucleases and other regulatory factors. Although viral RBPs often differ from cellular RBPs in architecture and structure, the molecular principles for RNA binding described above are preserved.

RNA viruses encode polymerases to multiply the viral genomes and/or produce mRNAs. RNA-dependent RNA polymerases (RDRP) can produce an RNA chain using RNA as a template and are encoded for most RNA viruses. Retroviruses, however, encode a reverse transcriptase (RT), that uses RNA or DNA as a template to produce DNA. RNA-dependent polymerases are error-prone and typically lack error correction activity, driving the high diversity and high evolution potential of RNA viruses. Viral RDRPs and RTs are supported by the activity of cellular and viral RNA helicases and chaperones that unwind and reconfigure local RNA structures to aid their processivity. This is illustrated by the RNA chaperone NS5A and helicase NS3 of the hepatitis C virus (HCV) (Li et al. 2021), the helicase NSP2 of alphavirus (Law et al. 2019), the nucleoprotein (NP) of influenza virus (Te Velthuis and Fodor 2016) and the nucleocapsid of human immunodeficiency virus (HIV) (Lyonnais et al. 2013).

Cellular mRNAs contain an m7G cap at the 5′ end that is important to promote translation initiation and prevent degradation via 5′-3′ exonucleases. The presence of a 5′ tri-phosphate end instead of an m7G cap activates a specialised set of antiviral sensors that specifically bind to this PAMP (Jensen and Thomsen 2012). To hinder their 5′ ends and minimise sensing, viruses typically add a cap or a protein to the 5′ end of their RNA. Picornaviruses, for example, covalently fuse a protein called Vpg to the 5′ end of their RNAs which is crucial for priming replication while hindering detrimental PAMPs at the 5′ end of the viral RNA (Paul and Wimmer 2015). Alternatively, viral RNA can be capped either by hijacking the host cell capping machinery (e.g. herpes virus), by 'stealing' cellular caps through cap snatching (e.g. influenza virus), or by using viral-encoded capping enzymes (e.g. alphaviruses and coronaviruses). The use of virus-encoded capping enzymes is a common strategy, the specifics of which vary depending on the virus. SARS-CoV-2 RNA capping is a multi-step process involving several viral proteins. It starts with the transfer of the RNA molecule to the N-terminus of the viral protein NSP9, creating a protein–RNA covalent intermediary (Park et al. 2022). This is followed by the transfer of the RNA to a terminal GDP molecule by NSP12 and the methylation of the resulting cap by NSP14 and NSP16 (Park et al. 2022). Recently, it was discovered that viral capping machinery can form complex pore structures, as is the case of NSP1 of alphavirus (Jones et al. 2021). NSP1 oligomerises into a dodecameric ring with 12 catalytic sites forming a pore on the membrane of the replication vesicles in the cytoplasm. Viral mRNA is capped as it traverses the membrane of the replication vesicle through the NSP1 pore (Jones et al. 2021). This mechanism ensures that only capped RNAs are transported to the cytosol, which, unlike the lumen of the replication vesicle, is enriched in antiviral factors. Influenza virus follows a different strategy to add the cap to their mRNAs. The viral RDRP is able to cleave the 5′ end of cellular mRNAs that are then used to prime the synthesis of viral mRNAs (De Vlugt et al. 2018; Te Velthuis and Fodor 2016; Walker and Fodor 2019). This mechanism, known as cap snatching, adds the cap structure together with a short sequence derived from the 5′ end of cellular mRNA to the beginning of the viral RNAs, making viral and cellular mRNAs virtually indistinguishable (Ho et al. 2020). While this mechanism is well understood for influenza A viruses in the nucleus, the mechanism of cap snatching remains poorly understood for other negative-stranded RNA viruses such as bunyaviruses (Olschewski et al. 2020).

Viral RBPs are also critical for the assembly of viral RNAs into viral particles. Recognition of viral RNA for packaging is typically initiated by the interaction of the viral capsid (e.g. alphaviruses and flaviviruses) or nucleocapsid (e.g. retroviruses and coronaviruses) with a cis-acting element, known as the packaging signal. In silico modelling and in vitro assays have revealed that this pioneering protein–RNA interaction is followed by a multimerisation of the viral capsid/nucleocapsid onto the viral RNA. It is believed that this multimerisation process is facilitated by secondary packaging signals placed downstream in the viral RNA (Borodavka et al. 2012; Chandler-Bostock et al. 2020; Twarock et al. 2018). The presence of multiple packaging signals in a viral genome has recently been shown in cellulo through the binding of the nucleocapsid region of the Gag polyprotein of HIV to the genomic RNA. This study shows that, in the cytosol, Gag not only engages with the known packaging signal at the 5′ end of the genomic RNA, but also with several discrete downstream regions (Kutluay and Bieniasz 2016). Interestingly, when the genomic RNA reaches the membrane, Gag's RNA-binding footprint begins to extend beyond these initial sites, posing a model in which few discrete interactions in the cytosol serve to recruit the genomic RNA to the virus assembly areas at the plasma membrane. Once there, the high local concentration of Gag triggers its multimerisation along the genomic RNA. Influenza virus follows an alternative approach for packaging, in which the nucleoprotein (NP) is deposited on the viral genomes concomitant to replication (probably via interaction with the cellular RDRP) without the participation of packaging signals (Te Velthuis and Fodor 2016; Walker and Fodor 2019). How the different genome segments are recruited and packaged into the virions remains poorly understood.

Viral RBPs can also establish crucial interactions with the host cell to bypass or hijack cellular processes. One of the best-characterised cases is the Rev protein of HIV (Truman et al. 2020). HIV expresses a plethora of viral mRNAs that are generated by host-mediated splicing of the genomic RNA. These spliced RNAs are exported from the nucleus to the cytoplasm using the canonical nuclear mRNA export pathway to produce the accessory and auxiliary proteins, including Rev. By contrast, unspliced (genomic) and partially spliced HIV RNAs accumulate in the nucleus as they are perceived as immature by the host cell since they still contain intronic-like sequences (Truman et al. 2020). Once expressed, Rev protein is imported into the nucleus, where it interacts with a cis-acting element present in the last one-third of the HIV genome called the Rev response element (RRE). Upon interaction with the RRE, Rev exports nuclear-retained unspliced and partially spliced HIV RNAs via the CRM1 pathway, which typically exports ribosomal subunits. HIV RNAs thus bypass nuclear retention through the engagement of a viral RBP with an alternative nuclear export pathway (Truman et al. 2020). Another example of a viral RBP that interferes with host processes is SARS-CoV-2 NSP1. NSP1 interacts with the mRNA channel of the ribosome to hamper the translation of cellular mRNAs (Banerjee et al. 2020; Schubert et al. 2020; Thoms et al. 2020). This process will be discussed in further detail later.

## Virus Infection Profoundly Remodels the Cellular RNA-Binding Proteome (RBPome)

Viral proteins are crucial for the viral life cycle. However, viruses rely on the host cell for a productive life cycle. This is due to the limited coding capacity of viral genomes and the need for complex machinery to mediate the metabolism of viral RNA and other processes of the viral life cycle. For example, the ribosome is formed of dozens of proteins and several rRNAs, which is well beyond the coding capacity of RNA (~10 Kb on average) and double-stranded DNA (~100 Kb on average) viruses. Ribosomal proteins are just a fraction of all RBPs required for RNA metabolism in the cell (Hentze et al. 2018; Jackson et al. 2010). Therefore, viruses are expected to rely on a wide range of cellular RBPs to aid their life cycle. Over the last decades, this idea has been proven true through individual cases, showing that

RBPs can repress or promote virus infection (Dicker et al. 2021; Garcia-Moreno et al. 2018).

Viral fitness will thus inevitably be conditioned by the configuration of RBPs present in the cell (Garcia-Moreno et al. 2019). For example, interferon treatment augments the levels of antiviral sensors, increasing the probability of these proteins engaging with the viral RNA early in infection (Shaw et al. 2021). The relative abundance of antiviral and 'pro-viral' RBPs at the moment of infection is thus expected to impact the ability of the virus to establish a productive infection, affecting cell tropism and susceptibility. The development of methods such as RNA interactome capture (RIC) will now enable us to approach this underexplored idea in a global and unbiased way.

To determine if virus infection affects the configuration of cellular RBPs in the cell, RIC (Fig. 13.2a) was applied to cells infected with Sindbis virus (SINV) or SARS-CoV-2 (Garcia-Moreno et al. 2019; Kamel et al. 2021a). RIC employs oligo (dT) to isolate RNA, and thus captures both cellular mRNAs as well as SINV and SARS-CoV-2 RNAs as they are polyadenylated. Strikingly, RIC revealed that both viruses induced a pervasive and largely overlapping remodelling of the RNA-binding proteome (RBPome). Analysis of the early times post-infection (4 h for SINV and 8 h for SARS-CoV-2) revealed discrete changes, whereas more than 200 RBPs exhibited differential association with RNA at late times post-infection (18 h for SINV and 24 h for SARS-CoV-2). One of the most exciting proteins upregulated at early times post-infection was the interferon-stimulated gene IFI16, which is a known ISG with antiviral activity. IFI16 was classified initially as a DNA sensor (Haronikova et al. 2016; Jonsson et al. 2017), but RIC analysis revealed that it is an early responder to RNA virus infection that interacts with RNA in infected cells (Garcia-Moreno et al. 2019; Kim et al. 2021). Other antiviral proteins such as GEMIN5, ZC3HAV1 (ZAP), TRIM25 and TRIM56 gradually increased their activity as infection progressed (Garcia-Moreno et al. 2019; Kamel et al. 2021a).

GEMIN5 is an unorthodox RBP that is involved to the assembly of spliceosomes. However, pioneering work from the Martinez-Salas lab showed that it has an additional function as a repressor of protein synthesis (Martinez-Salas et al. 2020). RIC revealed that GEMIN5 RNA-binding activity is strongly enhanced after SINV and SARS-CoV-2 infection, interacting with the 5′ end of viral RNA (Garcia-Moreno et al. 2019; Kamel et al. 2021a). Overexpression of this protein partially suppresses SINV gene expression suggesting that it is an antiviral factor. Interestingly, foot-and-mouth-disease virus (FMDV) protease L cleaves GEMIN5, disrupting its function as a repressor and repurposing its C-terminus to promote translation from the internal ribosome entry site (IRES) of its genomic RNA (Pineiro et al. 2013; Pineiro et al. 2012).

TRIM25 is an E3 ubiquitin ligase with known antiviral roles (Gack et al. 2008; Gack et al. 2007). However, the initial RIC studies revealed that it is an unorthodox RBP, providing a new angle on how this protein might exert its function in infected cells (Castello et al. 2012; Castello et al. 2016). Infection with both SINV and SARS-CoV-2 stimulates this protein in a time-dependent manner, reducing viral fitness (Garcia-Moreno et al. 2019; Kamel et al. 2021a). Recent work has shown that

TRIM25 E3 ubiquitin ligase activity is dependent on RNA binding, highlighting another example of riboregulation that is analogous to PKR (Fig. 13.2b and c) (Choudhury et al. 2017).

Activation of RBPs by infection is not only limited to antiviral proteins. For example, cyclophilin A (PPIA), the chaperone HSP90 and the exonuclease XRN1 are also activated as infection progresses, while being necessary for infection (Garcia-Moreno et al. 2019; Kamel et al. 2021a). PPIA is a peptidyl-prolyl cis-trans isomerase that plays important roles in HCV replication by regulating NS5A (Colpitts et al. 2020; Verdegem et al. 2011) and in HIV viral particle stability through its interaction with the capsid core (Kim et al. 2019; Liu et al. 2016). Inhibitors of PPIA such as cyclosporine and DEB025 have been explored as potential antivirals (Coelmont et al. 2010).

Interestingly, proteins upregulated by SINV and SARS-CoV-2 infection frequently redistribute from their native localisation (e.g. nucleus, cytosol, etc) to the membrane-bound compartments where these viruses replicate, suggesting that they directly engage with viral RNA (Garcia-Mo reno et al. 2019; Kamel et al. 2021a).

## *How Are Cellular RBPs Regulated by a Virus Infection?*

A plausible explanation for the observed RBP remodelling by infection is that the alterations in RBP activity are linked to changes in protein abundance. In other words, if there is more protein, there will be more binding and vice versa. To explore this possibility, the whole cell proteome of the SARS-CoV-2 and SINV-infected cells was analysed. Strikingly, no changes in abundance were observed for most regulated RBPs, suggesting that virus infection does not affect the RBPome by impacting the cellular proteome (Garcia-Moreno et al. 2019; Kamel et al. 2021a). RNA-binding activity can instead be modulated by the abundance of target RNAs (i.e. substrate). Supporting this idea, transcriptomic analysis of SARS-CoV-2 and SINV-infected cells revealed pervasive changes in RNA abundance. Particularly, housekeeping cellular mRNAs were progressively lost upon infection, while new RNAs emerged in the cell, including mRNAs encoding antiviral factors and the viral RNA. Indeed, SINV and SARS-CoV-2 RNAs become the most abundant polyadenylated RNAs in the cell, representing ~70% and ~ 20% of all the reads, respectively, at late times post-infection. The emergence of viral RNAs as predominant RNA species in the cell is likely to affect the configuration of active RBPs. In addition, cross-comparison of the responsive RBPs with analyses of phosphorylation and ubiquitination in SARS-CoV-2 infected cells suggested that post-translational modifications may also contribute to the regulation of cellular RBPs in virus-infected cells (Kamel et al. 2021a).

## New Methods to Uncover the Viral RNA Interactome

RIC revealed profound changes in the RBPome in response to virus infection, concomitant with pervasive alterations of the cellular transcriptome. However, it remained unclear which of the observed alterations involved a direct interaction with the viral RNA. Several methods were developed to examine the scope of cellular proteins that bind to viral RNA in virus-infected cells (Iselin et al. 2022). In brief, these methods have two key steps: (1) the covalent immobilisation of RBPs to RNA by crosslinking and (2) the capture of the RNA of interest (Fig. 13.3). Iterations of

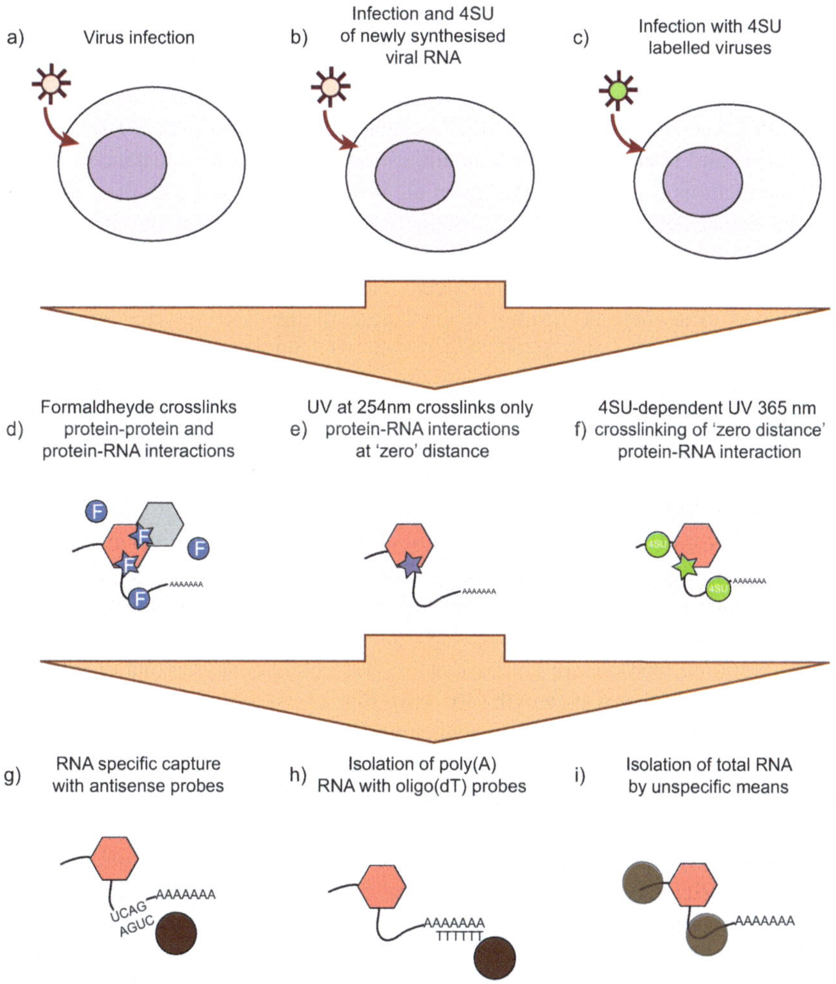


**Fig. 13.3** Different strategies to study the viral RNA interactome. Key steps of the protocol are described from a to h. The different approaches use precise combinations of these steps

this concept have led to complementary approaches to study the viral RNA interactome. These methods have been applied to several RNA viruses, including HIV, SARS-CoV-2, SINV, Zika and Dengue viruses, amongst others.

## *Methods to Uncover the Viral RNA Interactome*

The first step consists of the covalent immobilisation of RBPs onto the viral RNA to prevent loss or exchange with non-interactors upon lysis and following stringent washes. The most common approaches to this are UV irradiation at 254 nm ($UV_{254}$), UV irradiation at 365 nm $UV_{365}$, and formaldehyde crosslinking. $UV_{254}$ and formaldehyde crosslinking are not viral RNA specific; instead, they stabilise 'all' RNA–protein interactions in the cell, so they must be paired with an RNA capture method that selects specifically for the viral RNA (Fig. 13.4d and f with g and h). $UV_{254}$ crosslinking is employed by RAP-MS (Lee et al. 2021; Phillips et al. 2016; Schmidt et al. 2021) and results in covalent bonds forming between zero distance protein–RNA interactions (Fig. 13.3e) (Pashev et al. 1991). This, paired with the fact that it does not induce detectable levels of protein–protein crosslinking, makes it a valuable method for identifying only the proteins binding directly to RNA (Castello et al. 2016; Pashev et al. 1991). It is, however, relatively inefficient and has biases in crosslinking efficiency, according to the nature of the RNA–protein interaction. Formaldehyde, on the other hand, is a highly efficient crosslinker that works by forming methylene bridges between amino/imino groups that are present in amino acids and nucleic acids. It stabilises both protein–RNA and protein–protein interactions (Fig. 13.3d). Formaldehyde is employed by ChIRP-MS (Flynn et al. 2021; Labeau et al. 2022; Ooi et al. 2019), HyPR-MS (Knoener et al. 2021; Knoener et al. 2017) and CLAMP (Gebhart et al. 2020; LaPointe et al. 2018). It can be used to assess protein complexes interacting with vRNA, including indirect interactions. This comes at the cost of a higher rate of false positives due to the stabilisation of sub-stoichiometric, stochastic and long-distance interactions through protein–RNA and protein–protein bridges.

4-Thiouridine (4SU) is a photoactivatable nucleotide analogue that enables UV crosslinking at a higher wavelength (365 nm) (Fig. 13.3f) (Baltz et al. 2012; Castello et al. 2012; Hafner et al. 2010), a wavelength in which natural nucleotide exhibits negligible absorbance. 4SU is taken up by cells and incorporated into RNA, offering a useful tool to study newly synthesised RNA. In VIR-CLASP, viruses are produced in the presence of 4SU, so the virion-encapsidated genome contains the nucleotide analogue (Fig. 13.3c). These particles are used to infect cells that are cultured in the absence of 4SU. As 4SU is only present in the incoming RNA molecules, VIR-CLASP can be used to study the protein–RNA interactions occurring in the initial steps of infection (Kim et al. 2020). TUX-MS (Lenarcic et al. 2013; Viktorovskaya et al. 2016), vRIC (Kamel et al. 2021a; Kamel et al. 2021b) and CLAMP (Gebhart et al. 2020; LaPointe et al. 2018), by contrast, focus on the post-replication steps of the viral life cycle by exploiting the relative insensitivity of viral

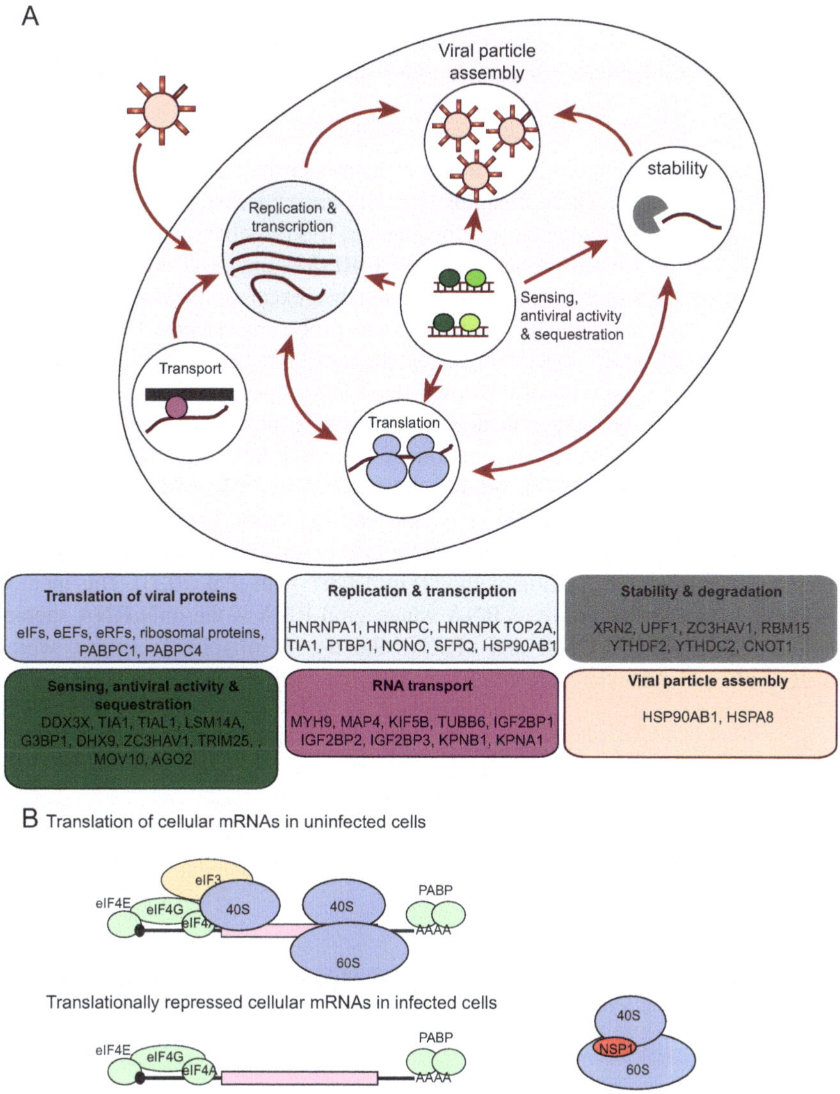


**Fig. 13.4** (**a**) The known regulatory roles of the core viral RNA interactome. The core viral RNA interactome is composed by proteins identified in the RNA interactomes of coronaviruses, alphaviruses and flaviviruses. The association of these proteins with a wide range of viral RNAs suggests that they are endowed with master regulatory roles in the viral life cycle. (**b**) NSP1 represses translation initiation in SARS-CoV-2 infected cells following a mechanism that is independent from the association of the cap and poly(A) binding complexes

RDRP to transcription inhibitors such as Actinomycin D and flavopiridol. By using host transcription inhibitors, it is possible to force the 4SU to be exclusively incorporated into viral RNA. TUX-MS and CLAMP use Actinomycin D, which can result in reduced viral replication and significant cytotoxicity (Sawicki and Godman 1971). vRIC uses flavopiridol instead, which is specific to RNA polymerase II and does not have any detectable effect on the cell or virus (Bensaude 2011; Kamel et al. 2021b).

For methods that do not use specific labelling and crosslinking of the viral RNA, it becomes essential to incorporate a downstream step that captures it specifically (Fig. 13.3e, f, g and h). These methods typically use antisense probes to hybridise with the viral RNA in a sequence-specific manner. It is possible to use a single probe as in HyPR-MS or a multi-probe set as in ChIRP-MS and RAP-MS (Iselin et al. 2022). As a rule of thumb, single-probe approaches excel at specificity at the cost of sensitivity, while the opposite is true for multi-probe approaches. The benefits of using one or the other will depend on the abundance of the target RNA, the quality of the probes and the sensitivity of the downstream mass spectrometry approach.

If viral RNA has been specifically labelled with a photoactivatable nucleotide such as 4SU (Fig. 13.3f), then isolation of the viral RNA can be non-specific (i.e. isolating all RNA species), semi-specific (e.g. using oligo (dT) to enrich for polyadenylated RNA) or specific (e.g. antisense probes) (Fig. 13.3g, h and i). VIR-CLASP, for example, relies on solid-phase reversible immobilisation beads, which is a total RNA purification method (Fig. 13.3i) (Kim et al. 2020). This method will purify a complex mixture of RNA where viral RNA is the only RNA species labelled with 4SU. TUX-MS and vRIC make use of oligo(dT) capture, which isolates only polyadenylated RNA (Fig. 13.3h) (Kamel et al. 2021a; Kamel et al. 2021b; Lenarcic et al. 2013; Viktorovskaya et al. 2016). As viral RNA is typically the most abundant polyadenylated RNA in the cell, this approach enriches for viral RNA first by 4SU-specific crosslinked followed by the subsequent poly (A) selection, boosting the signal-to-noise ratio. However, oligo(dT) capture would only work for polyadenylated viruses and should therefore be replaced by other RNA capture approaches if working with non-polyadenylated viral RNAs.

## Viral RNA Is a Hub for Critical and Conserved Host–Virus Interactions

Cellular RBPs are critical for infection (Fig. 13.4), however, the scope of RBPs involved in virus infection and the extent to which they overlap across viral models remains unknown. There have been 21 vRNA interactomes established for 11 viruses from six viral families, with higher coverage for the Togaviridae, Coronaviridae and Flaviviridae families. The combination of these datasets resulted in nearly 2000 proteins, half of which were only identified in a single experiment. Since these methods have been applied at least twice to each virus, on average, these uniquely identified proteins are likely experimental noise (Iselin et al. 2022). However, numerous RBPs were present in the dataset of several viruses belonging to different families, suggesting that they have pan-regulatory roles. Particularly, ~200 RBPs were present in viral RNA interactomes of the three best-represented viral families (i.e. Coronaviridae, Togaviridae and Flaviviridae), being referred to here as the 'core viral RNA interactome'. Importantly, 68% of the RBPs within the core interactome have been previously associated with virus infection, and 25% lack known RNA-binding domains and thus display unorthodox RNA binding modes. These include the chaperone HSP90AB1, whose RNA-binding surfaces have been recently identified (Castello et al. 2016; Dicker et al. 2021; Garcia-Moreno et al. 2019). 73 RBPs within the core RNA interactome lack connections with virus infection, offering new avenues to expand our understanding of the viral life cycle.

## *RBPs Involved in Innate Immunity*

The core viral RNA interactome includes several RBPs that have been involved in antiviral processes, including stress granule formation (G3BP1 and TIA1) and viral RNA sensing (DDX3, DHX9 and MOV10) (Fig. 13.4a). Stress granules (SGs) are membrane-less cellular organelles that are formed by liquid-to-liquid phase transition. SGs are assembled in infected cells and are known to sequester viral RNA in a repressed state, playing an antiviral role (Onomoto et al. 2014). However, viruses have developed mechanisms to prevent the formation of SGs, often mediated by inhibitory interactions with viral proteins (e.g. alphaviral NSP3) (Lu et al. 2021) and/or viral RNA (flavivirus) (Bidet et al. 2014). G3BP1 has been shown to interact with many viral RNAs suggesting that SG inhibition by riboregulation might be a broadly used mechanism across viruses.

## *RNA Stability and Degradation*

The core viral RNA interactome contains several RBPs that have been previously linked to the control of RNA stability and degradation (Fig. 13.4a). This is the case for ELAVL1 (also known as HuR) and PCBP2, which stabilise their target RNAs mitigating degradation. Conversely, the components of the core viral RNA interactome ZC3HAV1 (also known as ZAP), UPF1 and the exonuclease XRN2 are known to trigger RNA degradation. The balance of these stability and pro-degradation factors on the viral RNA is likely to define whether it is degraded or remains stable.

## *Protein Synthesis*

Many translation initiation factors, ribosomal proteins and ribosome-associated proteins are enriched in the core viral RNA interactomes (Fig. 13.4a). This is expected as viral RNA is heavily translated in virus-infected cells to produce the viral progeny. Interestingly, the cap-binding protein eIF4E is absent in SINV and SARS-CoV-2 RNA interactomes despite viral RNAs being capped (Iselin et al. 2022; Kamel et al. 2021a). Conversely, initiation factors involved in the downstream step such as bridging the viral mRNA and the ribosome (eIF4G, eIF3) and unwinding secondary structure (eIF4A and eIF4B) are detected. These findings suggest that SINV and SARS-CoV-2 mRNAs may initiate translation in an eIF4E-independent manner, probably using a non-canonical cap-binding protein. This agrees with previous results showing that SINV is insensitive to conditions that impair the eIF4E-dependent translation of capped mRNAs (Carrasco et al. 2018).

Interestingly, eIF4E, eIF4G, eIF4A and PABP displayed enhanced association with cellular mRNAs in SARS-CoV-2 infected cells, while eIF3 and the ribosome were downregulated (Fig. 13.4b) (Kamel et al. 2021a). These alterations correlate with the shut off of cellular protein synthesis, which is mediated by the association of the viral protein NSP1 with the mRNA binding channel of the ribosome and the induction of translation termination (Banerjee et al. 2020; Bujanic et al. 2022; Rao et al. 2021; Schubert et al. 2020; Simeoni et al. 2021; Thoms et al. 2020). Altogether, the data depicts a scenario in which cellular mRNAs associate with the cap (eIF4E-eIF4G-eIF4A) and poly(A) (PABP) binding machinery, but these fail to recruit the eIF3 and 40S ribosomal complex as ribosomes are 'locked' by NSP1, leading to a translationally repressed state. It has been proposed that NSP1 can associate with SARS-CoV-2 RNAs and promote their translation; however, this is controversial. While some studies described the interaction of NSP1 with SARS-CoV-2 RNAs (Banerjee et al. 2020; Flynn et al. 2021; Schmidt et al. 2021), others revealed that it mainly interacts with cellular mRNAs (Kamel et al. 2021a). Further experiments will be required to identify the substrates of NSP1 with molecular detail.

## Outlook

While our knowledge of the composition of viral RNPs is in its infancy, important developments have revealed a new universe of host–virus interactions involving cellular RBPs. In the future, the analysis of the RNA interactome with other viruses, different cell types and experimental conditions will provide new insights into the plasticity and dynamics of viral RNPs and how the cellular state and intrinsic properties of the viral RNA dictate composition. Moreover, the discovery of novel host–virus interactions that span virus species and families opens new avenues of research to improve our understanding of the molecular principles of virus infection and to identify novel antivirals with broad-spectrum potential.

**Acknowledgements** A.C. is supported by the ERC Consolidator Grant vRNP-capture number 101001634. L.I. is funded by the BBSRC DTP scholarship number BB/M011224/1.

## References

Bach-Pages M, Homma F, Kourelis J, Kaschani F, Mohammed S, Kaiser M, van der Hoorn RAL, Castello A, Preston GM (2020) Discovering the RNA-binding proteome of plant leaves with an improved RNA Interactome capture method. Biomol Ther 10(4):661

Baltz AG, Munschauer M, Schwanhausser B, Vasile A, Murakawa Y, Schueler M, Youngs N, Penfold-Brown D, Drew K, Milek M et al (2012) The mRNA-bound proteome and its global occupancy profile on protein-coding transcripts. Mol Cell 46:674–690

Banerjee AK, Blanco MR, Bruce EA, Honson DD, Chen LM, Chow A, Bhat P, Ollikainen N, Quinodoz SA, Loney C et al (2020) SARS-CoV-2 disrupts splicing, translation, and protein trafficking to suppress host defenses. Cell 183:1325–1339.e1321

Beckmann BM, Horos R, Fischer B, Castello A, Eichelbaum K, Alleaume AM, Schwarzl T, Curk T, Foehr S, Huber W et al (2015) The RNA-binding proteomes from yeast to man harbour conserved enigmRBPs. Nat Commun 6:10127

Bensaude O (2011) Inhibiting eukaryotic transcription: which compound to choose? How to evaluate its activity? Transcription 2:103–108

Bidet K, Dadlani D, Garcia-Blanco MA (2014) G3BP1, G3BP2 and CAPRIN1 are required for translation of interferon stimulated mRNAs and are targeted by a dengue virus non-coding RNA. PLoS Pathog 10:e1004242

Borodavka A, Tuma R, Stockley PG (2012) Evidence that viral RNAs have evolved for efficient, two-stage packaging. Proc Natl Acad Sci U S A 109:15769–15774

Bujanic L, Shevchuk O, von Kugelgen N, Kalinina A, Ludwik K, Koppstein D, Zerna N, Sickmann A, Chekulaeva M (2022) The key features of SARS-CoV-2 leader and NSP1 required for viral escape of NSP1-mediated repression. RNA 28:766–779

Carrasco L, Sanz MA, Gonzalez-Almela E (2018) The regulation of translation in alphavirus-infected cells. Viruses 10(2):70

Castello A, Fischer B, Eichelbaum K, Horos R, Beckmann BM, Strein C, Davey NE, Humphreys DT, Preiss T, Steinmetz LM et al (2012) Insights into RNA biology from an atlas of mammalian mRNA-binding proteins. Cell 149:1393–1406

Castello A, Hentze MW, Preiss T (2015) Metabolic enzymes enjoying new partnerships as RNA-binding proteins. Trends Endocrinol Metab 26:746–757

Castello A, Horos R, Strein C, Fischer B, Eichelbaum K, Steinmetz LM, Krijgsveld J, Hentze MW (2016) Comprehensive identification of RNA-binding proteins by RNA Interactome capture. Methods Mol Biol 1358:131–139

Chandler-Bostock R, Mata CP, Bingham RJ, Dykeman EC, Meng B, Tuthill TJ, Rowlands DJ, Ranson NA, Twarock R, Stockley PG (2020) Assembly of infectious enteroviruses depends on multiple, conserved genomic RNA-coat protein contacts. PLoS Pathog 16:e1009146

Choudhury NR, Heikel G, Trubitsyna M, Kubik P, Nowak JS, Webb S, Granneman S, Spanos C, Rappsilber J, Castello A et al (2017) RNA-binding activity of TRIM25 is mediated by its PRY/SPRY domain and is required for ubiquitination. BMC Biol 15:105

Coelmont L, Hanoulle X, Chatterji U, Berger C, Snoeck J, Bobardt M, Lim P, Vliegen I, Paeshuyse J, Vuagniaux G et al (2010) DEB025 (Alisporivir) inhibits hepatitis C virus replication by preventing a cyclophilin a induced cis-trans isomerisation in domain II of NS5A. PLoS One 5:e13687

Colpitts CC, Ridewood S, Schneiderman B, Warne J, Tabata K, Ng CF, Bartenschlager R, Selwood DL, Towers GJ (2020) Hepatitis C virus exploits cyclophilin a to evade PKR. elife 9:e52237

De Vlugt C, Sikora D, Pelchat M (2018) Insight into influenza: a virus cap-snatching. Viruses 10(11):641

Dias Junior AG, Sampaio NG, Rehwinkel J (2019) A balancing act: MDA5 in antiviral immunity and autoinflammation. Trends Microbiol 27:75–85

Dicker K, Jarvelin AI, Garcia-Moreno M, Castello A (2021) The importance of virion-incorporated cellular RNA-binding proteins in viral particle assembly and infectivity. Semin Cell Dev Biol 111:108–118

Flynn RA, Belk JA, Qi Y, Yasumoto Y, Wei J, Alfajaro MM, Shi Q, Mumbach MR, Limaye A, DeWeirdt PC et al (2021) Discovery and functional interrogation of SARS-CoV-2 RNA-host protein interactions. Cell 184:2394–2411e2316

Gack MU, Shin YC, Joo CH, Urano T, Liang C, Sun L, Takeuchi O, Akira S, Chen Z, Inoue S et al (2007) TRIM25 RING-finger E3 ubiquitin ligase is essential for RIG-I-mediated antiviral activity. Nature 446:916–920

Gack MU, Kirchhofer A, Shin YC, Inn KS, Liang C, Cui S, Myong S, Ha T, Hopfner KP, Jung JU (2008) Roles of RIG-I N-terminal tandem CARD and splice variant in TRIM25-mediated antiviral signal transduction. Proc Natl Acad Sci U S A 105:16743–16748

Garcia-Moreno M, Jarvelin AI, Castello A (2018) Unconventional RNA-binding proteins step into the virus-host battlefront. Wiley Interdiscip Rev RNA 9:e1498

Garcia-Moreno M, Noerenberg M, Ni S, Jarvelin AI, Gonzalez-Almela E, Lenz CE, Bach-Pages M, Cox V, Avolio R, Davis T et al (2019) System-wide profiling of RNA-binding proteins uncovers key regulators of virus infection. Mol Cell 74:196–211e111

Gebhart NN, Hardy RW, Sokoloski KJ (2020) Comparative analyses of alphaviral RNA:Protein complexes reveals conserved host-pathogen interactions. PLoS One 15:e0238254

Hafner M, Landthaler M, Burger L, Khorshid M, Hausser J, Berninger P, Rothballer A, Ascano M Jr, Jungkamp AC, Munschauer M et al (2010) Transcriptome-wide identification of RNA-binding protein and microRNA target sites by PAR-CLIP. Cell 141:129–141

Haronikova L, Coufal J, Kejnovska I, Jagelska EB, Fojta M, Dvorakova P, Muller P, Vojtesek B, Brazda V (2016) IFI16 preferentially binds to DNA with Quadruplex structure and enhances DNA Quadruplex formation. PLoS One 11:e0157156

He C, Sidoli S, Warneford-Thomson R, Tatomer DC, Wilusz JE, Garcia BA, Bonasio R (2016) High-resolution mapping of RNA-binding regions in the nuclear proteome of embryonic stem cells. Mol Cell 64:416–430

Hentze MW, Preiss T (2010) The REM phase of gene regulation. Trends Biochem Sci 35:423–426

Hentze MW, Muckenthaler MU, Galy B, Camaschella C (2010) Two to tango: regulation of mammalian iron metabolism. Cell 142:24–38

Hentze MW, Castello A, Schwarzl T, Preiss T (2018) A brave new world of RNA-binding proteins. Nat Rev Mol Cell Biol 19:327–341

Ho JSY, Angel M, Ma Y, Sloan E, Wang G, Martinez-Romero C, Alenquer M, Roudko V, Chung L, Zheng S et al (2020) Hybrid gene origination creates human-virus chimeric proteins during infection. Cell 181:1502–1517.e1523

Huppertz I, Perez-Perri JI, Mantas P, Sekaran T, Schwarzl T, Russo F, Ferring-Appel D, Koskova Z, Dimitrova-Paternoga L, Kafkia E et al (2022) Riboregulation of enolase 1 activity controls glycolysis and embryonic stem cell differentiation. Mol Cell 82:2666–2680.e2611

Iselin L, Palmalux N, Kamel W, Simmonds P, Mohammed S, Castello A (2022) Uncovering viral RNA-host cell interactions on a proteome-wide scale. Trends Biochem Sci 47:23–38

Jackson RJ, Hellen CU, Pestova TV (2010) The mechanism of eukaryotic translation initiation and principles of its regulation. Nat Rev Mol Cell Biol 11:113–127

Jarvelin AI, Noerenberg M, Davis I, Castello A (2016) The new (dis)order in RNA regulation. Cell Commun Signal 14:9

Jensen S, Thomsen AR (2012) Sensing of RNA viruses: a review of innate immune receptors involved in recognizing RNA virus invasion. J Virol 86:2900–2910

Jones R, Bragagnolo G, Arranz R, Reguera J (2021) Capping pores of alphavirus nsP1 gate membranous viral replication factories. Nature 589:615–619

Jonsson KL, Laustsen A, Krapp C, Skipper KA, Thavachelvam K, Hotter D, Egedal JH, Kjolby M, Mohammadi P, Prabakaran T et al (2017) IFI16 is required for DNA sensing in human macrophages by promoting production and function of cGAMP. Nat Commun 8:14391

Kamel W, Noerenberg M, Cerikan B, Chen H, Jarvelin AI, Kammoun M, Lee JY, Shuai N, Garcia-Moreno M, Andrejeva A et al (2021a) Global analysis of protein-RNA interactions in SARS-CoV-2-infected cells reveals key regulators of infection. Mol Cell 81:2851–2867.e2857

Kamel W, Ruscica V, Garcia-Moreno M, Palmalux N, Iselin L, Hannan M, Jaervelin AI, Noerenberg M, Moore S, Merits A et al (2021b) Compositional analysis of Sindbis virus ribonucleoproteins reveals an extensive co-opting of key nuclear RNA-binding proteins. BioRxiv

Kim K, Dauphin A, Komurlu S, McCauley SM, Yurkovetskiy L, Carbone C, Diehl WE, Strambio-De-Castillia C, Campbell EM, Luban J (2019) Cyclophilin a protects HIV-1 from restriction by human TRIM5alpha. Nat Microbiol 4:2044–2051

Kim B, Arcos S, Rothamel K, Jian J, Rose KL, McDonald WH, Bian Y, Reasoner S, Barrows NJ, Bradrick S et al (2020) Discovery of widespread host protein interactions with the pre-replicated genome of CHIKV using VIR-CLASP. Mol Cell 78:624–640.e627

Kim B, Arcos S, Rothamel K, Ascano M (2021) Viral crosslinking and solid-phase purification enables discovery of ribonucleoprotein complexes on incoming RNA virus genomes. Nat Protoc 16:516–531

Knoener RA, Becker JT, Scalf M, Sherer NM, Smith LM (2017) Elucidating the in vivo interactome of HIV-1 RNA by hybridization capture and mass spectrometry. Sci Rep 7:16965

Knoener R, Evans E 3rd, Becker JT, Scalf M, Benner B, Sherer NM, Smith LM (2021) Identification of host proteins differentially associated with HIV-1 RNA splice variants. elife 10: e62470

Kramer K, Sachsenberg T, Beckmann BM, Qamar S, Boon KL, Hentze MW, Kohlbacher O, Urlaub H (2014) Photo-cross-linking and high-resolution mass spectrometry for assignment of RNA-binding sites in RNA-binding proteins. Nat Methods 11:1064–1070

Kutluay SB, Bieniasz PD (2016) Analysis of HIV-1 gag-RNA interactions in cells and Virions by CLIP-seq. Methods Mol Biol 1354:119–131

Labeau A, Fery-Simonian L, Lefevre-Utile A, Pourcelot M, Bonnet-Madin L, Soumelis V, Lotteau V, Vidalain PO, Amara A, Meertens L (2022) Characterization and functional interrogation of the SARS-CoV-2 RNA interactome. Cell Rep 39:110744

LaPointe AT, Moreno-Contreras J, Sokoloski KJ (2018) Increasing the capping efficiency of the Sindbis virus nsP1 protein negatively affects viral infection. mBio 9(6):e02342-18

Law YS, Utt A, Tan YB, Zheng J, Wang S, Chen MW, Griffin PR, Merits A, Luo D (2019) Structural insights into RNA recognition by the chikungunya virus nsP2 helicase. Proc Natl Acad Sci U S A 116:9558–9567

Lee S, Lee YS, Choi Y, Son A, Park Y, Lee KM, Kim J, Kim JS, Kim VN (2021) The SARS-CoV-2 RNA interactome. Mol Cell 81:2838–2850.e2836

Lenarcic EM, Landry DM, Greco TM, Cristea IM, Thompson SR (2013) Thiouracil cross-linking mass spectrometry: a cell-based method to identify host factors involved in viral amplification. J Virol 87:8697–8712

Li HC, Yang CH, Lo SY (2021) Hepatitis C viral replication complex. Viruses 13(3):520

Liu C, Perilla JR, Ning J, Lu M, Hou G, Ramalho R, Himes BA, Zhao G, Bedwell GJ, Byeon IJ et al (2016) Cyclophilin a stabilizes the HIV-1 capsid through a novel non-canonical binding site. Nat Commun 7:10714

Lu X, Alam U, Willis C, Kennedy D (2021) Role of chikungunya nsP3 in regulating G3BP1 activity, stress granule formation and drug efficacy. Arch Med Res 52:48–57

Lunde BM, Moore C, Varani G (2007) RNA-binding proteins: modular design for efficient function. Nat Rev Mol Cell Biol 8:479–490

Lyonnais S, Gorelick RJ, Heniche-Boukhalfa F, Bouaziz S, Parissi V, Mouscadet JF, Restle T, Gatell JM, Le Cam E, Mirambeau G (2013) A protein ballet around the viral genome orchestrated by HIV-1 reverse transcriptase leads to an architectural switch: from nucleocapsid-condensed RNA to Vpr-bridged DNA. Virus Res 171:287–303

Martinez-Salas E, Embarc-Buh A, Francisco-Velilla R (2020) Emerging roles of Gemin5: from snRNPs assembly to translation control. Int J Mol Sci 21(11):3868

Matia-Gonzalez AM, Laing EE, Gerber AP (2015) Conserved mRNA-binding proteomes in eukaryotic organisms. Nat Struct Mol Biol 22:1027–1033

Muckenthaler MU, Rivella S, Hentze MW, Galy B (2017) A red carpet for iron metabolism. Cell 168:344–361

Olschewski S, Cusack S, Rosenthal M (2020) The cap-snatching mechanism of Bunyaviruses. Trends Microbiol 28:293–303

Onomoto K, Yoneyama M, Fung G, Kato H, Fujita T (2014) Antiviral innate immunity and stress granule responses. Trends Immunol 35:420–428

Ooi YS, Majzoub K, Flynn RA, Mata MA, Diep J, Li JK, van Buuren N, Rumachik N, Johnson AG, Puschnik AS et al (2019) An RNA-centric dissection of host complexes controlling flavivirus infection. Nat Microbiol 4:2369–2382

Park GJ, Osinski A, Hernandez G, Eitson JL, Majumdar A, Tonelli M, Henzler-Wildman K, Pawlowski K, Chen Z, Li Y et al (2022) The mechanism of RNA capping by SARS-CoV-2. Nature 609:793–800

Pashev IG, Dimitrov SI, Angelov D (1991) Crosslinking proteins to nucleic acids by ultraviolet laser irradiation. Trends Biochem Sci 16:323–326

Paul AV, Wimmer E (2015) Initiation of protein-primed picornavirus RNA synthesis. Virus Res 206:12–26

Phillips SL, Soderblom EJ, Bradrick SS, Garcia-Blanco MA (2016) Identification of proteins bound to dengue viral RNA in vivo reveals new host proteins important for virus replication. mBio 7: e01865–e01815

Pineiro D, Ramajo J, Bradrick SS, Martinez-Salas E (2012) Gemin5 proteolysis reveals a novel motif to identify L protease targets. Nucleic Acids Res 40:4942–4953

Pineiro D, Fernandez N, Ramajo J, Martinez-Salas E (2013) Gemin5 promotes IRES interaction and translation control through its C-terminal region. Nucleic Acids Res 41:1017–1028

Queiroz RML, Smith T, Villanueva E, Marti-Solano M, Monti M, Pizzinga M, Mirea DM, Ramakrishna M, Harvey RF, Dezi V et al (2019) Comprehensive identification of RNA-protein interactions in any organism using orthogonal organic phase separation (OOPS). Nat Biotechnol 37:169–178

Rao S, Hoskins I, Tonn T, Garcia PD, Ozadam H, Sarinay Cenik E, Cenik C (2021) Genes with 5′ terminal oligopyrimidine tracts preferentially escape global suppression of translation by the SARS-CoV-2 Nsp1 protein. RNA 27:1025–1045

Rehwinkel J, Gack MU (2020) RIG-I-like receptors: their regulation and roles in RNA sensing. Nat Rev Immunol 20:537–551

Sawicki SG, Godman GC (1971) On the differential cytotoxicity of actinomycin D. J Cell Biol 50: 746–761

Schmidt N, Lareau CA, Keshishian H, Ganskih S, Schneider C, Hennig T, Melanson R, Werner S, Wei Y, Zimmer M et al (2021) The SARS-CoV-2 RNA-protein interactome in infected human cells. Nat Microbiol 6:339–353

Schubert K, Karousis ED, Jomaa A, Scaiola A, Echeverria B, Gurzeler LA, Leibundgut M, Thiel V, Muhlemann O, Ban N (2020) SARS-CoV-2 Nsp1 binds the ribosomal mRNA channel to inhibit translation. Nat Struct Mol Biol 27:959–966

Shaw AE, Rihn SJ, Mollentze N, Wickenhagen A, Stewart DG, Orton RJ, Kuchi S, Bakshi S, Collados MR, Turnbull ML et al (2021) The antiviral state has shaped the CpG composition of the vertebrate interferome to avoid self-targeting. PLoS Biol 19:e3001352

Simeoni M, Cavinato T, Rodriguez D, Gatfield D (2021) I(nsp1)ecting SARS-CoV-2-ribosome interactions. Commun Biol 4:715

Tagawa T, Serquina A, Kook I, Ziegelbauer J (2021) Viral non-coding RNAs: stealth strategies in the tug-of-war between humans and herpesviruses. Semin Cell Dev Biol 111:135–147

Taylor SS, Haste NM, Ghosh G (2005) PKR and eIF2alpha: integration of kinase dimerization, activation, and substrate docking. Cell 122:823–825

Te Velthuis AJ, Fodor E (2016) Influenza virus RNA polymerase: insights into the mechanisms of viral RNA synthesis. Nat Rev Microbiol 14:479–493

Thoms M, Buschauer R, Ameismeier M, Koepke L, Denk T, Hirschenberger M, Kratzat H, Hayn M, Mackens-Kiani T, Cheng J et al (2020) Structural basis for translational shutdown and immune evasion by the Nsp1 protein of SARS-CoV-2. Science 369:1249–1255

Trendel J, Schwarzl T, Horos R, Prakash A, Bateman A, Hentze MW, Krijgsveld J (2019) The human RNA-binding proteome and its dynamics during translational arrest. Cell 176:391–403. e319

Truman CT, Jarvelin A, Davis I, Castello A (2020) HIV Rev-isited. Open Biol 10:200320

Twarock R, Bingham RJ, Dykeman EC, Stockley PG (2018) A modelling paradigm for RNA virus assembly. Curr Opin Virol 31:74–81

Verdegem D, Badillo A, Wieruszeski JM, Landrieu I, Leroy A, Bartenschlager R, Penin F, Lippens G, Hanoulle X (2011) Domain 3 of NS5A protein from the hepatitis C virus has intrinsic alpha-helical propensity and is a substrate of cyclophilin a. J Biol Chem 286:20441–20454

Viktorovskaya OV, Greco TM, Cristea IM, Thompson SR (2016) Identification of RNA binding proteins associated with dengue virus RNA in infected cells reveals temporally distinct host factor requirements. PLoS Negl Trop Dis 10:e0004921

Vladimer GI, Gorna MW, Superti-Furga G (2014) IFITs: emerging roles as key anti-viral proteins. Front Immunol 5:94

Walker AP, Fodor E (2019) Interplay between influenza virus and the host RNA polymerase II transcriptional machinery. Trends Microbiol 27:398–407

# Chapter 14
# Influenza A Virus: Cellular Entry

**Yasuyuki Miyake, Yuya Hara, Miki Umeda, and Indranil Banerjee**

**Abstract** The frequent emergence of pathogenic viruses with pandemic potential has posed a significant threat to human health and economy, despite enormous advances in our understanding of infection mechanisms and devising countermeasures through developing various prophylactic and therapeutic strategies. The recent coronavirus disease (COVID-19) pandemic has re-emphasised the importance of rigorous research on virus infection mechanisms and highlighted the need for our preparedness for potential pandemics. Although viruses cannot self-replicate, they tap into host cell factors and processes for their entry, propagation and dissemination. Upon entering the host cells, viruses ingeniously utilise the innate biological functions of the host cell to replicate themselves and maintain their existence in the hosts. Influenza A virus (IAV), which has a negative-sense, single-stranded RNA as its genome, is no exception. IAVs are enveloped viruses with a lipid bilayer derived from the host cell membrane and have a surface covered with the spike glycoprotein haemagglutinin (HA) and neuraminidase (NA). Viral genome is surrounded by an M1 shell, forming a "capsid" in the virus particle. IAV particles use HA to recognise sialic acids on the cell surface of lung epithelial cells for their attachment. After attachment to the cell surface, IAV particles are endocytosed and sorted into the early endosomes. Subsequently, as the early endosomes mature into late endosomes, the endosomal lumen becomes acidified, and the low pH of the late endosomes induces conformational reaggangements in the HA to initiate fusion between the

Y. Miyake (✉)
Department of Virology, Nagoya University Graduate School of Medicine, Nagoya, Japan

Institute for Advanced Research (IAR), Nagoya University, Nagoya, Japan
e-mail: yasumiyake@med.nagoya-u.ac.jp

Y. Hara · M. Umeda
Department of Virology, Nagoya University Graduate School of Medicine, Nagoya, Japan
e-mail: yuya.hara0403@med.nagoya-u.ac.jp; m.umeda@med.nagoya-u.ac.jp

I. Banerjee (✉)
Department of Biological Sciences, Indian Institute of Science Education and Research, Mohali (IISER Mohali), Mohali, India
e-mail: indranil@iisermohali.ac.in

 
S. Vijayakrishnan et al. (eds.), *Virus Infected Cells*, Subcellular Biochemistry 106,
https://doi.org/10.1007/978-3-031-40086-5_14

endosomal and viral membranes. Upon fusion, the viral capsid disintegrates and the viral ribonucleoprotein (vRNP) complexes containing the viral genome are released into the cytosol. The process of viral capsid disintegration is called "uncoating". After successful uncoating, the vRNPs are imported into the nucleus by importin α/β (IMP α/β), where viral replication and transcription take place and the new vRNPs are assembled. Recently, we have biochemically elucidated the molecular mechanisms of the processes of viral capsid uncoating subsequent viral genome dissociation. In this chapter, we present the molecular details of the viral uncoating process.

**Keywords** Influenza A virus · In Vitro uncoating · HDAC6 · Ubiquitin · TNPO1

## Molecular Mechanisms of Influenza A Virus Entry

To establish successful infection in the target cell, virus particles need to deliver their genome and the accessory proteins for replication and transcription. Due to small viral particle sizes, viruses can only accommodate a limited number of protein-producing genes to help with their multiplication and spread. Therefore, they largely rely on the host proteins and processes to carry out essential functions. Virus particles produced from infected cells need to be stable enough to endure the challenges in the hostile extracellular environment for their survival. The stability of the virion is imparted by strong macromolecular interactions among the viral genome and accessory proteins during viral packaging, ensuring highly ordered condensation of the viral components into an extremely compact structure. However, when a highly stable virus in the extracellular environment finds its target cell, the macromolecular interactions that lead to its assembly need to be reversed to release its genome into the cell for replication and viral protein synthesis. Growing evidence suggests that viruses alone are unable to undergo self-disassembly in the host cell, therefore, they exploit various cellular factors to accomplish uncoating (Edinger et al. 2015; He et al. 2013; Huotari et al. 2012; Konig et al. 2010; Pohl et al. 2014; Su et al. 2013; Yanguez et al. 2018; Banerjee et al. 2014; Gschweitl et al. 2016).

Upon binding to the cell surface, IAV enters the target cell via either receptor-mediated endocytosis (Eierhoff et al. 2010; Fujioka et al. 2018) or macropinocytosis (de Vries et al. 2011). Receptor-mediated endocytosis is a process by which cells internalise specific proteins, metabolites, hormones, viruses, etc., by inwardly budding the plasma membrane and forming endocytic vesicles, whereas in macropinocytosis, cells engulf non-specific soluble materials in bulk into large uncoated vesicles called macropinosomes (Mercer et al. 2010). After entry into the host cells, IAV is sorted into the mildly acidic, Rab5-positive early endosomes (Chavrier et al. 1990; Gorvel et al. 1991). The activity of vacuolar ATPase (vATPase) present on the early endosomes causes a gradual drop in the luminal pH, and as the early endosomes mature into Rab7-positive late endosomes, the pH drops to 5.0 (Liang et al. 2016). The low pH in the late endosomes induces rapid structural rearrangements in the HA, exposing the fusion peptide which mediates the fusion between the viral membrane and the limiting membrane of the late endosome (Harrison 2008, 2015; Chlanda et al. 2016; Kanaseki et al. 1997; Lee 2010).

Although some viruses can enter the host cells by penetrating at the plasma membrane, the vast majority of enveloped viruses depend on the host cell endocytic pathways for entry and subsequent penetration at endosomes for transmission of their genome into the host cells (Yamauchi and Greber 2016; Mercer et al. 2010). Upon fusion at the endosomes, the uncoating process may differ from various viruses and cell types; however, the general processes of capsid disassembly and genome release are still not well understood. Previously it was believed that the viral fusion process is coupled to the capsid disassembly process, but recent studies indicate that host molecules play a critical role in capsid disassembly after fusion (Hao et al. 2013; Banerjee et al. 2014; Miyake et al. 2019; Kawaguchi et al. 2003). Uncoating is a highly regulated, step-wise process, which is dependent on host cell cues; dysregulation of the endocytic pathways or absence of critical host factors may lead to abortive uncoating.

After initial attachment to the plasma membrane, IAV, is internalised by either clathrin-mediated endocytosis or macropinocytosis, and the virus is sorted into early endosomes. The vATPase present in the early endosome causes an accumulation of protons ($H^+$) in the lumen of the early endosome, as a result of which the luminal pH gradually drops. As the early endosome matures to late endosome while travelling toward the microtubule organising centre (MTOC) from the cell periphery, the pH drops to 5.0–5.5, which serves as the cue for HA conformational change to make the virus fusion-competent (Alvarado-Facundo et al. 2015; Grambas and Hay 1992). The low pH also opens up the viral M2 channel through which, $H^+$ and potassium ions ($K^+$) enter the viral core, loosening up the macromolecular interactions and priming the virus for uncoating (Stauffer et al. 2014). The pH-induced fusogenic HA brings the viral and endosomal membranes in close proximity, which is followed by the merging of the two membranes (Harrison 2015). Fusion of the virus in the late endosome is followed by uncoating, which ensures the escape of the virus from the endosome and the release of the vRNPs into the cytosol (Yamauchi and Helenius 2013; Yamauchi and Greber 2016; Staring et al. 2018; Yamauchi 2020). For example, IAV uncoating requires free ubiquitin chains in the virus and the host histone deacetylase 6 (HDAC6) (Banerjee et al. 2014).

In influenza virions, the vRNPs are attached to the viral envelope by nucleoprotein (NP) and M1 interactions (Martin and Helenius 1991). M1 also interacts with the cytoplasmic tails of HA and NA (Sempere Borau and Stertz 2021). The influx of $H^+$ and $K^+$ triggers conformational rearrangements in M1, leading to the disruption in the M1-NP interactions. Although priming of the virus in the late endosome is a prerequisite for uncoating, it is not sufficient for efficient capsid disassembly; the virus needs assistance from the host to complete the process of uncoating to release the vRNPs into the cytosol. One of the host proteins found to play a critical role in IAV uncoating is HDAC6, which deacetylases tubulin, Hsp90, cortactin and DDX3X (Zhang et al. 2006; Zhang et al. 2008; Hubbert et al. 2002; Kovacs et al.

2005; Zhang et al. 2007; Saito et al. 2019). HDAC6 plays a role in the transport of ubiquitinated protein aggregates to the aggresome for clearance via recruitment of the motor protein dynein (Fig. 14.1) (Hao et al. 2013; Kawaguchi et al. 2003; Banerjee et al. 2014). HDAC6 has two catalytic domains whose functions have been reported to involve tubulin deacetylation and the regulation of liquid phase separation of target proteins (Miyake et al. 2016; Saito et al. 2019). On the other hand, the ZnF domain of HDAC6 at the C-terminus recognises and strongly binds to the—LRGG sequence at the free C-terminus of the ubiquitin chain (Hao et al. 2013; Banerjee et al. 2014). The ubiquitin-binding activity of HDAC6 has been shown to be important for IAV uncoating. Although IAV could successfully internalise and fuse at the late endosome in the HDAC6-knockout (KO) mouse embryonic fibroblasts (MEFs), it was unable to undergo uncoating, and the virus particles remained trapped in the endosomal compartments. A similar phenotype was observed in the lung alveolar cell line (A549) deficient in HDAC6. The block in uncoating resulted in the inhibition of viral replication. Further, it was found that uncoating was independent of the mutations abrogating the enzymatic functions of HDAC6 but required the ubiquitin-binding activity; a point mutation in the ubiquitin-binding ZnF motif suppressed uncoating to the level observed in the HDAC6 KO cells. Biochemical investigations and super-resolution microscopy of purified IAVs revealed the presence of host-derived free ubiquitin chains within the virus particles. After confirming the presence of free ubiquitin chains within virions, the localisation of the ubiquitin chains was examined; if ubiquitin molecules were present on the viral surface, they should have been degraded by protease K in the absence of detergent (Triton). But no degradation of the ubiquitin molecules was observed upon protease K treatment. On the other hand, viral particles wrapped in the lipid bilayers showed membrane breakdown in the presence of Triton, and the ubiquitins present inside the particles were degraded by proteases. This suggests that ubiquitin molecules are present beneath the viral membrane and inside the core (Fig. 14.1). An in vitro assay with purified HDAC6 and IAV revealed an interaction of the viral ubiquitin with the ZnF domain of HDAC6. HDAC6 is a factor that responds to various cellular stresses and is known to transport untranslated mRNAs to stress granules (SGs). Misfolded and ubiquitinated proteins that have not moved to microtubule organising centres (MTOCs) along microtubules form aggregates corresponding to membraneless organelles called aggresomes. HDAC6 has been reported to interact with dynein motor proteins that move on microtubules in the retrograde direction (Kawaguchi et al. 2003), and HDAC6 mutants lacking the dynein binding motif also showed reduced viral infection. In addition to dynein, HDAC6 was found to interact with the actin-based motor protein myosin 10 (MYO10). We found that after fusion at the late endosome, the free ubiquitin chains present in the viral particles are detected by HDAC6, which, after binding to the ubiquitin molecules, recruit dynein and MYO10 to the fusion site. The activity of the motor proteins generates a pulling force on the softened virus capsid, which further helps in the dissociation of the capsid and release of the vRNPs into the cytosol. This suggests that IAV, with its packaged ubiquitin chains, make clever use of the HDAC6-dependent ubiquitin-sensing pathways lined to aggresome

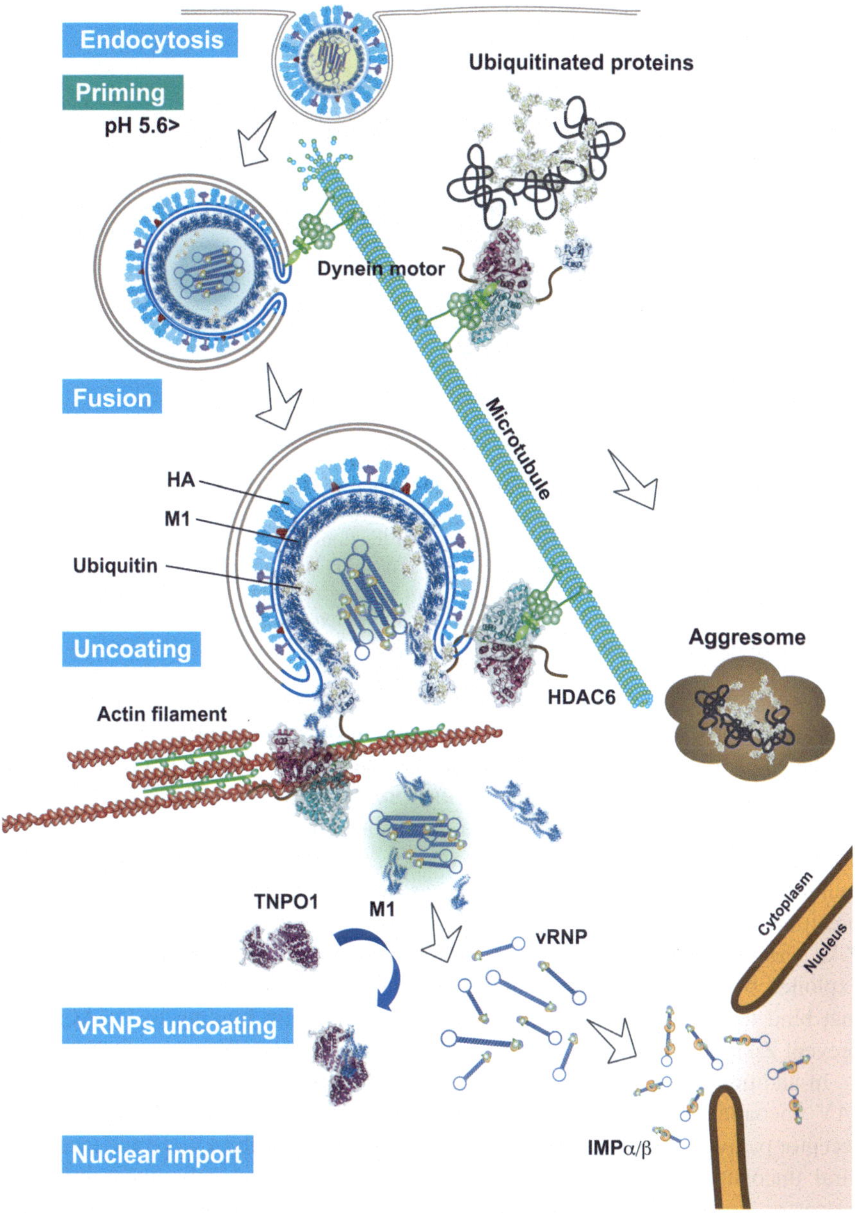


**Fig. 14.1** Schematic of influenza A virus entry. Viral particles are taken up into the cell by endocytosis and become acidified in the endosomes as they mature (Priming). Within matured endosomes, HA undergoes structural conformation change and fuses to the endosomal membrane. Then, host histone deacetylase HDAC6 is recruited by the ubiquitin chains derived from the virus. The ubiquitin-binding domain of HDAC6 tightly binds to unanchored ubiquitin chains. By protein quality control mechanisms, unfold or misfold proteins that are no longer required are ubiquitinated for its degradation and transport to the aggresome by binding to HDAC6. As a stress response, HDAC6 bound to ubiquitin chains in the virus interacts with microtubules and actin fibres to promote uncoating response. After uncoating, the viral nucleoprotein complexes (vRNPs) are released into the cytoplasm. Host factor TNPO1 pulls off the M1 protein on the vRNPs bundle, and the eight bundles of vRNPs are separated into each vRNP. Dissociated vRNPs bind to IMPα/β and migrate into the nucleus for replication and transcription

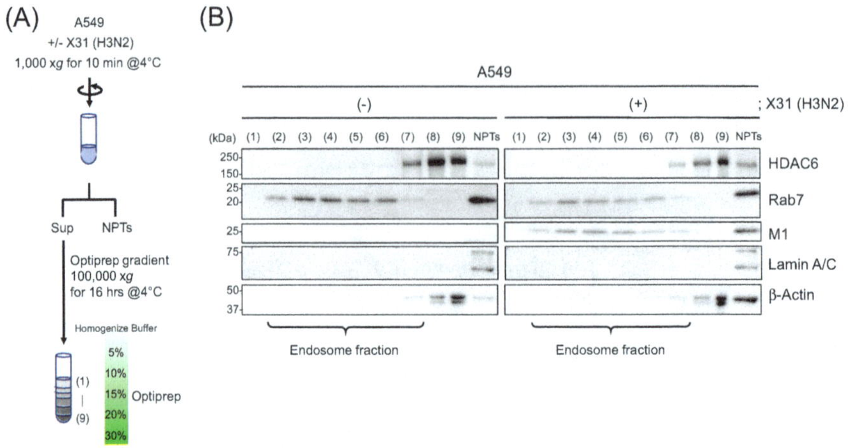


**Fig. 14.2** Biochemical analysis of the endosomal fraction of virus-infected cells. (**a**) After collection of the virus-infected cells, the supernatant of the cell lysate was fractionated by Optiprep-density gradient centrifugation. NPTs; Nuclear pellet fractions. (**b**) Analysis of each fraction after density gradient centrifugation by Western blotting. Fraction (2) up to fraction (7) is positive for the late endosomal marker Rab7. The M1 matrix protein, used as a marker for IAV, is also co-localised in the endosomal fraction (right panel). Fraction (7) is considered to be the endosomal fraction from which HDAC6 is recruited immediately after uncoating

processing for its entry and establishment of a successful infection (Rudnicka and Yamauchi 2016). Biochemical fractionation of endosomal fractions of IAV-infected cells by ultracentrifugation revealed the presence of HDAC6 in one fraction containing viral component M1 and a late endosome marker Rab7 in the soluble endosomal fraction (Fig. 14.2; see fraction (7)). As a fraction containing Rab7 is also found in the pellet fraction, the endosomes that are undergoing uncoating may be precipitated in complex with various other proteins (Fig. 14.2). A recent study on Zika virus has shown how a similar HDAC6-dependent uncoating mechanism is exploited by the virus for uncoating. Designed ankyrin repeat proteins (DARPins) that bind to the ZnF motif of HDAC6 disrupt ubiquitin binding to HDAC6 and prevent Zika virus infection by suppressing viral uncoating (Wang et al. 2022).

In addition to HDAC6, several other host factors have been found to promote IAV uncoating. Larson and colleagues reported that the epidermal growth factor receptor pathway substrate 8 (EPS8) physically interacts with IAV NP and facilitates viral uncoating at a post-fusion step. IAV uncoating and subsequent infection processes were significantly compromised in the EPS8-deficient cells, and complementing EPS8 in the cells restored infection (Larson et al. 2019). IAV uncoating is also blocked in cells with endosomal maturation defect. Houtari and colleagues found that Cullin-3, which serves as a scaffolding subunit in a large family of E3 ubiquitin ligases, plays a critical role in endosome maturation. Upon Cullin-3 depletion, although IAV can enter the cell and the HA can assume an acid-induced conformation in the acidic lumen of the late endosome, the virus fails to uncoat due to impaired endosomal maturation (Huotari et al. 2012). Another study found that the lymphocyte antigen 6 family member E (LY6E) promotes IAV uncoating; ectopic expression of LY6E increases capsid disassembly, suggesting the supporting role of the protein in uncoating at a post-fusion step (Mar et al. 2018). A genome-wide pooled shRNA screen identified an E3 ubiquitin ligase ITCH, which participates in the release of IAV from late endosomes. In the early stage of IAV infection, ITCH is activated by phosphorylation and is recruited to endosomes. In ITCH-depleted cells, the amount of mono- and poly-ubiquitinated M1 was reduced and the M1 co-immunoprecipitated with ITCH. Together, the results suggested that ITCH ubiquitinates M1 in the endosomes and facilitates IAV uncoating after fusion (Su et al. 2013). It is clear from the above studies that IAV uncoating is not a spontaneous event that follows fusion at the late endosome, but requires the participation of the host molecules. Additional uncoating factors could be involved, and future studies may illuminate new molecules that promote IAV uncoating.

After uncoating, the IAV genome is transferred into the nucleus, where the genome is transcribed and replicated. The IAV genome is divided into eight vRNPs segments, which encode 11 structural and non-structural proteins. These eight vRNPs are linked to each other by RNA-RNA interactions or protein-mediated interactions (Hutchinson et al. 2010; Noda et al. 2012; Dadonaite et al. 2019; Noda 2021). When the vRNPs are released into the cytosol following uncoating, they are bundled together and, therefore, cannot travel into the nucleus due to size limitation for the passage of the vRNPs through the nuclear pores. To identify the host factors promoting vRNP nuclear import, we conducted an RNAi screen targeting 70 potential genes involved in nucleocytoplasmic transport and identified transportin 1 (TNPO1) as one of the top hits (Miyake et al. 2019). RNAi-mediated knockdown of TNPO1 significantly arrested vRNP nuclear import. In the TNPO1-deficient cells, both viral capsid uncoating and vRNP nuclear import were impeded, and strong NP and M1 signals were observed in the cytoplasm. We reasoned that reduced uncoating and the block in vRNP nuclear import was possibly due to clustered vRNPs that needed to be debundled into individual segments in order to pass through the nuclear pores. Close examination of the amino acid sequence of the M1 protein revealed the presence of a sequence similar to the existing Proline-Tyrosine nuclear localisation signal (PY-NLS) at the N-terminus of the protein. TNPO1 was previously documented to interact with PY-NLS. Hydrophobic-rich amino acid sequences, including proline, tyrosine and glycine and their neighbouring basic sequences are widely conserved in the sequences of proteins interacting with TNPO1. This TNPO1-binding sequence is also conserved in the capsid protein CA of HIV-1, and it was reported that TNPO1 is an essential factor that facilitates HIV-1 uncoating (Fernandez et al. 2019). IAVs are exposed to acidic conditions in late endosomes for fusion, but upon penetration, the matrix protein M1, vRNPs and the accessory proteins are released into the cytosol at an almost neutral pH. Mimicking both endosomal and cytosolic pH conditions, we examined in vitro whether TNPO1 interacts with M1. Extracts were prepared after treatment of the virus in a buffer solution of pH 5.6 containing potassium salts mimicking the late endosome luminal environment, and immunoprecipitation with purified TNPO1 protein was attempted.

Interestingly, TNPO1 interacted with M1 exclusively at pH 5.6 but failed to bind to the protein at pH 7.0, indicating the interaction of TNPO1 with M1 was dependent on the acidic pH environment found in the lumen of late endosomes. On the other hand, in a mutant (G18A) virus in which the glycine at position 18 of M1, the most conserved protein in the TNPO1-binding sequence, was replaced by alanine, the interaction with TNPO1 was abrogated, and the infectivity of the virus was also greatly reduced. X-ray crystallography of the M1 protein with mutation at glycine 18 revealed a significant structural change around the mutated amino acid. This conformational change is expected to have a significant impact on the interaction with TNPO1, but molecular interaction details are still unclear and are currently being investigated. Taken together, these results suggest that TNPO1 acts on the eight-bundle vRNPs complex of IAVs and binds to the M1 protein, dissociating them into individual vRNPs to be subsequently imported into the nucleus for replication and transcription (vRNPs uncoating, Fig. 14.1).

## Uncoating Reaction In Vitro

To further understand the mechanism of IAV uncoating, we performed an in vitro assay, exposing purified viruses to buffer solutions at neutral or acidic pH conditions to mimic reactions within late endosomes, and analysed the properties of ubiquitin molecules interacting with HDAC6 in the viral extracts. Virus particles exposed to pH 7.0 or pH 5.4 were disrupted by lysis buffer. Viral lysates were reacted with ZnF, the ubiquitin-binding domain of HDAC6, and purification was attempted by histidine tagging of the ZnF domain. Samples that reacted with viral extracts and were exposed to acidic conditions resulted in smears on the gel, corresponding to high molecular weight ubiquitin chains (Fig. 14.3; lane (12)). This suggests that upon acidification, the ubiquitin chains in the virus can increase to higher molecular weight. A detailed molecular mechanism of viral uncoating is currently being investigated.

Several attempts have been made to reproduce this uncoating reaction in vitro using biochemical experimental methods so far (Zhirnov 1990; Stauffer et al. 2014; Stauffer et al. 2016). To mimic the maturation process in endosomes, viral particles were exposed to different pH conditions. The lipid bilayer containing surface glycoproteins was biochemically removed by treatment with non-ionic detergents such as NP-40. In a follow-up study, we attempted to extract vRNP complexes from viral particles using this method. The surface glycoproteins of IAV particles that passed through a buffer containing NP-40 at pH 7.4 (near neutral) were removed, and only the viral core was purified in the pellet fraction (Fig. 14.4a, b). In contrast, under acidic conditions (pH 5.4) mimicking late endosomes, the shell consisting of a lipid bilayer and M1 was removed and the eight segments of vRNPs, were observed in the lower fractions by glycerol density gradient centrifugation (Fig. 14.4c). Thus, this process allowed the extraction of vRNPs with rod-like octameric segments of different lengths from the viral particles without shells, as previously demonstrated

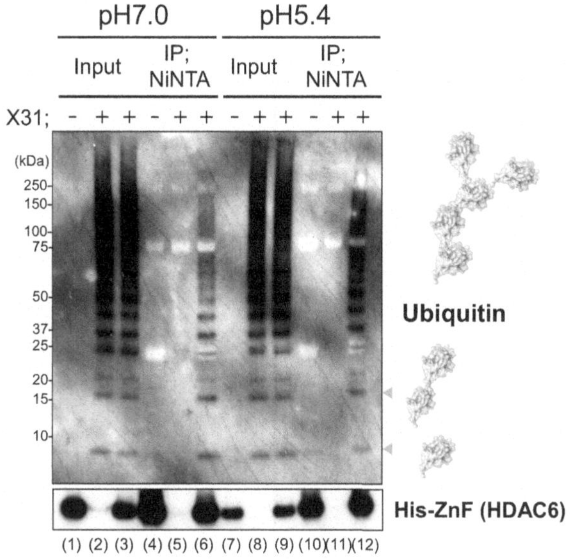


**Fig. 14.3** High molecular weight ubiquitin chains accumulate within viral particles exposed to acidic conditions. The purified ubiquitin-binding domain of HDAC6 with viral extracts exposed to acidic conditions immunoprecipitate high molecular weight ubiquitin chains

by Noda et al. (2006) in ultrathin sections. Negative staining electron microscopy of the vRNPs fraction containing NPs extracted from this intermediate layer revealed individually dispersed vRNPs as well as multiple segments of vRNPs. Since many octameric vRNPs have been observed, it is likely that IAV vRNPs are organised by RNA-RNA interactions and/or RNA-protein interactions (Fig. 14.4d). In future, such purified vRNPs could be used as substrates to reveal the molecular mechanisms of dispersion and assembly of vRNPs.

## Application to Coronavirus Endosomal Uncoating

The in vitro uncoating approach to extracting vRNPs of IAVs can be applied to other viruses. For example, coronaviruses, such as SARS-CoV-2, are also RNA viruses covered with a lipid bilayer. As with SARS-CoV-2, common cold coronaviruses have spike proteins on their viral surface and a 30 kb positive-sense single-stranded RNA genome inside the virus (V'Kovski et al. 2021; Evans and Liu 2021). Coronaviruses have been reported to have an envelope protein (E protein) on the lipid bilayer membrane, which functions as an ion channel (Ruch and Machamer 2012). Since the 229E and OC43 strains have been reported to be taken up into cells in a Rab7-dependent manner (Schneider et al. 2021), the seasonal coronaviruses also internalise via endosocytosis and uncoat after fusion at the late endosome. However, the detailed mechanisms of coronavirus uncoating are still unclear. Similar to IAV, intact viral particles were precipitated by in vitro uncoating by ultracentrifugation under neutral conditions without NP-40 (Fig. 14.5; left panel). On the other hand, when ultracentrifugation was performed at neutral and/or acidic pH conditions in a

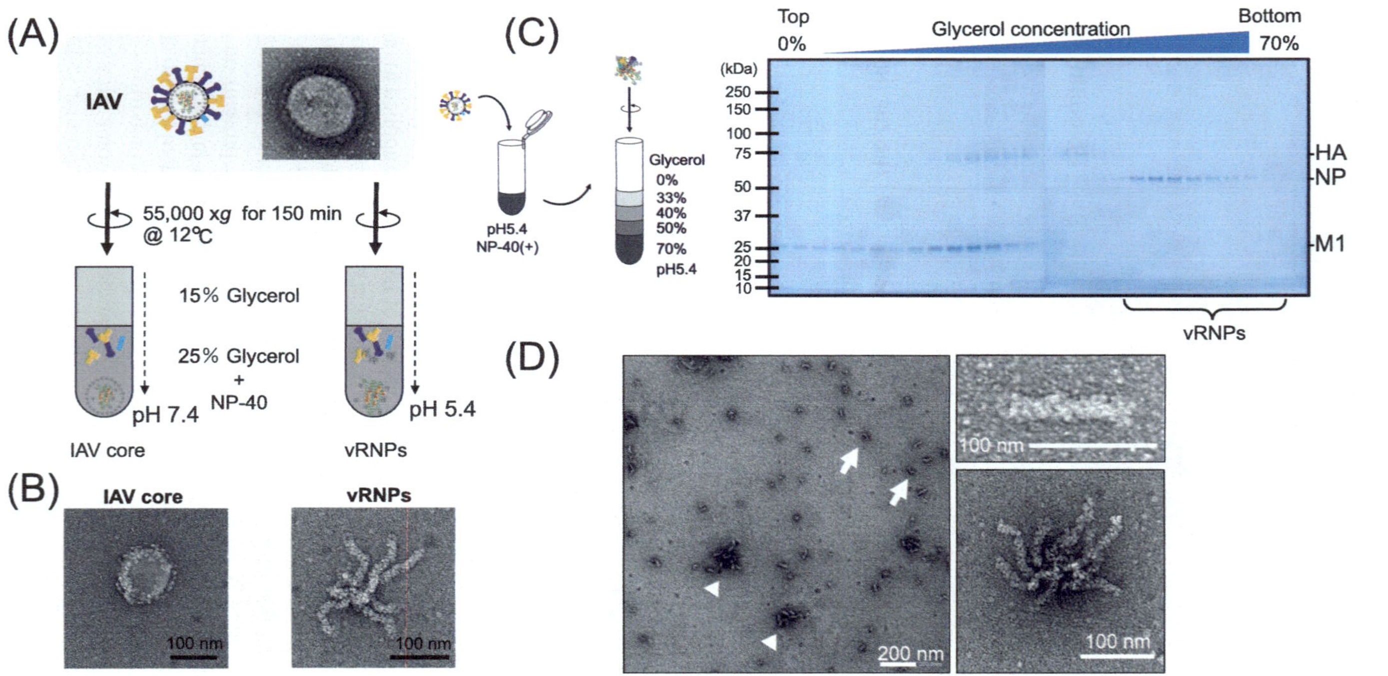


**Fig. 14.4** In vitro uncoating of IAV. (**a**) Two-layer glycerol ultracentrifugation with NP-40 at different pH precipitated viral cores under neutral conditions and vRNPs genomes under acidic conditions, respectively. (**b**) IAV core without HA and NA and lipid bilayer (left); vRNPs without M1 shell (right). (**c**) Viral fractionation by glycerol density gradient centrifugation. IAV particles uncoated in buffer solution with acidic conditions pH 5.4 containing NP-40 were layered on top of buffer containing 0–70% glycerol and fractionated by ultracentrifugation. The fractions containing NPs were collected and observed under an electron microscope. (**d**) Electron microscopic image of the vRNPs fraction. In addition to individually isolated rod-shaped vRNPs (arrow), bundled vRNPs complexes (arrowhead) were also observed. Enlarged images are shown on the right panels

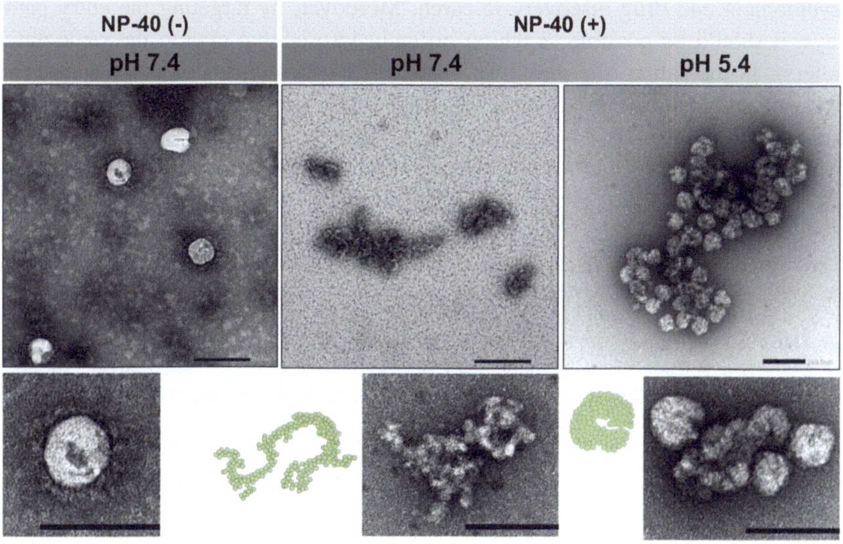


**Fig. 14.5** In vitro uncoating of coronavirus 229E strain. Purified coronavirus strain 229E was ultracentrifuged in two-layer glycerol buffer as in IAV. Virus particles precipitate in the buffer without NP-40 (left panel). Viral genome vRNPs are purified when ultracentrifuged in the buffer containing NP-40. Under neutral conditions, the structure was dissociated (centre panel). Under acidic conditions, vRNPs from each virus particle condensed (right panel). Viral genomes incorporated within endosomes in the cell may form condensates under acidic conditions as the endosomes mature. Scale bar; 200 nm

buffer containing NP-40, cleaved vRNP fragments were observed under neutral conditions at pH 7.4. Interestingly, individual vRNPs form condensates at pH 5.4, mimicking the late endosomal luminal environment, and some condensates were also observed with the neighbouring virus-derived vRNPs (Fig. 14.5). The aggregation of these N-protein-wrapped RNA complexes under acidic conditions may favour the efficient release of the genome into the cytoplasm during uncoating from the endosome. It has also been suggested that coronavirus RNA-N-protein complexes may undergo phase separation (Carlson et al. 2020; Iserman et al. 2020; Jack et al. 2021), and there may be a mechanism by which, the phenomenon of phase separation can be successfully exploited to facilitate genome packaging and uncoating reactions (Etibor et al. 2021).

## Conclusions

As viral entry represents the first step in the life cycle of a virus, inhibition of these early infection events can block the downstream infection processes. Therefore, a mechanistic understanding of virus entry has important implications for vaccine development and drug discovery research. Moreover, by targeting the entry pathways, it should be possible to inhibit different viruses that exploit the same entry route(s). In the future, the design of host-targeted molecular compounds that can block the intracellular entry of emerging and re-emerging viruses is expected to lead to the development of prophylactic and therapeutic agents common to different viruses. Alternatively, if efficient uncoating can be inhibited by drugs, viral infection can also be attenuated.

**Acknowledgements** The authors wish to acknowledge Division for Medical Research Engineering, Nagoya University Graduate School of Medicine, for technical support of electron microscopy; Dr. Koji Itakura, analysis with Mass spec facility; Dr. Kentaro Taki, technical support for microscopy; Dr. Eri Yorifuji. We also thank Prof. Yohei Yamauchi (ETH Zürich), Prof. Hiroshi Kimura and Mr. Atzin Bolanos-Ceron for comments on the manuscript, Prof. Jiro Usukura (Nagoya University Graduate School of Science) and Ms. Tomoko Kunogi for technical assistance.

**Funding** This work was supported by JSPS RPD research fellowship (Grant Number 18J40214), JSPS KAKENHI (Grant-in-Aid for Scientific Research (C)) Grant Number JP20K07513, JST FOREST Program (Grant Number JPMJFR206B, Japan), and MRC-AMED SICORP (Grant Number JP22jm0210070, Japan) to Y.M.

## References

Alvarado-Facundo E, Gao Y, Ribas-Aparicio RM, Jimenez-Alberto A, Weiss CD, Wang W (2015) Influenza virus M2 protein ion channel activity helps to maintain pandemic 2009 H1N1 virus hemagglutinin fusion competence during transport to the cell surface. J Virol 89(4):1975–1985. https://doi.org/10.1128/JVI.03253-14

Banerjee I, Miyake Y, Nobs SP, Schneider C, Horvath P, Kopf M, Matthias P, Helenius A, Yamauchi Y (2014) Influenza A virus uses the aggresome processing machinery for host cell entry. Science 346(6208):473–477. https://doi.org/10.1126/science.1257037

Carlson CR, Asfaha JB, Ghent CM, Howard CJ, Hartooni N, Safari M, Frankel AD, Morgan DO (2020) Phosphoregulation of phase separation by the SARS-CoV-2 N protein suggests a biophysical basis for its dual functions. Mol Cell 80(6):1092–1103. e1094. https://doi.org/10.1016/j.molcel.2020.11.025

Chavrier P, Parton RG, Hauri HP, Simons K, Zerial M (1990) Localization of low molecular weight GTP binding proteins to exocytic and endocytic compartments. Cell 62(2):317–329. https://doi.org/10.1016/0092-8674(90)90369-p

Chlanda P, Mekhedov E, Waters H, Schwartz CL, Fischer ER, Ryham RJ, Cohen FS, Blank PS, Zimmerberg J (2016) The hemifusion structure induced by influenza virus haemagglutinin is determined by physical properties of the target membranes. Nat Microbiol 1(6):16050. https://doi.org/10.1038/nmicrobiol.2016.50

Dadonaite B, Gilbertson B, Knight ML, Trifkovic S, Rockman S, Laederach A, Brown LE, Fodor E, Bauer DLV (2019) The structure of the Influenza A virus genome. Nat Microbiol 4(11):1781–1789. https://doi.org/10.1038/s41564-019-0513-7

de Vries E, Tscherne DM, Wienholts MJ, Cobos-Jimenez V, Scholte F, Garcia-Sastre A, Rottier PJ, de Haan CA (2011) Dissection of the Influenza A virus endocytic routes reveals macropinocytosis as an alternative entry pathway. PLoS Pathog 7(3):e1001329. https://doi.org/10.1371/journal.ppat.1001329

Edinger TO, Pohl MO, Yanguez E, Stertz S (2015) Cathepsin W is required for escape of Influenza A virus from late endosomes. mBio 6(3):e00297. https://doi.org/10.1128/mBio.00297-15

Eierhoff T, Hrincius ER, Rescher U, Ludwig S, Ehrhardt C (2010) The epidermal growth factor receptor (EGFR) promotes uptake of Influenza A viruses (IAV) into host cells. PLoS Pathog 6(9):e1001099. https://doi.org/10.1371/journal.ppat.1001099

Etibor TA, Yamauchi Y, Amorim MJ (2021) Liquid biomolecular condensates and viral lifecycles: review and perspectives. Viruses 13(3):366. https://doi.org/10.3390/v13030366

Evans JP, Liu SL (2021) Role of host factors in SARS-CoV-2 entry. J Biol Chem 297(1):100847. https://doi.org/10.1016/j.jbc.2021.100847

Fernandez J, Machado AK, Lyonnais S, Chamontin C, Gartner K, Leger T, Henriquet C, Garcia C, Portilho DM, Pugniere M, Chaloin L, Muriaux D, Yamauchi Y, Blaise M, Nisole S, Arhel NJ (2019) Transportin-1 binds to the HIV-1 capsid via a nuclear localization signal and triggers uncoating. Nat Microbiol 4(11):1840–1850. https://doi.org/10.1038/s41564-019-0575-6

Fujioka Y, Nishide S, Ose T, Suzuki T, Kato I, Fukuhara H, Fujioka M, Horiuchi K, Satoh AO, Nepal P, Kashiwagi S, Wang J, Horiguchi M, Sato Y, Paudel S, Nanbo A, Miyazaki T, Hasegawa H, Maenaka K, Ohba Y (2018) A Sialylated voltage-dependent Ca(2+) channel binds hemagglutinin and mediates Influenza A virus entry into mammalian cells. Cell Host Microbe 23(6):809–818. e805. https://doi.org/10.1016/j.chom.2018.04.015

Gorvel JP, Chavrier P, Zerial M, Gruenberg J (1991) rab5 controls early endosome fusion in vitro. Cell 64(5):915–925. https://doi.org/10.1016/0092-8674(91)90316-q

Grambas S, Hay AJ (1992) Maturation of Influenza A virus hemagglutinin--estimates of the pH encountered during transport and its regulation by the M2 protein. Virology 190(1):11–18. https://doi.org/10.1016/0042-6822(92)91187-y

Gschweitl M, Ulbricht A, Barnes CA, Enchev RI, Stoffel-Studer I, Meyer-Schaller N, Huotari J, Yamauchi Y, Greber UF, Helenius A, Peter M (2016) A SPOPL/Cullin-3 ubiquitin ligase complex regulates endocytic trafficking by targeting EPS15 at endosomes. elife 5:e13841. https://doi.org/10.7554/eLife.13841

Hao R, Nanduri P, Rao Y, Panichelli RS, Ito A, Yoshida M, Yao TP (2013) Proteasomes activate aggresome disassembly and clearance by producing unanchored ubiquitin chains. Mol Cell 51(6):819–828. https://doi.org/10.1016/j.molcel.2013.08.016

Harrison SC (2008) Viral membrane fusion. Nat Struct Mol Biol 15(7):690–698. https://doi.org/10.1038/nsmb.1456

Harrison SC (2015) Viral membrane fusion. Virology 479-480:498–507. https://doi.org/10.1016/j.virol.2015.03.043

He J, Sun E, Bujny MV, Kim D, Davidson MW, Zhuang X (2013) Dual function of CD81 in influenza virus uncoating and budding. PLoS Pathog 9(10):e1003701. https://doi.org/10.1371/journal.ppat.1003701

Hubbert C, Guardiola A, Shao R, Kawaguchi Y, Ito A, Nixon A, Yoshida M, Wang XF, Yao TP (2002) HDAC6 is a microtubule-associated deacetylase. Nature 417(6887):455–458. https://doi.org/10.1038/417455a

Huotari J, Meyer-Schaller N, Hubner M, Stauffer S, Katheder N, Horvath P, Mancini R, Helenius A, Peter M (2012) Cullin-3 regulates late endosome maturation. Proc Natl Acad Sci U S A 109(3):823–828. https://doi.org/10.1073/pnas.1118744109

Hutchinson EC, von Kirchbach JC, Gog JR, Digard P (2010) Genome packaging in Influenza A virus. J Gen Virol 91(Pt 2):313–328. https://doi.org/10.1099/vir.0.017608-0

Iserman C, Roden CA, Boerneke MA, Sealfon RSG, McLaughlin GA, Jungreis I, Fritch EJ, Hou YJ, Ekena J, Weidmann CA, Theesfeld CL, Kellis M, Troyanskaya OG, Baric RS, Sheahan TP, Weeks KM, Gladfelter AS (2020) Genomic RNA elements drive phase separation of the SARS-CoV-2 nucleocapsid. Mol Cell 80(6):1078–1091. e1076. https://doi.org/10.1016/j.molcel.2020.11.041

Jack A, Ferro LS, Trnka MJ, Wehri E, Nadgir A, Nguyenla X, Fox D, Costa K, Stanley S, Schaletzky J, Yildiz A (2021) SARS-CoV-2 nucleocapsid protein forms condensates with viral genomic RNA. PLoS Biol 19(10):e3001425. https://doi.org/10.1371/journal.pbio.3001425

Kanaseki T, Kawasaki K, Murata M, Ikeuchi Y, Ohnishi S (1997) Structural features of membrane fusion between influenza virus and liposome as revealed by quick-freezing electron microscopy. J Cell Biol 137(5):1041–1056. https://doi.org/10.1083/jcb.137.5.1041

Kawaguchi Y, Kovacs JJ, McLaurin A, Vance JM, Ito A, Yao TP (2003) The deacetylase HDAC6 regulates aggresome formation and cell viability in response to misfolded protein stress. Cell 115(6):727–738. https://doi.org/10.1016/s0092-8674(03)00939-5

Konig R, Stertz S, Zhou Y, Inoue A, Hoffmann HH, Bhattacharyya S, Alamares JG, Tscherne DM, Ortigoza MB, Liang Y, Gao Q, Andrews SE, Bandyopadhyay S, De Jesus P, Tu BP, Pache L, Shih C, Orth A, Bonamy G, Miraglia L, Ideker T, Garcia-Sastre A, Young JA, Palese P, Shaw ML, Chanda SK (2010) Human host factors required for influenza virus replication. Nature 463(7282):813–817. https://doi.org/10.1038/nature08699

Kovacs JJ, Murphy PJ, Gaillard S, Zhao X, Wu JT, Nicchitta CV, Yoshida M, Toft DO, Pratt WB, Yao TP (2005) HDAC6 regulates Hsp90 acetylation and chaperone-dependent activation of glucocorticoid receptor. Mol Cell 18(5):601–607. https://doi.org/10.1016/j.molcel.2005.04.021

Larson GP, Tran V, Yu S, Cai Y, Higgins CA, Smith DM, Baker SF, Radoshitzky SR, Kuhn JH, Mehle A (2019) EPS8 facilitates uncoating of Influenza A virus. Cell Rep 29(8):2175–2183. e2174. https://doi.org/10.1016/j.celrep.2019.10.064

Lee KK (2010) Architecture of a nascent viral fusion pore. EMBO J 29(7):1299–1311. https://doi.org/10.1038/emboj.2010.13

Liang R, Swanson JMJ, Madsen JJ, Hong M, DeGrado WF, Voth GA (2016) Acid activation mechanism of the Influenza A M2 proton channel. Proc Natl Acad Sci U S A 113(45):E6955–E6964. https://doi.org/10.1073/pnas.1615471113

Mar KB, Rinkenberger NR, Boys IN, Eitson JL, McDougal MB, Richardson RB, Schoggins JW (2018) LY6E mediates an evolutionarily conserved enhancement of virus infection by targeting a late entry step. Nat Commun 9(1):3603. https://doi.org/10.1038/s41467-018-06000-y

Martin K, Helenius A (1991) Nuclear transport of influenza virus ribonucleoproteins: the viral matrix protein (M1) promotes export and inhibits import. Cell 67(1):117–130. https://doi.org/10.1016/0092-8674(91)90576-k

Mercer J, Schelhaas M, Helenius A (2010) Virus entry by endocytosis. Annu Rev Biochem 79:803–833. https://doi.org/10.1146/annurev-biochem-060208-104626

Miyake Y, Keusch JJ, Wang L, Saito M, Hess D, Wang X, Melancon BJ, Helquist P, Gut H, Matthias P (2016) Structural insights into HDAC6 tubulin deacetylation and its selective inhibition. Nat Chem Biol 12(9):748–754. https://doi.org/10.1038/nchembio.2140

Miyake Y, Keusch JJ, Decamps L, Ho-Xuan H, Iketani S, Gut H, Kutay U, Helenius A, Yamauchi Y (2019) Influenza virus uses transportin 1 for vRNP debundling during cell entry. Nat Microbiol 4(4):578–586. https://doi.org/10.1038/s41564-018-0332-2

Noda T (2021) Selective genome packaging mechanisms of Influenza A viruses. Cold Spring Harb Perspect Med 11(7):a038497. https://doi.org/10.1101/cshperspect.a038497

Noda T, Sagara H, Yen A, Takada A, Kida H, Cheng RH, Kawaoka Y (2006) Architecture of ribonucleoprotein complexes in Influenza A virus particles. Nature 439(7075):490–492. https://doi.org/10.1038/nature04378

Noda T, Sugita Y, Aoyama K, Hirase A, Kawakami E, Miyazawa A, Sagara H, Kawaoka Y (2012) Three-dimensional analysis of ribonucleoprotein complexes in Influenza A virus. Nat Commun 3:639. https://doi.org/10.1038/ncomms1647

Pohl MO, Edinger TO, Stertz S (2014) Prolidase is required for early trafficking events during Influenza A virus entry. J Virol 88(19):11271–11283. https://doi.org/10.1128/JVI.00800-14

Ruch TR, Machamer CE (2012) The coronavirus E protein: assembly and beyond. Viruses 4(3): 363–382. https://doi.org/10.3390/v4030363

Rudnicka A, Yamauchi Y (2016) Ubiquitin in influenza virus entry and innate immunity. Viruses 8(10):293. https://doi.org/10.3390/v8100293

Saito M, Hess D, Eglinger J, Fritsch AW, Kreysing M, Weinert BT, Choudhary C, Matthias P (2019) Acetylation of intrinsically disordered regions regulates phase separation. Nat Chem Biol 15(1):51–61. https://doi.org/10.1038/s41589-018-0180-7

Schneider WM, Luna JM, Hoffmann HH, Sanchez-Rivera FJ, Leal AA, Ashbrook AW, Le Pen J, Ricardo-Lax I, Michailidis E, Peace A, Stenzel AF, Lowe SW, MacDonald MR, Rice CM, Poirier JT (2021) Genome-scale identification of SARS-CoV-2 and pan-coronavirus host factor networks. Cell 184(1):120–132. e114. https://doi.org/10.1016/j.cell.2020.12.006

Sempere Borau M, Stertz S (2021) Entry of Influenza A virus into host cells - recent progress and remaining challenges. Curr Opin Virol 48:23–29. https://doi.org/10.1016/j.coviro.2021.03.001

Staring J, Raaben M, Brummelkamp TR (2018) Viral escape from endosomes and host detection at a glance. J Cell Sci 131(15):jcs216259. https://doi.org/10.1242/jcs.216259

Stauffer S, Feng Y, Nebioglu F, Heilig R, Picotti P, Helenius A (2014) Stepwise priming by acidic pH and a high K+ concentration is required for efficient uncoating of Influenza A virus cores after penetration. J Virol 88(22):13029–13046. https://doi.org/10.1128/JVI.01430-14

Stauffer S, Nebioglu F, Helenius A (2016) In vitro disassembly of Influenza A virus capsids by gradient centrifugation. J Vis Exp 109:e53909. https://doi.org/10.3791/53909

Su WC, Chen YC, Tseng CH, Hsu PW, Tung KF, Jeng KS, Lai MM (2013) Pooled RNAi screen identifies ubiquitin ligase Itch as crucial for Influenza A virus release from the endosome during virus entry. Proc Natl Acad Sci U S A 110(43):17516–17521. https://doi.org/10.1073/pnas.1312374110

V'Kovski P, Kratzel A, Steiner S, Stalder H, Thiel V (2021) Coronavirus biology and replication: implications for SARS-CoV-2. Nat Rev Microbiol 19(3):155–170. https://doi.org/10.1038/s41579-020-00468-6

Wang L, Moreira EA, Kempf G, Miyake Y, Oliveira Esteves BI, Fahmi A, Schaefer JV, Dreier B, Yamauchi Y, Alves MP, Pluckthun A, Matthias P (2022) Disrupting the HDAC6-ubiquitin interaction impairs infection by influenza and Zika virus and cellular stress pathways. Cell Rep 39(4):110736. https://doi.org/10.1016/j.celrep.2022.110736

Yamauchi Y (2020) Influenza A virus uncoating. Adv Virus Res 106:1–38. https://doi.org/10.1016/bs.aivir.2020.01.001

Yamauchi Y, Greber UF (2016) Principles of virus uncoating: cues and the snooker ball. Traffic 17(6):569–592. https://doi.org/10.1111/tra.12387

Yamauchi Y, Helenius A (2013) Virus entry at a glance. J Cell Sci 126(Pt 6):1289–1295. https://doi.org/10.1242/jcs.119685

Yanguez E, Hunziker A, Dobay MP, Yildiz S, Schading S, Elshina E, Karakus U, Gehrig P, Grossmann J, Dijkman R, Schmolke M, Stertz S (2018) Phosphoproteomic-based kinase profiling early in influenza virus infection identifies GRK2 as antiviral drug target. Nat Commun 9(1):3679. https://doi.org/10.1038/s41467-018-06119-y

Zhang Y, Gilquin B, Khochbin S, Matthias P (2006) Two catalytic domains are required for protein deacetylation. J Biol Chem 281(5):2401–2404. https://doi.org/10.1074/jbc.C500241200

Zhang X, Yuan Z, Zhang Y, Yong S, Salas-Burgos A, Koomen J, Olashaw N, Parsons JT, Yang XJ, Dent SR, Yao TP, Lane WS, Seto E (2007) HDAC6 modulates cell motility by altering the acetylation level of cortactin. Mol Cell 27(2):197–213. https://doi.org/10.1016/j.molcel.2007.05.033

Zhang Y, Kwon S, Yamaguchi T, Cubizolles F, Rousseaux S, Kneissel M, Cao C, Li N, Cheng HL, Chua K, Lombard D, Mizeracki A, Matthias G, Alt FW, Khochbin S, Matthias P (2008) Mice lacking histone deacetylase 6 have hyperacetylated tubulin but are viable and develop normally. Mol Cell Biol 28(5):1688–1701. https://doi.org/10.1128/MCB.01154-06

Zhirnov OP (1990) Solubilization of matrix protein M1/M from virions occurs at different pH for orthomyxo- and paramyxoviruses. Virology 176(1):274–279. https://doi.org/10.1016/0042-6822(90)90253-n

# Chapter 15
# Human Endogenous Retroviruses in Diseases

**Tian-Jiao Fan and Jie Cui**

**Abstract** Human endogenous retroviruses (HERVs), which are conserved sequences of ancient retroviruses, are widely distributed in the human genome. Although most HERVs have been rendered inactive by evolution, some have continued to exhibit important cytological functions. HERVs in the human genome perform dual functions: on the one hand, they are involved in important physiological processes such as placental development and immune regulation; on the other hand, their aberrant expression is closely associated with the pathological processes of several diseases, such as cancers, autoimmune diseases, and viral infections. HERVs can also regulate a variety of host cellular functions, including the expression of protein-coding genes and regulatory elements that have evolved from HERVs. Here, we present recent research on the roles of HERVs in viral infections and cancers, including the dysregulation of HERVs in various viral infections, HERV-induced epigenetic modifications of histones (such as methylation and acetylation), and the potential mechanisms of HERV-mediated antiviral immunity. We also describe therapies to improve the efficacy of vaccines and medications either by directly or indirectly targeting HERVs, depending on the HERV.

**Keywords** Human endogenous retroviruses (HERVs) · Viral infection · Tumor · Antiviral immunity · Histone epigenetics · Therapies

## HERV Functions

### *Introduction to ERVs*

Endogenous retroviruses (ERVs) are retroviral sequences in vertebrate genomes derived from the invasion and integration of ancient exogenous retroviruses. When infecting host cells, retroviruses are first retrotranscribed into double-stranded DNA

T.-J. Fan · J. Cui (✉)
CAS Key Laboratory of Molecular Virology & Immunology, Shanghai Institute of Immunity and Infection, Chinese Academy of Sciences, Shanghai, China
e mail: tjfan@ips.ac.cn; jcui@ips.ac.cn

 
S. Vijayakrishnan et al. (eds.), *Virus Infected Cells*, Subcellular Biochemistry 106,
https://doi.org/10.1007/978-3-031-40086-5_15

and then integrated into host genome DNA. Exogenous retroviral DNA integrated into germ cells may become ERVs. Over the long course of evolution, random genetic drift and natural selection have led to the loss or alteration of most retroviruses integrated into the human genome (Johnson 2019). As a result, few ERVs express viral genes. However, full-length ERVs contain the viral genes *gag, pol,* and sometimes *env*, flanked by two LTRs (long noncoding terminal repeats). Moreover, the vast majority of ERV-derived sequences in the human genome are solitary LTRs (Bannert and Kurth 2006), which most likely arose through homologous recombination between two LTRs, resulting in the loss of their retroviral genes (Gemmell et al. 2019; Mao et al. 2021). LTRs carry important gene regulatory sequence elements, such as promoters and enhancers (Bannert and Kurth 2006; Dolei 2018). When inserted in close proximity to a host gene, LTRs can influence the gene expression pattern (Mager and Stoye 2015; Hurst and Magiorkinis 2017). However, little is known about the mechanisms that modulate LTR promoter activity or the functional implications of LTR activities in human cells.

The replication cycle of all retroviruses consists of two typical steps (Freed and Mouland 2006; Novikova et al. 2019). (a) reverse transcription of the true viral RNA genome into dsDNA and (b) integration of a newly synthesized viral DNA into the chromosomal DNA of a host cell. The resulting provirus is typically 8–12 kb long (depending on the type of retrovirus) and flanked by short repeats (~4–6 bp) of the host genome sequence. The provirus includes all retroviral genes that together encode the structural and nonstructural proteins of a virus, as well as the cis-acting elements necessary to regulate viral gene expression and synthesize new copies of the viral RNA genome.

## *Classification of HERVs*

ERVs have been found in the human genome and are thus called human endogenous retroviruses (HERVs). ERVs were discovered in the late 1960s and early 1970s, when independent groups reported the identification of three types of ERVs within a few years: avian leukemia virus (ALV), murine leukemia virus (MLV) (Aaronson et al. 1971) and murine mammary tumor virus (MMTV) (Bentvelzen et al. 1970). Since their initial identification in the 1980s (Hayward et al. 2015), HERVs have been found in many loci of the human genome and occupy approximately 8% of human genomic sequences (Griffiths 2001). Some reports have shown that HERVs are carried in vertebrates in addition to humans (Bannert and Kurth 2006; Mager and Stoye 2015; Escalera-Zamudio and Greenwood 2016). These findings suggest that some HERVs undergo vertical transmission, similar to nonhuman ERVs (Jern and Coffin 2008; Garcia-Montojo et al. 2018). HERVs are classified into 3 classes and 11 supergroups: Class I (gamma- and epsilon-retrovirus-like) HERVs comprise the MLLV, HERVERI, HERVFRDLIKE, HEPSI, HUERSP, HERVW9, HERVIPADP, MER50-like, and HERVHF supergroups; Class II (beta-retrovirus-like) comprise the HML (HERV-K) supergroup; and Class III (Spuma-retroviruses).

### Human Endogenous Retrovirus-K (HML-2)

Among the aforementioned supergroups, multiple copies of the HERV-K retrovirus family have recently been identified in the human genome; some of these copies carry intact open reading frames and can be transcribed and translated, especially in early embryogenesis. Phylogenetically, HERV-K viruses comprise a super virus group. Among these viruses, the HML-2 (human endogenous MMTV-like-2) subtype shows the most transcriptional activity; therefore, it has been the most intensively HERV-K virus studied to date. Hundreds of HERV-K copies have been integrated into the human genome (Wildschutte et al. 2016). Although all HML-2 proviruses are defective in at least one gene (Subramanian et al. 2011), many carry intact open reading frames (ORFs) with coding capacity, and aberrant expression of HML-2 in adult tissues has been associated with certain types of cancer and neurodegenerative diseases (Garcia-Montojo et al. 2018). Studies have shown that HERV-K plays a key role in embryogenesis and is silently expressed in most cell types in healthy adults (Medstrand and Mager 1998). Moreover, its reactivation has been associated with several types of cancer (Agoni et al. 2013; Argaw-Denboba et al. 2017) and with the neurodegenerative disease amyotrophic lateral sclerosis (ALS) (Douville et al. 2011).

Phylogenetically, the HERV-K group belongs to the ERV2 (or Class II) or beta retrovirus-like supergroup. To date, 10 groups (from HML-1 to HML-10) constitute the HERV-K evolutionary branch (Andersson et al. 1996), and each group is classified according to the reverse transcriptase (RT) sequence of the member. Notably, J. M. Coffin's group revealed that 2 proviruses (located in chromosome bands 17p13.1 and 8p22) that had been previously classified in the HML-2 evolutionary branch, carried gene sequences that are very dissimilar to those of other viruses in this group. Therefore, this research group proposed reclassifying the viruses with these motifs to a new group named HML-11 (Subramanian et al. 2011). Based on phylogenetic analysis of the LTR sequences, HML-2 can be categorized into three subgroups: LTR5Hs, LTR5A, and LTR5B (Buzdin et al. 2003). LTR5H viruses were recently acquired by humans, while the LTR5A or LTR5B subgroups were acquired via earlier integration (Buzdin et al. 2003). Class I proviruses are found only in the LTR5H subgroup and are scattered among all subclasses, indicating frequent recombination (Subramanian et al. 2011). To clearly identify HML-2 proviruses, Hughes and Coffin (2001) proposed naming a virus on the basis of its position of a chromosomal band. In cases of multiple proviral loci in the same chromosomal band, the Coffin laboratory suggested labeling each provirus with "a," "b," "c," etc., depending on the order in the band (Subramanian et al. 2011).

HML-2 is usually constitutively expressed at low levels in human tissues and may confer benefit to the host (Kurth and Bannert 2010). HML-2 is primarily involved in the following cytological functions: a) Maintenance of cellular pluripotency: High levels of HML-2 transcripts and proteins have been found in undifferentiated embryonic stem cells (ESCs) and induced pluripotent stem cells (IPSCs). Induction of cell differentiation rapidly suppressed HML-2 expression (Fuchs et al. 2011), and the oncogene p53 may be involved in regulating HML-2 expression. b) Neurotrophic effects: Human neural cell lines transfected with HML-2 Env expression led to increased expression of nerve growth factor (NGF) and brain-derived neurotrophic factor (BDNF), which enhanced cell viability and prevented neurotoxicity (Bhat et al. 2014). c) Antiviral effects: Expression of truncated Gag derived from Fv-1 ERV protected against infection of certain MLV strains (Best et al. 1997), and activation of HML-2 in HIV-infected patients produced a similar protective effect. However, the potential antiviral role of HML-2 has not been fully investigated. Interestingly, no exogenous counterparts to any of the three classes of HERVs (γ, β, and spuma-like retroviruses) are infectious to humans; notably, spuma-like viruses infect humans only through zoonoses. Similarly, exogenous retroviruses capable of infecting human cells (lentiviruses and delta retroviruses) are not represented by endogenous viruses in humans (van der Kuyl 2012). These patterns of integration support the hypothesis that HERVs are antiviral agents with activity against related exogenous retroviruses. (d) Placental formation; The HML-2 envelope does not form in syncytia. In contrast, it is expressed in placental villi and extravillous trophoblast cells but not in syncytial trophoblast cells (Kammerer et al. 2011).

HML-2 has been associated with the development of cancer because its expression has been associated with many types of cancer, such as teratomas, germ cell tumors, melanoma, ovarian cancer, and prostate cancer (Buscher et al. 2005, Kurth and Bannert 2010). HML-2 also exhibits various malignant cell characteristics (Oricchio et al. 2007; Reis et al. 2013; Bhardwaj et al. 2014; Garcia-Montojo et al. 2018).

## HERV-H

HERV-H is one of the most prolific HERV families, with approximately 1000 copies, including those with full-length, truncated, and individual LTRs. Previous studies on the function of HERVs have focused on the envelope protein encoded by the *env* gene, which is thought to be the most important gene in disease-causing HERVs due to its potential immunosuppressive properties (Wang-Johanning et al. 2007). The *env* gene fragment in the HERV-H family has been detected in various human tissues and cancer cells; for example, RT–PCR of several human tissues (placenta, skeletal muscle, spleen, and thymus) and cancer cells (RT4, BT-474, HCT-116, TE-1, UO-31, Jurkat, HepG2 A549, MCF7, OVCAR-3, MIA-PaCa-2, PC3, LOX-IMVI, AZ521, 2F7, U-937, and C-33A cells) has led to the identification of *env* fragments in mRNA (Yi et al. 2006). However, few HERV-H family members have been shown to encode intact envelope proteins. Only three HERV-H integrins retain an *env* ORF: HERV-H/*env*62 (or HERV-H19), HERV-H/*env*60, and HERV-H/*env* (Kelleher et al. 1996; Lindeskog et al. 1999). HERV-H19 has been reported to be a code-competent family member carrying an *env* ORF comprising 1752 bp that encodes an envelope protein of 77 kDa via an in vitro translational reaction (Hirose et al. 1993). Performing a transcriptome analysis of specific copies of HERV-H is challenging due to the high similarity among the 1000 members of the HERV-H family distributed across the human genome. However, multiple studies focusing on activating the expression of HERV-H elements, particularly spliced noncoding transcripts, have been performed.

Transcripts for HERV-H have been detected in cell lines derived from different types of cancer, including teratocarcinoma, bladder cancer, testicular tumors, lung cancer, and colorectal cancer (Yi et al. 2006). Expression of a HERV-H *env* gene, however, has not been detected in normal tissues of the brain, prostate, kidney, liver, and lung, although it has been detected in cancer cells. Using massively parallel signature sequencing, for example, it has been found that HERV-H elements were preferentially expressed in the colon, small bowel, and gastric cancer tissues of the various samples of cancer tissues tested (Kitsou et al. 2021). Alves et al. (2008) reported that specific HERV-H elements on the X chromosome were selectively transcribed in 60% of colon cancers and were transcribed at a higher rate in metastatic colon cancers. Wentzensen et al. (2007) also reported overexpression of HERV-H RNA sequences in a range of gastrointestinal cancers, which was associated with HERV-H LTR demethylation. Pérot and colleagues detected colorectal cancer-specific HERV-H transcripts at significantly higher levels in tumors with lymph node invasion throughout disease progression from adenoma to liver metastasis (Perot et al. 2015). Shu Zheng et al. performed microarray expression profiling to search for genes differentially expressed in colorectal tumors and adjacent nontumor tissues, and they identified a HERV-H element located at Xp22, named HERV-HX, that was specifically upregulated in colon tumor tissues, and this HERV-H locus was found to be more active in colon tumors than in adjacent normal tissues. Leonardo A. Sechi and colleagues found that patients with prostate cancer who carried antibodies to HERV-H *env*-su229–241 exhibited a significant response, showing the possible involvement of HERV-H in prostate cancer development and suggesting that this gene may be used as a biomarker for this disease (Manca et al. 2022).

## *Replication of HERVs*

The replication of HERVs is very similar to that of exogenous retroviruses such as HIV. In the case of HERVK, two HERVKs encoding different auxiliary protein genes are first transcribed into several different transcripts, such as a transcript encoding the *gag-pro-pol* polyprotein, auxiliary Rec/NP9 protein or the ENV protein. In addition, the termination of HERV transcription due to the accumulation of mutations leads to the production of noncoding RNA molecules. Transcribed mRNA molecules are translated by ribosomes into individual HERVE proteins. First, the HERV protein protease (PR) undergoes self-cleavage to form the active form of PR, which in turn cleaves the polyprotein precursor of other HERV protcins to produce the structural proteins of the matrix (MA), capsid (CA) and nucleocapsid

(NC), as well as enzymatically active reverse transcriptase (RT) and integrase (IN). The transcript (the genome of a daughter virus) is then encircled with the IN and RT proteins in a viral capsid formed by the three structural proteins. In the case of ENV proteins, synthesis proceeds via the secretory pathway: ENV proteins are first guided into the endoplasmic reticulum via a signal peptide in their head, and then, the signal peptide is excised. Subsequently, the ENV protein is cleaved in the Golgi apparatus by host furin to form surface units and transmembrane units, forming dimers by noncovalent binding. Then, three dimers form a trimer. The mature ENV protein is anchored to the cell membrane via the TM structural domain and attaches to the surface of the viral capsid, which in turn forms new daughter viral particles. The newly generated daughter viral particles carry properties similar to those of exogenous retroviruses: They can enter cells in a manner similar to exogenous retroviruses, and thus, the viral genome is integrated into the cellular genome (Garcia-Montojo, et al. 2018).

## *HERVs as Transcriptional Regulators*

Epigenomic studies in recent years have revealed that ERVs are unexpectedly large sources of cell type-specific regulatory elements, including promoters, enhancers, chromatin boundary elements, and regulatory RNAs. These regulatory sequences are often derived from the LTR region of an ERV, suggesting that ancestral regulatory integrated proviral LTRs may have been leveraged for genomic regulation at the cellular level. Regulatory activity derived from ERVs has recently been demonstrated in a variety of tissues and cell types, including many immune cells, whereas the sequence of the MIR retrotransposon insulates the human genome. Together, these studies suggest that ERVs and other repetitive elements are important drivers of gene regulatory network evolution (Schmidt et al. 2012; Ohnuki et al. 2014).

HERVs regulate gene transcription through different mechanisms. At least 4 different mechanisms have been reported: (1) HERVs function as promoters or enhancer elements in genes, (2) HERVs encode proteins involved in the regulation of gene transcription, (3) noncoding RNA (ncRNA) or double-stranded RNA (dsRNA) molecules derived from that HERVs regulate gene transcription, and (4) HERVs alter the structure of genes or induce an antisense sequence to modulate gene transcription.

### HERVs as Promoters or Enhancers

HERVs may regulate genes by functioning as promoters or enhancers. For example, Wang et al. analyzed the transcriptome of human cells infected with a variety of viruses, including influenza viruses, flaviviruses, herpesviruses, and coronaviruses, and found that certain HERVs likely promoted the expression of certain interferon-stimulated genes (ISGs) [1], and this mechanistic finding was consistent with previous findings obtained through HIV-1 research [2]. HERVs may not only act as ISG promoters but also as ISG enhancers. Chuong et al. found that MER41 (HERVW9) elements close to a signal transducer and activator of transcription 1 (STAT1) gene showed a marked increase in histone H3 lysine 27 acetylation (H3K27ac) after IFN-gamma (IFN-γ) stimulation, and this H3K27ac marked abundance suggested cis-regulatory enhancer activity. Subsequently, this research group demonstrated that deleting MER41 from the human genome altered IFN-induced adjacent gene expression [3].

## HERV-Encoded Proteins

Some HERVs with an intact open reading frame have been reported to produce virus-like particles with peptides and functional proteins involved in tumorigenesis due to epigenetic dysregulation; in other words, HERVs can regulate gene expression through the translation of their own proteins. The pathogenic properties of envelope (Env) proteins, among the protein products of HERVs, are currently subjects of intense investigation and are likely involved in a number of diseases with complex etiologies, particularly in the context of autoimmunity and cancer.

In fact, envelope proteins of HERVs have been shown to initiate both innate and adaptive immunity, contributing to inflammatory, cytotoxic, and apoptotic responses and, conversely, to prevent the activation of immune responses. Notably, immunosuppressive properties and functions through immune negative regulators, namely Env proteins (i.e. Syncytin-1, Syncytin-2, Np9, and Rec) in different HERV groups, such as the HK2, HERV-W, HERV-V, HERV-H and HERV-P groups (Lee et al. 2020), have been reported to suppress the immune system (Shekarian et al. 2017) and affect cell signaling pathways (Yu et al. 2012). HERV Env proteins have also been shown to induce abnormal cell-cell fusion, which may contribute to tumor progression and metastasis. The expression of ENV proteins, therefore, leads to oncogenic consequences. Specifically, it promotes cell proliferation, growth, migration, invasion, metastasis, and stemness in a variety of cancer types, such as breast cancer (Sorek et al. 2022), melanoma (Song et al. 2021), leukemia (Wang-Johanning et al. 2012), Kaposi's sarcoma (Levet et al. 2019) and pancreatic cancer (Curtin et al. 2012). Notably, even HERVs with profoundly defective envelope proteins and alternative envelope splice variants may contribute to alternative pathogenic mechanisms. One of the best-characterized examples of a disease-mediating mutant HERV-K (HML2), the *env* gene, with or without a 292-bp deletion, may produce a protein with a different length (either Np9 or Rec) that is thought to be oncogenic. Understanding their involvement in complex pathologic diseases has indicated that HERV Env proteins may be potential targets for therapeutic intervention. Notably, a monoclonal antibody against HERV-W Env is currently in clinical trials for the treatment for multiple sclerosis, making it the first HERV-based therapy. Hence, the HERV Env protein is an exciting target for anticancer therapy.

A small number of studies have demonstrated that the expression of multiple HERV clusters, such as HERV-W and HERV-K, contributes to the cancer stemness or pluripotency of colorectal cancer cells (Krishnamurthy et al. 2015) and melanoma cells (Song et al. 2021), respectively. However, the underlying mechanisms remain to be investigated. In addition, nonallelic recombination of HERV sequences resulted in their translocation to different regions of the genome. The novel proximity of these genes may be related to their expression, thus resulting in the activation of specific oncogenes (Diebold and Derfuss 2019) or the disrupted expression of tumor suppressor genes (Kornmann and Curtin 2020).

HERV LTRs can function as alternative promoters that regulate cellular gene expression leading to abnormal gene expression, such as proto-oncogene expression, ultimately contributing to tumorigenesis (Krone et al. 2005). In addition, in cases of T-cell acute lymphoblastic leukemia (T-ALL), the E1B exon derived from HERV-E has been observed to attenuate the expression of CD5 on the T-cell plasma membrane, thereby inhibiting CD5 function and causing uncontrolled leukemic T-cell proliferation (Curtin et al. 2020). Hence, HERVs seem to play important roles in tumorigenesis, with Env proteins from different HERV subgroups showing particularly crucial activity. Ariza and Williams demonstrated that overexpression of wild-type or mutant HERV-K dUTPase proteins upregulated the expression of genes related to the NF-κB pathway, such as IL-6, -8, -10, -17, -23, and TNF-α, and that this process did not rely on the enzymatic activity of dUTPase but on Toll-like receptor 2 (TLR2) activity (Escalera-Zamudio and Greenwood 2016).

## HERV-Derived ncRNAs or dsRNAs

HERV-derived ncRNAs or dsRNAs may regulate gene expression. Zhou et al. discovered a long noncoding RNA molecule (lncRNA) derived from the ERVs of mice that competitively bound and displace splicing factor proline glutamine-rich (SFPQ), which is a transcriptional repressor of v-rel reticuloendotheliosis viral oncogene homolog A (avian) (RELA). Interestingly, the human SFPQ protein has been shown to bind to the RNAs of three representative HERV transcripts (MER9a2, LTR5A, and MLT2A1), and the human SFPQ protein shares greater than 95% identity with the murine SFPQ protein. This finding implied that although these lncRNAs are not evolutionarily conserved, the mechanism of SPFQ interactions with ERV-derived lncRNAs is thought to be conserved in both mice and humans (Mager and Stoye 2015).

In addition to lncRNAs, HERVs can generate dsRNAs to regulate gene expression. For example, Lee et al. showed that radiation-induced transcription of HERV-related dsRNA and the subsequent activation of an innate antiviral pathway led to the transcription of downstream ISGs (Bannert and Kurth 2006). Moreover, it has been reported that different HERV-encoded proteins and nucleic acids are identified as pathogen-associated molecular patterns (PAMPs) or danger-associated molecular patterns (DAMPs) by cellular sensors located on the plasma or endosome membrane (Jern and Coffin 2008). Detection of these HERV-related molecules by transmembrane or cytosolic pattern recognition receptors (PRRs) initiates a signaling cascade that results in the transcription of immune genes encoding proinflammatory effectors, particular cytokines, and type I IFNs (Garcia-Montojo et al. 2018).

#### HERVs May Alter Gene Structure or Yield Antisense Sequences

HERVs can ultimately influence gene transcription by altering the gene structure or producing an antisense gene sequence. In Hodgkin lymphoma, colony-stimulating factor 1 receptor (CSF1R) is transcribed from single hypomethylated LTRs in the MaLR family (THE1B), which are aberrantly activated in Hodgkin lymphoma cells (Vargiu et al. 2016). Conley et al. found that LTRs located downstream of a gene promoted the transcription of antisense sequences at that gene, leading to regulatory effects at the transcriptional level (Hayward et al. 2015).

Similar to other ERVs and TEs, HERVs can insert randomly into host genomes. These insertions can alter chromosome structure and gene expression. The acquisition of these altered phenotypes was recapitulated by Payer and Burns, who pointed out that TE insertions (LINE-1) caused genetic disease in humans. Mechanistically, a TE insertion alters the splicing of a normal transcript and leads to aberrant protein translation by disrupting exons, providing additional splice sites, and driving homologous recombination that alters the gene structure (Mao et al. 2021). Although no evidence has indicated that HERVs are involved in genetic disease, the insertion of ERVs resulted in mutations in the genomes of inbred mice (Gemmell et al. 2019). Many studies have shown that HERVs are abnormally expressed in many kinds of human diseases, but the function of these HERVs is still unclear. These HERVs may be involved in the regulation of certain key genes in a disease process. Further investigation is required to explore HERV function in disease.

## *HERV Biological Functions*

Several studies have demonstrated that HERVs are functionally complex and have revealed both a "light side" and a "dark side" of HERVs (Suntsova et al. 2015).

### Physiological Functions Involving HERVs

HERVs are involved in the regulation of a variety of important physiological processes in the human body. One of the best-characterized examples of HERV involvement is the development of the early human embryo and placenta. In addition, HERVs may be closely related to neuronal and immune gene regulation.

#### HERVs Participate in Early Embryonic Development

HERVs are beneficial in the maintenance of normal bodily functions and in the promotion of human evolution because they provide adaptive material derived from syncytin proteins of HERV-W and HERV-H (Young et al. 2013) during, for example, placental formation and embryogenesis (Maliniemi et al. 2013; Zhang et al. 2013).

Syncytin-1 (syncytin-1) and syncytin-2 (syncytin-2) are both HERV envelope proteins, with the former belonging to the HERV-W family and the latter to the HERV-FRD family (Blaise et al. 2003). Syncytin proteins have been shown to be expressed mainly in placental syncytiotrophoblasts and multinucleated cells derived from fetal trophectoderm and mediate cell fusion: Recombinant syncytin proteins induced the formation of giant syncytia from multiple cell types, and in the presence of anti-syncytin antiserum, the fusion of human trophoblasts was significantly inhibited (Brooks et al. 2001). Furthermore, during early human embryonic development, a number of LTR elements are synchronously transcribed at each stage of embryonic development, with the transcribed LTR elements differing during different stages of development. Researchers have suggested that these LTR elements are markers that can be used to identify cell types, determine cell differentiation potential and distinguish cell groups during early embryonic development (Goke et al. 2015). In a systematic study, researchers identified a class of LTR elements in HERV-K that were transcribed during human embryonic development, starting at the activation of the embryonic genome at the 8-cell stage, continuing throughout the emergence of ectodermal cells in the preimplantation blastocyst and eventually ceasing when the ectodermal cells differentiated into embryonic stem cells. More importantly, the researchers detected the expression of HERV-K-derived viral-like particles and coat proteins in ectodermal cells, suggesting that HERV-K-derived proteins may function in the development of ectodermal cells. In addition, the team demonstrated that the HERV-K-encoded Rec (regulator of expression encoded by corf) protein increased IFITM1 (interferon-induced transmembrane protein 1) expression on the surface of pluripotent cells and inhibited the suppression of viral infection. These findings suggest that HERVK may be involved in antiviral activity during infection in embryonic development (Grow et al. 2015).

**HERVs Involved in Cellular Reprogramming**

Expression of HERVs is repressed in pluripotent stem cells. Methylation of DNA inhibits the expression of HERVs in differentiated cell types, whereas cis-methylation of ERVs and transactivation of transcriptional repressors mediate HERV repression in pluripotent stem cells (Gaudet et al. 2004; Wolf et al. 2008). Many studies have focused on the transcriptional and epigenetic regulation of HERVs and their roles in cell fate commitment and establishment. The ubiquitination-modified Kruppel-associated box-zinc finger proteins (KRAB-ZFPs) recognize specific DNA sequences to recruit TRIM28 (tripartite motif-containing 28), which in turn recruits histone modification complexes to catalyze the addition of repressive marks on HERVs (Margolin et al. 1994; Schultz et al. 2001; Wolf and Goff 2009).

The transcriptional regulation of HERV-H (ERVH) LTRs may be one of the main mediators of cell fate reprogramming. For example, the "Yamanaka cocktail''(overexpressing Oct3/4, SOX2, and KLF4 proteins) induces the generation of pluripotent stem cells, and it seems highly likely that this effect is mediated by long intergenic noncoding regulatory RNAs that are driven by HERV-H (ERVH) (Lu et al. 2014; Ohnuki et al. 2014; Ko et al. 2021). In addition, the Env protein encoded by HERV-R (ERV3-1) has been suggested to be a possible mediator in development, as it is overexpressed in developing tissues such as the kidney, tongue, heart, liver, and brain (Andersson et al. 2002; Ko et al. 2021). The LTR7 repeats are flanked by HERV-H elements that are essential for the maintenance of pluripotency (Friedli et al. 2014). When human-reprogrammed cells enter the iPSC stage, HERV-H is transiently overactivated. However, when the cells reach full pluripotency, the expression level of HERV-H is partially suppressed. Thus, transient reactivation of HERV-H has been associated with the reprogramming of somatic cells toward pluripotency and the establishment of their differentiation potential. Genes located in neighboring loci were also activated after HERVH de-repression (Lim and Knowles 2015). During reprogramming, HERVs may drive gene expression.

**HERVs Involved in Human Evolution**
The long-term effects of retroviruses on vertebrate evolution are twofold: On the one hand, as with other viruses, retroviruses drive the evolution of host genes that impede viral infection or mitigate their pathogenic effects; on the other hand, through endogenization, retroviruses contribute a large number of new genetic features to a host genome, including unique protein-coding genes and cis-regulatory elements.

The estimated age of syncytin ranges from approximately 12 million to more than 80 million years (Lavialle et al. 2013), but the receptor-blocking envelope is younger, as evidenced by a narrower taxonomic distribution, lower degree insertion polymorphism, and an estimated integration time in the past 20 million years (Jern and Coffin 2008; Blanco-Melo et al. 2017). It is conceivable that many defective env gene sequences in the genome of humans and other vertebrates may have originally blocked the entry of viruses but were eliminated due to the extinction of selectors. This possibility suggests that viruses, which play important roles in the development of organisms, likely evolved via continual purifying selection. In contrast, genes that inhibit the replication of exogenous viruses may have been exposed to a shorter selection process; that is, when exoviruses became extinct or were replaced by resistant variants, exogenes were lost via drift (Aswad and Katzourakis 2012). Recently, certain evidence has suggested that HERVs may have facilitated the coevolution of genomes coordinated by gene regulatory networks (GRNs) (Khodosevich et al. 2002; Lynch 2016; Simonti et al. 2017). Coregulation of genes may involve shared cis-regulatory elements (CREs), and the evolutionary recombination of GRNs may be a source of phenotypic variation and species diversity (Carroll 2008).

**HERVs Involved in Human Evolution**
HERVs can also act as master transcriptional regulators of human genes by direct enhancement or promotion or via other mechanisms, sometimes involving long noncoding transcripts (Pi et al. 2010). Human hippocampus-specific transcription, for example, is regulated by enhancer elements generated by the insertion of HERV-

K/HML-2 (ERVK) LTRs, which may show significant implications in the evolution of humans (Suntsova et al. 2013).

**HERVs Involved in Immune System Regulation**

HERVs may be involved in the regulation of the human immune system and may be influenced by a variety of environmental factors (Gosenca et al. 2012).

On one hand, highly immunogenic HERVs are readily recognized due the responses of endogenous T cells and B cells and cleared by the immune system. Some HERV epitopes generate immune responses that are too weak to promote antitumor immunity; these responses can be classified into two categories: nonspecific responses (checkpoint blockade therapies and innate immune agonists) or epitope-specific responses (vaccination and T-cell therapies). However, the reactivation and expression of HERVs generate antigens that stimulate the immune system through an upregulated viral defense mechanism pathway, which positively correlates with the presence of infiltrated immune CD8+ T cells. In contrast, HERVs encode the Rec and NP9 oncogenes, which upregulate the expression of the immunosuppressive β-catenin pathway by interacting with promyelocytic leukemia zinc finger protein (PLZF).

However, HERVs are also involved in antiviral immune processes. Schmidt et al. demonstrated that influenza A virus (IAV) infection can reduce SUMO-TRIM28 activity (small ubiquitin-like modifier-modified tripartite motif-containing 28), leading to the desuppression of HERVs, which may elicit interferon-mediated natural immunity and inhibit influenza virus replication.

The abovementioned examples illustrate the roles played by HERVs in important human physiological processes. The molecular mechanisms involved in these functions enable HERVs to act as cis-acting elements or encoding proteins to regulate many important genetic and cytological processes in the human body.

## HERVs Involved in Human Diseases

Nonetheless, numerous reports have implicated HERVs in different types of diseases, including autoimmune diseases, neurological diseases, infectious diseases, and cancers. For example, the dysregulation of HERVs has been detected in a variety of cancer types, such as breast cancer (BC), melanoma, renal cancer, lung cancer and ovarian cancer (OC) (Alcazer et al. 2020; Geis and Goff 2020; Gao et al. 2021; Jansz and Faulkner 2021; Matteucci et al. 2022). In addition, HERVs have been associated with other diseases, such as type 1 diabetes (Levet et al. 2019), neurological disorders (Kury et al. 2018, Groger et al. 2021), and other autoimmune diseases (Kremer et al. 2020). Furthermore, due to their ability to move in the genome and their own cis-element functions, HERVs are capable of altering the normal structure of genes after their insertion, leading to the development of certain diseases (Geis and Goff 2020; Jakimovski et al. 2020).

**Autoimmune Diseases**
Biases in the expression of proviral proteins in human tissues can initiate autoimmune diseases (Karimi et al. 2019; Jin et al. 2022). This outcome was first shown by the high levels of proviral transcripts (Ehlhardt et al. 2006) and the presence of anti-HERV protein antibodies in the sera of several groups of patients with these systemic diseases (Bannert and Kurth 2004). Immune responses to HERV products typically arise spontaneously in infections or cancers and are believed to drive several autoimmune diseases in humans and in mice. The marked overexpression of HERV genes may be associated with the global hypomethylation of T-cell DNA in patients with systemic lupus erythematosus (SLE). The behavior of different HERV families in SLE varies. HERV-E mRNA expression, for example, is increased in CD4+ T cells from lupus patients, while members of the HERV-K/HML-2 (ERVK) and HERV-W (ERVW-1) families are expressed to a similar degree in patients with SLE and healthy controls. HERV-E mRNA expression levels are positively correlated with disease activity in SLE. HERV-E LTR methylation levels were reduced in lupus patients and negatively correlated with HERV-E mRNA expression (Wu et al. 2015). HRES-1, ERV-3, HERV-E4-1 (ERVE 1–4), HERV-K10 (ERVK-10), and HERV-K18 (ERVK-18) were overexpressed and may possibly be associated with SLE (Nelson et al. 2014a, b). A statistically significant increase in HERVK10 (ERVK-10) Gag protein, HERV-W (ERVW-1) transcripts, and protein products (isoforms of syncytial proteins) was found in patients with rheumatoid arthritis compared with the increase in normal control subjects (Nelson et al. 2014a, b). The presence of HERV-K/HML-2 (ERVK) components in autoimmune disease samples may be related to multiple binding sites for a variety of inflammation-associated transcription factors in LTRs. Inflammatory diseases, however, may also be associated with a reduction in HERV gene expression (Nogueira et al. 2015); for example, psoriasis is a multifactorial chronic skin disease in which patients exhibit a significantly reduced antibody response to HERV-K/HML-2 (ERVK) Gag and Env protein products detected in plasma (Gupta et al. 2014).

**Neurological Disorders**
The increased expression of HERV-encoded proteins is a candidate biomarker for several neurologic diseases (Manghera et al. 2015). IFNc is a proinflammatory cytokine that plays a key role in HERV-related neurologic disease pathology; for example, IFNc signaling has been shown to significantly increase HERV-K/HML-2 (ERVK) protein expression levels in astrocytes and human neurons. These findings suggest that HERV expression may be induced under inflammatory conditions (Christensen 2010). For example, the expression of the gene encoding HERV-K/HML-2 (ERVK) was significantly upregulated in the postmortem brains of amyotrophic lateral sclerosis (ALS) patients (Alfahad and Nath 2013). In the case of multiple sclerosis (MS), the slow progression of disease after neurodegeneration may be directly induced by reactivation of HERV expression and polymorphisms within the HERV-Fc1 gene on chromosome X as well as the HERV-K18 gene locus (ERVK-18) on chromosome 1 have been associated with susceptibility to MS (de la Hera et al. 2014). Infectious mononucleosis can lead to high levels of HERV-W

envelope protein and mRNA (ERVW-1) in blood mononuclear cells (Mameli et al. 2013). The expression of HERV-W elements (ERVW-1) and ERV9 in CNS cells was highly upregulated by treatment with neuroleptics and antidepressants (e.g., valproic acid, haloperidol, risperidone, and clozapine) (Diem et al. 2012). HERV-W expression (ERVW-1) was also significantly upregulated by certain prevalent nutrients as well as drugs such as caffeine and aspirin (Liu et al. 2013).

**Infectious Diseases**

The spontaneous long-term evolution of human and animal viruses has led to the development of a variety of mechanisms involving the cooperation or interference of endogenous and exogenous retroviruses.

Endogenous retroviral Env production may provide partial resistance to infection by exogenous pathogenic retroviruses in host cells. Interference with the Env receptor confers host cells with partial resistance to infection with exogenous pathogenic retroviruses or related retroviruses (Buzdin et al. 2003; Ponferrada et al. 2003), as in the case of the endogenous sheep retrovirus Jaagsiekte (JSRV), which blocks the entry of the exogenously associated viruses. In both infections, the same protein receptor is leveraged for viral entry, implying interference between endogenous and exogenous viruses (Spencer et al. 2003). Endogenous Gag proteins might also be involved in antiviral host cell protection. For example, expression of the endogenous Gag sequence Fv1 in mice blocked infection with some murine leukemia virus (MLV) strains, most likely as a result of direct encounters with the MLV virus (Bamunusinghe et al. 2017). In a similar manner, endogenous retroviral elements have been reported in both ground squirrels and cats (Best et al. 1997; Fujino et al. 2014). To date, there is no direct evidence for a human element, but it is theoretically possible because human DNA contains hundreds of intact or largely intact retroviral *env* genes (Malfavon-Borja and Feschotte 2015). In a recent association study with 23,000 participants, susceptibility to herpes zoster caused by varicella zoster virus (VZV) was shown to be associated with the noncoding gene HCP5 (HLA complex P5), which is also known as varicella zoster, an endogenous herpesvirus that can suppress viral activity through its regulatory function. Variants in this region of the gene have been associated with a delay in the development of AIDS in HIV-1-infected patients (Crosslin et al. 2015). Below, in the section HERV involvement in viral diseases, we outline the specific relationship between HERVs and viruses.

**Tumors**

HERVs are aberrantly expressed in cancer. For instance, the HERV-K/HML-2 element (ERVK) is overexpressed in germ cell tumors and melanomas (Muster et al. 2003; Oricchio et al. 2007). Positive regulation of HERVs can be mediated by combinations of cancer-specific transcription factors, as evidenced by activation of HERV-K/HML-2 (ERVK) via the melanoma-specific transcription factor MITF-M (Katoh et al. 2011). HERV-K/HML-2 (ERVK) mRNA and antibodies against proviral proteins have been found in primary breast cancer patients and testicular cancer patients, and serum levels of provirus mRNA tend to be higher in patients who eventually develop metastatic disease (Wang-Johanning et al. 2014). In addition, aberrant overexpression of the HERV-K/HML-2 (ERVK) provirus has been found in prostate cancer samples (Wallace et al. 2014), and the transcript or protein encoded by HERV-K/HML-2 (ERVK) is considered to be a possible biomarker of malignancy and is overexpressed in patients with a poor prognosis (Reis et al. 2013). However, the expression of these genes has not been reported in the literature. Bone marrow transplantation led to tumor regression in a group of patients with renal cell carcinoma, possibly as a result of the antitumor effect of the transplant. In these patients, antitumor cytotoxic lymphocytes targeted the HERV-E (ERVE)-encoded epitope (Takahashi et al. 2008), demonstrating the importance of immune responses to anti-HERVs in the development of or the cure for human diseases. Syncytium protein-1 is an envelope protein of the HERV-W family of proteins (ERVW-1), and its expression is usually limited to the placenta. However, it has been found in tumors, including melanoma, neuroblastoma, and breast cancer, where cell fusion mediated by syncytium proteins may be involved in the transformation or metastasis of cancer cells (Pothlichet et al. 2006; Manghera et al. 2015). The expression of immunosuppressive structural domains of the HERV-W envelope (ERVW-1) and HERV-K/HML-2 envelope (ERVK) in tumors suppressed immune responses and thus prevented the rejection of tumors and embryos (Morozov et al. 2013). In conclusion, the discovery of HTLV-1 (human T-lymphotropic virus-1) has provided clear evidence for a tumor-inducing exogenous human retrovirus (Kassiotis 2014).

## HERV Involvement in Viral Diseases

### *HERVs Associated with Retroviral Infections*

Several lines of evidence suggest that HERVs are associated with viral infections, especially retroviral infections.

Nearly all infectious retroviruses contain LTR sequences that carry dense clusters of binding sites that hijack cellular transcription factors and directly integrate the transcription of a provirus. To defend against retroviruses, mammalian cells have evolved mechanisms to epigenetically target and silence the expression of proviruses. As a counter adaptation, retroviruses, such as human immunodeficiency virus 1 (HIV-1), have been reported to have evolved multiple interferon-binding sites and inducible NF-κB-binding sites that can reactivate proviral transcription in response to innate immunity stimulation (Tchasovnikarova et al. 2015; Yang et al. 2015; Zhang et al. 2022). These switches in transcriptional activity depend on the epigenetic remodeling of an LTR, in which acetylation of a nucleosome activation mark drives the reactivation of proviral transcription (Asin et al. 2001; Chan and Greene 2012). At the cellular level, the balance between epigenetic repression and activation contributes to the stochastic reactivation of a latent retrovirus. HIV-1 infection can lead to the transactivation of the HERV-W family (ERVW-1) members, including their Env gene and syncytial proteins (Uleri et al. 2014).

A large body of evidence has indicated that HERVs are linked to viral infections. Notable studies have shown a correlation between HERVs and infection with *Retroviridae*. Nearly all infectious retroviruses carry long noncoding terminal repeat (LTR) sequences with dense clusters of binding sites that hijack cellular transcription factors and directly integrate transcription of the provirus.

Contreras-Galindo and colleagues showed that infection with human immunodeficiency virus type-1 (HIV-1) resulted in the expression of a novel HERV-K (HML-2) provirus called K111(Contreras-Galindo et al. 2007; Contreras-Galindo et al. 2013); Young et al. found that expression of certain other HERV-K HML-2 loci was increased in primary human CD4+ T cells following infection with HIV-1 (Young et al. 2018). Garrison et al. (2007) reported that HERV-specific CD8+ T cells present features consistent with a prominent function in the response to HIV-1 infection, and they proposed to use HERV-K peptides to treat HIV-1 infection. Also, Srinivasachar Badarinarayan et al. (2020) reported that HIV-1 infection resulted in the upregulation of the expression of LTR12C, which is a member of the ERVW9 superfamily and can serve as a promoter through which the expression of antiviral genes is regulated.

HERVs have also been reported to be involved not only in HIV infection but also in human T-lymphotropic virus 1 (HTLV-1) infection. For example, Toufaily et al. (2011) demonstrated that the Tax protein of HTLV-1 activated several terminal repeats that were HERVs LTRs, including HERV-W8 and HERV-H (MC16) LTRs, in vitro. Perzova et al. (2013) found that antibody titers of HERV-K10 peptides were significantly increased in patients with myelopathy associated with HTLV-I infection.

Interestingly, the vast majority of ERVs are HIV-1-like viruses and are normally targeted for silencing by epigenetic repressors at the intracellular level (Manghera et al. 2016). For example, the KRAB-ZNF family of transcriptional repressor proteins has extensively evolved to bind specific sequence motifs associated with different families of TEs. The binding of KRAB-ZNF proteins results in the recruitment of KAP1/TRIM28 and SETDB1 to repressive epigenetic marks, including H3K9me3 and are eventually involved in DNA methylation. These targeted inhibitory effects on an ERV have persisted long after these genes lost their coding ability (Yang et al. 2017). The KRAB-ZNF pathway is arguably the most highly evolved epigenetic repression mechanism of TEs identified to date, but multiple additional mechanisms, such as the piRNA pathway and the HUSH pathway, also lead to the epigenetic silencing of TEs.

## HERVs Associated with RNA Virus Infections

Reports have suggested that HERVs are functional not only in retroviruses but also in many other types of RNA virus infections.

Schmidt et al. (2019) provided evidence that influenza A virus (IAV) triggered the loss of the small ubiquitin-like modifier (SUMO)-modified tripartite motif-containing 28 (TRIM28), leading to depression of HERV expression, which may have triggered interferon (IFN)-mediated innate immune defenses to inhibit the replication of IAV. Notably, Tovo et al. demonstrated that HERV-H-*pol* and HERV-K-*pol* were overexpressed during chronic infection with hepatitis C virus (HCV) (Machlowska et al. 2020), and Wang et al. (2020) confirmed the activation of HERVs in infection with dengue virus serotype-2 (DENV-2). These studies validated previous observations suggesting differential expression of HERVs in positive or negative single-strand RNA virus infections. Interestingly, HERVs have also been found to be differentially expressed in severe acute respiratory syndrome coronavirus type 2 (SARS-CoV-2) infection, which has recently become a matter of global public health concern. Notably, HERV-W expression in COVID-19 patients has been found to correlate with disease outcome, suggesting the use of HERVs as biomarkers for viral infection. In addition, HERVs were upregulated in a human cell line following infection with SARS-CoV-2 and in bronchoalveolar lavage fluid obtained from COVID-19 patients (Kitsou et al. 2021; Sorek et al. 2021).

### *HERVs Associated with DNA Virus Infections*

In addition to those of HERVs and RNA viruses, relationships between HERVs and DNA virus infections, such as herpesviruses, have been reported.

Dai et al. demonstrated that Kaposi's sarcoma-associated herpesvirus (KSHV) transactivated HERV-K and that the HERV-K accessory gene NP9 was transcribed, playing a significant role in KSHV42 pathogenesis and tumorigenesis. Bello-Morales et al. (2021) concluded that herpes simplex virus type-1 (HSV-1) was involved in the regulation of a variety of HERVs, including HERV-H, HEV-W, and HERV-K viruses. Moreover, Assinger et al. (2013) and Bergallo et al. (2015) reported the upregulation of HERV-K and HERV-W in vitro and in vivo, respectively, following infection with human cytomegalovirus (HCMV).

## HERVs and Epigenetic Modifications

HERVs consist of a large number of heterogeneous retroviral integrated sequences. Successive waves of HERVs have proliferated and diversified throughout evolution, providing fertile ground for genomic innovation. HERVs, however, can compromise genome integrity by disrupting genome structure and expression; therefore, hosts have evolved a variety of mechanisms to limit the general activity of HERVs; these mechanisms include inhibitory small RNAs and histone modifications. These modifications can reduce the promoter activity of retroviral LTRs and interfere with the transcription of downstream viral genes, thereby preventing the spread of HERVs across the genome. Regulation of HERV homeostasis is achieved by cellular monitoring of different stages of the HERV life cycle. In particular, chromatin silencing due to DNA methylation and histone modifications as well as posttranscriptional control via RNA editing and RNA interference, has been extensively studied, and its involvement in regulated cellular processes such as chromatin remodeling, the cell cycle, splicing, nuclear transport, and actin nucleation has been examined (Schmidt et al. 2012; Ohnuki et al. 2014, Wang et al. 2015).

## *Retroviruses and Epigenetics*

During retroviral incubation, the cellular balance between repressive and activating epigenetic states contributes to random reactivation.

Despite the fact that most ERVs in cells are in a repressed epigenetic state, some biochemical evidence suggests that hundreds of ERVs exhibit epigenetic or transcriptional activity in mammalian cells. Similar to the silencing of modern proviruses, the epigenetic activation of HERVs is primarily driven by cellular transcription factors that bind to sequences within an LTR and recruit additional transcriptional machinery to the cell. HERVs that are epigenetically activated display all the characteristics typical of enhancer elements: increased accessibility, an abundance of histone marks associated with cancer (including H3K27ac and H3K4me3), p300 coactivator binding, and nascent POL2 RNA transcription. Different ERV families present with distinct transcription factor-binding-site repertoires, resulting in distinct cell type-specific or context-dependent patterns of activation, and individual insertions with a given HERV family can be mutated, duplicated, or truncated to expand or reduce HERV enhancer capacity (Ito et al. 2017; Tokuyama et al. 2018).

## *HERVs and DNA Methylation*

The methylation of DNA, the most abundant and stably inherited DNA modification that leads to 5-methylcytosine, occurs predominantly in CpG dinucleotides and has been shown to regulate transcription (Law and Jacobsen 2010; Jones 2012). The establishment and patterns of DNA methylation in different tissues and cells are critical for gene regulation, transposon silencing, and gene imprinting. DNA methylation is catalyzed throughout the cell cycle by members of the DNA methyltransferase family (DNMTs), which are highly conserved (Goll and Bestor 2005). DNMT1 maintains patterns of DNA methylation, whereas DNMT3A and DNMT3B methylate DNA. Epigenetics is intimately linked to the regulation of transcription. DNA methylation in a promoter region, for example, has been associated with repression of transcription (Esteller 2007), and dimethylation and trimethylation marks on histone 3 lysine 9 (H3K9me2/3) are repressive modifications, whereas histone 3 lysine 4 single and trimethylation (H3K4me1/3) marks indicate transcriptionally active sites (Vallianatos and Iwase 2015). Most human

ERVs are highly methylated in somatic cells, and demethylation has been associated with transcription or enhancers(Lavie et al. 2005, Schulz et al. 2006, Reiss et al. 2007, Stengel et al. 2010). CpG substitution rates in the primate genome are significantly higher in repeat sequences such as retroelements and the ERVs of L1 compared to nonrepeat DNA sequences, suggesting a mechanism of homology-dependent methylation(Aravin and Bourc'his 2008). Evidence suggests that small RNA pathways promote DNA methylation and repress human LINE1 non-LTR retrotransposon activity(Yang and Kazazian Jr. 2006), but their role in human ERV repression is not clear.

## Epigenetic Suppression of HERV Expression

Several studies have shown that DNA methylation has primarily evolved to silence ERVs and other TEs and that DNMTi silences ERVs (Chiappinelli et al. 2015; Amabile et al. 2016; Kogan et al. 2022).

LTR expression occurs in somatic cells due to the loss of DNA methylation. Although DNA methylation is important for ERV silencing in somatic stem cell tissues, histone modifications have also been shown to play important roles, particularly in undifferentiated and/or stem cells. Interactions of as many as hundreds of proteins may be required, in addition to DNA methylation early, in embryonic development, as the genome is largely unmethylated during this stage. Several of these proteins have been implicated in chromatin remodeling or a variety of histone posttranslational modifications. For example, H3K9me2 and H3K9me3 have been associated with the repression of transcription (Shilatifard 2006). SETDB1, a methyltransferase that catalyzes the addition of methyl groups to H3K9, was necessary for silencing LTR in mouse embryonic stem cells (ESCs) but not in mouse embryonic fibroblasts (Matsui et al. 2010). Many of the LTRs in mouse ESCs overlap with H3K9me3, and LTRs in differentiated cells mostly lacked this histone mark (Mikkelsen et al. 2007).

## Epigenetic Activation of HERVs

Most studies have focused on epigenetic repression of HERV expression, but some studies have shown that elements of HERV enhancer or promoter activity seem to be associated with activating epigenetic marks (Huda et al. 2011).

For example, tissue-specific hypermethylated LTRs and another retrovirally acquired mark, H3K4me1, on enhancers, have been shown to exhibit enhancer activity in reporter assays (Xie et al. 2013). In mouse ES cells with Kap1 deficiency, the repressive mark H3K9me3 was lost pm IAP ERVs, and the enhancer marks H3K27ac and H3K4me1, which are associated with the upregulation and induction of nearby genes, were acquired (Rowe et al. 2013). Mouse LTRs close to actively transcribed genes can acquire the active promoter mark H3K4me3, but when unmethylated, these LTRs may serve as alternative promoters of nearby genes

(Rebollo et al. 2012). Mice harboring a large number of human chromosomes are transcribed with many LTRs that are normally silenced on this chromosome and are associated with lower levels of DNA methylation and activating epigenetic marks (Ward et al. 2013), suggesting that the absence of species-specific repression mechanisms may result in the widespread upregulation of ERV sequence transcription (Mager and Stoye 2015).

## *HERVs and DNA Acetylation*

Histone deacetylases (HDACs) constitute an important class of "erasers" that antagonize histone acetyltransferases (HATs) by removing acyl groups from histones and nonhistones (Dokmanovic et al. 2007). A total of 18 HDAC proteins are encoded in the human genome, and they show sequence similarity to yeast homologs and require Zn2+ (Classes I, II, and IV) or the NAD+ cofactor (Class III) for enzymatic activity. Deletion of several HDACs leads to embryonic lethality or severe defects in development. HDACs have been shown to play important functions in appropriate downregulated genes (Haberland et al. 2009). The removal of acyl groups from H3K27 at enhancers inhibited proximal transcription (Mendenhall et al. 2013), whereas H3K27 acetylation at a fluoride promoter increased transcriptional activity (Hilton et al. 2015).

Inhibitors of histone deacetylases (HDACis) activate HERVs primarily through (i) activation of HERVs with the simultaneous synthesis of double-stranded RNAs (dsRNAs) and (ii) massive promoter activation at long terminal repeat sequences (LTRs) derived from previous instances of HERV invasion. This activation results in the synthesis of unannotated transcripts that may encode new or cryptic proteins. The dsRNA activates an antiviral response pathway, followed by apoptosis, and ultimately may be used for the treatment of tumors or viral infections (Daskalakis et al. 2018).

Several studies have demonstrated that HDAC inhibitors can increase the transcription of HERV elements in malignant cells such as ovarian cancer cells. HERV-V2, for example, encodes an envelope protein similar to syncytium. This protein is significantly overexpressed after HDACi exposure and can be effectively targeted by the concurrent application of TLR7/8 agonists to trigger intrinsic apoptosis. These synergistic cytotoxic effects are accompanied by the functional disruption of the TLR7/8-NFκB, Akt/PKB, and Ras-MEK-ERK signaling pathways (Diaz-Carballo et al. 2021). In lung cancer cell lines, DNMTi and HDACi have been shown to aggregate at the same site to induce unannotated TSSs (TINATs) because TINATs are encoded in long unique stretches of terminal repeats in the ERV9/LTR12 family and are epigenetically silenced in nearly all normal cells (Brocks et al. 2017). When testicular cancer cells were treated with HDACi and the death ligand TNF-related apoptosis-inducing ligand (TRAIL), rapid cell death dependent on ERV9-LTR-driven TNFRSF10B expression was observed (Beyer et al. 2016). HDAC1–3 inhibition resulted in the activation of LTR12. Notably, HDAC inhibitors induced

LTR12 activity not only in testicular cancer cells but also in cells of many other tumor types. In addition, the transcription factor NF-Y mediated LTR12 promoter activity and apoptosis induced by HDAC inhibitors. Therefore, activation of LTR12 by HDAC inhibitors is a generally applicable method to induce proapoptotic genes in human cancer cells (Kronung et al. 2016).

Many HDACi drugs that induce the proviral expression of HIV-1 may show the potential to promote HERV expression and cause unpredictable pathological changes, but this possibility is debated (Magiorkinis et al. 2013). HERV-K (HML-2) is the most recently discovered active HERV family (it shows unfixed endogenous retroviral insertion in the population) with intact open reading frames and intact long terminal repeats (LTRs) in transcriptional enhancers. Two genes encoding HERVs *env* (HERV-W *env* and HERV-FRD *env*) have been shown to be controlled via host selection and to encode proteins (syncytin-1 and syncytin-2, respectively) critical to placentation and immune tolerance of a fetus (Dupressoir et al. 2011). For example, Gkikas Magiorkinis et al. analyzed HERV genes in HIV-1 latent T-cell (J-LAT 8.4) and monocyte (U1) models after treatment with the HDAC inhibitors vorinostat, panobinostat, and romidepsin transcription. They found that HDAC inhibitors did not significantly upregulate the transcription of HERV-K *env, pol* or HERV-W *env* (encoding syncytin-1) or HERV-FRD *env* (encoding syncytin-2), suggesting that the acetylation of histones was not involved in these experiments. Thus, acetylation was not critical for the control of HERV expression in primary human T cells in a model or ex vivo (Hurst et al. 2016).

## The Mechanisms of HERVs in Antiviral Immunity

Although numerous studies have been conducted to evaluate HERV-mediated antiviral immunity (Hurst and Magiorkinis 2015; Srinivasachar Badarinarayan and Sauter 2021), the precise mechanism of HERV activity in human viral infection remains poorly understood. Investigators have discovered that HERVs can be transcribed into RNA and can produce proteins (Paces et al. 2013). Thus, expressed HERVs may produce nucleic acids or proteins with viral features, such as pathogen-associated molecular patterns of foreign viruses that are sensed by the innate immune system. In the present study, expressed HERVs induce innate immune activation and exert four distinct antiviral effects: (1) HERVs function as promoters or enhancers of antiviral genes; (2) HERV-derived nucleic acids or proteins are involved in antiviral responses; (3) HERV-derived noncoding RNA molecules (ncRNA) or dsRNA molecules (dsRNA) regulate the antiviral response; and (4) the endogenous retroviral proteins significantly and negatively supplement the effects of the exogenous virus particles. Therefore, HERVs may be associated with multiple immune signaling pathways.

The NF-κB signaling pathway has been linked to HERVs in viral infections. Ariza and Williams (2011) demonstrated that both wild-type and mutant HERV-K dUTPase proteins induced NF-κB activation and reported that this process did not depend on dUTPase enzyme activity but on Toll-like receptor 2 (TLR2) activity. In mice, Zhou et al. (2019) found that a full-length ERV was a positive regulator of NF-κB signaling; from this ERV, a long noncoding RNA molecule (lncRNA) was derived, and the lncRNA promoted RELA expression by competitively binding to and displacing the SFPQ protein.

Furthermore, HERVs may affect other signaling pathways related to immune responses. HERVs are closely related to cyclic GMP-AMP (cGAS) and cyclic GMP-AMP receptor stimulator of the interferon gene (STING) signaling pathway and have been shown to be involved in the development of the immune system. The intrinsic keratinocyte response to an ERV depended on STING signaling and promoted the induction of commensal-specific T cells (Lima-Junior et al. 2021). In another study, radiation-induced ERV activation played an important role in regulating the antitumor immune response2. Radiation-induced transcription of HERV-related dsRNAs and the subsequent activation of the innate antiviral MDA5/MAVS/TBK1 pathway led to the transcription of genes stimulated by downstream IFNs (Lee et al. 2020).

In addition, HERVs have been reported to be involved in innate immune responses. Pattern recognition receptors (PRRs) are germline-encoded receptors that play pivotal roles in antiviral defenses, as well as in response to self-harm, through recognition of pathogen-associated molecular patterns (PAMPs; i.e., lipids, proteins, glycans and nucleic acids) (Lee et al. 2020) and danger-associated molecular patterns (DAMPs) (Shekarian et al. 2017). A study revealed that the innate immune pathways activated by HERV-derived products were involved in the first-line antiviral defense against exogenous retroviruses. Upregulation of HERV transcription, including upregulation not caused by the initiation of immune-related pathogenesis, led to the expression of many HERV-derived PAMPs or DAMPs that drove further inflammation, contributing to the development of symptomatology.

In summary, HERVs are closely associated with different types of viruses, such as retroviruses, RNA viruses, and DNA viruses. HERVs regulate immune-related genes through a variety of mechanisms. It is, therefore, important to understand the connections between HERVs and human viruses, which may provide novel insights into therapeutics for diseases.

## The Roles of HERVs in Therapy

### *Therapies Targeting HERVs*

As mentioned herein, HERVs have been reported to be aberrantly expressed in several types of disease. As a result, therapies have been developed to target these HERVs and thus control their abnormal expression. HERVs are markers in biomedicine for the early diagnosis of diseases such as cancers, providing a new perspective for the clinical translation of HERVs (Song et al. 2021).

### Specific Monoclonal Antibodies (mAbs)

Notably, Wang-Johanning et al. (2012) reported that anti-HERV-K-specific monoclonal antibodies (mAbs) inhibited the growth of BC cells and induced their apoptosis in vitro by targeting highly expressed HERV-K. They also indicated that mice treated with anti-HERV-K mAbs showed significantly decreased xenograft tumor growth compared with that in control groups. Another example of the application of anti-HERV mAbs is GNbAC1 (also called temelimab), a humanized immunoglobulin G4 (IgG4) mAb that targets the retroviral envelope protein (MSRV-Env), which has been associated with multiple sclerosis (e.g., HERVW9) (Curtin et al. 2012). GNbAC1 has been reported to reduce the severity of cortical and thalamic atrophy in patients with multiple sclerosis (Diebold and Derfuss 2019) and did not produce serious adverse events during treatment (Kornmann and Curtin 2020). In addition, pharmacodynamic effects of GNbAC1 have been observed in patients with type I diabetes (Curtin et al. 2020), pointing to the possibility of using antibodies derived from a family of HERVs to treat multiple diseases. It has also been reported that HERVs are actionable vaccine antigens. For example, Kraus et al. (2013) generated a recombinant vaccinia virus (MVA-HK *env*) based on a vaccinia virus Ankara (MVA)-engineered construct and expressed the envelope glycoprotein HERV-K (HKenv). These authors found that a single vaccination in MVA-HKenv markedly reduced the number of lung RLZ-HKenv tumor nodules (derived from murine renal carcinoma cells that had been genetically transformed to express HKenv) compared with the effect of a vaccine on murine models with wild-type MVA. In another study, the same group used the same method and found that treatment with the HERV-K *gag* protein (HK*gag*) decreased the number of murine pulmonary RLZ-HK *gag* tumor nodules (Kraus et al. 2014).

### Chimeric Antigen Receptor T Cells (CAR-T)

Chimeric antigen receptor T-cell (CAR-T) therapy has been developed for targeting HERVs. Krishnamurthy et al. (2015) found that most genetically engineered T cells expressing CAR acquired a memory phenotype and lysed HERV-K envelope-positive (melanoma) tumor targets in an antigen-specific manner. HERV-K envelope-specific CAR-T cells have also been used effectively to treat mice with disseminated HERV-K envelope-positive tumors. In addition, HERV-K has been used for the construction of CAR-T-cell therapy against BC. For example, Zhou et al. (2015) reported that HERVK-CAR-T cells derived from peripheral blood mononuclear cells (PBMCs) obtained from both BC patients and normal donors inhibited the growth of BC cells and induced significant cytotoxicity in these cells. These findings, however, were not observed in normal breast cells, indicating that HERV-K is a tumor-specific antigen of BC. In addition, Zhou and colleagues found a significant decrease in tumor growth and tumor weight in xenograft mice, and HERVK-CAR-T cells prevented the metastasis of tumors to other organs.

Furthermore, the downregulation of HERV-K expression in the aforementioned murine model was correlated with the upregulation of tumor protein P53 (p53) expression and the downregulation of mouse double minute 2 (MDM2) and phosphorylated extracellular signal-regulated kinase (p-ERK) expression. For example, the HERV-K envelope protein in metastatic tumor tissues was synchronously reduced with the expression of the viral oncogene homolog (H-Ras) in V-Ha-Ras Harvey rat sarcoma 66. Collectively, these observations linked HERV-K to the p53 and Ras signaling pathways in tumorigenesis and provided mechanistic insights into HERVs, showing that they can be utilized as therapeutic targets in cancer. Saini et al. (2020) recently assessed the potential of T cells to recognize HERVs in myeloid malignancies and found that populations of CD8+ T cells recognized 29 HERV-derived peptides encoded at 18 different HERV loci, and some of these genes (HERVH-5, HERVW-1, and HERVE-3) were found to induce strong T-cell reactivity. In addition, HERV-specific T cells were found in 50% of patients (17/34) but were found less commonly in healthy donors. This study confirmed that HERVs are prospective targets for immunotherapies.

## Epigenetic Modifications

An alternative therapy targeting HERVs has been shown to modulate HERV expression via epigenetic modification. Epigenetic regulation involves the modification of DNA and histones with DNA to generate chromatin. Closely packed chromatin is related to gene silencing, and loose chromatin is related to increased gene expression. Moreover, nucleotide modification can be achieved through the addition of chemical groups (such as a methyl or acetyl group). DNA methyltransferase inhibitors (DNMTis) induced the expression of dsRNAs that had been transcribed from HERVs to trigger IFN-linked pathways (Dawson and Kouzarides 2012). Two different research groups concurrently reported this mechanism. Briefly, Chiappinelli et al. (2015) and Roulois et al. (2015) showed that HERV (HERVHF and HERVW9) transcription was upregulated in OC epithelial cells following treatment with DNMTis and that these HERVs formed dsRNAs. Further analyses revealed that the expression of these HERVs was significantly correlated with viral defense genes expressed in primary OC. Roulois and colleagues also demonstrated that DNMTs targeted colorectal cancer-initiating cells (CICs) through the induction of viral mimicry, which was associated with induction of dsRNAs derived from HERVs (HSERVIII), activation of the melanoma differentiation-associated protein 5 (MDA5)/mitochondrial antiviral signaling protein (MAVS) RNA recognition pathway and activation of INF regulatory factor 7 (IRF7) downstream of this pathway (Chiappinelli et al. 2015). Furthermore, transfection of dsRNA into CIC mimicked the effects of DNMTis, and knocking down the expression of MDA5, MAVS, or IRF7 inhibited these effects (Roulois et al. 2015). Histone deacetylase inhibitors (HDACis) have also been reported to regulate the expression of HERVs in addition to DNMTi proteins. These findings led to better mechanistic insights: The LTR promoter retained from a previous invasion of HERVs was broadly activated. Activation of this LTR promoter led to the synthesis of unannotated transcripts, which may encode previously unknown or novel proteins (Hurst and Magiorkinis 2017). Brocks et al. (2017) observed the transcription of thousands of nonannotated transcription start sites induced by HDACi treatment (TINATs), and HDACis have been shown to specifically induce a series of TINATs in association with acetylation of histone H2AK9 (H2AK9ac), H3K14ac, and H3K23ac proteins. Most importantly, this group found that TINATs were encoded in single LTRs of the ERV9/LTR12 family of proteins, which were repressed by epigenetic modifications in almost all normal cells, and in an in silico analysis, TINATs were predicted to encode potential immunoregulatory peptides. Another study demonstrated that TINATs were upregulated not only in cancer cell lines but also in +PBMCs of cancer patients following treatment with vorinostat (suberoylanilide hydroxamic acid, SAHA). However, this was not the case in the present study (Daskalakis et al. 2018). This discovery revealed the feasibility of using epigenetic modification drugs to regulate the expression of HERVs and treat diseases.

## *Therapies Utilizing HERV Tools*

### HERVs as Vaccine Carriers

In contrast to therapies that directly target HERVs, many therapies indirectly utilize HERVs as tools. For example, envelope glycoproteins of HERVs have been used to increase vaccine delivery efficiency. Notably, Lee et al. (2012) constructed a recombinant baculovirus system named AcHERV that was based on the baculoviral vector pFastBac1 and a synthetic envelope gene optimized with codons of W-type HERVs, and they used AcHERV as the nanocarrier for a DNA vaccine against human papillomavirus (HPV) (AcHERV-HP16L1) (Hurst and Magiorkinis 2017). Compared with the marketed vaccine, their DNA vaccine induced similar humoral immune responses and neutralized antibody titers but maintained significantly higher levels of cell-mediated immune responses in mice. Recently, the AcHERV system has been applied to develop a vaccine against Zika virus (AcHERV-ZIKV), and AcHERV-ZIKV induced elevated IgG, neutralizing antibody, and IFN-$\gamma$ levels in mice. In addition, AcHERV-ZIKV conferred complete protection to female IFN-knockout mice (A129), pregnant mice, and ZIKV-challenged fetal mice (Lee et al. 2012). HERVs have been used not only as vaccine vectors but also as vaccine adjuvants. Choi et al. (2015) reported that female mice immunized with an inactivated influenza vaccine in conjunction with a murine granulocyte-macrophage colony-stimulating factor (GmCSF) gene delivered via the AcHERV system exhibited greater humoral and cellular immune responses than mice immunized with the inactivated vaccine alone. The influenza-GmCSF-vaccinated mice were completely protected from the lethal challenge of the influenza virus. These findings showed that HERVs exhibit multiple functions in vaccine design.

### LTR as a Retrotransposon Homolog

Recently, Segel et al. (2021) identified a retrotransposon homolog of an LTR, paternally expressed gene 10 (PEG10), which preferentially bound and facilitated the vesicle-mediated secretion of messenger RNA (mRNA).

Surprisingly, this group showed that random messenger RNA was transferred in a manner similar to PEG10 mRNA after the mRNA sequence was flanked by untranslated regions (UTRs) of PEG10, and they developed a system called selective endogenous encapsidation for cell delivery (SEND) by remodeling both murine and human PEG10 to package, secrete, and deliver specific RNAs (Segel et al. 2021). Although PEG10 is a Ty3/Gypsy retrotransposon that cannot be classified as an HERV, PEG10 has been shown to form virus-like particles (VLPs) and efficiently transport RNA into cells. Similar results had been previously observed in neurons by Ashley et al. (2018), and notably, mammalian activity-regulated cytoskeleton-associated protein (Arc1), which they used in their study, is a retrovirus-like GAG protein of the Ty3/Gypsy family of retrotransposons. Given the close genetic relationship between Ty3/Gypsy retrotransposons and retroviruses, it is reasonable to hypothesize that a few HERV proteins perform similar functions.

In general, HERVs can be widely used either directly or indirectly. However, the clinical outcomes of HERV-related therapies have been inadequately evaluated, and therefore, further studies are urgently required to assess the efficacy and safety of these therapies.

**Acknowledgments** This work was supported by the National Natural Science Foundation of China (31970176), a grant from the CAS Interdisciplinary Innovation Team, and research grants from the Innovation Capacity Building Project of Jiangsu province (BM2020019).

## References

Aaronson SA, Todaro GJ, Scolnick EM (1971) Induction of murine C-type viruses from clonal lines of virus-free BALB-3T3 cells. Science 174(4005):157–159

Agoni L, Guha C, Lenz J (2013) Detection of human endogenous retrovirus K (HERV-K) transcripts in human prostate cancer cell lines. Front Oncol 3:180

Alcazer V, Bonaventura P, Depil S (2020) Human endogenous retroviruses (HERVs): shaping the innate immune response in cancers. Cancers (Basel) 12(3):610

Alfahad T, Nath A (2013) Retroviruses and amyotrophic lateral sclerosis. Antivir Res 99(2): 180–187

Alves PM, Levy N, Stevenson BJ, Bouzourene H, Theiler G, Bricard G, Viatte S, Ayyoub M, Vuilleumier H, Givel JC, Rimoldi D, Speiser DE, Jongeneel CV, Romero PJ, Levy F (2008) Identification of tumor-associated antigens by large-scale analysis of genes expressed in human colorectal cancer. Cancer Immun 8:11

Amabile A, Migliara A, Capasso P, Biffi M, Cittaro D, Naldini L, Lombardo A (2016) Inheritable silencing of endogenous genes by hit-and-run targeted epigenetic editing. Cell 167(1): 219–232. e214

Andersson ML, Medstrand P, Yin H, Blomberg J (1996) Differential expression of human endogenous retroviral sequences similar to mouse mammary tumor virus in normal peripheral blood mononuclear cells. AIDS Res Hum Retrovir 12(9):833–840

Andersson AC, Venables PJ, Tonjes RR, Scherer J, Eriksson L, Larsson E (2002) Developmental expression of HERV-R (ERV3) and HERV-K in human tissue. Virology 297(2):220–225

Aravin AA, Bourc'his D (2008) Small RNA guides for de novo DNA methylation in mammalian germ cells. Genes Dev 22(8):970–975

Argaw-Denboba A, Balestrieri E, Serafino A, Cipriani C, Bucci I, Sorrentino R, Sciamanna I, Gambacurta A, Sinibaldi-Vallebona P, Matteucci C (2017) HERV-K activation is strictly required to sustain CD133+ melanoma cells with stemness features. J Exp Clin Cancer Res 36(1):20

Ariza ME, Williams MV (2011) A human endogenous retrovirus K dUTPase triggers a TH1, TH17 cytokine response: does it have a role in psoriasis? J Invest Dermatol 131(12):2419–2427

Ashley J, Cordy B, Lucia D, Fradkin LG, Budnik V, Thomson T (2018) Retrovirus-like gag protein Arc1 binds RNA and traffics across synaptic boutons. Cell 172(1–2):262–274. e211

Asin S, Bren GD, Carmona EM, Solan NJ, Paya CV (2001) NF-kappaB cis-acting motifs of the human immunodeficiency virus (HIV) long terminal repeat regulate HIV transcription in human macrophages. J Virol 75(23):11408–11416

Assinger A, Yaiw KC, Gottesdorfer I, Leib-Mosch C, Soderberg-Naucler C (2013) Human cytomegalovirus (HCMV) induces human endogenous retrovirus (HERV) transcription. Retrovirology 10:132

Aswad A, Katzourakis A (2012) Paleovirology and virally derived immunity. Trends Ecol Evol 27(11):627–636

Bamunusinghe D, Liu Q, Plishka R, Dolan MA, Skorski M, Oler AJ, Yedavalli VRK, Buckler-White A, Hartley JW, Kozak CA (2017) Recombinant origins of pathogenic and nonpathogenic mouse gammaretroviruses with polytropic host range. J Virol 91(21):e00855-17

Bannert N, Kurth R (2004) Retroelements and the human genome: new perspectives on an old relation. Proc Natl Acad Sci U S A 101(Suppl 2):14572–14579

Bannert N, Kurth R (2006) The evolutionary dynamics of human endogenous retroviral families. Annu Rev Genomics Hum Genet 7:149–173

Bello-Morales R, Andreu S, Ripa I, Lopez-Guerrero JA (2021) HSV-1 and endogenous retroviruses as risk factors in demyelination. Int J Mol Sci 22(11):5738

Bentvelzen P, Daams JH, Hageman P, Calafat J (1970) Genetic transmission of viruses that incite mammary tumor in mice. Proc Natl Acad Sci U S A 67(1):377–384

Bergallo M, Galliano I, Montanari P, Gambarino S, Mareschi K, Ferro F, Fagioli F, Tovo PA, Ravanini P (2015) CMV induces HERV-K and HERV-W expression in kidney transplant recipients. J Clin Virol 68:28–31

Best S, Le Tissier PR, Stoye JP (1997) Endogenous retroviruses and the evolution of resistance to retroviral infection. Trends Microbiol 5(8):313–318

Beyer U, Kronung SK, Leha A, Walter L, Dobbelstein M (2016) Comprehensive identification of genes driven by ERV9-LTRs reveals TNFRSF10B as a re-activatable mediator of testicular cancer cell death. Cell Death Differ 23(1):64–75

Bhardwaj N, Maldarelli F, Mellors J, Coffin JM (2014) HIV-1 infection leads to increased transcription of human endogenous retrovirus HERV-K (HML-2) proviruses in vivo but not to increased virion production. J Virol 88(19):11108–11120

Bhat RK, Rudnick W, Antony JM, Maingat F, Ellestad KK, Wheatley BM, Tonjes RR, Power C (2014) Human endogenous retrovirus-K(II) envelope induction protects neurons during HIV/AIDS. PLoS One 9(7):e97984

Blaise S, de Parseval N, Benit L, Heidmann T (2003) Genomewide screening for fusogenic human endogenous retrovirus envelopes identifies syncytin 2, a gene conserved on primate evolution. Proc Natl Acad Sci U S A 100(22):13013–13018

Blanco-Melo D, Gifford RJ, Bieniasz PD (2017) Co-option of an endogenous retrovirus envelope for host defense in hominid ancestors. elife 6:e22519

Brocks D, Schmidt CR, Daskalakis M, Jang HS, Shah NM, Li D, Li J, Zhang B, Hou Y, Laudato S, Lipka DB, Schott J, Bierhoff H, Assenov Y, Helf M, Ressnerova A, Islam MS, Lindroth AM, Haas S, Essers M, Imbusch CD, Brors B, Oehme I, Witt O, Lubbert M, Mallm JP, Rippe K, Will R, Weichenhan D, Stoecklin G, Gerhauser C, Oakes CC, Wang T, Plass C (2017) DNMT and HDAC inhibitors induce cryptic transcription start sites encoded in long terminal repeats. Nat Genet 49(7):1052–1060

Brooks EM, Branda RF, Nicklas JA, O'Neill JP (2001) Molecular description of three macro-deletions and an Alu-Alu recombination-mediated duplication in the HPRT gene in four patients with Lesch-Nyhan disease. Mutat Res 476(1–2):43–54

Buscher K, Trefzer U, Hofmann M, Sterry W, Kurth R, Denner J (2005) Expression of human endogenous retrovirus K in melanomas and melanoma cell lines. Cancer Res 65(10):4172–4180

Buzdin A, Ustyugova S, Khodosevich K, Mamedov I, Lebedev Y, Hunsmann G, Sverdlov E (2003) Human-specific subfamilies of HERV-K (HML-2) long terminal repeats: three master genes were active simultaneously during branching of hominoid lineages. Genomics 81(2):149–156

Carroll SB (2008) Evo-devo and an expanding evolutionary synthesis: a genetic theory of morphological evolution. Cell 134(1):25–36

Chan JK, Greene WC (2012) Dynamic roles for NF-kappaB in HTLV-I and HIV-1 retroviral pathogenesis. Immunol Rev 246(1):286–310

Chiappinelli KB, Strissel PL, Desrichard A, Li H, Henke C, Akman B, Hein A, Rote NS, Cope LM, Snyder A, Makarov V, Budhu S, Slamon DJ, Wolchok JD, Pardoll DM, Beckmann MW, Zahnow CA, Merghoub T, Chan TA, Baylin SB, Strick R (2015) Inhibiting DNA methylation causes an interferon response in cancer via dsRNA including endogenous retroviruses. Cell 162(5):974–986

Choi HJ, Gwon YD, Jang Y, Cho Y, Heo YK, Lee HJ, Kim KC, Choi J, Lee JB, Kim YB (2015) Effect of AcHERV-GmCSF as an influenza virus vaccine adjuvant. PLoS One 10(6):e0129761

Christensen T (2010) HERVs in neuropathogenesis. J Neuroimmune Pharmacol 5(3):326–335

Contreras-Galindo R, Lopez P, Velez R, Yamamura Y (2007) HIV-1 infection increases the expression of human endogenous retroviruses type K (HERV-K) in vitro. AIDS Res Hum Retrovir 23(1):116–122

Contreras-Galindo R, Kaplan MH, He S, Contreras-Galindo AC, Gonzalez-Hernandez MJ, Kappes F, Dube D, Chan SM, Robinson D, Meng F, Dai M, Gitlin SD, Chinnaiyan AM, Omenn GS, Markovitz DM (2013) HIV infection reveals widespread expansion of novel centromeric human endogenous retroviruses. Genome Res 23(9):1505–1513

Crosslin DR, Carrell DS, Burt A, Kim DS, Underwood JG, Hanna DS, Comstock BA, Baldwin E, de Andrade M, Kullo IJ, Tromp G, Kuivaniemi H, Borthwick KM, McCarty CA, Peissig PL, Doheny KF, Pugh E, Kho A, Pacheco J, Hayes MG, Ritchie MD, Verma SS, Armstrong G, Stallings S, Denny JC, Carroll RJ, Crawford DC, Crane PK, Mukherjee S, Bottinger E, Li R, Keating B, Mirel DB, Carlson CS, Harley JB, Larson EB, Jarvik GP (2015) Genetic variation in the HLA region is associated with susceptibility to herpes zoster. Genes Immun 16(1):1–7

Curtin F, Lang AB, Perron H, Laumonier M, Vidal V, Porchet HC, Hartung HP (2012) GNbAC1, a humanized monoclonal antibody against the envelope protein of multiple sclerosis-associated endogenous retrovirus: a first-in-humans randomized clinical study. Clin Ther 34(12): 2268–2278

Curtin F, Champion B, Davoren P, Duke S, Ekinci EI, Gilfillan C, Morbey C, Nathow T, O'Moore-Sullivan T, O'Neal D, Roberts A, Stranks S, Stuckey B, Vora P, Malpass S, Lloyd D, Maestracci-Beard N, Buffet B, Kornmann G, Bernard C, Porchet H, Simpson R (2020) A safety and pharmacodynamics study of temelimab, an antipathogenic human endogenous retrovirus type W envelope monoclonal antibody, in patients with type 1 diabetes. Diabetes Obes Metab 22(7):1111–1121

Daskalakis M, Brocks D, Sheng YH, Islam MS, Ressnerova A, Assenov Y, Milde T, Oehme I, Witt O, Goyal A, Kuhn A, Hartmann M, Weichenhan D, Jung M, Plass C (2018) Reactivation of endogenous retroviral elements via treatment with DNMT- and HDAC-inhibitors. Cell Cycle 17(7):811–822

Dawson MA, Kouzarides T (2012) Cancer epigenetics: from mechanism to therapy. Cell 150(1): 12–27

de la Hera B, Varade J, Garcia-Montojo M, Alcina A, Fedetz M, Alloza I, Astobiza I, Leyva L, Fernandez O, Izquierdo G, Antiguedad A, Arroyo R, Alvarez-Lafuente R, Vandenbroeck K, Matesanz F, Urcelay E (2014) Human endogenous retrovirus HERV-Fc1 association with multiple sclerosis susceptibility: a meta-analysis. PLoS One 9(3):e90182

Diaz-Carballo D, Saka S, Acikelli AH, Homp E, Erwes J, Demmig R, Klein J, Schroer K, Malak S, D'Souza F, Noa-Bolano A, Menze S, Pano E, Andrioff S, Teipel M, Dammann P, Klein D, Nasreen A, Tannapfel A, Grandi N, Tramontano E, Ochsenfarth C, Strumberg D (2021) Enhanced antitumoral activity of TLR7 agonists via activation of human endogenous retroviruses by HDAC inhibitors. Commun Biol 4(1):276

Diebold M, Derfuss T (2019) The monoclonal antibody GNbAC1: targeting human endogenous retroviruses in multiple sclerosis. Ther Adv Neurol Disord 12:1756286419833574

Diem O, Schaffner M, Seifarth W, Leib-Mosch C (2012) Influence of antipsychotic drugs on human endogenous retrovirus (HERV) transcription in brain cells. PLoS One 7(1):e30054

Dokmanovic M, Clarke C, Marks PA (2007) Histone deacetylase inhibitors: overview and perspectives. Mol Cancer Res 5(10):981–989

Dolei A (2018) The aliens inside us: HERV-W endogenous retroviruses and multiple sclerosis. Mult Scler 24(1):42–47

Douville R, Liu J, Rothstein J, Nath A (2011) Identification of active loci of a human endogenous retrovirus in neurons of patients with amyotrophic lateral sclerosis. Ann Neurol 69(1):141–151

Dupressoir A, Vernochet C, Harper F, Guegan J, Dessen P, Pierron G, Heidmann T (2011) A pair of co-opted retroviral envelope syncytin genes is required for formation of the two-layered murine placental syncytiotrophoblast. Proc Natl Acad Sci U S A 108(46):E1164–E1173

Ehlhardt S, Seifert M, Schneider J, Ojak A, Zang KD, Mehraein Y (2006) Human endogenous retrovirus HERV-K(HML-2) Rec expression and transcriptional activities in normal and rheumatoid arthritis synovia. J Rheumatol 33(1):16–23

Escalera-Zamudio M, Greenwood AD (2016) On the classification and evolution of endogenous retrovirus: human endogenous retroviruses may not be 'human' after all. APMIS 124(1–2): 44–51

Esteller M (2007) Epigenetic gene silencing in cancer: the DNA hypermethylome. Hum Mol Genet 16. Spec No 1:R50–R59

Freed EO, Mouland AJ (2006) The cell biology of HIV-1 and other retroviruses. Retrovirology 3:77

Friedli M, Turelli P, Kapopoulou A, Rauwel B, Castro-Diaz N, Rowe HM, Ecco G, Unzu C, Planet E, Lombardo A, Mangeat B, Wildhaber BE, Naldini L, Trono D (2014) Loss of transcriptional control over endogenous retroelements during reprogramming to pluripotency. Genome Res 24(8):1251–1259

Fuchs NV, Kraft M, Tondera C, Hanschmann KM, Lower J, Lower R (2011) Expression of the human endogenous retrovirus (HERV) group HML-2/HERV-K does not depend on canonical promoter elements but is regulated by transcription factors Sp1 and Sp3. J Virol 85(7): 3436–3448

Fujino K, Horie M, Honda T, Merriman DK, Tomonaga K (2014) Inhibition of Borna disease virus replication by an endogenous bornavirus-like element in the ground squirrel genome. Proc Natl Acad Sci U S A 111(36):13175–13180

Gao Y, Yu XF, Chen T (2021) Human endogenous retroviruses in cancer: expression, regulation and function. Oncol Lett 21(2):121

Garcia-Montojo M, Doucet-O'Hare T, Henderson L, Nath A (2018) Human endogenous retrovirus-K (HML-2): a comprehensive review. Crit Rev Microbiol 44(6):715–738

Garrison KE, Jones RB, Meiklejohn DA, Anwar N, Ndhlovu LC, Chapman JM, Erickson AL, Agrawal A, Spotts G, Hecht FM, Rakoff-Nahoum S, Lenz J, Ostrowski MA, Nixon DF (2007) T cell responses to human endogenous retroviruses in HIV-1 infection. PLoS Pathog 3(11):e165

Gaudet F, Rideout WM 3rd, Meissner A, Dausman J, Leonhardt H, Jaenisch R (2004) Dnmt1 expression in pre- and postimplantation embryogenesis and the maintenance of IAP silencing. Mol Cell Biol 24(4):1640–1648

Geis FK, Goff SP (2020) Silencing and transcriptional regulation of endogenous retroviruses: an overview. Viruses 12(8):884

Gemmell P, Hein J, Katzourakis A (2019) The exaptation of HERV-H: evolutionary analyses reveal the genomic features of highly transcribed elements. Front Immunol 10:1339

Goke J, Lu X, Chan YS, Ng HH, Ly LH, Sachs F, Szczerbinska I (2015) Dynamic transcription of distinct classes of endogenous retroviral elements marks specific populations of early human embryonic cells. Cell Stem Cell 16(2):135–141

Goll MG, Bestor TH (2005) Eukaryotic cytosine methyltransferases. Annu Rev Biochem 74:481–514

Gosenca D, Gabriel U, Steidler A, Mayer J, Diem O, Erben P, Fabarius A, Leib-Mosch C, Hofmann WK, Seifarth W (2012) HERV-E-mediated modulation of PLA2G4A transcription in urothelial carcinoma. PLoS One 7(11):e49341

Griffiths DJ (2001) Endogenous retroviruses in the human genome sequence. Genome Biol 2(6): 1017

Groger V, Emmer A, Staege MS, Cynis H (2021) Endogenous retroviruses in nervous system disorders. Pharmaceuticals (Basel) 14(1):70

Grow EJ, Flynn RA, Chavez SL, Bayless NL, Wossidlo M, Wesche DJ, Martin L, Ware CB, Blish CA, Chang HY, Pera RA, Wysocka J (2015) Intrinsic retroviral reactivation in human preimplantation embryos and pluripotent cells. Nature 522(7555):221–225

Gupta R, Michaud HA, Zeng X, Debbaneh M, Arron ST, Jones RB, Ormsby CE, Nixon DF, Liao W (2014) Diminished humoral responses against and reduced gene expression levels of human endogenous retrovirus-K (HERV-K) in psoriasis. J Transl Med 12:256

Haberland M, Montgomery RL, Olson EN (2009) The many roles of histone deacetylases in development and physiology: implications for disease and therapy. Nat Rev Genet 10(1):32–42

Hayward A, Cornwallis CK, Jern P (2015) Pan-vertebrate comparative genomics unmasks retrovirus macroevolution. Proc Natl Acad Sci U S A 112(2):464–469

Hilton IB, D'Ippolito AM, Vockley CM, Thakore PI, Crawford GE, Reddy TE, Gersbach CA (2015) Epigenome editing by a CRISPR-Cas9-based acetyltransferase activates genes from promoters and enhancers. Nat Biotechnol 33(5):510–517

Hirose Y, Takamatsu M, Harada F (1993) Presence of env genes in members of the RTVL-H family of human endogenous retrovirus-like elements. Virology 192(1):52–61

Huda A, Bowen NJ, Conley AB, Jordan IK (2011) Epigenetic regulation of transposable element derived human gene promoters. Gene 475(1):39–48

Hughes JF, Coffin JM (2001) Evidence for genomic rearrangements mediated by human endogenous retroviruses during primate evolution. Nat Genet 29(4):487–489

Hurst TP, Magiorkinis G (2015) Activation of the innate immune response by endogenous retroviruses. J Gen Virol 96(Pt 6):1207–1218

Hurst TP, Magiorkinis G (2017) Epigenetic control of human endogenous retrovirus expression: focus on regulation of long-terminal repeats (LTRs). Viruses 9(6):130

Hurst T, Pace M, Katzourakis A, Phillips R, Klenerman P, Frater J, Magiorkinis G (2016) Human endogenous retrovirus (HERV) expression is not induced by treatment with the histone deacetylase (HDAC) inhibitors in cellular models of HIV-1 latency. Retrovirology 13:10

Ito J, Sugimoto R, Nakaoka H, Yamada S, Kimura T, Hayano T, Inoue I (2017) Systematic identification and characterization of regulatory elements derived from human endogenous retroviruses. PLoS Genet 13(7):e1006883

Jakimovski D, Weinstock-Guttman B, Ramanathan M, Dwyer MG, Zivadinov R (2020) Infections, vaccines and autoimmunity: a multiple sclerosis perspective. Vaccines (Basel) 8(1):50

Jansz N, Faulkner GJ (2021) Endogenous retroviruses in the origins and treatment of cancer. Genome Biol 22(1):147

Jern P, Coffin JM (2008) Effects of retroviruses on host genome function. Annu Rev Genet 42:709–732

Jin X, Li X, Guan F, Zhang J (2022) Human endogenous retroviruses and toll-like receptors. Viral Immunol 36(2):73–82

Johnson WE (2019) Origins and evolutionary consequences of ancient endogenous retroviruses. Nat Rev Microbiol 17(6):355–370

Jones PA (2012) Functions of DNA methylation: islands, start sites, gene bodies and beyond. Nat Rev Genet 13(7):484–492

Kammerer U, Germeyer A, Stengel S, Kapp M, Denner J (2011) Human endogenous retrovirus K (HERV-K) is expressed in villous and extravillous cytotrophoblast cells of the human placenta. J Reprod Immunol 91(1–2):1–8

Karimi A, Esmaili N, Ranjkesh M, Zolfaghari MA (2019) Expression of human endogenous retroviruses in pemphigus vulgaris patients. Mol Biol Rep 46(6):6181–6186

Kassiotis G (2014) Endogenous retroviruses and the development of cancer. J Immunol 192(4): 1343–1349

Katoh I, Mirova A, Kurata S, Murakami Y, Horikawa K, Nakakuki N, Sakai T, Hashimoto K, Maruyama A, Yonaga T, Fukunishi N, Moriishi K, Hirai H (2011) Activation of the long terminal repeat of human endogenous retrovirus K by melanoma-specific transcription factor MITF-M. Neoplasia 13(11):1081–1092

Kelleher CA, Wilkinson DA, Freeman JD, Mager DL, Gelfand EW (1996) Expression of novel-transposon-containing mRNAs in human T cells. J Gen Virol 77(Pt 5):1101–1110

Khodosevich K, Lebedev Y, Sverdlov E (2002) Endogenous retroviruses and human evolution. Comp Funct Genomics 3(6):494–498

Kitsou K, Kotanidou A, Paraskevis D, Karamitros T, Katzourakis A, Tedder R, Hurst T, Sapounas S, Kotsinas A, Gorgoulis V, Spoulou V, Tsiodras S, Lagiou P, Magiorkinis G (2021) Upregulation of human endogenous retroviruses in bronchoalveolar lavage fluid of COVID-19 patients. Microbiol Spectr 9(2):e0126021

Ko EJ, Song KS, Ock MS, Choi YH, Kim S, Kim HS, Cha HJ (2021) Expression profiles of human endogenous retrovirus (HERV)-K and HERV-R Env proteins in various cancers. BMB Rep 54(7):368–373

Kogan AA, Topper MJ, Dellomo AJ, Stojanovic L, McLaughlin LJ, Creed TM, Eberly CL, Kingsbury TJ, Baer MR, Kessler MD, Baylin SB, Rassool FV (2022) Activating STING1-dependent immune signaling in TP53 mutant and wild-type acute myeloid leukemia. Proc Natl Acad Sci U S A 119(27):e2123227119

Kornmann G, Curtin F (2020) Temelimab, an IgG4 anti-human endogenous retrovirus monoclonal antibody: an early development safety review. Drug Saf 43(12):1287–1296

Kraus B, Fischer K, Buchner SM, Wels WS, Lower R, Sliva K, Schnierle BS (2013) Vaccination directed against the human endogenous retrovirus-K envelope protein inhibits tumor growth in a murine model system. PLoS One 8(8):e72756

Kraus B, Fischer K, Sliva K, Schnierle BS (2014) Vaccination directed against the human endogenous retrovirus-K (HERV-K) gag protein slows HERV-K gag expressing cell growth in a murine model system. Virol J 11:58

Kremer D, Weyers V, Gruchot J, Gottle P, Hartung HP, Perron H, Kury P (2020) Meeting report: "human endogenous retroviruses: HERVs or transposable elements in autoimmune, chronic inflammatory and degenerative diseases or cancer", Lyon, France, November 5th and 6th 2019 - an MS scientist's digest. Mult Scler Relat Disord 42:102068

Krishnamurthy J, Rabinovich BA, Mi T, Switzer KC, Olivares S, Maiti SN, Plummer JB, Singh H, Kumaresan PR, Huls HM, Wang-Johanning F, Cooper LJ (2015) Genetic engineering of T cells to target HERV-K, an ancient retrovirus on melanoma. Clin Cancer Res 21(14):3241–3251

Krone B, Kolmel KF, Henz BM, Grange JM (2005) Protection against melanoma by vaccination with Bacille Calmette-Guerin (BCG) and/or vaccinia: an epidemiology-based hypothesis on the nature of a melanoma risk factor and its immunological control. Eur J Cancer 41(1):104–117

Kronung SK, Beyer U, Chiaramonte ML, Dolfini D, Mantovani R, Dobbelstein M (2016) LTR12 promoter activation in a broad range of human tumor cells by HDAC inhibition. Oncotarget 7(23):33484–33497
Kurth R, Bannert N (2010) Beneficial and detrimental effects of human endogenous retroviruses. Int J Cancer 126(2):306–314
Kury P, Nath A, Creange A, Dolei A, Marche P, Gold J, Giovannoni G, Hartung HP, Perron H (2018) Human endogenous retroviruses in neurological diseases. Trends Mol Med 24(4): 379–394
Lavialle C, Cornelis G, Dupressoir A, Esnault C, Heidmann O, Vernochet C, Heidmann T (2013) Paleovirology of 'syncytins', retroviral env genes exapted for a role in placentation. Philos Trans R Soc Lond Ser B Biol Sci 368(1626):20120507
Lavie L, Kitova M, Maldener E, Meese E, Mayer J (2005) CpG methylation directly regulates transcriptional activity of the human endogenous retrovirus family HERV-K(HML-2). J Virol 79(2):876–883
Law JA, Jacobsen SE (2010) Establishing, maintaining and modifying DNA methylation patterns in plants and animals. Nat Rev Genet 11(3):204–220
Lee HJ, Hur YK, Cho YD, Kim MG, Lee HT, Oh YK, Kim YB (2012) Immunogenicity of bivalent human papillomavirus DNA vaccine using human endogenous retrovirus envelope-coated baculoviral vectors in mice and pigs. PLoS One 7(11):e50296
Lee AK, Pan D, Bao X, Hu M, Li F, Li CY (2020) Endogenous retrovirus activation as a key mechanism of anti-tumor immune response in radiotherapy. Radiat Res 193(4):305–317
Levet S, Charvet B, Bertin A, Deschaumes A, Perron H, Hober D (2019) Human endogenous retroviruses and type 1 diabetes. Curr Diab Rep 19(12):141
Lim AK, Knowles BB (2015) Controlling endogenous retroviruses and their chimeric transcripts during natural reprogramming in the oocyte. J Infect Dis 212(Suppl 1):S47–S51
Lima-Junior DS, Krishnamurthy SR, Bouladoux N, Collins N, Han SJ, Chen EY, Constantinides MG, Link VM, Lim AI, Enamorado M, Cataisson C, Gil L, Rao I, Farley TK, Koroleva G, Attig J, Yuspa SH, Fischbach MA, Kassiotis G, Belkaid Y (2021) Endogenous retroviruses promote homeostatic and inflammatory responses to the microbiota. Cell 184(14):3794–3811. e3719
Lindeskog M, Mager DL, Blomberg J (1999) Isolation of a human endogenous retroviral HERV-H element with an open env reading frame. Virology 258(2):441–450
Liu C, Chen Y, Li S, Yu H, Zeng J, Wang X, Zhu F (2013) Activation of elements in HERV-W family by caffeine and aspirin. Virus Genes 47(2):219–227
Lu X, Sachs F, Ramsay L, Jacques PE, Goke J, Bourque G, Ng HH (2014) The retrovirus HERVH is a long noncoding RNA required for human embryonic stem cell identity. Nat Struct Mol Biol 21(4):423–425
Lynch VJ (2016) GENETICS. A copy-and-paste gene regulatory network. Science 351(6277): 1029–1030
Machlowska J, Baj J, Sitarz M, Maciejewski R, Sitarz R (2020) Gastric cancer: epidemiology, risk factors, classification, genomic characteristics and treatment strategies. Int J Mol Sci 21(11): 4012
Mager DL, Stoye JP (2015) Mammalian endogenous retroviruses. Microbiol Spectr 3(1):MDNA3-0009-2014
Magiorkinis G, Belshaw R, Katzourakis A (2013) 'There and back again': revisiting the pathophysiological roles of human endogenous retroviruses in the post-genomic era. Philos Trans R Soc Lond Ser B Biol Sci 368(1626):20120504
Malfavon-Borja R, Feschotte C (2015) Fighting fire with fire: endogenous retrovirus envelopes as restriction factors. J Virol 89(8):4047–4050
Maliniemi P, Vincendeau M, Mayer J, Frank O, Hahtola S, Karenko L, Carlsson E, Mallet F, Seifarth W, Leib-Mosch C, Ranki A (2013) Expression of human endogenous retrovirus-w including syncytin-1 in cutaneous T-cell lymphoma. PLoS One 8(10):e76281

Mameli G, Madeddu G, Mei A, Uleri E, Poddighe L, Delogu LG, Maida I, Babudieri S, Serra C, Manetti R, Mura MS, Dolei A (2013) Activation of MSRV-type endogenous retroviruses during infectious mononucleosis and Epstein-Barr virus latency: the missing link with multiple sclerosis? PLoS One 8(11):e78474

Manca MA, Solinas T, Simula ER, Noli M, Ruberto S, Madonia M, Sechi LA (2022) HERV-K and HERV-H Env proteins induce a humoral response in prostate cancer patients. Pathogens 11(1): 95

Manghera M, Ferguson J, Douville R (2015) ERVK polyprotein processing and reverse transcriptase expression in human cell line models of neurological disease. Viruses 7(1):320–332

Manghera M, Ferguson-Parry J, Lin R, Douville RN (2016) NF-kappaB and IRF1 induce endogenous retrovirus K expression via interferon-stimulated response elements in its 5′ long terminal repeat. J Virol 90(20):9338–9349

Mao J, Zhang Q, Cong YS (2021) Human endogenous retroviruses in development and disease. Comput Struct Biotechnol J 19:5978–5986

Margolin JF, Friedman JR, Meyer WK, Vissing H, Thiesen HJ, Rauscher FJ 3rd (1994) Kruppel-associated boxes are potent transcriptional repression domains. Proc Natl Acad Sci U S A 91(10):4509–4513

Matsui T, Leung D, Miyashita H, Maksakova IA, Miyachi H, Kimura H, Tachibana M, Lorincz MC, Shinkai Y (2010) Proviral silencing in embryonic stem cells requires the histone methyltransferase ESET. Nature 464(7290):927–931

Matteucci C, Balestrieri E, Hurst TP, Magiorkinis G, Strick R (2022) Editorial: unravelling the role of HERVs in cancer: insights and new targets for therapy. Front Oncol 12:874245

Medstrand P, Mager DL (1998) Human-specific integrations of the HERV-K endogenous retrovirus family. J Virol 72(12):9782–9787

Mendenhall EM, Williamson KE, Reyon D, Zou JY, Ram O, Joung JK, Bernstein BE (2013) Locus-specific editing of histone modifications at endogenous enhancers. Nat Biotechnol 31(12):1133–1136

Mikkelsen TS, Ku M, Jaffe DB, Issac B, Lieberman E, Giannoukos G, Alvarez P, Brockman W, Kim TK, Koche RP, Lee W, Mendenhall E, O'Donovan A, Presser A, Russ C, Xie X, Meissner A, Wernig M, Jaenisch R, Nusbaum C, Lander ES, Bernstein BE (2007) Genome-wide maps of chromatin state in pluripotent and lineage-committed cells. Nature 448(7153): 553–560

Morozov VA, Dao Thi VL, Denner J (2013) The transmembrane protein of the human endogenous retrovirus--K (HERV-K) modulates cytokine release and gene expression. PLoS One 8(8): e70399

Muster T, Waltenberger A, Grassauer A, Hirschl S, Caucig P, Romirer I, Fodinger D, Seppele H, Schanab O, Magin-Lachmann C, Lower R, Jansen B, Pehamberger H, Wolff K (2003) An endogenous retrovirus derived from human melanoma cells. Cancer Res 63(24):8735–8741

Nelson P, Rylance P, Roden D, Trela M, Tugnet N (2014a) Viruses as potential pathogenic agents in systemic lupus erythematosus. Lupus 23(6):596–605

Nelson PN, Roden D, Nevill A, Freimanis GL, Trela M, Ejtehadi HD, Bowman S, Axford J, Veitch AM, Tugnet N, Rylance PB (2014b) Rheumatoid arthritis is associated with IgG antibodies to human endogenous retrovirus gag matrix: a potential pathogenic mechanism of disease? J Rheumatol 41(10):1952–1960

Nogueira MA, Gavioli CF, Pereira NZ, de Carvalho GC, Domingues R, Aoki V, Sato MN (2015) Human endogenous retrovirus expression is inversely related with the up-regulation of interferon-inducible genes in the skin of patients with lichen planus. Arch Dermatol Res 307(3):259–264

Novikova M, Zhang Y, Freed EO, Peng K (2019) Multiple roles of HIV-1 capsid during the virus replication cycle. Virol Sin 34(2):119–134

Ohnuki M, Tanabe K, Sutou K, Teramoto I, Sawamura Y, Narita M, Nakamura M, Tokunaga Y, Nakamura M, Watanabe A, Yamanaka S, Takahashi K (2014) Dynamic regulation of human

endogenous retroviruses mediates factor-induced reprogramming and differentiation potential. Proc Natl Acad Sci U S A 111(34):12426–12431
Oricchio E, Sciamanna I, Beraldi R, Tolstonog GV, Schumann GG, Spadafora C (2007) Distinct roles for LINE-1 and HERV-K retroelements in cell proliferation, differentiation and tumor progression. Oncogene 26(29):4226–4233
Paces J, Huang YT, Paces V, Ridl J, Chang CM (2013) New insight into transcription of human endogenous retroviral elements. New Biotechnol 30(3):314–318
Perot P, Mullins CS, Naville M, Bressan C, Huhns M, Gock M, Kuhn F, Volff JN, Trillet-Lenoir V, Linnebacher M, Mallet F (2015) Expression of young HERV-H loci in the course of colorectal carcinoma and correlation with molecular subtypes. Oncotarget 6(37):40095–40111
Perzova R, Graziano E, Sanghi S, Welch C, Benz P, Abbott L, Lalone D, Glaser J, Loughran T, Sheremata W, Poiesz BJ (2013) Increased seroreactivity to HERV-K10 peptides in patients with HTLV myelopathy. Virol J 10:360
Pi W, Zhu X, Wu M, Wang Y, Fulzele S, Eroglu A, Ling J, Tuan D (2010) Long-range function of an intergenic retrotransposon. Proc Natl Acad Sci U S A 107(29):12992–12997
Ponferrada VG, Mauck BS, Wooley DP (2003) The envelope glycoprotein of human endogenous retrovirus HERV-W induces cellular resistance to spleen necrosis virus. Arch Virol 148(4): 659–675
Pothlichet J, Heidmann T, Mangeney M (2006) A recombinant endogenous retrovirus amplified in a mouse neuroblastoma is involved in tumor growth in vivo. Int J Cancer 119(4):815–822
Rebollo R, Miceli-Royer K, Zhang Y, Farivar S, Gagnier L, Mager DL (2012) Epigenetic interplay between mouse endogenous retroviruses and host genes. Genome Biol 13(10):R89
Reis BS, Jungbluth AA, Frosina D, Holz M, Ritter E, Nakayama E, Ishida T, Obata Y, Carver B, Scher H, Scardino PT, Slovin S, Subudhi SK, Reuter VE, Savage C, Allison JP, Melamed J, Jager E, Ritter G, Old LJ, Gnjatic S (2013) Prostate cancer progression correlates with increased humoral immune response to a human endogenous retrovirus GAG protein. Clin Cancer Res 19(22):6112–6125
Reiss D, Zhang Y, Mager DL (2007) Widely variable endogenous retroviral methylation levels in human placenta. Nucleic Acids Res 35(14):4743–4754
Roulois D, Loo Yau H, Singhania R, Wang Y, Danesh A, Shen SY, Han H, Liang G, Jones PA, Pugh TJ, O'Brien C, De Carvalho DD (2015) DNA-demethylating agents target colorectal cancer cells by inducing viral mimicry by endogenous transcripts. Cell 162(5):961–973
Rowe HM, Kapopoulou A, Corsinotti A, Fasching L, Macfarlan TS, Tarabay Y, Viville S, Jakobsson J, Pfaff SL, Trono D (2013) TRIM28 repression of retrotransposon-based enhancers is necessary to preserve transcriptional dynamics in embryonic stem cells. Genome Res 23(3): 452–461
Saini SK, Orskov AD, Bjerregaard AM, Unnikrishnan A, Holmberg-Thyden S, Borch A, Jensen KV, Anande G, Bentzen AK, Marquard AM, Tamhane T, Treppendahl MB, Gang AO, Dufva IH, Szallasi Z, Ternette N, Pedersen AG, Eklund AC, Pimanda J, Gronbaek K, Hadrup SR (2020) Human endogenous retroviruses form a reservoir of T cell targets in hematological cancers. Nat Commun 11(1):5660
Schmidt D, Schwalie PC, Wilson MD, Ballester B, Goncalves A, Kutter C, Brown GD, Marshall A, Flicek P, Odom DT (2012) Waves of retrotransposon expansion remodel genome organization and CTCF binding in multiple mammalian lineages. Cell 148(1–2):335–348
Schmidt N, Domingues P, Golebiowski F, Patzina C, Tatham MH, Hay RT, Hale BG (2019) An influenza virus-triggered SUMO switch orchestrates co-opted endogenous retroviruses to stimulate host antiviral immunity. Proc Natl Acad Sci U S A 116(35):17399–17408
Schultz DC, Friedman JR, Rauscher FJ 3rd (2001) Targeting histone deacetylase complexes via KRAB-zinc finger proteins: the PHD and bromodomains of KAP-1 form a cooperative unit that recruits a novel isoform of the Mi-2alpha subunit of NuRD. Genes Dev 15(4):428–443
Schulz WA, Steinhoff C, Florl AR (2006) Methylation of endogenous human retroelements in health and disease. Curr Top Microbiol Immunol 310:211–250

Segel M, Lash B, Song J, Ladha A, Liu CC, Jin X, Mekhedov SL, Macrae RK, Koonin EV, Zhang F (2021) Mammalian retrovirus-like protein PEG10 packages its own mRNA and can be pseudotyped for mRNA delivery. Science 373(6557):882–889
Shekarian T, Valsesia-Wittmann S, Brody J, Michallet MC, Depil S, Caux C, Marabelle A (2017) Pattern recognition receptors: immune targets to enhance cancer immunotherapy. Ann Oncol 28(8):1756–1766
Shilatifard A (2006) Chromatin modifications by methylation and ubiquitination: implications in the regulation of gene expression. Annu Rev Biochem 75:243–269
Simonti CN, Pavlicev M, Capra JA (2017) Transposable element exaptation into regulatory regions is rare, influenced by evolutionary age, and subject to pleiotropic constraints. Mol Biol Evol 34(11):2856–2869
Song Y, Li X, Wei X, Cui J (2021) Human endogenous retroviruses as biomedicine markers. Virol Sin 36(5):852–858
Sorek M, Meshorer E, Schlesinger S (2021) Impaired activation of transposable elements in SARS-CoV-2 infection. EMBO Rep 23(9):e55101
Sorek M, Meshorer E, Schlesinger S (2022) Impaired activation of transposable elements in SARS-CoV-2 infection. EMBO Rep 23(9):e55101
Spencer TE, Mura M, Gray CA, Griebel PJ, Palmarini M (2003) Receptor usage and fetal expression of ovine endogenous betaretroviruses: implications for coevolution of endogenous and exogenous retroviruses. J Virol 77(1):749–753
Srinivasachar Badarinarayan S, Sauter D (2021) Switching sides: how endogenous retroviruses protect us from viral infections. J Virol 95(12):e02299-20
Srinivasachar Badarinarayan S, Shcherbakova I, Langer S, Koepke L, Preising A, Hotter D, Kirchhoff F, Sparrer KMJ, Schotta G, Sauter D (2020) HIV-1 infection activates endogenous retroviral promoters regulating antiviral gene expression. Nucleic Acids Res 48(19): 10890–10908
Stengel S, Fiebig U, Kurth R, Denner J (2010) Regulation of human endogenous retrovirus-K expression in melanomas by CpG methylation. Genes Chromosomes Cancer 49(5):401–411
Subramanian RP, Wildschutte JH, Russo C, Coffin JM (2011) Identification, characterization, and comparative genomic distribution of the HERV-K (HML-2) group of human endogenous retroviruses. Retrovirology 8:90
Suntsova M, Gogvadze EV, Salozhin S, Gaifullin N, Eroshkin F, Dmitriev SE, Martynova N, Kulikov K, Malakhova G, Tukhbatova G, Bolshakov AP, Ghilarov D, Garazha A, Aliper A, Cantor CR, Solokhin Y, Roumiantsev S, Balaban P, Zhavoronkov A, Buzdin A (2013) Human-specific endogenous retroviral insert serves as an enhancer for the schizophrenia-linked gene PRODH. Proc Natl Acad Sci U S A 110(48):19472–19477
Suntsova M, Garazha A, Ivanova A, Kaminsky D, Zhavoronkov A, Buzdin A (2015) Molecular functions of human endogenous retroviruses in health and disease. Cell Mol Life Sci 72(19): 3653–3675
Takahashi Y, Harashima N, Kajigaya S, Yokoyama H, Cherkasova E, McCoy JP, Hanada K, Mena O, Kurlander R, Tawab A, Srinivasan R, Lundqvist A, Malinzak E, Geller N, Lerman MI, Childs RW (2008) Regression of human kidney cancer following allogeneic stem cell transplantation is associated with recognition of an HERV-E antigen by T cells. J Clin Invest 118(3): 1099–1109
Tchasovnikarova IA, Timms RT, Matheson NJ, Wals K, Antrobus R, Gottgens B, Dougan G, Dawson MA, Lehner PJ (2015) GENE SILENCING. Epigenetic silencing by the HUSH complex mediates position-effect variegation in human cells. Science 348(6242):1481–1485
Tokuyama M, Kong Y, Song E, Jayewickreme T, Kang I, Iwasaki A (2018) ERVmap analysis reveals genome-wide transcription of human endogenous retroviruses. Proc Natl Acad Sci U S A 115(50):12565–12572
Toufaily C, Landry S, Leib-Mosch C, Rassart E, Barbeau B (2011) Activation of LTRs from different human endogenous retrovirus (HERV) families by the HTLV-1 tax protein and T-cell activators. Viruses 3(11):2146–2159

Uleri E, Mei A, Mameli G, Poddighe L, Serra C, Dolei A (2014) HIV Tat acts on endogenous retroviruses of the W family and this occurs via toll-like receptor 4: inference for neuroAIDS. AIDS 28(18):2659–2670

Vallianatos CN, Iwase S (2015) Disrupted intricacy of histone H3K4 methylation in neurodevelopmental disorders. Epigenomics 7(3):503–519

van der Kuyl AC (2012) HIV infection and HERV expression: a review. Retrovirology 9:6

Vargiu L, Rodriguez-Tome P, Sperber GO, Cadeddu M, Grandi N, Blikstad V, Tramontano E, Blomberg J (2016) Classification and characterization of human endogenous retroviruses; mosaic forms are common. Retrovirology 13:7

Wallace TA, Downey RF, Seufert CJ, Schetter A, Dorsey TH, Johnson CA, Goldman R, Loffredo CA, Yan P, Sullivan FJ, Giles FJ, Wang-Johanning F, Ambs S, Glynn SA (2014) Elevated HERV-K mRNA expression in PBMC is associated with a prostate cancer diagnosis particularly in older men and smokers. Carcinogenesis 35(9):2074–2083

Wang J, Vicente-Garcia C, Seruggia D, Molto E, Fernandez-Minan A, Neto A, Lee E, Gomez-Skarmeta JL, Montoliu L, Lunyak VV, Jordan IK (2015) MIR retrotransposon sequences provide insulators to the human genome. Proc Natl Acad Sci U S A 112(32):E4428–E4437

Wang M, Qiu Y, Liu H, Liang B, Fan B, Zhou X, Liu D (2020) Transcription profile of human endogenous retroviruses in response to dengue virus serotype 2 infection. Virology 544:21–30

Wang-Johanning F, Liu J, Rycaj K, Huang M, Tsai K, Rosen DG, Chen DT, Lu DW, Barnhart KF, Johanning GL (2007) Expression of multiple human endogenous retrovirus surface envelope proteins in ovarian cancer. Int J Cancer 120(1):81–90

Wang-Johanning F, Rycaj K, Plummer JB, Li M, Yin B, Frerich K, Garza JG, Shen J, Lin K, Yan P, Glynn SA, Dorsey TH, Hunt KK, Ambs S, Johanning GL (2012) Immunotherapeutic potential of anti-human endogenous retrovirus-K envelope protein antibodies in targeting breast tumors. J Natl Cancer Inst 104(3):189–210

Wang-Johanning F, Li M, Esteva FJ, Hess KR, Yin B, Rycaj K, Plummer JB, Garza JG, Ambs S, Johanning GL (2014) Human endogenous retrovirus type K antibodies and mRNA as serum biomarkers of early-stage breast cancer. Int J Cancer 134(3):587–595

Ward MC, Wilson MD, Barbosa-Morais NL, Schmidt D, Stark R, Pan Q, Schwalie PC, Menon S, Lukk M, Watt S, Thybert D, Kutter C, Kirschner K, Flicek P, Blencowe BJ, Odom DT (2013) Latent regulatory potential of human-specific repetitive elements. Mol Cell 49(2):262–272

Wentzensen N, Coy JF, Knaebel HP, Linnebacher M, Wilz B, Gebert J, von Knebel Doeberitz M (2007) Expression of an endogenous retroviral sequence from the HERV-H group in gastrointestinal cancers. Int J Cancer 121(7):1417–1423

Wildschutte JH, Williams ZH, Montesion M, Subramanian RP, Kidd JM, Coffin JM (2016) Discovery of unfixed endogenous retrovirus insertions in diverse human populations. Proc Natl Acad Sci U S A 113(16):E2326–E2334

Wolf D, Goff SP (2009) Embryonic stem cells use ZFP809 to silence retroviral DNAs. Nature 458(7242):1201–1204

Wolf D, Cammas F, Losson R, Goff SP (2008) Primer binding site-dependent restriction of murine leukemia virus requires HP1 binding by TRIM28. J Virol 82(9):4675–4679

Wu Z, Mei X, Zhao D, Sun Y, Song J, Pan W, Shi W (2015) DNA methylation modulates HERV-E expression in CD4+ T cells from systemic lupus erythematosus patients. J Dermatol Sci 77(2): 110–116

Xie M, Hong C, Zhang B, Lowdon RF, Xing X, Li D, Zhou X, Lee HJ, Maire CL, Ligon KL, Gascard P, Sigaroudinia M, Tlsty TD, Kadlecek T, Weiss A, O'Geen H, Farnham PJ, Madden PA, Mungall AJ, Tam A, Kamoh B, Cho S, Moore R, Hirst M, Marra MA, Costello JF, Wang T (2013) DNA hypomethylation within specific transposable element families associates with tissue-specific enhancer landscape. Nat Genet 45(7):836–841

Yang N, Kazazian HH Jr (2006) L1 retrotransposition is suppressed by endogenously encoded small interfering RNAs in human cultured cells. Nat Struct Mol Biol 13(9):763–771

Yang BX, El Farran CA, Guo HC, Yu T, Fang HT, Wang HF, Schlesinger S, Seah YF, Goh GY, Neo SP, Li Y, Lorincz MC, Tergaonkar V, Lim TM, Chen L, Gunaratne J, Collins JJ, Goff SP,

Daley GQ, Li H, Bard FA, Loh YH (2015) Systematic identification of factors for provirus silencing in embryonic stem cells. Cell 163(1):230–245

Yang P, Wang Y, Macfarlan TS (2017) The role of KRAB-ZFPs in transposable element repression and mammalian evolution. Trends Genet 33(11):871–881

Yi JM, Kim HM, Kim HS (2006) Human endogenous retrovirus HERV-H family in human tissues and cancer cells: expression, identification, and phylogeny. Cancer Lett 231(2):228–239

Young JM, Whiddon JL, Yao Z, Kasinathan B, Snider L, Geng LN, Balog J, Tawil R, van der Maarel SM, Tapscott SJ (2013) DUX4 binding to retroelements creates promoters that are active in FSHD muscle and testis. PLoS Genet 9(11):e1003947

Young GR, Terry SN, Manganaro L, Cuesta-Dominguez A, Deikus G, Bernal-Rubio D, Campisi L, Fernandez-Sesma A, Sebra R, Simon V, Mulder LCF (2018) HIV-1 infection of primary CD4 (+) T cells regulates the expression of specific human endogenous retrovirus HERV-K (HML-2) elements. J Virol 92(1):e01507-17

Yu P, Lubben W, Slomka H, Gebler J, Konert M, Cai C, Neubrandt L, Prazeres da Costa O, Paul S, Dehnert S, Dohne K, Thanisch M, Storsberg S, Wiegand L, Kaufmann A, Nain M, Quintanilla-Martinez L, Bettio S, Schnierle B, Kolesnikova L, Becker S, Schnare M, Bauer S (2012) Nucleic acid-sensing Toll-like receptors are essential for the control of endogenous retrovirus viremia and ERV-induced tumors. Immunity 37(5):867–879

Zhang Y, Babaian A, Gagnier L, Mager DL (2013) Visualized computational predictions of transcriptional effects by intronic endogenous retroviruses. PLoS One 8(8):e71971

Zhang Q, Pan J, Cong Y, Mao J (2022) Transcriptional regulation of endogenous retroviruses and their misregulation in human diseases. Int J Mol Sci 23(17):10112

Zhou F, Krishnamurthy J, Wei Y, Li M, Hunt K, Johanning GL, Cooper LJ, Wang-Johanning F (2015) Chimeric antigen receptor T cells targeting HERV-K inhibit breast cancer and its metastasis through downregulation of Ras. Onco Targets Ther 4(11):e1047582

Zhou B, Qi F, Wu F, Nie H, Song Y, Shao L, Han J, Wu Z, Saiyin H, Wei G, Wang P, Ni T, Qian F (2019) Endogenous retrovirus-derived long noncoding RNA enhances innate immune responses via derepressing RELA expression. MBio 10(4):e00937-19

# Chapter 16
# Cholesterol and M2 Rendezvous in Budding and Scission of Influenza A Virus

**Jesper J. Madsen and Jeremy S. Rossman**

**Abstract** The cholesterol of the host cell plasma membrane and viral M2 protein plays a crucial role in multiple stages of infection and replication of the influenza A virus. Cholesterol is required for the formation of heterogeneous membrane microdomains (or rafts) in the budozone of the host cell that serves as assembly sites for the viral components. The raft microstructures act as scaffolds for several proteins. Cholesterol may further contribute to the mechanical forces necessary for membrane scission in the last stage of budding and help to maintain the stability of the virus envelope. The M2 protein has been shown to cause membrane scission in model systems by promoting the formation of curved lipid bilayer structures that, in turn, can lead to membrane vesicles budding off or scission intermediates. Membrane remodeling by M2 is intimately linked with cholesterol as it affects local lipid composition, fluidity, and stability of the membrane. Thus, both cholesterol and M2 protein contribute to the efficient and proper release of newly formed influenza viruses from the virus-infected cells.

**Keywords** Influenza virus · Budding · Scission · Membrane curvature · Cholesterol · M2 protein · Lipid raft · Lipid phase boundary

---

J. J. Madsen (✉)
Global and Planetary Health, Center for Global Health and Infectious Diseases Research, College of Public Health, University of South Florida, Tampa, FL, USA

Department of Molecular Medicine, Morsani College of Medicine, University of South Florida, Tampa, FL, USA
e-mail: jespermadsen@usf.edu

J. S. Rossman
School of Biosciences, University of Kent, Canterbury, Kent, UK

Research-Aid Networks, Chicago, IL, USA
e-mail: jeremy@researchaidnetworks.org

 
S. Vijayakrishnan et al. (eds.), *Virus Infected Cells*, Subcellular Biochemistry 106,
https://doi.org/10.1007/978-3-031-40086-5_16

## Introduction

Crossing the membrane is a significant physical barrier for viruses, and different strategies are employed to facilitate this process. The newly synthesized components of influenza A virus (IAV) in the infected host cell will come together and organize laterally in a region of the plasma membrane called the "budozone" (Fig. 16.1) (Leser and Lamb 2017). After this assembly occurs, the budozone becomes the site from which budding occurs. Importantly, the lipid composition in this region differs markedly from the average stoichiometric ratios found in the host cell as it is enriched in cholesterol and certain lipids (e.g., saturated phospholipids and sphingolipids), which help drive the assembly of viral components at the phase boundary of lipid microdomains and alter membrane biomechanics (Fig. 16.1). Some of the viral proteins (e.g., hemagglutinin (HA) and nucleoprotein) appear to concentrate spontaneously in the cholesterol-rich membrane environment whereas matrix protein 2 (M2) association with the budozone materializes in a fashion dependent on interactions with other proteins (matrix protein 1 (M1) and HA in the cluster (Leser and Lamb 2017). Notably, the lateral enrichment of certain lipid components persists through the entire process of budding and virus release, making the emerging viral envelope likewise enriched in cholesterol and sphingolipids (Lenard and Compans 1974; Nayak and Barman 2002). Lateral heterogeneity of lipids and (host as well as viral) proteins enable thermodynamically or dynamically stabilized microdomains of functional relevance to exist under the control of cellular regulatory processes (Mouritsen and Jorgensen 1992; Mouritsen 2016). The popular term membrane/lipid "raft" is frequently used to signify such structures of the membrane. Rafts appear to be an essential platform for viral and host components to interact with one another.

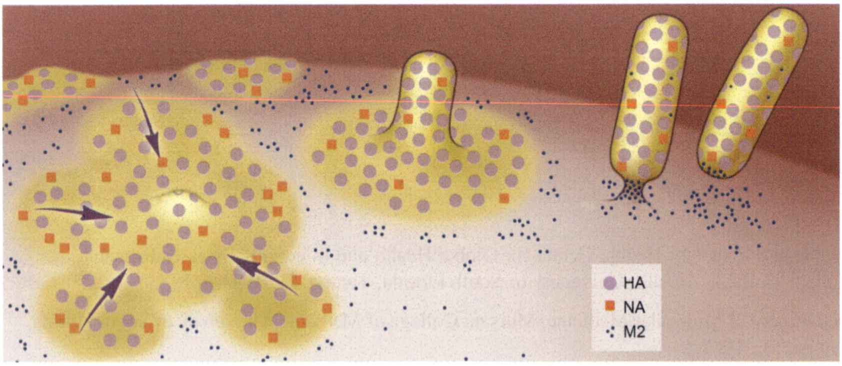


**Fig. 16.1** "Overview of the budding of influenza viruses, showing the coalescence of HA and NA containing lipid rafts (shown in yellow), the formation of a filamentous virion and membrane scission caused by M2 clustered at the neck of the budding virus." Reprinted from Rossman and Lamb (2011), Copyright 2011, with permission from Elsevier

Viruses are compact and small by necessity. Their genetic material will typically contain a minimal set of genes encoding proteins strictly needed for functions they cannot hijack from the host cell. Therefore, the scission of enveloped viruses, the process in which the budding virus cuts the membrane in order to release from the host, is normally performed by using the endosomal sorting complex required for transport (ESCRT) machinery belonging to the host (Votteler and Sundquist 2013). Compelling evidence has led to the proposal that IAV is an exception to this principle. Microscopy data show that M2 is localized at the budding neck of the virus. Additionally, evidence from Nuclear magnetic resonance (NMR) and Electron paramagnetic resonance (EPR) spectroscopy, as well as computational studies, supports this hypothesis at the molecular level, although there are still many unresolved aspects.

This chapter will describe the roles and interactions between two of the key biomolecular components, cholesterol and M2, of the IAV-infected host cell lipid environment during budding and scission. Cholesterol is important in facilitating infection, replication, and budding of IAV (Sun and Whittaker 2003). M2 is important because of its evident role in facilitating the scission of the membrane to complete budding (Rossman et al. 2010b) in addition to its essential role as a proton channel during viral entry (Lamb et al. 1994; Helenius 1992; Pinto et al. 1992). We will set the frame by sketching out the key players and steps of IAV budding and scission (see below) and work toward a contemporary hypothesis for the roles of cholesterol and M2 in budding and scission based on recent discoveries. The essential biophysical techniques used to uncover the knowledge we have in this area are listed in the following section.

## Methods

### *Experimental Techniques*

Microscopic techniques are used to visualize objects. Cells, microbes, and viruses are typically studied using light, fluorescence, or electron microscopy (Harry et al. 1995; Swanson et al. 1969; Goldsmith and Miller 2009; Scaffidi et al. 2021). Light microscopy illuminates the sample with visible light and uses arrangements of optical lenses to magnify the image. Fluorescence microscopy uses a high-intensity light source to excite a molecular tag (the fluorophore) attached to components of the sample (Fig. 16.2). The tag reemits light on a different wavelength once the excitation relaxes to a low-energy state, allowing for image magnification and effective resolution beyond what is possible with conventional light microscopy. This technique is also especially suited for highlighting certain components in order to see where they localize, including specific organelles of a cell or even individual proteins. In electron microscopy, on the other hand, the sample is exposed to a beam of accelerated electrons. The electron beam has a wavelength of around five orders of magnitude shorter than the photons of visible light and can therefore resolve much

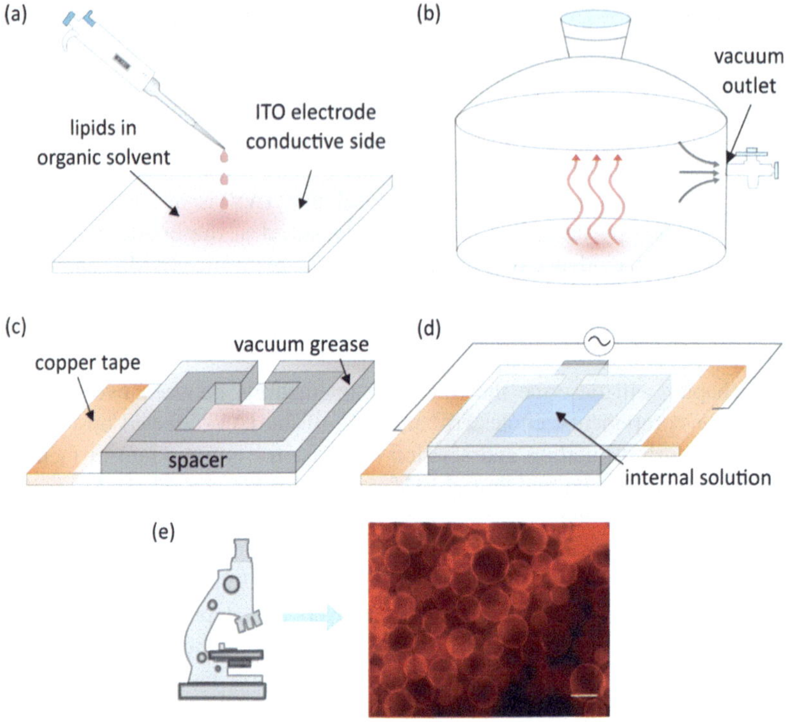


**Fig. 16.2** Vesicle formation "(**a**) Deposition of lipid droplets onto the electrode surface. (**b**) Evaporation of organic solvent under vacuum. (**c**) Construction of the electroformation chamber. (**d**) Electroformation chamber filled with an internal solution and connected to an alternating current function generator. (**e**) An image of fluorescently labeled giant unilamellar vesicles (GUVs) obtained using fluorescence microscopy. The scale bar denotes 50 μm." Reprinted from Membranes, Vol 11:11, Zvonimir Boban, Ivan Mardešić, Witold Karol Subczynski, Marija Raguz, Giant Unilamellar Vesicle Electroformation: What to Use, What to Avoid, and How to Quantify the Results, Article 860, Copyright 2021, under CC BY 4.0 license from MDPI (Boban et al. 2021)

smaller structures. Often the technique of immunogold labeling is used in conjunction with electron microscopy to determine the localization of specific components. This labeling works by attaching gold particles to a secondary antibody that will bind to a primary antibody, which itself attaches to the target component (e.g., protein) (Slot and Geuze 1985; Leser et al. 1996). The high electron density of the gold particle gives rise to high levels of electron scattering, making them appear as distinct dark spots in the image. However, in contrast to light microscopy techniques, imaging by electron microscopy requires fixation and positive staining of the sample, preventing the examination of living samples (Leser et al. 1996).

In spectroscopic methods such as NMR and EPR, the goal is to measure quantities that will reveal properties of the sample by radiating it, for instance, with a strong magnetic field. The magnet excites the behavior of the atomic nuclei (in NMR) on the electron (in EPR). The results are quantifiable spectral lines or energies, which will often allow for making inferences about correlations or distances between components of the sample, for example, the distance between two amino acid residues of a protein (Klug and Feix 2008; Marion 2013; Fan and Lane 2016).

## *Theoretical Modeling and Simulation*

The structural resolving of biomolecular components by experimental means such as X-ray crystallography, NMR, or cryo-electron microscopy enables interrogation of the molecules and their interplay using computational methods. Molecular dynamics (MD) simulation, affectionately called "the computational microscope," is now routinely employed due to a relatively low entry barrier, both monetary and technical, and because the method enables investigation with unparalleled spatiotemporal resolution. MD simulations are used to study the behavior of complex molecular systems over time. A minimal, well-defined system called the simulation "box" containing all desired components (proteins, nucleic acids, lipids, ions, water, etc.) is constructed (Fig. 16.3). The interactions between all atoms in the system are described by physics-based potentials collectively known as the "force field" and the time evolution of the system can be simulated by integrating Newton's equations of motion for the atoms in the system subject to the force field. From these data, a

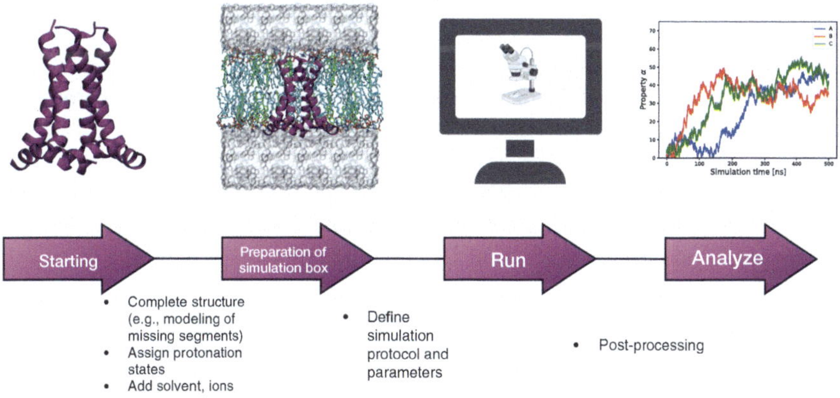


**Fig. 16.3** The general workflow of molecular dynamics simulation. A model structure of the M2 protein is selected together with the initial system setup configuration. We show a ribbon diagram of M2 resolved by solid-state NMR (PDB ID: 2L0J) (Sharma et al. 2010) embedded in a phospholipid bilayer consisting of 1-palmitoyl-2-oleoyl-*sn*-glycero-3-phosphocholine (POPC) and cholesterol. Water and ions fill the remainder of the box. The system is assembled, for example, using the CHARMM-GUI tool (Jo et al. 2008; Wu et al. 2014). From here, Newton's equations of motion are solved, propagating the dynamics until the system's properties stabilize over time. The resulting data are analyzed for the desired properties. Visualizations of the protein structure and simulation box components are created with VMD (Humphrey et al. 1996). Graph data graph is plotted with Matplotlib (Hunter 2007). The Fig. is created, in part, using BioRender (https://BioRender.com)

wide range of properties can be computed, including structural properties (e.g., distribution of distances and angles between atoms, and the organization of molecules), thermodynamic properties (e.g., temperature, entropy, and heat capacity), and kinetic properties (e.g., rates of reactions and transformations) (Allen and Tildesley 1989; Frenkel et al. 2002). For an accessible account of the more detailed aspects of the MD technique to nonexperts, we can refer to our recent review article (Ohkubo and Madsen 2021). In general, modern force fields demonstrate satisfactory accuracy, and there is a growing recognition of limitations in certain situations, such as those pertaining to cholesterol in lipid mixtures (Javanainen et al. 2023).

In certain studies, "coarse-grained" computational techniques are employed, which refers to methods where the individual biomolecular constituents of the simulation are represented at a lower level of detail than the atomistic level. These techniques can range from intermediate resolution methods (e.g., the Martini force field (Marrink et al. 2007)) to highly coarse-grained techniques (e.g., elastic network models (Madsen et al. 2017; Grime and Madsen 2019)) and are used to accelerate computational speed, enabling the exploration of larger systems over longer time scales. However, it is important to note that using such methods can exacerbate inaccuracies in the system (Ingolfsson et al. 2014; Jarin et al. 2021; Pak et al. 2019). Comprehensive validation to ensure that the results are reliable is required for any application of theoretical modeling and simulation.

## Biomolecules and Interactions Driving Budding and Scission

IAV proteins and RNA interact intricately with each other and with components of the host cell in and around the plasma membrane in a complex, multistep process to assemble and bud-off progeny viruses (Lakadamyali et al. 2004; Noda and Kawaoka 2010). The key players in this process are the viral protein HA, neuraminidase (NA), M1, M2, the ribonucleoprotein (RNP) complex, and the arena of interaction at the heterogeneous plasma membrane environment. In this section, we review the fundamentals of budding and scission of IAV.

### *Initiation of Budding*

Viral proteins HA and NA are transported to the plasma membrane where they cluster with cholesterol, sphingolipids, and saturated phospholipids, forming the budozone, a heterogeneous lipid domain on the plasma membrane where the virus assembles (Fig. 16.4A) (Leser and Lamb 2017). The budozone region serves as a scaffold for the recruitment of other viral proteins and the viral RNA. It is thought that this initial clustering and lipid domain formation initiates the process of viral budding by creating a lipid phase boundary, resulting in a local change in membrane curvature (Fig. 16.4A) (Wang et al. 2012). This change in curvature leads to bud formation, which subsequently grows and matures into a new virus particle in a

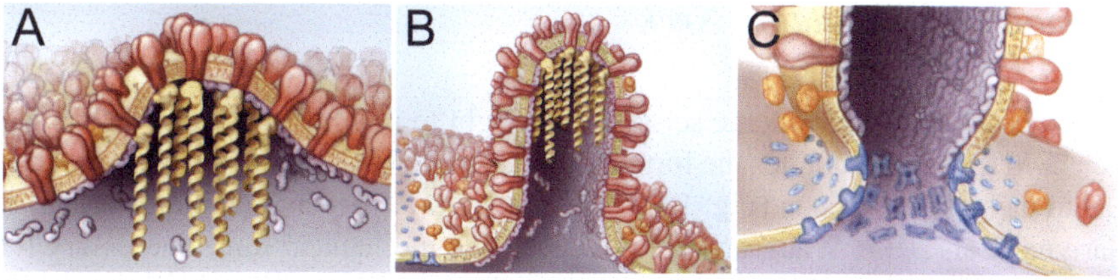


**Fig. 16.4** "Model of influenza virus budding. (**a**) The initiation of virus budding caused by clustering of HA (shown in red) and NA (shown in orange) in lipid raft domains. M1 (shown in purple) is seen binding to the cytoplasmic tails of HA and NA and serves as a docking site for the vRNPs (shown in yellow). (**b**) Elongation of the budding virion caused by polymerization of the M1 protein, resulting in a polarized localization of the vRNPs. M2 (shown in blue) is recruited to the periphery of the budding virus though interactions with M1. (**c**) Membrane scission caused by the insertion of the M2 amphipathic helix at the lipid phase boundary, altering membrane curvature at the neck of the budding virus and leading to release of the budding virus." Reprinted from Rossman and Lamb (2011), Copyright 2011, with permission from Elsevier

process driven by the mutual interactions between the viral proteins (in particular, M1 polymerization), host cell proteins, and the plasma membrane.

## *Virion Assembly*

The cytoplasmic tails of clustered HA and NA, coupled with the lipid environment of the budozone, attract the internal structural components of the virus, in particular, M1 and the RNP complex (Fig. 16.4B). Polymerization of M1 at the budozone is thought to be responsible for the formation and elongation of the nascent virus, which can form spherical or filamentous virion morphologies depending on specific sequence motifs within certain viral proteins (Bourmakina and Garcia-Sastre 2005; Elleman and Barclay 2004; Roberts et al. 1998). The resulting virions contain HA and NA on their surface, a helical layer of polymerized M1 protein underlying the membrane and an internal collection of RNP complexes, each packaging a different segment of the viral genome (Fig. 16.4B) (Waterson et al. 1963; Nermut and Frank 1971; Schulze 1972). The presence of these different virus morphologies can have significant implications for viral budding, entry, host interactions, and may affect the severity of the disease caused by the virus (Taubenberger and Morens 2008; Kalil and Thomas 2019). Despite the formation of fully assembled virus particles, the virions remain attached to the plasma membrane by a small membrane neck that requires scission for their complete release (Fig. 16.4C) (McCown and Pekosz 2006; Rossman et al. 2010b).

### *Scission of the Budding Neck*

The process of IAV membrane scission is thought to involve the action of the viral protein M2, which promotes the formation of curved lipid bilayer structures that can lead to membrane scission (Fig. 16.4C) (Rossman et al. 2010b; Schmidt et al. 2013; Martyna et al. 2017; Wang and Hong 2015). The mechanical forces necessary for scission may be contributed by cholesterol (Hubert et al. 2020; Anderson et al. 2021), which fittingly is also required for the stability of the virus envelope (Lenard and Compans 1974; Nayak and Barman 2002; Bajimaya et al. 2017). During the initial process of viral budding, the M2 protein is conspicuously excluded from the core of the budozone and may not be actively involved in virion formation (Leser and Lamb 2017). The first two steps of budding can even occur in viruses where the M2 protein is deleted or otherwise dysfunctional (Rossman et al. 2010b; Herneisen et al. 2017; McCown and Pekosz 2006). Intriguingly, the process comes to a complete stall in the absence of M2, leaving behind the uncut constricted budding neck from which a secondary budding can occur at the same site. This gives rise to a characteristic phenotype when several unsuccessful budding attempts occur in succession and a "beads-on-a-string" morphology emerges (Fig. 16.5, *top panel*) (McCown and Pekosz 2006; Rossman et al. 2010b). As virion assembly progresses, the planar budozone becomes folded and incorporated into the emerging virus. The M2 protein, located at the periphery of the budozone is then drawn to the constricted membrane neck of the budding virus (Rossman et al. 2010b). There is evidence that interactions between the M2 amphipathic helix M2(AH), which is a highly conserved region of the protein, and the plasma membrane can alter membrane curvature at the virion neck and are responsible, at least in part, for the role that M2 directly plays in membrane scission (Rossman et al. 2010b; Roberts et al. 2013; Schmidt et al. 2013; Rossman et al. 2010a; Martyna et al. 2017; Wang and Hong 2015; Andreas et al. 2015). During the process of scission, some M2 protein gets incorporated into the newly released virion, generating an infectious virus capable of starting the next round of infection (Fig. 16.5, *top panel*).

## A Contemporary Hypothesis on the Roles of Cholesterol and M2

M2 protein can form clusters in the plasma membrane and viral envelope (Sutherland et al. 2022; Madsen et al. 2018; Paulino et al. 2019; Elkins et al. 2017). Clustering is thought to be driven by hydrophobic interaction between M2 transmembrane (TM) domains, as well as the specific arrangements of the amino acids within the protein (Bao et al. 2022). The clustering process is believed to be highly dependent on lipid composition (Sutherland et al. 2022; Elkins et al. 2017; Elkins et al. 2018) and can be influenced by the microenvironment or the action of other proteins (Leser and Lamb 2017). Curvature sorting plays a role in membrane

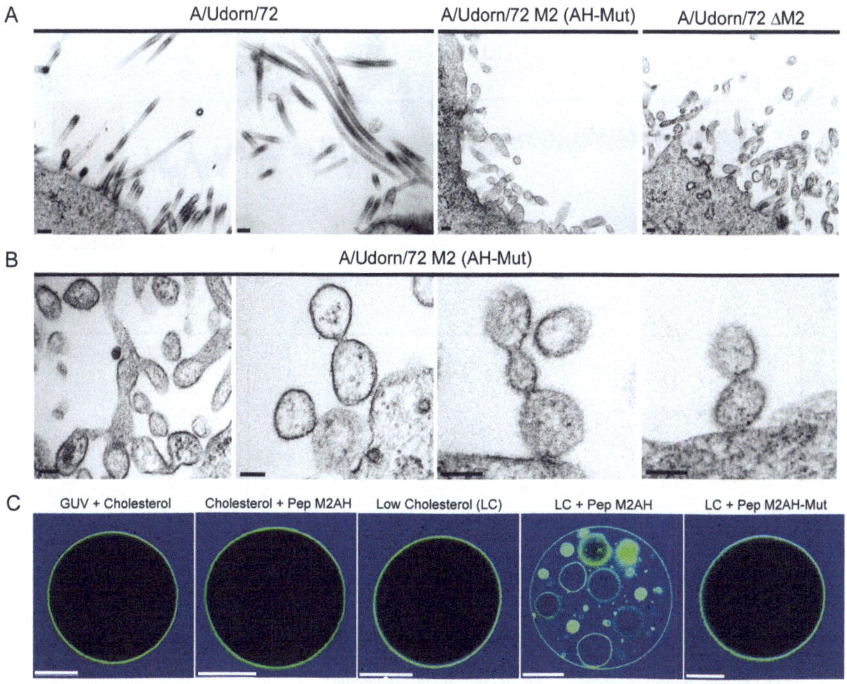


**Fig. 16.5** "The M2 Amphipathic Helix is Necessary for Membrane Scission and Virion Release. MDCK cells were infected with an MOI of 3 pfu/cell of (**a**) A/Udorn/72" or (**b**) "A/Udorn/72 M2 (AH-Mut) for 18 hr and thin sections were analyzed by electron microscopy. The scale bars indicate 100 nm." (**c**) "GUVs electroformed with 30 or 0.5 molar % of cholesterol, with 0.5 mg/ml of lucifer yellow (shown in blue) added to the resuspension buffer, were treated with 10 mM of the indicated peptide and imaged within 1 hr." Reprinted from Cell, Vol 142:6, Jeremy S. Rossman, Xianghong Jing, George P. Leser, Robert A. Lamb, Influenza Virus M2 Protein Mediates ESCRT-Independent Membrane Scission, Pages 902–913, Copyright 2010, with permission from Cell Press (Rossman et al. 2010b)

geometries where the curvature is commensurate with the wedge-shaped M2 (Madsen et al. 2018; Martyna et al. 2016; Ho et al. 2016). Proximity among multiple M2 proteins is thought to play a role in the functioning of the protein, with clustering thought to occur in a stepwise, hierarchical fashion (Fig. 16.6) (Sutherland et al. 2022; Chen et al. 2008).

At the molecular level, M2 protein interacts with cholesterol in several ways to facilitate the budding and release of IAV particles. They are both integrated into the plasma membrane and can therefore interact by direct contact. Cholesterol molecules are embedded within the bilayer membrane, though typically asymmetrically distributed among the two leaflets in a nonstatic fashion that is subject to regulation by cellular processes (Wood et al. 2011). The functional consequences of cholesterol asymmetry likely include changes in membrane fluidity, alterations in lipid domains, lateral and transbilayer ("flip-flop") diffusion, lipid packing, and perhaps even the function of some proteins (Wood et al. 2011). The M2 protein is seen to associate

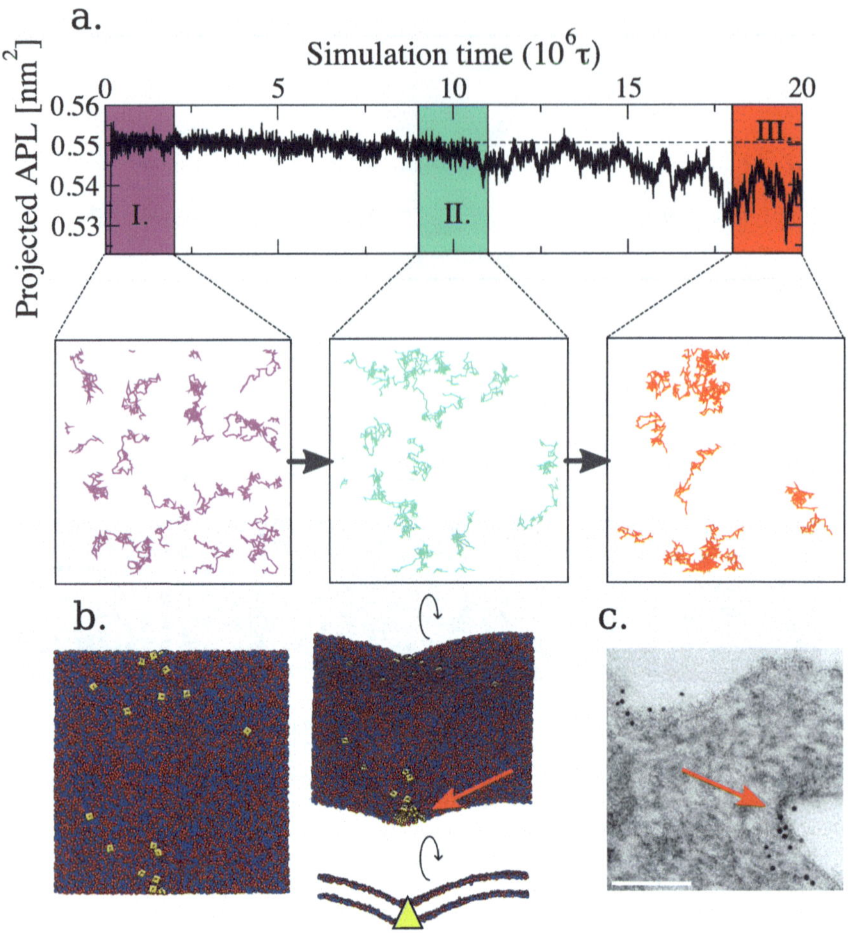


**Fig. 16.6** "(**a**) M2 in a planar bilayer that is exposed to a constant, external compressive stress. (*Top*) (x, y)-plane projected area per lipid in the simulation box. Three time blocks are defined for further analysis, I, II, and III. (*Bottom*) Top view of protein tracer lines show how proteins gradually come together in the three time blocks (I, II, and III). (**b**) Final top view and side view snapshots after the membrane significantly deviates from planarity at t = 20 Mτ. Red arrows indicate the analogy with the in vitro system. (**c**) MDCK cells were infected with 3 pfu per cell of A/Udorn/72 for 18 h before fixation, immunogold labeling of M2, thin sectioning, and analysis by electron microscopy. (Scale bar: 100 nm.)" Reprinted from PNAS, Vol 115:37, Jesper J. Madsen, John M. A. Grime, Jeremy S. Rossman, Gregory A. Voth, Entropic forces drive clustering and spatial localization of influenza A M2 during viral budding, Pages E8595-E8603, with permission from PNAS (Madsen et al. 2018)

with cholesterol-enriched rafts in the plasma membrane primarily through association with other proteins such as M1 (Leser and Lamb 2017). Since the transmembrane domain of M2 traverses the membrane, it is tempting to speculate that cholesterol molecules can get in direct contact with this region of the M2 protein.

Support for this idea comes from sequence analysis of M2, which reveals a cholesterol recognition/interaction amino acid consensus (CRAC) sequence motif thought to be involved with cholesterol recognition in other proteins, as well as the observation that M2 co-purifies with cholesterol (Schroeder et al. 2005). However, extensive mutagenesis studies have shown that the association between cholesterol and M2, and the proper functioning of M2 in facilitating virus replication, are not affected when the CRAC motif is deleted (Thaa et al. 2012; Stewart et al. 2010; Thaa et al. 2014; Thaa et al. 2011). In fact, some strains of IAV do not contain a M2 CRAC motif at all. We must therefore conclude that the CRAC motif is of no major functional importance in this context. Instead, the way that cholesterol and the M2 protein interact is likely through their ability to participate in lipid rafts or through the affinity of the M2 TM domain with cholesterol, positioning it to facilitate scission (Fig. 16.6) (Pan et al. 2019; Kolokouris et al. 2021; Rossman and Lamb 2010; Elkins et al. 2017). The presence of cholesterol in the plasma membrane might further have an impact on the structure and orientation of the M2 protein (Ekanayake et al. 2016; Wright et al. 2022; Claridge et al. 2013). Cholesterol has been shown to increase the insertion depth of the M2 protein in the lipid bilayer, bringing it closer to the hydrophobic core of the bilayer (Elkins et al. 2017; Martyna et al. 2020).

The key to elucidating the interplay between cholesterol and M2 in budding and scission will be to understand how their individual effects on membrane fluidity and mechanical properties add up when they colocalize in the budding neck. Is the sum greater than the parts, and how does the presence of M2 in low copy number (14 to 68 molecules per virion (Zebedee and Lamb 1988)) tip the balance to facilitate membrane scission? There is evidence that lipid phase separation phenomena and membrane scission are closely related (Roux et al. 2005; Ryu et al. 2014). Lipid phase separation refers to the spontaneous segregation of different species in the plasma membrane into distinct domains, which can form spontaneously due to differences in lipid composition or physical properties such as size, shape, and charge. We remark that this is distinct from the occurrence of lipid rafts, which are dynamic in nature and extend over much smaller spatial and temporal scales (and therefore cannot be resolved by microscopic methods) (Rajendran and Simons 2005; Lozano et al. 2016). The phase separation process results in the formation of nonuniform lipid structures, including liquid-ordered and liquid-disordered domains (Heberle and Feigenson 2011). The interface between the distinct phases is the focal point of mechanical forces such as tension, shear, and bending stresses potentially causing membrane scission (Liu et al. 2006). A variety of factors may contribute to membrane scission: Changes in lipid composition, to which both cholesterol and M2 protein contribute, changes in mechanical stress, likely caused by lipid phase boundary dynamics, and the action of additional specific proteins. Lipid raft structures have also been proposed to play a role in scission (Schroeder et al. 2005; Barman and Nayak 2007).

The AH domain within M2 acts as a membrane-sculpting tool (Huang et al. 2015). An AH is characterized by a Janus-faced physical/chemical surface consisting of both hydrophobic and hydrophilic amino acids residues concentrated on opposing sides of the α-helix structure. This arrangement allows the AH to insert into the plasma membrane right at the hydrophilic/hydrophobic interface region so that the hydrophobic residues interact with the lipid tails while the hydrophilic residues interact with the aqueous environment. The M2AH can facilitate budding and scission by inducing changes, primarily in composition, curvature, and stability, that help generate the appropriate lipid environment for budding and membrane scission to occur in model membrane systems, even in the absence of the other M2 domains (Fig. 16.5, *lower panel*) (Rossman et al. 2010b; Roberts et al. 2013; Martyna et al. 2017; Martyna et al. 2020). The effect appears most dependent on the chemical/physical properties of the AH domain and not dependent on the conserved amino acid motif per se, as one might have been tempted to assume (Hu et al. 2020). Furthermore, the conformation of the M2 protein is pH-dependent and can be altered when there are changes in pH or lipid composition, particularly when cholesterol is present, an effect that is likely facilitated by the protein's AH domain (Torabifard et al. 2020; Liao et al. 2013; Liao et al. 2015; Kim et al. 2015; Thomaston et al. 2013; Saotome et al. 2015; Martyna et al. 2020; Kwon et al. 2015; Kyaw et al. 2023). While proteins, in general, can adapt conformations to suit different lipid environments, the opposite may be an important additional factor in some situations (Bacle et al. 2021).

Several other proteins and cellular processes have been shown to play a role in the processes of budding and membrane scission mediated by cholesterol and the M2 protein of IAV (Mao et al. 2022; Bruce et al. 2010; Han et al. 2021; Zhu et al. 2017; Wohlgemuth et al. 2018; Beale et al. 2014; Bhowmick et al. 2017; Vahey and Fletcher 2019). It is important to note that the exact roles of these factors are not fully understood, and additional research is needed to reveal the complex network of interactions that drive these processes.

## Conclusion

The sophisticated interplay between cholesterol and the viral M2 protein plays a crucial role in multiple stages of the influenza virus life cycle. Cholesterol is essential for the formation of membrane raft microdomains that serve as virus assembly sites and contribute to the mechanical forces necessary for membrane scission during the last stage of virus budding. Furthermore, cholesterol helps to maintain the stability of the virus envelope. M2 protein also plays a critical role in influenza virus budding by promoting the formation of curved lipid bilayer structures that facilitate the process of membrane scission. The interaction between cholesterol and the M2 protein affects the local lipid composition, fluidity, and stability of the membrane, thus contributing to the efficient and proper release of newly formed influenza viruses from virus-infected cells.

In closing, investigation of the intricacies of the mechanistic steps of the influenza virus life cycle is critical because of the ongoing threat of influenza epidemics and pandemics. The knowledge gained from studying the interplay between cholesterol, the viral M2 protein and other associated components in influenza virus budding can contribute to the development of novel antiviral strategies and strengthen our efforts to combat viral infections and protect public health.

## References


Allen MP, Tildesley DJ (1989) Computer simulation of liquids. Oxford science publications, Pbk. (with corr.). edn. Clarendon Press, Oxford

Anderson RH, Sochacki KA, Vuppula H, Scott BL, Bailey EM, Schultz MM, Kerkvliet JG, Taraska JW, Hoppe AD, Francis KR (2021) Sterols lower energetic barriers of membrane bending and fission necessary for efficient clathrin-mediated endocytosis. Cell Rep 37(7):110008. https://doi.org/10.1016/j.celrep.2021.110008

Andreas LB, Reese M, Eddy MT, Gelev V, Ni QZ, Miller EA, Emsley L, Pintacuda G, Chou JJ, Griffin RG (2015) Structure and mechanism of the Influenza A M218-60 dimer of dimers. J Am Chem Soc 137(47):14877–14886. https://doi.org/10.1021/jacs.5b04802

Bacle A, Buslaev P, Garcia-Fandino R, Favela-Rosales F, Mendes Ferreira T, Fuchs PFJ, Gushchin I, Javanainen M, Kiirikki AM, Madsen JJ, Melcr J, Milan Rodriguez P, Miettinen MS, Ollila OHS, Papadopoulos CG, Peon A, Piggot TJ, Pineiro A, Virtanen SI (2021) Inverse conformational selection in lipid-protein binding. J Am Chem Soc 143(34):13701–13709. https://doi.org/10.1021/jacs.1c05549

Bajimaya S, Frankl T, Hayashi T, Takimoto T (2017) Cholesterol is required for stability and infectivity of influenza A and respiratory syncytial viruses. Virology 510:234–241. https://doi.org/10.1016/j.virol.2017.07.024

Bao D, Lu C, Ma T, Xu G, Mao Y, Xin L, Niu S, Wu Z, Li X, Teng Q, Li Z, Liu Q (2022) Hydrophobic residues at the intracellular domain of the M2 protein play an important role in budding and membrane integrity of influenza virus. J Virol 96(9):e0037322. https://doi.org/10.1128/jvi.00373-22

Barman S, Nayak DP (2007) Lipid raft disruption by cholesterol depletion enhances influenza A virus budding from MDCK cells. J Virol 81(22):12169–12178. https://doi.org/10.1128/JVI.00835-07

Beale R, Wise H, Stuart A, Ravenhill BJ, Digard P, Randow F (2014) A LC3-interacting motif in the influenza A virus M2 protein is required to subvert autophagy and maintain virion stability. Cell Host Microbe 15(2):239–247. https://doi.org/10.1016/j.chom.2014.01.006

Bhowmick S, Chakravarty C, Sellathamby S, Lal SK (2017) The influenza A virus matrix protein 2 undergoes retrograde transport from the endoplasmic reticulum into the cytoplasm and bypasses cytoplasmic proteasomal degradation. Arch Virol 162(4):919–929. https://doi.org/10.1007/s00705-016-3153-8

Boban Z, Mardesic I, Subczynski WK, Raguz M (2021) Giant unilamellar vesicle electroformation: what to use, what to avoid, and how to quantify the results. Membranes (Basel) 11(11):860. https://doi.org/10.3390/membranes11110860

Bourmakina SV, Garcia-Sastre A (2005) The morphology and composition of influenza A virus particles are not affected by low levels of M1 and M2 proteins in infected cells. J Virol 79(12): 7926–7932. https://doi.org/10.1128/JVI.79.12.7926-7932.2005

Bruce EA, Digard P, Stuart AD (2010) The Rab11 pathway is required for influenza A virus budding and filament formation. J Virol 84(12):5848–5859. https://doi.org/10.1128/JVI.00307-10

Chen BJ, Leser GP, Jackson D, Lamb RA (2008) The influenza virus M2 protein cytoplasmic tail interacts with the M1 protein and influences virus assembly at the site of virus budding. J Virol 82(20):10059–10070. https://doi.org/10.1128/JVI.01184-08

Claridge JK, Aittoniemi J, Cooper DM, Schnell JR (2013) Isotropic bicelles stabilize the juxtamembrane region of the influenza M2 protein for solution NMR studies. Biochemistry 52(47):8420–8429. https://doi.org/10.1021/bi401035m

Ekanayake EV, Fu R, Cross TA (2016) Structural influences: cholesterol, drug, and proton binding to full-length Influenza A M2 protein. Biophys J 110(6):1391–1399. https://doi.org/10.1016/j.bpj.2015.11.3529

Elkins MR, Williams JK, Gelenter MD, Dai P, Kwon B, Sergeyev IV, Pentelute BL, Hong M (2017) Cholesterol-binding site of the influenza M2 protein in lipid bilayers from solid-state NMR. Proc Natl Acad Sci U S A 114(49):12946–12951. https://doi.org/10.1073/pnas.1715127114

Elkins MR, Sergeyev IV, Hong M (2018) Determining cholesterol binding to membrane proteins by cholesterol (13) C labeling in yeast and dynamic nuclear polarization NMR. J Am Chem Soc 140(45):15437–15449. https://doi.org/10.1021/jacs.8b09658

Elleman CJ, Barclay WS (2004) The M1 matrix protein controls the filamentous phenotype of influenza A virus. Virology 321(1):144–153. https://doi.org/10.1016/j.virol.2003.12.009

Fan TW, Lane AN (2016) Applications of NMR spectroscopy to systems biochemistry. Prog Nucl Magn Reson Spectrosc 92-93:18–53. https://doi.org/10.1016/j.pnmrs.2016.01.005

Frenkel D, Smit B, ProQuest (2002) Understanding molecular simulation: from algorithms to applications, Computational science series, vol 1, 2nd edn. Academic Press, San Diego

Goldsmith CS, Miller SE (2009) Modern uses of electron microscopy for detection of viruses. Clin Microbiol Rev 22(4):552–563. https://doi.org/10.1128/CMR.00027-09

Grime JMA, Madsen JJ (2019) Efficient simulation of tunable lipid assemblies across scales and resolutions. arXiv:191005362. https://doi.org/10.48550/arXiv.1910.05362

Han J, Ganti K, Sali VK, Twigg C, Zhang Y, Manivasagam S, Liang CY, Vogel OA, Huang I, Emmanuel SN, Plung J, Radoshevich L, Perez JT, Lowen AC, Manicassamy B (2021) Host factor Rab11a is critical for efficient assembly of influenza A virus genomic segments. PLoS Pathog 17(5):e1009517. https://doi.org/10.1371/journal.ppat.1009517

Harry EJ, Pogliano K, Losick R (1995) Use of immunofluorescence to visualize cell-specific gene expression during sporulation in Bacillus subtilis. J Bacteriol 177(12):3386–3393. https://doi.org/10.1128/jb.177.12.3386-3393.1995

Heberle FA, Feigenson GW (2011) Phase separation in lipid membranes. Cold Spring Harb Perspect Biol 3(4):a004630. https://doi.org/10.1101/cshperspect.a004630

Helenius A (1992) Unpacking the incoming influenza virus. Cell 69(4):577–578. https://doi.org/10.1016/0092-8674(92)90219-3

Herneisen AL, Sahu ID, McCarrick RM, Feix JB, Lorigan GA, Howard KP (2017) A budding-defective M2 mutant exhibits reduced membrane interaction, insensitivity to cholesterol, and perturbed interdomain coupling. Biochemistry 56(44):5955–5963. https://doi.org/10.1021/acs.biochem.7b00924

Ho CS, Khadka NK, She F, Cai J, Pan J (2016) Influenza M2 transmembrane domain senses membrane heterogeneity and enhances membrane curvature. Langmuir 32(26):6730–6738. https://doi.org/10.1021/acs.langmuir.6b00150

Hu B, Siche S, Moller L, Veit M (2020) Amphipathic helices of cellular proteins can replace the helix in M2 of Influenza A virus with only small effects on virus replication. J Virol 94:3. https://doi.org/10.1128/JVI.01605-19

Huang S, Green B, Thompson M, Chen R, Thomaston J, DeGrado WF, Howard KP (2015) C-terminal juxtamembrane region of full-length M2 protein forms a membrane surface associated amphipathic helix. Protein Sci 24(3):426–429. https://doi.org/10.1002/pro.2631

Hubert M, Larsson E, Vegesna NVG, Ahnlund M, Johansson AI, Moodie LW, Lundmark R (2020) Lipid accumulation controls the balance between surface connection and scission of caveolae. elife 9:e55038. https://doi.org/10.7554/eLife.55038

Humphrey W, Dalke A, Schulten K (1996) VMD: visual molecular dynamics. J Mol Graph 14(1): 33–38, 27-38. https://doi.org/10.1016/0263-7855(96)00018-5

Hunter JD (2007) Matplotlib: a 2D graphics environment. Comput Sci Eng 9(3):90–95

Ingolfsson HI, Lopez CA, Uusitalo JJ, de Jong DH, Gopal SM, Periole X, Marrink SJ (2014) The power of coarse graining in biomolecular simulations. Wiley Interdiscip Rev Comput Mol Sci 4(3):225–248. https://doi.org/10.1002/wcms.1169

Jarin Z, Newhouse J, Voth GA (2021) Coarse-grained force fields from the perspective of statistical mechanics: better understanding of the origins of a MARTINI hangover. J Chem Theory Comput 17(2):1170–1180. https://doi.org/10.1021/acs.jctc.0c00638

Javanainen M, Heftberger P, Madsen JJ, Miettinen MS, Pabst G, Ollila OHS (2023) Quantitative comparison against experiments reveals imperfections in force fields' descriptions of popc–cholesterol interactions. J Chem Theory Comput. In press. https://doi.org/10.1021/acs.jctc.3c00648

Jo S, Kim T, Iyer VG, Im W (2008) CHARMM-GUI: a web-based graphical user interface for CHARMM. J Comput Chem 29(11):1859–1865. https://doi.org/10.1002/jcc.20945

Kalil AC, Thomas PG (2019) Influenza virus-related critical illness: pathophysiology and epidemiology. Crit Care 23(1):258. https://doi.org/10.1186/s13054-019-2539-x

Kim SS, Upshur MA, Saotome K, Sahu ID, McCarrick RM, Feix JB, Lorigan GA, Howard KP (2015) Cholesterol-dependent conformational exchange of the C-terminal domain of the Influenza A M2 protein. Biochemistry 54(49):7157–7167. https://doi.org/10.1021/acs.biochem.5b01065

Klug CS, Feix JB (2008) Methods and applications of site-directed spin labeling EPR spectroscopy. Methods Cell Biol 84:617–658. https://doi.org/10.1016/S0091-679X(07)84020-9

Kolokouris D, Kalenderoglou IE, Kolocouris A (2021) Inside and out of the pore: comparing interactions and molecular dynamics of Influenza A M2 viroporin complexes in standard lipid bilayers. J Chem Inf Model 61(11):5550–5568. https://doi.org/10.1021/acs.jcim.1c00264

Kwon B, Tietze D, White PB, Liao SY, Hong M (2015) Chemical ligation of the influenza M2 protein for solid-state NMR characterization of the cytoplasmic domain. Protein Sci 24(7):1087–1099. https://doi.org/10.1002/pro.2690

Kyaw A, Roepke K, Arthur T, Howard KP (2023) Conformation of influenza AM2 membrane protein in nanodiscs and liposomes. Biochim Biophys Acta Biomembr 1865(5):184152. https://doi.org/10.1016/j.bbamem.2023.184152

Lakadamyali M, Rust MJ, Zhuang X (2004) Endocytosis of influenza viruses. Microbes Infect 6(10):929–936. https://doi.org/10.1016/j.micinf.2004.05.002

Lamb RA, Holsinger LJ, Pinto LH (1994) Receptor-mediated virus entry into cells. Cold Spring Harbor Laboratory Press, Cold Spring Harbor

Lenard J, Compans RW (1974) The membrane structure of lipid-containing viruses. Biochim Biophys Acta 344(1):51–94. https://doi.org/10.1016/0304-4157(74)90008-2

Leser GP, Lamb RA (2017) Lateral organization of influenza virus proteins in the Budozone region of the plasma membrane. J Virol 91(9):e02104-16. https://doi.org/10.1128/JVI.02104-16

Leser GP, Ector KJ, Lamb RA (1996) The paramyxovirus simian virus 5 hemagglutinin-neuraminidase glycoprotein, but not the fusion glycoprotein, is internalized via coated pits and enters the endocytic pathway. Mol Biol Cell 7(1):155–172. https://doi.org/10.1091/mbc.7.1.155

Liao SY, Fritzsching KJ, Hong M (2013) Conformational analysis of the full-length M2 protein of the influenza A virus using solid-state NMR. Protein Sci 22(11):1623–1638. https://doi.org/10.1002/pro.2368

Liao SY, Yang Y, Tietze D, Hong M (2015) The influenza m2 cytoplasmic tail changes the proton-exchange equilibria and the backbone conformation of the transmembrane histidine residue to facilitate proton conduction. J Am Chem Soc 137(18):6067–6077. https://doi.org/10.1021/jacs.5b02510

Liu J, Kaksonen M, Drubin DG, Oster G (2006) Endocytic vesicle scission by lipid phase boundary forces. Proc Natl Acad Sci U S A 103(27):10277–10282. https://doi.org/10.1073/pnas.0601045103

Lozano MM, Hovis JS, Moss FR 3rd, Boxer SG (2016) Dynamic reorganization and correlation among lipid raft components. J Am Chem Soc 138(31):9996–10001. https://doi.org/10.1021/jacs.6b05540

Madsen JJ, Sinitskiy AV, Li J, Voth GA (2017) Highly coarse-grained representations of transmembrane proteins. J Chem Theory Comput 13(2):935–944. https://doi.org/10.1021/acs.jctc.6b01076

Madsen JJ, Grime JMA, Rossman JS, Voth GA (2018) Entropic forces drive clustering and spatial localization of influenza A M2 during viral budding. Proc Natl Acad Sci U S A 115(37):E8595–E8603. https://doi.org/10.1073/pnas.1805443115

Mao H, Cao L, Xu T, Xia X, Ren P, Han P, Li C, Hui X, Lin X, Huang K, Jin M (2022) YWHAG inhibits influenza a virus replication by suppressing the release of viral M2 protein. Front Microbiol 13:951009. https://doi.org/10.3389/fmicb.2022.951009

Marion D (2013) An introduction to biological NMR spectroscopy. Mol Cell Proteomics 12(11): 3006–3025. https://doi.org/10.1074/mcp.O113.030239

Marrink SJ, Risselada HJ, Yefimov S, Tieleman DP, de Vries AH (2007) The MARTINI force field: coarse grained model for biomolecular simulations. J Phys Chem B 111(27):7812–7824. https://doi.org/10.1021/jp071097f

Martyna A, Gomez-Llobregat J, Linden M, Rossman JS (2016) Curvature sensing by a viral scission protein. Biochemistry 55(25):3493–3496. https://doi.org/10.1021/acs.biochem.6b00539

Martyna A, Bahsoun B, Badham MD, Srinivasan S, Howard MJ, Rossman JS (2017) Membrane remodeling by the M2 amphipathic helix drives influenza virus membrane scission. Sci Rep 7: 44695. https://doi.org/10.1038/srep44695

Martyna A, Bahsoun B, Madsen JJ, Jackson F, Badham MD, Voth GA, Rossman JS (2020) Cholesterol alters the orientation and activity of the influenza virus M2 amphipathic helix in the membrane. J Phys Chem B 124(31):6738–6747. https://doi.org/10.1021/acs.jpcb.0c03331

McCown MF, Pekosz A (2006) Distinct domains of the influenza a virus M2 protein cytoplasmic tail mediate binding to the M1 protein and facilitate infectious virus production. J Virol 80(16): 8178–8189. https://doi.org/10.1128/JVI.00627-06

Mouritsen OG (2016) Life - as a matter of fat. Softcover reprint of the original, 2nd 2016. edn. Springer International Publishing A+, Cham

Mouritsen OG, Jorgensen K (1992) Dynamic lipid-bilayer heterogeneity: a mesoscopic vehicle for membrane function? BioEssays 14(2):129–136. https://doi.org/10.1002/bies.950140211

Nayak DP, Barman S (2002) Role of lipid rafts in virus assembly and budding. Adv Virus Res 58: 1–28. https://doi.org/10.1016/s0065-3527(02)58001-5

Nermut MV, Frank H (1971) Fine structure of influenza A2 (Singapore) as revealed by negative staining, freeze-drying and freeze-etching. J Gen Virol 10(1):37–51. https://doi.org/10.1099/0022-1317-10-1-37

Noda T, Kawaoka Y (2010) Structure of influenza virus ribonucleoprotein complexes and their packaging into virions. Rev Med Virol 20(6):380–391. https://doi.org/10.1002/rmv.666

Ohkubo YZ, Madsen JJ (2021) Uncovering membrane-bound models of coagulation factors by combined experimental and computational approaches. Thromb Haemost 121(9):1122–1137. https://doi.org/10.1055/s-0040-1722187

Pak AJ, Dannenhoffer-Lafage T, Madsen JJ, Voth GA (2019) Systematic coarse-grained lipid force fields with semiexplicit solvation via virtual sites. J Chem Theory Comput 15(3):2087–2100. https://doi.org/10.1021/acs.jctc.8b01033

Pan J, Dalzini A, Song L (2019) Cholesterol and phosphatidylethanolamine lipids exert opposite effects on membrane modulations caused by the M2 amphipathic helix. Biochim Biophys Acta Biomembr 1861(1):201–209. https://doi.org/10.1016/j.bbamem.2018.07.013

Paulino J, Pang X, Hung I, Zhou HX, Cross TA (2019) Influenza A M2 channel clustering at high protein/lipid ratios: viral budding implications. Biophys J 116(6):1075–1084. https://doi.org/10.1016/j.bpj.2019.01.042

Pinto LH, Holsinger LJ, Lamb RA (1992) Influenza virus M2 protein has ion channel activity. Cell 69(3):517–528. https://doi.org/10.1016/0092-8674(92)90452-i

Rajendran L, Simons K (2005) Lipid rafts and membrane dynamics. J Cell Sci 118(Pt 6): 1099–1102. https://doi.org/10.1242/jcs.01681

Roberts PC, Lamb RA, Compans RW (1998) The M1 and M2 proteins of influenza A virus are important determinants in filamentous particle formation. Virology 240(1):127–137. https://doi.org/10.1006/viro.1997.8916
Roberts KL, Leser GP, Ma C, Lamb RA (2013) The amphipathic helix of influenza A virus M2 protein is required for filamentous bud formation and scission of filamentous and spherical particles. J Virol 87(18):9973–9982. https://doi.org/10.1128/JVI.01363-13
Rossman JS, Lamb RA (2010) Swine-origin influenza virus and the 2009 pandemic. Am J Respir Crit Care Med 181(4):295–296. https://doi.org/10.1164/rccm.200912-1876ED
Rossman JS, Jing X, Leser GP, Balannik V, Pinto LH, Lamb RA (2010a) Influenza virus m2 ion channel protein is necessary for filamentous virion formation. J Virol 84(10):5078–5088. https://doi.org/10.1128/JVI.00119-10
Rossman JS, Jing X, Leser GP, Lamb RA (2010b) Influenza virus M2 protein mediates ESCRT-independent membrane scission. Cell 142(6):902–913. https://doi.org/10.1016/j.cell.2010.08.029
Roux A, Cuvelier D, Nassoy P, Prost J, Bassereau P, Goud B (2005) Role of curvature and phase transition in lipid sorting and fission of membrane tubules. EMBO J 24(8):1537–1545. https://doi.org/10.1038/sj.emboj.7600631
Ryu YS, Lee IH, Suh JH, Park SC, Oh S, Jordan LR, Wittenberg NJ, Oh SH, Jeon NL, Lee B, Parikh AN, Lee SD (2014) Reconstituting ring-rafts in bud-mimicking topography of model membranes. Nat Commun 5:4507. https://doi.org/10.1038/ncomms5507
Saotome K, Duong-Ly KC, Howard KP (2015) Influenza A M2 protein conformation depends on choice of model membrane. Biopolymers 104(4):405–411. https://doi.org/10.1002/bip.22617
Scaffidi SJ, Shebes MA, Yu W (2021) Tracking the subcellular localization of surface proteins in Staphylococcus aureus by immunofluorescence microscopy. Bio Protoc 11(10):e4038. https://doi.org/10.21769/BioProtoc.4038
Schmidt NW, Mishra A, Wang J, DeGrado WF, Wong GC (2013) Influenza virus A M2 protein generates negative Gaussian membrane curvature necessary for budding and scission. J Am Chem Soc 135(37):13710–13719. https://doi.org/10.1021/ja400146z
Schroeder C, Heider H, Moncke-Buchner E, Lin TI (2005) The influenza virus ion channel and maturation cofactor M2 is a cholesterol-binding protein. Eur Biophys J 34(1):52–66. https://doi.org/10.1007/s00249-004-0424-1
Schulze IT (1972) The structure of influenza virus. II. A model based on the morphology and composition of subviral particles. Virology 47(1):181–196. https://doi.org/10.1016/0042-6822(72)90251-6
Sharma M, Yi M, Dong H, Qin H, Peterson E, Busath DD, Zhou HX, Cross TA (2010) Insight into the mechanism of the influenza A proton channel from a structure in a lipid bilayer. Science 330(6003):509–512. https://doi.org/10.1126/science.1191750
Slot JW, Geuze HJ (1985) A new method of preparing gold probes for multiple-labeling cytochemistry. Eur J Cell Biol 38(1):87–93
Stewart SM, Wu WH, Lalime EN, Pekosz A (2010) The cholesterol recognition/interaction amino acid consensus motif of the influenza A virus M2 protein is not required for virus replication but contributes to virulence. Virology 405(2):530–538. https://doi.org/10.1016/j.virol.2010.06.035
Sun X, Whittaker GR (2003) Role for influenza virus envelope cholesterol in virus entry and infection. J Virol 77(23):12543–12551. https://doi.org/10.1128/jvi.77.23.12543-12551.2003
Sutherland M, Tran N, Hong M (2022) Clustering of tetrameric influenza M2 peptides in lipid bilayers investigated by (19)F solid-state NMR. Biochim Biophys Acta Biomembr 1864(7):183909. https://doi.org/10.1016/j.bbamem.2022.183909
Swanson J, Hsu KC, Gotschlich EC (1969) Electron microscopic studies on streptococci. I. M antigen. J Exp Med 130(5):1063–1091. https://doi.org/10.1084/jem.130.5.1063
Taubenberger JK, Morens DM (2008) The pathology of influenza virus infections. Annu Rev Pathol 3:499–522. https://doi.org/10.1146/annurev.pathmechdis.3.121806.154316

Thaa B, Levental I, Herrmann A, Veit M (2011) Intrinsic membrane association of the cytoplasmic tail of influenza virus M2 protein and lateral membrane sorting regulated by cholesterol binding and palmitoylation. Biochem J 437(3):389–397. https://doi.org/10.1042/BJ20110706

Thaa B, Tielesch C, Moller L, Schmitt AO, Wolff T, Bannert N, Herrmann A, Veit M (2012) Growth of influenza A virus is not impeded by simultaneous removal of the cholesterol-binding and acylation sites in the M2 protein. J Gen Virol 93(Pt 2):282–292. https://doi.org/10.1099/vir.0.038554-0

Thaa B, Siche S, Herrmann A, Veit M (2014) Acylation and cholesterol binding are not required for targeting of influenza A virus M2 protein to the hemagglutinin-defined budozone. FEBS Lett 588(6):1031–1036. https://doi.org/10.1016/j.febslet.2014.02.014

Thomaston JL, Nguyen PA, Brown EC, Upshur MA, Wang J, DeGrado WF, Howard KP (2013) Detection of drug-induced conformational change of a transmembrane protein in lipid bilayers using site-directed spin labeling. Protein Sci 22(1):65–73. https://doi.org/10.1002/pro.2186

Torabifard H, Panahi A, Brooks CL 3rd (2020) M2 amphipathic helices facilitate pH-dependent conformational transition in influenza A virus. Proc Natl Acad Sci U S A 117(7):3583–3591. https://doi.org/10.1073/pnas.1913385117

Vahey MD, Fletcher DA (2019) Low-fidelity assembly of Influenza A virus promotes escape from host cells. Cell 176(1–2):281–294. e219. https://doi.org/10.1016/j.cell.2018.10.056

Votteler J, Sundquist WI (2013) Virus budding and the ESCRT pathway. Cell Host Microbe 14(3): 232–241. https://doi.org/10.1016/j.chom.2013.08.012

Wang T, Hong M (2015) Investigation of the curvature induction and membrane localization of the influenza virus M2 protein using static and off-magic-angle spinning solid-state nuclear magnetic resonance of oriented bicelles. Biochemistry 54(13):2214–2226. https://doi.org/10.1021/acs.biochem.5b00127

Wang T, Cady SD, Hong M (2012) NMR determination of protein partitioning into membrane domains with different curvatures and application to the influenza M2 peptide. Biophys J 102(4):787–794. https://doi.org/10.1016/j.bpj.2012.01.010

Waterson AP, Hurrell JM, Jensen KE (1963) The fine structure of influenza A, B and C viruses. Arch Gesamte Virusforsch 12:487–495. https://doi.org/10.1007/BF01242156

Wohlgemuth N, Lane AP, Pekosz A (2018) Influenza A virus M2 protein apical targeting is required for efficient virus replication. J Virol 92(22):e01425-18. https://doi.org/10.1128/JVI.01425-18

Wood WG, Igbavboa U, Muller WE, Eckert GP (2011) Cholesterol asymmetry in synaptic plasma membranes. J Neurochem 116(5):684–689. https://doi.org/10.1111/j.1471-4159.2010.07017.x

Wright AK, Paulino J, Cross TA (2022) Emulating membrane protein environments horizontal line how much lipid is required for a native structure: influenza S31N M2. J Am Chem Soc 144(5): 2137–2148. https://doi.org/10.1021/jacs.1c10174

Wu EL, Cheng X, Jo S, Rui H, Song KC, Davila-Contreras EM, Qi Y, Lee J, Monje-Galvan V, Venable RM, Klauda JB, Im W (2014) CHARMM-GUI membrane builder toward realistic biological membrane simulations. J Comput Chem 35(27):1997–2004. https://doi.org/10.1002/jcc.23702

Zebedee SL, Lamb RA (1988) Influenza A virus M2 protein: monoclonal antibody restriction of virus growth and detection of M2 in virions. J Virol 62(8):2762–2772. https://doi.org/10.1128/JVI.62.8.2762-2772.1988

Zhu P, Liang L, Shao X, Luo W, Jiang S, Zhao Q, Sun N, Zhao Y, Li J, Wang J, Zhou Y, Zhang J, Wang G, Jiang L, Chen H, Li C (2017) Host cellular protein TRAPPC6ADelta interacts with Influenza A virus M2 protein and regulates viral propagation by modulating M2 trafficking. J Virol 91(1):e01757-16. https://doi.org/10.1128/JVI.01757-16

## Further Reading

Chlanda P, Zimmerberg J (2016) Protein-lipid interactions critical to replication of the influenza A virus. FEBS Lett 590:1940–1954

Heaton NS, Randall G (2011) Multifaceted roles for lipids in viral infection. Trends Microbiol 19: 368–375

Lamb RA (2020) The structure, function, and pathobiology if the Influenza A and B virus ion channels. Cold Spring Harb Perspect Med 10:a038505. https://doi.org/10.1101/cshperspect.a038505

Martyna A, Rossman JS (2014) Alterations of membrane curvature during influenza virus budding. Biochem Soc Trans 42:1425–1428

Rossman JS, Lamb RA (2011) Influenza virus assembly and budding. Virology 411:229–236

Rossman JS, Lamb RA (2013) Viral membrane scission. Annu Rev. Cell Dev Biol 29:551–569

Schroeder C (2010) Cholesterol-binding viral proteins in virus entry and morphogenesis. Subcell Biochem 51:77–108

Veit M, Thaa B (2011) Association of influenza virus proteins with membrane rafts. Adv Virol 2011:370606

Zhou H-X, Cross TA (2013) Modeling the membrane environment has implications for membrane protein structure and function: Influenza A M2 protein. Protein Sci 22:381–394

# Correction to: In Situ Imaging of Virus-Infected Cells by Cryo-Electron Tomography: An Overview

**Swetha Vijayakrishnan**

## Correction to:
## Chapter 1 in: S. Vijayakrishnan et al. (eds.), *Virus Infected Cells*, Subcellular Biochemistry 106, https://doi.org/10.1007/978-3-031-40086-5_1

The following reference citation **(Li S, 2022)** has been cited on page 4, and the respective reference **(Li S. (2022). Cryo-electron tomography of enveloped viruses. *Trends Biochem Sci*, *47*(2), 173–186. doi:10.1016/j.tibs.2021.08.005)** is included in References list at the end of the chapter.

The following reference citation **(Vankadari et al., 2002)** has been cited on page 15, and the respective reference **(Vankadari N, Shepherd DC, Carter SD, Ghoshal D. (2002). Three-dimensional insights into human enveloped viruses in vitro and in situ. Biochem Soc Trans, 50(1), 95–105. doi:10.1042/BST20210433)** is included in References list at the end of the chapter.

The following figure **(Fig. 1.1)** caption has been updated with the following missing text "**Figure adapted from Li S (2022) under CC BY 4.0.**". The updated figure caption is as follows:

**Fig. 1.1 Correlation of structural biology and imaging methods with sample size and resolution**. Sample length, resolution, and the extent of nativeness (in vitro or in situ) are the main determinants when choosing structural biology and imaging methods. Sample size and resolution ranges of the most common methods are shown for comparison, highlighting the suitability of cryo-ET for imaging a wide range of molecular and cellular samples including protein assemblies, viruses, and cells. In combination with correlative light-electron microscopy (CLEM) and cryo-focused

The updated version for this chapter can be found at
https://doi.org/10.1007/978-3-031-40086-5_1

 
S. Vijayakrishnan et al. (eds.), *Virus Infected Cells*, Subcellular Biochemistry 106,
https://doi.org/10.1007/978-3-031-40086-5_17

ion beam milling (cryo-FIB), cryo-ET resolves native structures of protein at relatively high resolution as well as provides abundant information on protein–protein, protein–membrane, or protein–nucleic acids interactions in an in situ setting. Figure adapted from Li S (2022) under CC BY 4.0. The various sample types used in this figure were created with Biorender.com

The following figure **(Fig. 1.5)** caption has been updated from the following text **"Figures were made using publicly available atomic coordinates and maps from the protein data bank (PDB) and Electron Microscopy Data Bank (EMDB) using UCSF ChimeraX (Pettersen et al. 2021)" to the following text "Figure was adapted from Vankadari et al. (2002) under CC BY 4.0 using publicly available atomic coordinates and maps from the protein data bank (PDB) and Electron Microscopy Data Bank (EMDB) and UCSF ChimeraX (Pettersen et al., 2021)"**. The updated figure caption is as follows:

**Fig. 1.5 Structures of purified enveloped viruses resolved by cryo-ET and STA.** A selection of high-resolution structures of representative enveloped purified viruses determined by cryo-ET and STA is shown. (a) Structure of mature HIV-1 capsid (PDB: 3J3Y) colour coded with CA hexamers (brown) and pentamers (blue). (b) Ultrastructure of the SARS-CoV-2 virion (EMD-30430); lipid envelope (grey), S protein (blue), and RNPs (yellow). (c) The STA reconstructed prefusion LASV glycoprotein (GP) trimers (yellow) embedded on the viral envelope (blue) and filled with internal material (grey) segmented from the tomogram. Figure reproduced from Li et al. (2016) under CC BY 4.0. (d) Structure of the helical IAV M1 matrix protein (grey and blue) (EMD-22384). (e) Structure of RVFV prefusion GP pentamers (blue) and hexamers (brown) reprojected on the viral envelope (EMDB-4197). (f) Structure of mature HSV-1 capsid (EMD-5452) denoting the five-fold portal structure (blue) and surrounding hexamers (cyan). Figure was adapted from Vankadari et al. (2002) under CC BY 4.0 using publicly available atomic coordinates and maps from the protein data bank (PDB) and Electron Microscopy Data Bank (EMDB) and UCSF ChimeraX (Pettersen et al., 2021)

The manufacturer's authorised representative in the EU is Springer Nature Customer Service Centre GmbH, Europaplatz 3, 69115 Heidelberg, Germany. If you have any concerns regarding our products, please contact ProductSafety@springernature.com

Printed and bound by CPI Group (UK) Ltd, Croydon, CR0 4YY
01/10/2026
02225892-0015